Environmental Pollution: Causes, Impacts and Assessment

Environmental Pollution: Causes, Impacts and Assessment

Edited by Bruce Horak

SYRAWOOD
PUBLISHING HOUSE

New York

Published by Syrawood Publishing House,
750 Third Avenue, 9th Floor,
New York, NY 10017, USA
www.syrawoodpublishinghouse.com

Environmental Pollution: Causes, Impacts and Assessment
Edited by Bruce Horak

International Standard Book Number: 978-1-68286-415-9 (Hardback)

Cataloging-in-publication Data

Environmental pollution : causes, impacts and assessment / edited by Bruce Horak.
    p. cm.
Includes bibliographical references and index.
ISBN 978-1-68286-415-9
1. Pollution. 2. Pollution--Environmental aspects. 3. Pollutants. 4. Environmental quality. I. Horak, Bruce.
TD174 .E58 2017
628.5--dc23

Printed in the United States of America.

TABLE OF CONTENTS

PREFACE

Environment pollution is defined as the introduction of any pollutants into the earth's atmosphere which has adverse effects on the environment. This book on environmental pollution discusses the consequences and damage-assessment of hundreds of years of waste disposal and technological progress that has occurred. This book sheds light on the causes and impacts of environmental pollution. It unravels the recent studies and unfolds some innovative aspects in the field. This text presents researches and case studies contributed by experts from across the globe. It attempts to understand the multiple branches that fall under the discipline of environmental pollution. Those in search of information to further their knowledge will be greatly assisted by this book. Researchers and students in the fields of urban studies, environmental and ecological engineering, conservations, habitat management and wildlife preservation will find this book especially helpful.

This book has been an outcome of determined endeavour from a group of educationists in the field. The primary objective was to involve a broad spectrum of professionals from diverse cultural background involved in the field for developing new researches. The book not only targets students but also scholars pursuing higher research for further enhancement of the theoretical and practical applications of the subject.

It was an honour to edit such a profound book and also a challenging task to compile and examine all the relevant data for accuracy and originality. I wish to acknowledge the efforts of the contributors for submitting such brilliant and diverse chapters in the field and for endlessly working for the completion of the book. Last, but not the least; I thank my family for being a constant source of support in all my research endeavours.

Editor

Environmental Risk Score as a New Tool to Examine Multi-Pollutants in Epidemiologic Research: An Example from the NHANES Study Using Serum Lipid Levels

Sung Kyun Park[1,2]*, Yebin Tao[3], John D. Meeker[2], Siobán D. Harlow[1], Bhramar Mukherjee[3]

1 Department of Epidemiology, University of Michigan School of Public Health, Ann Arbor, Michigan, United States of America, 2 Department of Environmental Health Sciences, University of Michigan School of Public Health, Ann Arbor, Michigan, United States of America, 3 Department of Biostatistics, University of Michigan School of Public Health, Ann Arbor, Michigan, United States of America

Abstract

Objective: A growing body of evidence suggests that environmental pollutants, such as heavy metals, persistent organic pollutants and plasticizers play an important role in the development of chronic diseases. Most epidemiologic studies have examined environmental pollutants individually, but in real life, we are exposed to multi-pollutants and pollution mixtures, not single pollutants. Although multi-pollutant approaches have been recognized recently, challenges exist such as how to estimate the risk of adverse health responses from multi-pollutants. We propose an "Environmental Risk Score (ERS)" as a new simple tool to examine the risk of exposure to multi-pollutants in epidemiologic research.

Methods and Results: We examined 134 environmental pollutants in relation to serum lipids (total cholesterol, high-density lipoprotein cholesterol (HDL), low-density lipoprotein cholesterol (LDL) and triglycerides) using data from the National Health and Nutrition Examination Survey between 1999 and 2006. Using a two-stage approach, stage-1 for discovery (n = 10818) and stage-2 for validation (n = 4615), we identified 13 associated pollutants for total cholesterol, 9 for HDL, 5 for LDL and 27 for triglycerides with adjustment for sociodemographic factors, body mass index and serum nutrient levels. Using the regression coefficients (weights) from joint analyses of the combined data and exposure concentrations, ERS were computed as a weighted sum of the pollutant levels. We computed ERS for multiple lipid outcomes examined individually (single-phenotype approach) or together (multi-phenotype approach). Although the contributions of ERS to overall risk predictions for lipid outcomes were modest, we found relatively stronger associations between ERS and lipid outcomes than with individual pollutants. The magnitudes of the observed associations for ERS were comparable to or stronger than those for socio-demographic factors or BMI.

Conclusions: This study suggests ERS is a promising tool for characterizing disease risk from multi-pollutant exposures. This new approach supports the need for moving from a single-pollutant to a multi-pollutant framework.

Editor: Jaymie Meliker, Stony Brook University, Graduate Program in Public Health, United States of America

Funding: The research was supported by NIEHS (National Institute of Environmental Health Sciences) grant ES20811 (BM), and K01-ES016587 (SKP), and R01-ES021465 and P01-ES022844 (JDM). Additional support was provided by NIEHS Grant P30-ES017885 entitled "Lifestage Exposure and Adult Disease" and NIEHS Grant P42-ES017198. The funders had no role in study design, data collection and analysis, decision to publish, or preparation of the manuscript.

Competing Interests: The authors have declared that no competing interests exist.

* E-mail: sungkyun@umich.edu

Introduction

Over the last several decades, numerous environmental pollutants have been examined as potential risk factors for various diseases and health responses. Most studies have focused on single pollutants, that is, examining a single factor or a set of species (e.g., arsenic species; polychlorinated biphenyl (PCB) congeners). However, in real life we are exposed to multiple pollutants and pollutant mixtures, not single pollutants. This complex exposure profile may have additive, synergistic or antagonistic effects which are not being detected by single pollutant approaches. In addition, the impact of combined exposures to multiple pollutants may differ from the sum of the impacts from single pollutant assessments [1].

A main issue of the single pollutant approach in epidemiologic research is that it is prone to confounding. For example, the health effects of PCBs are subject to confounding by methylmercury if participants were co-exposed to both toxicants from fish consumption. This example also suggests that beneficial nutrients such as omega-3 fatty acids may confound the toxic effects by PCBs and methylmercury [2,3]. Therefore, a positive association in a single pollutant approach may be observed if the single pollutant is a proxy for other co-pollutants or a mixture of pollutants. Alternatively, if individual pollutants have relatively small effects but multiple pollutants as a whole influence the disease risk, the single-pollutant approach may not capture the true effects [4].

Recently, several studies have examined multiple pollutants. Patel and colleagues adopted an approach widely used in analyzing high-throughput genotype data, genome-wide association study (GWAS), and proposed an *Environment-Wide Association Study (EWAS)* to examined wide ranges of environmental factors including toxic chemicals as well as nutrients in relation to type-2 diabetes [5], lipid profiles [6], blood pressure [7] and all-cause mortality [8] using data from the National Health and Nutrition Examination Survey (NHANES). This systematic approach avoided a potential bias from selective reporting of subsets of analyses, outcomes, and adjustments [6]. Another EWAS approach which examined 76 environmental and lifestyle factors in relation to metabolic syndrome was conducted in Sweden [9]. Although these EWAS studies have yielded intriguing results, the statistical analyses were still based on single pollutant approaches. Multi-pollutant models were not considered. Of note, unlike GWAS with millions of markers, current EWAS studies have a moderate number of exposures and are not really comprehensive or "ultra high-dimensional" in nature. Similarly, misclassification, measurement error, temporal variations, and incomplete exposure data are inherent challenges to an EWAS study that modern genotyping techniques have overcome in GWAS.

Sun et al. [10] considered a number of statistical strategies to examine multiple pollutants and their interactions using regression methods for high-dimensional covariates, such as least absolute shrinkage and selection operator (LASSO) [11], Bayesian model averaging (BMA) [12] or supervised principal component analysis (SPCA) [13]. This study showed that LASSO and other dimension reduction techniques worked well for estimating risk models when a large number of candidate pollutants exist. Elastic-net method [14] or the adaptive elastic-net method [15] were proposed to take into account the issue of multi-collinearity when highly correlated predictors are fit simultaneously.

Another challenge in quantifying the health effects of multi-pollutant exposure is how to estimate the risk of adverse health responses from multiple pollutants. As stated above, single pollutant approaches and even EWAS in which the unit of analysis is based on a single pollutant have had small to modest effect sizes. The challenge is to construct the disease risk from exposure to multiple environmental risk factors [16–18]. Some advances have been made in the air pollution area (air pollution mixtures). For example, in the indicator approach one pollutant represents the combined exposure to several pollutants [19,20]; or, in the source apportionment approach particle constituents are assigned to emission sources using principal component analysis and hierarchical clustering [21,22]. However, these approaches do not account for a wide range of environmental pollutants.

In the general context of risk factor epidemiology, risk prediction models, such as the Framingham risk score for coronary heart disease [23] and genetic risk scores (a.k.a Genetic Risk Prediction Studies (GRIPS)) [24–29], have been widely used. Following from these ideas, it would be interesting to assess the predictive ability of an "*Environmental Risk Score*" as a follow-up to an EWAS study after identifying environmental pollutants significantly associated with health outcomes. A risk score may also facilitate targeting of preventive interventions [27].

Here, we propose an "Environmental Risk Score (ERS)" as a new tool to examine the risk of exposure to multi-pollutants in epidemiologic research. As a "proof of concept", we used environmental biomonitoring data from NHANES to illustrate our methodology because it includes a wide range of environmental pollutants from representative U.S. populations and independent data from different cycles enabled us to discover and validate our findings. As outcomes, we examined serum lipid levels including total cholesterol, high-density lipoprotein cholesterol (HDL), low-density lipoprotein cholesterol (LDL) and

triglycerides, because these are continuous measures that can be dichotomized at clinically relevant cutoff points, allowing us to evaluate both continuous and binary outcomes. These outcomes were used in the previous EWAS by Patel et al. [6]. We focused on environmental pollutants in this study rather than a broader array of environmental exposures including dietary, behavioral, psycho-social, socioeconomic and neighborhood, and microorganismic factors, which may limit the feasibility and applicability of ERS. Instead, we treated important determinants of lipid outcomes such as age, sex, race/ethnicity, education (an indicator of socioeconomic factor), body mass index (BMI), and selected dietary nutrients as covariates and confounding factors. The methodology can of course be generalized when the agnostic search for important predictors is expanded to a broader set of exposures capturing personal and community environment.

As the primary goal of the present study is to introduce this novel approach rather than to estimate and generalize actual risks in the U.S. population, and as some of the statistical procedures used in our approach are not equipped with automated handling of survey weights, we did not account for the complex sampling design and used conventional regression modeling. Biomonitoring data in NHANES were not measured in all participants; some pollutants were measured only in a subset (e.g., one third) and different kinds (classes) of pollutants were measured in different subsets in order to reduce the burden of examinations, which limits the sample size for this multi-pollutant model. To maximize the power of the proposed approach, we imputed unmeasured or missing pollutant data. For these reasons, our findings should be cautiously interpreted as potential associations. Another new feature of the present study is that we examined 4 lipid outcomes separately (single-phenotype approach) as well as all 4 lipid outcomes together as a whole (multi-phenotype approach). This multi-phenotype approach can also help improve the power to detect modest individual effects of environmental pollutants and reduce the burden of multiple testing [30–32].

Methods

Ethics Statement

NHANES is a publicly available data set and all participants in NHANES provide written informed consent, consistent with approval by the National Center for Health Statistics Institutional Review Board.

Data

We obtained all publicly available data from the NHANES website (http://www.cdc.gov/nchs/nhanes.htm). Following the two-stage design as in genome-wide association studies [33], we selected three NHANES cycles, 1999–2000, 2001–2002, and 2005–2006 as stage 1 samples and NHANES 2003–2004 as stage 2 samples, because not all measures of environmental pollutants are available in all cycles and the 2003–2004 cycle had the largest number of shared pollutants. We restricted the sample to adults aged 20 years or older and did not include children in this study.

We focused on the 149 environmental pollutant variables that were measured in both stage 1 and 2 samples. The basic idea of an EWAS, like GWAS, is to conduct an agnostic search in a broad set of environmental compounds without any prior belief or hypothesis regarding the effects related to a given outcome. As our study was based on such a non-targeted approach and had no *a priori* assumption of the association directions, chemicals known to be less toxic, such as arsenosugars, were not screened out. For the concentrations below the National Centers for Health Statistics (NCHS) documented limit of detection (LOD), the values of each pollutant's $LOD/\sqrt{2}$ were replaced. We eliminated 15 variables that had more than 90% of the observations missing (including missing due to below LOD), leaving 134 pollutants available for our analysis (Table S1). As stated above the four outcome variables included total cholesterol, HDL, LDL and triglycerides. Important covariates were chosen *a priori* and included age, sex, race/ethnicity (Mexican American, Other Hispanic, non-Hispanic white, non-Hispanic black, Other), education (categorized to less than high school diploma, high school diploma, and greater than high school diploma), BMI, and NHANES cycle. We selected education as an indicator of socioeconomic status because it is widely used and has less missing data than other proxies, such as household income or poverty income ratio. We also considered 21 blood measures of micronutrients (vitamins and isoflavone compounds), some of which were identified to predict serum lipids in the previous EWAS [6]. We imputed our data with a sequential imputation strategy using IVEWARE where the variables to be imputed were treated as the outcomes and all other variables were used as predictors [34,35]. Since we used the data solely for an illustrative purpose, we used only one imputed dataset. The distributions of the data before and after imputation were similar (see File S1 for more details). The sample sizes after imputation were 10818 for the stage 1 sample and 4615 for the stage 2 sample. We applied logarithmic transformation with base 10 to the continuous outcomes and pollutant levels because of skewness in the distributions of the raw values.

Discovery Process of Environmental Factors Contributing to ERS for Single Phenotype

1. Choice of covariates and micronutrients. Our base model included age, gender, race/ethnicity, education and BMI as was also done by Patel et al. [5,6]. Then we selected important micronutrients corresponding to each phenotype using the full data (stage 1 and 2 samples combined). Specifically, we first regressed each phenotype on the set of covariates in the base model to obtain the residuals, and then used the residuals as the outcome to select the micronutrients. For micronutrient selection we applied the Bayesian model averaging technique (BMA) to jointly analyze all micronutrients and select the ones with posterior inclusion probability greater than 0.8 (see Sun et al. [10] for details). Other simpler methods (e.g., best subset regression) may also be used at this step.

2. Single-pollutant models. We selected environmental pollutants for each lipid outcome with adjustment for base covariates and outcome-specific micronutrients. Specifically, for subject i ($i = 1, ..., N$), let Y_i represent one given phenotype, E_i be one given environmental pollutant, and Z_i ($k \times 1$) be the vector of base covariates and micronutrients. The fitted single-pollutant model was

$$Y_i = \beta_0 + \beta_1 E_i + \beta_2' Z_i + \varepsilon_i, \qquad (1)$$

where $\varepsilon_i \sim N(0, \sigma^2)$. We adopted a two-stage analyses strategy following Skol et al. [36] using the model in (1). In stage 1, we analyzed the single-pollutant model for every pollutant using stage 1 samples and calculated the standard Wald test statistic z_1 corresponding to $\hat{\beta}_1$. In stage 2, we only included pollutants with $|z_1| > C_1$ (pre-defined significance threshold). For each of these chosen pollutants, we repeated the same regression analysis using stage 2 samples, and calculated Wald test statistics z_2 corresponding to $\hat{\beta}_1$. Finally, we conducted joint analysis to combine z_1 and z_2 and get a new statistic that allows for between-stage heterogeneity [36],

$$z_{joint} = \sqrt{\pi_{samples}} z_1 + \sqrt{1 - \pi_{samples}} z_2, \qquad (2)$$

where $\pi_{samples}$ was the proportion of samples in stage 1 (0.7 in our case). z_{joint} was compared with a significance threshold C_{joint}. Thresholds C_1 and C_{joint} were selected to control for the false positive rate. Details for the calculation can be found in Skol et al. [36]. Pollutants with $|z_1| > C_1$ and $|z_{joint}| > C_{joint}$ were selected for ERS and in our study, we chose C_1 and C_{joint} to be 2.58 and 3.57, respectively (corresponding to a significance level of 0.01 for the Wald test in both stage 1 and stage 2 analyses). The choice of these thresholds can be optimized for enhanced power at a given false positive rate; however, we wanted to be liberal in the choice of these thresholds. Our primary goal was to identify pollutants to be included in the construction of the ERS that can be used for prediction of health risks, not just identification of individual pollutants, thus, we are less concerned about the false positive rate of the discovery process at this step. We denote the set of pollutants selected in this step as E^s.

3. Conditional analysis via multi-pollutant models. Motivated by the discovery strategy of additional genetic loci via conditioning on the loci identified through marginal association in GWAS [37], we further explored the possibility of identifying additional pollutants not selected in the previous two-stage analysis, in the presence of the previously selected ones in a multivariate model. Specifically, for subject i, let E_i^+ denote a pollutant not belonging to E^s. The conditional model is given by.

$$Y_i = \gamma_0 + \gamma_1 E_i^+ + \gamma_2' E_i^s + \gamma_3' Z_i + e_i, \qquad (3)$$

where $e_i \sim N(0, \tau^2)$. We repeated the two-stage analysis with this conditional model for each pollutant not belonging to E^s. We calculated the same Wald test statistics and compared them to the same thresholds to select additional exposures for ERS. We denoted the set of pollutants selected in this step as E^c, denoting pollutants identified based on conditional analysis.

Construction of ERS and Assessment of its Predictive Power

We conceptualized the ERS as a weighted sum of the exposures identified by marginal and conditional analysis, namely, E^s and E^c i.e., for subject i, $ERS_i = w^s E_i^s + w^c E_i^c$, where w^s and w^c are vectors of weights corresponding to E^s and E^c, respectively. Given that all exposure variables were log-transformed in the present study, the weights (regression coefficients) are on a relative (ratio) scale, not an absolute (difference) scale, and therefore the weights did not need to be scaled. For comparability of the weights on an absolute scale if exposure variables are linearly fit, they need to be scaled (by either standard deviation or IQR).

To estimate the weights and evaluate the performance of ERS, we randomly split the full data (all cycles combined) by a 3:1 ratio: the larger part (n = 11586) used for estimation/training and the smaller part (n = 3847) for validation/testing. We considered two types of weights. ERS1 used regression coefficients from single-pollutant models for each pollutant in the E^s and E^c sets as weights, while ERS2 used regression coefficients from a multi-pollutant model that included all members of E^s and E^c simultaneously. The weights of ERS1 and ERS2 were both adjusted for base covariates and phenotype-specific micronutrients. ERS1 and ERS2 differ in terms of the weights corresponding to each pollutant, in particular, the weights in ERS2 are taking into account correlation among the pollutants in the entire E^s and

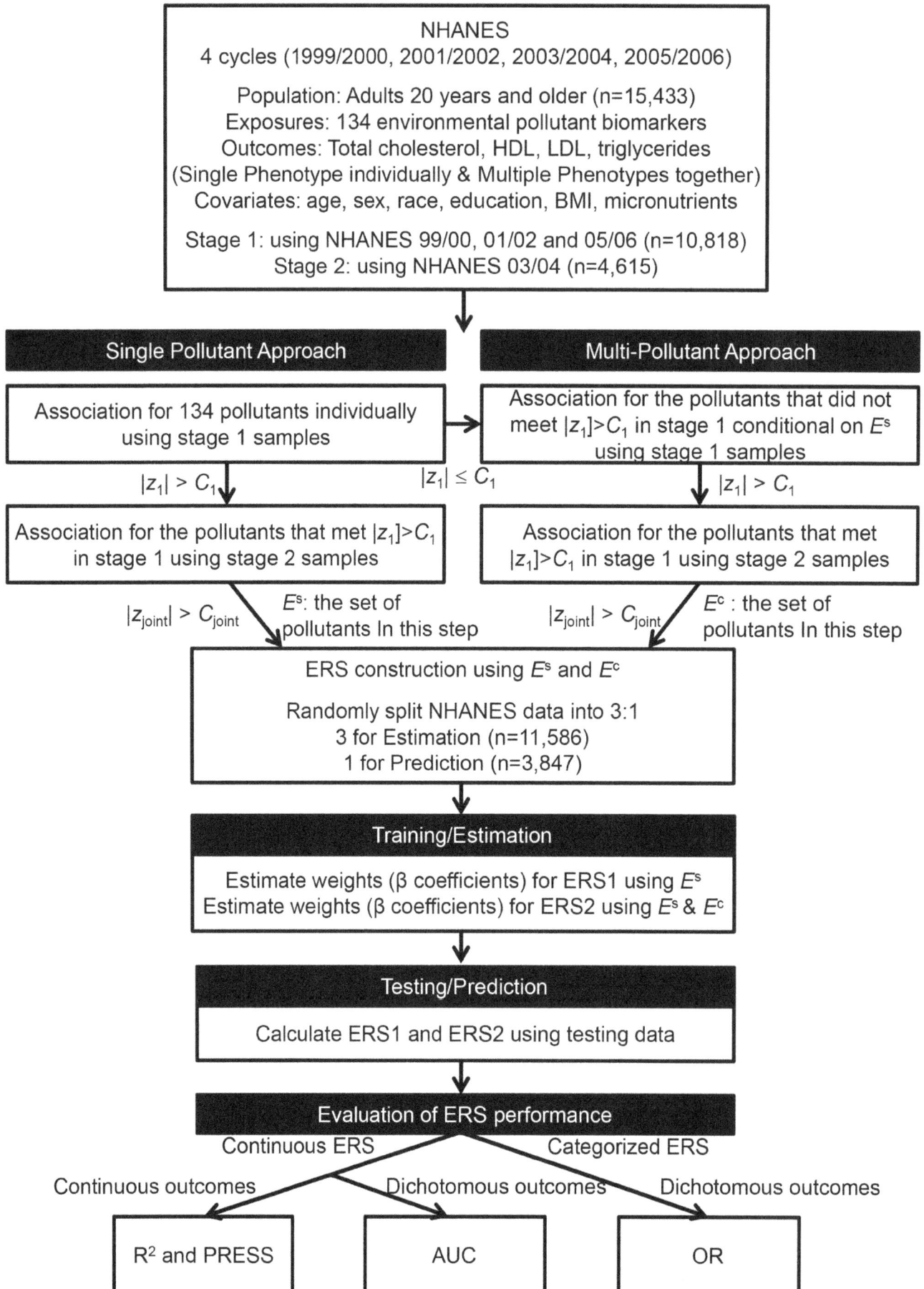

Figure 1. Schematic plot of statistical methods for Environmental Risk Score.

E^c sets. We estimated the weights using the training data and calculated the ERS in the validation data based on those weights to avoid issues of over-fitting. We realize that the multiple regression model that includes both E^s and E^c with adjustment for base covariates and phenotype-specific micronutrients may have some redundant variables in terms of statistical significance, and a further variable selection step may lead to a smaller model and a more concise measure of ERS. We wanted to retain all the identified pollutants in both versions of ERS and thus refrained from applying this additional model selection step in constructing the weights from the multivariate model.

We evaluated the performance of ERS using three metrics. In each case, the contribution of ERS was measured in the presence of base covariates and micronutrients retained in the model. First, we used linear regression with the continuous phenotype outcome and continuous version of the ERS, with R^2 and the predicted residual sums of squares (PRESS) statistic measuring model fit. Second, we dichotomized the levels of the phenotypes as high vs. low (200 mg/dL for total cholesterol; 40 mg/dL (male) or 50 mg/dL (female) for HDL; 130 mg/dL for LDL; and 150 mg/dL for triglycerides [38]), and conducted logistic regression analysis with this dichotomized outcome and with continuous ERS as predictor. We used area under the receiver operating characteristic (ROC) curve or AUC to assess predictive ability of the ERS with these binary endpoints. In each of the above two metrics we compared a sequence of models, with only base covariates, base covariates + micronutrients, base covariates + micronutrients + ERS. Note that the above two metrics measure overall prediction, aggregated over all subjects. A bootstrap resampling (2000 iterations) was used to compute 95% confidence intervals of AUCs for different models [39] (the ci.auc() function in the pROC package in R [40]).

In order to assess risk stratification/discrimination power of the ERS we further categorized ERS by its quintiles and conducted logistic regression for the binary phenotype and categorical ERS. We used the odds ratio (OR) for the highest quintile vs. the lowest quintile of ERS to measure the risk stratification properties of ERS.

Extension to Multiple Phenotypes

Since we are dealing with multiple lipid outcomes that are correlated, a natural question may be to investigate whether simultaneously analyzing the phenotypes lead to methods with superior/different performance. In this step we used four phenotypes together to select environmental pollutants by multivariate regression. The micronutrients adjusted for were the union of all phenotype-specific micronutrients selected in Section 1. Specifically, for subject i, the multivariate single-pollutant model is.

$$\tilde{Y}_i = \tilde{\alpha}_0 + \tilde{\alpha}_1 E_i + \tilde{\alpha}_2 W_i + \tilde{\varepsilon}_i, \qquad (4)$$

where $\tilde{Y}_i$ is the 4×1 vector of phenotypes, $\tilde{\alpha}_0$ and $\tilde{\alpha}_1$ are 4×1 vectors of intercepts and regression coefficients for one given pollutant, respectively, $\tilde{\alpha}_2$ is the $4 \times m$ matrix of regression coefficients for base covariates and micronutrients W, $(m \times 1)$ and $\tilde{\varepsilon}_i \tilde{N}(0, \Sigma_{4 \times 4})$. Similar to the single-phenotype method, we also applied the two-stage analysis. In stage 1, we analyzed the multivariate single-pollutant model for every pollutant using stage 1 samples and calculated the likelihood ratio Chi-squared test statistic with 4 degrees of freedom, namely, χ_1 comparing the multivariate single-pollutant model with the base model ($\tilde{\alpha}_1 = 0$). In stage 2, we repeated the same analysis using stage 2 samples, but only for pollutants with $|\chi_1| > C_1^*$ (pre-defined significance

threshold), and calculated the same likelihood ratio test statistic χ_2. We also used equation (2) (replace z with χ) to calculate χ_{joint} which was compared with a significance threshold C_{joint}^*. Again, thresholds C_1^* and C_{joint}^* were selected to control the false positive rate and we set them to be 13.3 and 18.4, respectively (corresponding to a significance level of 0.01 for the chi-squared test with 4 degrees of freedom in each stage).

Similarly, we also conduct the conditional analysis using the multivariate multi-pollutant model adjusted for pollutants selected in the previous step, base covariates and micronutrients. We calculated the same likelihood ratio test statistics and compared them to the same thresholds to select additional exposures for ERS.

The ERS consists of pollutants selected in the multivariate single- or multi-pollutant analyses. Its construction and assessment steps were the same as in Section 2. A schematic representation of the procedures is presented in Figure 1.

Results

Table 1 shows population characteristics of the stage 1 and 2 samples. Mean (SD) age and the proportion female were 48 (18.7) years and 53.5% in Stage 1 and 50 (19.5) years and 51.9% in Stage 2, respectively. The mean BMI was 28.4 kg/m^2 in both Stages. The Stage 1 samples included more Mexican American and other Hispanic and were less educated than the Stage 2 samples. Participants in the Stage 1 had lower HDL (53.0 vs. 54.7 mg/dL) and higher triglycerides (150.2 vs. 140.0 mg/dL) than those in the Stage 2. Total cholesterol was highly correlated with LDL (Spearman correlation coefficient (rho) = 0.86) but modestly correlated with HDL (rho = 0.16) and triglycerides (rho = 0.37) (Table S2). HDL was inversely correlated with triglycerides (rho = –0.42).

Of 31 micronutrient measures in blood, we identified 12 significant predictors for total cholesterol, 9 for HDL, 9 for LDL and 11 for triglycerides (Table S3). Measures of B vitamins (folate, B12, methylmalonic acid), vitamin A (retinol, retinyl palmitate, retinyl stearate), carotenoids (α-carotene, β-carotene, β-cryptoxanthin, lutein/zeaxanthin, lycopene), and/or vitamin E (α- and γ-tocopherol) were selected for each lipid outcome. These phenotype-specific nutrient variables along with the pre-selected base covariates were adjusted for when identifying environmental pollutants for ERS.

Discovery of Environmental Pollutants for ERS

Table 2 shows environmental pollutants that reached the significance threshold (C_{joint} of 0.01) for each lipid outcome and their estimated weights (regression coefficients) for ERS from single-pollutant models (ERS1) and a multi-pollutant model (ERS2). Figure S1 presents visual distributions of the P values for the individual environmental pollutants examined in the Stage-1 samples (Manhattan plot [41]). Out of 134 environmental pollutants, 11, 9, 5 and 23 pollutants were significantly associated with total cholesterol, HDL, LDL, and triglycerides, respectively, in single pollutant models (marginal analyses) with adjustment for the base covariates and phenotype-specific nutrients. Note that the weights in Table 2 are the regression coefficients for each log-transformed exposure in relation to the log-transformed lipid outcome, which are not directly interpretable. Generally, percent changes for a two-fold increase in exposure concentrations are presented as $[\exp(\text{regression coefficient} \times \log(2)) - 1] \times 100\%$. For example, a two-fold increase in blood lead was associated with a 19% higher levels of total cholesterol ($[\exp(1.71 \times \log(2)) - 1] \times 100\% = 19\%$). Since we used these weights to construct

Table 1. Population characteristics by two stage samples.

Variable	Stage 1 Samples (n = 10818)	Stage 2 Samples (n = 4615)
Continuous (Mean (SD))		
Age (years)	48.0 (18.7)	50.3 (19.5)
BMI (kg/m^2)	28.4 (6.4)	28.4 (6.3)
Total cholesterol (mg/dL)	201.8 (43.9)	202.0 (44.0)
HDL (mg/dL)	53.0 (16.3)	54.7 (16.3)
LDL (mg/dL)	118.9 (37.8)	119.9 (38.1)
Triglycerides (mg/dL)	150.2 (135)	140.0 (139)
Categorical (N (%))		
Gender		
Male	5029 (46.5)	2220 (48.1)
Female	5789 (53.5)	2395 (51.9)
Race/Ethnicity		
Non-Hispanic White	5397 (49.9)	2447 (53.0)
Mexican American	2433 (22.5)	925 (20.0)
Non-Hispanic Black	2121 (19.6)	905 (19.6)
Other Hispanic	498 (4.6)	139 (3.0)
Others	369 (3.4)	199 (4.3)
Education		
< High School	3383 (31.3)	1356 (29.4)
High School	2522 (23.3)	1159 (25.1)
College or Above	4913 (45.4)	2100 (45.5)
Study Year		
1999–2000	3089 (28.5)	-
2001–2002	4736 (43.8)	-
2003–2004	-	4615 (100)
2005–2006	2993 (27.7)	-

HDL, high-density lipoprotein cholesterol; LDL, low-density lipoprotein cholesterol.

ERS rather than interpret the associations of individual pollutants, we presented the direct weights rather than more interpretable estimates (percent changes). Also note that less significant associations in ERS2 compared with ERS1 are mainly due to lower power due to fitting of a larger model with larger number of parameters and with multiple pollutants that are potentially correlated. Two pollutants (1,2,3,4,6,7,8-HpCDD and PCB 177) for total cholesterol and 4 pollutants (PCB 118, PCB 138, PCB 153 and 3,3,4,4,5,5-PnCB) for triglycerides were additionally identified in conditional analyses in which the pollutants selected in the previous two-stage analyses were included as covariates. No further pollutants were identified in relation to HDL and LDL in the conditional analyses. Therefore, a total of 13 pollutants for total cholesterol, 9 for HDL, 5 for LDL and 27 for triglycerides were identified and used to construct ERS for each outcome. Various persistent organic pollutants (POPs) were positively associated with total cholesterol and triglycerides and inversely associated with HDL in single-pollutant models but the association directions for some POPs (2,3,4,7,8-PnCDF, 3,3,4,4,5-HxCB, PCB 138, PCB 146, PCB 156, PCB 177, PCB 180, and PCB 183) changed in the multi-pollutant model, probably due to multi-collinearity. Phthalates were inversely associated with HDL. Cadmium and lead were associated with lipid outcomes in expected directions, that is, higher concentrations of cadmium and lead were associated with higher levels of lipid outcomes except the

association between lead and HDL (good cholesterol) which was positive. Interestingly, the mercury (blood total and urinary) and arsenobetaine measures were inversely associated with triglycerides; as were perfluoroheptanoic acid and diethylphosphate with LDL.

Risk Prediction by ERS and its Associations with Lipid Outcomes

The ERS's from single-pollutant models ranged from −0.068 to 0.239 (mean±SD = 0.090±0.043) for total cholesterol (fit as a continuous outcome (log-transformation). Same for other outcomes); −0.226 to 0.205 (0.030±0.057) for HDL; −0.059 to 0.195 (0.088±0.029) for LDL; and −1.278 to 0.563 (−0.445±0.228) for triglycerides. Those from a multi-pollutant model ranged from −0.009 to 0.135 (0.058±0.019) for total cholesterol; −0.013 to 0.152 (0.061±0.022) for HDL; −0.054 to 0.183 (0.086±0.027) for LDL; and −0.291 to 0.339 (−0.009±0.082) for triglycerides (Table S4). The ERS2 were generally smaller than the ERS1 because of more inverse associations in ERS2.

Table 3 presents risk prediction measures by ERS when outcomes were continuous (R^2 and PRESS) and dichotomized (AUC). Base covariates and micronutrients explained approximately 13% of the variation for LDL, 26% for HDL, 33% for total cholesterol and 37% for triglycerides. ERS constructed with coefficients from single-pollutant models (ERS1) additionally explained variations from 0.33% for LDL to 0.72% for triglycerides. Addition of ERS1 decreased the PRESS by from 0.33% [(539.62–537.84)/539.62] for LDL to 1.1% [(967.24–956.76)/967.24] for triglycerides. When the dichotomous outcomes were used, the addition of the ERS1 only minimally modestly improved the AUC for each lipid outcome (Table 3 and Figure S2). Similar results were found with the ERS constructed with coefficients from multi-pollutant models (ERS2). Similar risk predictions were observed in the multi-phenotype approach although six new pollutants were identified in the multi-phenotype approach (Table S5).

Table 4 shows ORs of having adverse levels of lipid outcomes comparing the highest vs. the lowest quintiles of ERS. After controlling for base covariates and micronutrients, ORs of total cholesterol comparing the highest vs. the lowest quintiles were from 1.45 (95% confidence interval (CI), 1.11, 1.89) for ERS1 and single-phenotype approach to 1.78 (95% CI, 1.34, 2.37) for ERS1 and multi-phenotype approach. For HDL, ORs ranged from 1.37 (95% CI, 1.08, 1.75) for ERS1 and single-phenotype approach to 1.57 (95% CI, 1.23, 1.99) for ERS2 and multi-phenotype approach. For LDL, the highest quintile had a 82% higher odds of having high LDL levels (95% CI, 1.39, 2.38) compared with the lowest quintile in single-phenotype approaches, whereas the associations were relatively weak in multi-phenotype approaches (OR = 1.36 (95% CI, 1.06, 1.74) for ERS1 and 1.26 (95% CI, 0.97, 1.64) for ERS2). For triglycerides, ORs ranged from 1.54 (95% CI, 1.15, 2.06) for ERS2 and single-phenotype approach to 2.03 (95% CI, 1.52, 2.70) for ERS2 and the multi-phenotype. These ORs were comparable to or even stronger than those for socio-demographic factors or BMI (Table S6). For example, the OR of the association between total cholesterol and ERS from single-pollutant models (1.45) was consistent with ORs for females vs. males (1.47); for non-Hispanic blacks vs. non-Hispanic white (1.42); and for a 30 kg/m^2 increase in BMI (1.47); and stronger than ORs for <high school vs. college or higher (1.20).

Figure 2 shows ORs of having adverse levels of HDL and LDL for individual pollutants that compose the ERS. Three out of the 9 pollutants (antimony, mono-benzyl phthalate, mono-(3-carboxylpropyl) phthalate) had significant positive associations with the odds of HDL, the rest except for blood lead had weak non-significant positive associations and blood lead had a weak non-significant inverse association. One of the 5 pollutants (blood lead) had a significant positive association with the odds of LDL and the rest had weak non-significant associations. In particular, the effect sizes of ERS's in relation to LDL were larger than any of the effect sizes of individual pollutants. Here we present ORs of HDL and LDL because their ERS's comprise the smaller number of pollutants (9 and 5 pollutants each). The plots for total cholesterol and triglycerides are shown in Figure S3.

Discussion

In this study, we propose an Environmental Risk Score (ERS) as a novel approach that integrates information on the health effects of multiple pollutant exposures. We used serum lipid measures and various classes of pollutant biomonitoring data from NHANES to illustrate and validate this approach. Important environmental risk factors for lipid outcomes were identified individually (single-phenotype approach) or together (multi-phenotype approach) while controlling for socio-demographic risk factors and nutrients. Although the contributions of ERS to overall risk predictions for lipid outcomes (i.e., R^2, PRESS and AUC) were modest after accounting for important socio-demographic factors and nutrients,

we found relatively stronger associations between ERS and lipid outcomes than with individual pollutants. The magnitudes of the observed associations between ERS and lipid outcomes were comparable to or stronger than those for socio-demographic factors or BMI.

Although the importance of evaluating the health effects of multi-pollutant exposures has recently been recognized [18,42], only a few studies have been conducted, mostly focused on multiple air pollutants [10,21,43–46], probably due to methodological challenges, such as collinearity, measurement errors, potential interaction between pollutants and potential non-linear exposure-health relationships [16]. Patel et al. adopted newer techniques used in genomics and proposed an Environment-Wide Association Study (EWAS) [5,6]. This approach provided excellent insight to identify 'top hit' pollutants. However, few epidemiologic studies have provided methods to estimate combined effects or to predict risks from multi-pollutant exposure [43,47].

Hong et al. examined the combined effects of 4 air pollutants (particulate matter<10 μm (PM$_{10}$), nitrogen dioxide (NO$_2$), sulfur dioxide (SO$_2$), and ozone) by summing each pollutant concentration divided by its mean (i.e., relative concentrations) and then fitting this index as an independent variable [47]. They found that the combined index had a stronger association with mortality than individual pollutants. In a study of indoor exposure to volatile organic compounds (VOCs) and respiratory health, Billionnet et al. computed a global VOC score of 20 VOCs by dichotomizing individual VOC as 1 if greater than the 75th percentile and otherwise 0 and then summing the 20 dichotomous VOCs, which indicates the number of VOCs whose concentrations were relatively high within the study population (range 0–17) [43]. Each additional VOC with a concentration higher than the 75th percentile was associated with 7% (95% CI, 1.00–1.13) and 4% (95% CI, 1.00–1.08) higher odds of asthma and rhinitis, respectively. Although these studies evaluated the combined effects of multi-pollutants, their approaches did not account for the relative effects of individual pollutants on the phenotype of interest, that is, each pollutant was not weighted depending on its relative effect size. Our study aimed to obtain a more precise relative effect size of each pollutant on each lipid outcome by estimating the weights (regression coefficients) from a randomly split training dataset and then computed ERS in an independent validation dataset.

In the real-world, we are exposed to multiple pollutants which may contribute to disease susceptibility in combination or as mixtures. In contrast, individual pollutants may have relatively small effects. Our study supports this notion that only a few pollutants were significantly associated with serum lipids levels while many individual pollutants had relatively weak associations (Figure 2 and Figure S3). The ERS as a multi-pollutant approach allows us to integrate those relatively small effects from multiple pollutants and provides a better opportunity to identify subpopulations that are at higher risk for diseases. We used multi-pollutant information at different steps of our process. Our discovery approach is different from Patel's [5,6] as we performed analysis with single pollutant models and then evaluated additional pollutants conditional on the identified pollutants. We then formed ERS using the set of all pollutants identified via this process using the weights from assessing them one at a time (ERS1) and jointly (ERS2). It appears that in terms of overall prediction, ERS1 and ERS2 were very similar in performance (Table 3), however, ERS2 was often slightly better in terms of risk stratification (Table 4). It is not possible to conclude definitively, without extensive and exhaustive simulation studies, which one performs better. Also,

Table 2. Estimated environmental risk score (ERS) weights for environmental pollutants selected for each phenotype.

Class	Variable name in NHANES	Pollutant Name	Weight[a] (10^-2) Total cholesterol ERS1[b]	Total cholesterol ERS2[c]	HDL ERS1[b]	HDL ERS2[c]	LDL ERS1[b]	LDL ERS2[c]	Triglyceride ERS1[b]	Triglyceride ERS2[c]
Heavy metals	LBXBPB	Lead in blood	1.71#	1.36#	1.62#	1.95#	2.54#	2.31#		
	LBXBCD	Cadmium in blood	1.18#	0.84#					4.69#	4.73#
	URXUCD	Cadmium in urine			-1.32#	-1.22#	0.98^	0.78		
	LBXTHG	Total mercury in blood							-2.95#	-1.65*
	LBXUHG	Mercury in urine							-2.15#	-1.58#
	URXUAB	Arsenobetaine in urine							-0.93#	-0.51^
	URXUSB	Antimony in urine			-1.23#	-0.43^				
Phthalates	URXMZP	Mono-benzyl phthalate			-0.62^	-0.09				
	URXMIB	Mono-isobutyl phthalate			-0.80#	-0.33				
	URXMBP	Mono-n-butyl phthalate			-0.75#	-0.09				
	URXMC1	Mono-(3-carboxylpropyl) phthalate			-0.70*	-0.17				
PAHs	URXP07	2-phenanthrene							1.41#	1.32#
PFCs	LBXPFHP	Perfluoroheptanoic acid					-3.99#	-3.84*		
Dioxins and Furans	LBXTCD	2,3,7,8-TCDD	0.64^	0.51^			1.55^	1.49^		
	LBXF03	2,3,4,7,8-PnCDF							1.72*	-0.24
	LBXF07	2,3,4,6,7,8-HxCDF							5.18#	4.71#
	LBXF08	1,2,3,4,6,7,8-HpCDF	0.82#	0.75*						
Dioxin-like PCBs	LBX066	PCB 066							2.44^	2.12^
	LBX105	PCB 105							2.05*	0.96
	LBX118	PCB 118							1.79#	0.34
	LBX156	PCB 156	0.54*	-0.36					1.59*	-0.90
	LBXPCB	3,3,4,5,5-PnCB							1.57#	0.70
	LBXHXC	3,3,4,5-HxCB	0.61*	-0.17					2.71#	2.15^
Non-dioxin-like PCBs	LBX099	PCB 099							1.76*	1.82
	LBX138	PCB 138							1.26^	-2.48
	LBX146	PCB 146	0.56^	-0.12					1.68^	-0.13
	LBX153	PCB 153							1.31^	1.41
	LBX156	PCB 156	0.54*	-0.36					1.59*	-0.90
	LBX170	PCB 170	0.79#	0.75					2.39#	3.36
	LBX177	PCB 177	0.46^	0.19					0.78	-1.41
	LBX180	PCB 180	0.69#	0.42					2.00#	-3.45
		PCB 183	0.48^	0.07					0.88	-1.27
	LBX187	PCB 187	0.69#	0.05					2.34#	2.41

Table 2. Cont.

Class	Variable name in NHANES	Pollutant Name	Weight^a (10^{-2})							
			Total cholesterol		HDL		LDL		Triglyceride	
			ERS1^b	ERS2^c	ERS1^b	ERS2^c	ERS1^b	ERS2^c	ERS1^b	ERS2^c
Organo-chlorine pesticides	LBXPDT	*p,p*-DDT							1.74$^\#$	0.78
	LBXOXY	Oxychlordane							2.64$^\#$	1.53*
	LBXHPE	Heptachlor Epoxide			−1.36$^\#$	−0.98$^\wedge$			3.18$^\#$	1.93$^\wedge$
	LBXDIE	Dieldrin			−1.36$^\#$	−0.58			3.03$^\#$	0.78
Dialkyl metabolites	URXOP2	Diethylphosphate					−0.35	−0.34		
Total number identified			13		9		5		27	

HDL, high-density lipoprotein cholesterol; LDL, low-density lipoprotein cholesterol; PAHs, polycyclic aromatic hydrocarbons; PFCs, perfluorinated compounds; PCBs, polychlorinated biphenyls; TCDD, tetrachlorodibenzodioxin; PnCDF, pentachlorodibenzofuran; HxCDF, hexachlorodibenzofuran; HpCDF, heptachlorodibenzofuran; PnCB, pentachlorobiphenyl; HxCB, hexachlorobiphenyl; DDT, dichlorodiphenyltrichloroethane.

All models were adjusted for age, gender, race/ethnicity, education, BMI and phenotype-specific nutrients shown in Table S3.

^{a}Weights were estimated using the training data (n = 11586).

^{b}ERS constructed with coefficient estimates from single-pollutant models as weights.

^{c}ERS constructed with coefficient estimates from multi-pollutant models as weights.

$^\#$*p*-value<0.001,

*0.001≤*p*-value<0.01, and $^\wedge$0.01≤*p*-value<0.05.

Table 3. Risk prediction by continuous environmental risk score (ERS) using single-phenotype approach[a] (n = 3847).

Phenotype	Continuous Outcome						Dichotomized[b] Outcome		
	Model 1[c]		ERS1[d]		ERS2[e]		Model 1[c]	ERS1[d]	ERS2[e]
	R²	PRESS[f]	R²	PRESS[f]	R²	PRESS[f]	AUC[g]	AUC[g]	AUC[g]
Total cholesterol	0.3270	122.46	0.3306	121.88	0.3308	121.85	0.7672 (0.7523, 0.7820)	0.7695 (0.7547, 0.7842)	0.7691 (0.7543, 0.7838)
HDL	0.2636	231.70	0.2677	230.52	0.2665	230.91	0.7193 (0.7024, 0.7362)	0.7217 (0.7050, 0.7385)	0.7208 (0.7040, 0.7376)
LDL	0.1342	539.52	0.1375	537.84	0.1376	537.80	0.7213 (0.7050, 0.7376)	0.7255 (0.7093, 0.7416)	0.7253 (0.7091, 0.7414)
Triglyceride	0.3709	967.24	0.3781	956.76	0.3775	957.70	0.8164 (0.8021, 0.8306)	0.8178 (0.8036, 0.8320)	0.8183 (0.8041, 0.8324)

[a]Pollutants selected by single-phenotype regression (n = 13, 9, 5 and 27 for total cholesterol, HDL, LDL and triglyceride, respectively) to construct ERS which was computed in the validation data (n = 3847), with adjustment for base covariates and phenotype-specific micronutrients.
[b]Continuous phenotypes dichotomized to be high vs. low by thresholds: 200 mg/dL for CHOL, 40 mg/dL (male) or 50 mg/dL (female) for HDL, 130 mg/dL for LDL and 150 mg/dL for TRIG.
[c]adjusted for base covariates and phenotype-specific micronutrients.
[d]Model 1 plus ERS constructed with coefficient estimates from single-pollutant models as weights.
[e]Model 1 plus ERS constructed with coefficient estimates from multi-pollutant models as weights.
[f]Predicted residual sums of squares.
[g]Area under the receiver operating characteristic (ROC) curve and its 95% confidence interval computed with 2000 stratified bootstrap replicates.

one could modify ERS2 by filtering potentially correlated predictors through variable selection, and reducing its variability. Although high risk groups were identified by the ERS in the present study, the ERS showed only modest improvement in lipid-related risk prediction of above and beyond the effect of traditional risk factors including sociodemographic and dietary factors (e.g., AUC improvements of 0.72 to 0.82, Table 3 and Table S5). This finding may not be surprising because a marker with an OR of 3 or lower is usually a poor tool for classifying or predicting risk for individuals [48]. In fact, the improvements of risk prediction/classification by the ERS are similar to the AUC improvements for coronary heart disease risk prediction by genetic risk scores (GRS) found in the Atherosclerosis Risk in Communities (ARIC) (from 0.742 to 0.749), Rotterdam Study (from 0.729 to 0.734) and Framingham Offspring Study (from 0.773 to 0.775) [49]. We also point out that for GWAS studies, a polygenic risk score has also contributed very modestly to risk prediction as measured by increment in AUC or R^2, however, similar risk stratification properties across the quintiles of genetic risk scores have been noted [23]. Nonetheless, our findings imply that ERS can better determine potential risk stratification where individuals are at increased risk of high lipid levels and related cardio-metabolic diseases than single pollutant approaches. The proposed ERS may allow us to identify susceptible subpopulations where targeted interventions are necessary and could have the greatest benefits [27].

Theoretically, a multiple phenotype approach always reduces the number of tests that are conducted, and also increases power by exploiting correlation across phenotypes. In our study, we discovered that the multi-phenotype approach leads to elevated ORs in Table 4, aiding with risk stratification. In general, if there is correlation among the pollutants, the discovery approach based on conditional associations may yield new results. If there is correlation among the outcomes or different phenotypes, the multi-phenotype approach, in spite of being a test with higher degrees of freedom, will yield a more powerful analysis. For example, six new pollutants were discovered with the multi-phenotype approach in our case study.

Our study has numerous limitations. The individual pollutants used to construct the ERS were identified in linear regression models with log-transformation due to skewed distributions, which assumes linear (in fact, log-linear) exposure-outcome relationships for all individual pollutants. However, not all pollutants are linearly associated with health outcomes, for example, some pesticides and/or other endocrine disrupting chemicals may have thresholds or non-monotonic dose-responses [50,51]. Pollutants whose dose-responses were misspecified may not be selected and not contribute to the ERS. Examining non-linearity in each of the single pollutant models may identify new pollutants but construction of a simple weighted risk score like ERS would no longer be possible, which led us to a linear regression based screening strategy in this initial paper. Moreover the ERS itself may have a non-linear association with the outcome when treated as a single predictor. We used quintiles of ERS to somewhat address this issue in the association models but a completely flexible generalized additive model will be more appropriate from a statistical point of view. We tried to retain simplicity in our approach for usability and thus compromised on some finer points that may be expanded upon in the future.

We did not consider pollutant-pollutant or pollutant-nutrient interactions when important individual pollutants were selected. Some pollutants may interact and have synergy. A well-known example is cigarette smoking and asbestos on lung cancer [52,53]. On the other hand, beneficial nutrients may mitigate toxic effects

Table 4. Odds ratios (95% CIs) for environmental risk score (ERS) categorized by quintile[a] (n = 3847).

Phenotype[b]	Single-phenotype Approach[c]		Multi-phenotype Approach[d]	
	ERS1[e]	ERS2[f]	ERS1[e]	ERS2[f]
Total cholesterol	1.450 (1.112, 1.892)	1.722 (1.317, 2.252)	1.781 (1.337, 2.374)	1.564 (1.191, 2.054)
HDL	1.372 (1.077, 1.748)	1.450 (1.144, 1.838)	1.471 (1.142, 1.894)	1.565 (1.230, 1.990)
LDL	1.824 (1.394, 2.386)	1.820 (1.391, 2.381)	1.357 (1.061, 1.735)	1.262 (0.973, 1.637)
Triglyceride	1.843 (1.366, 2.487)	1.536 (1.147, 2.056)	1.758 (1.275, 2.424)	2.027 (1.521, 2.703)

[a]Odds ratios for dichotomized phenotype (high vs. low) comparing subjects with ERS in the top 20% to those in the bottom 20%, adjusted for covariates and micronutrients.
[b]Dichotomization thresholds: 200 mg/dL for total cholesterol, 40 mg/dL (male) or 50 mg/dL (female) for HDL, 130 mg/dL for LDL and 150 mg/dL for triglyceride.
[c]Pollutants selected by single-phenotype regression (n = 13, 9, 5 and 27 for total cholesterol, HDL, LDL and triglyceride, respectively) to construct ERS, adjusted for phenotype-specific micronutrients.
[d]Pollutants selected by multi-phenotype regression (n = 45) to construct ERS, adjusted for union of selected micronutrients (n = 14).
[e]ERS constructed with coefficient estimates from single-pollutant models as weights.
[f]ERS constructed with coefficient estimates from multi-pollutant models as weights.

of pollutants. For example, people with higher intake of antioxidant vitamins, B-vitamins (folate and vitamin B12) or omega-3 fatty acids had lower effects of air pollution [54–56]. Conventional statistical approach that includes cross-product terms of two interacting factors may have low power and therefore effect estimates would be unstable. A recent study by Sun et al. [10] proposed statistical strategies to examine multi-pollutants and their interactions using a two-stage model. Other dimension reduction techniques may also work for estimating risk models when a large number of pollutants and their interactions exist. A planned future study accounting for pollutant-pollutant and pollutant-nutrient interactions is expected to improve the model prediction, and therefore, potentially the utility of ERS.

We used an arbitrary significance level of 0.01 to account for false positive rate. One reason is that we wanted to allow environmental pollutants that had even modest associations to be included in the ERS. We conducted sensitivity analyses using

significance levels of 0.05 and 0.001 and applied these different thresholds to the AUC as shown in Table 3. Under the significance level of 0.05, 30 pollutants (vs. 13 under the significance of 0.01) for total cholesterol; 16 (vs. 9) for HDL; 5 (vs. 5) for LDL; and 34 (vs. 27) for triglycerides were identified. However, the improvement in the AUC and OR were minimal. Using a significance level of 0.001, the number of pollutants identified decreased substantially, especially for LDL. The decrease in AUC was mainly for LDL while the decrease in OR was found for all phenotypes. Therefore we chose the intermediate threshold of 0.01. Even higher significance levels (e.g., alpha of 0.1) have been used as "pruning criteria" in genetic risk scores [57,58], therefore, genetic markers conferring only modest levels of disease risk could be aggregated in the risk score. In general, a liberal threshold is often noted to perform better for prediction as compared to controlling false discovery rate for identification of variables [59].

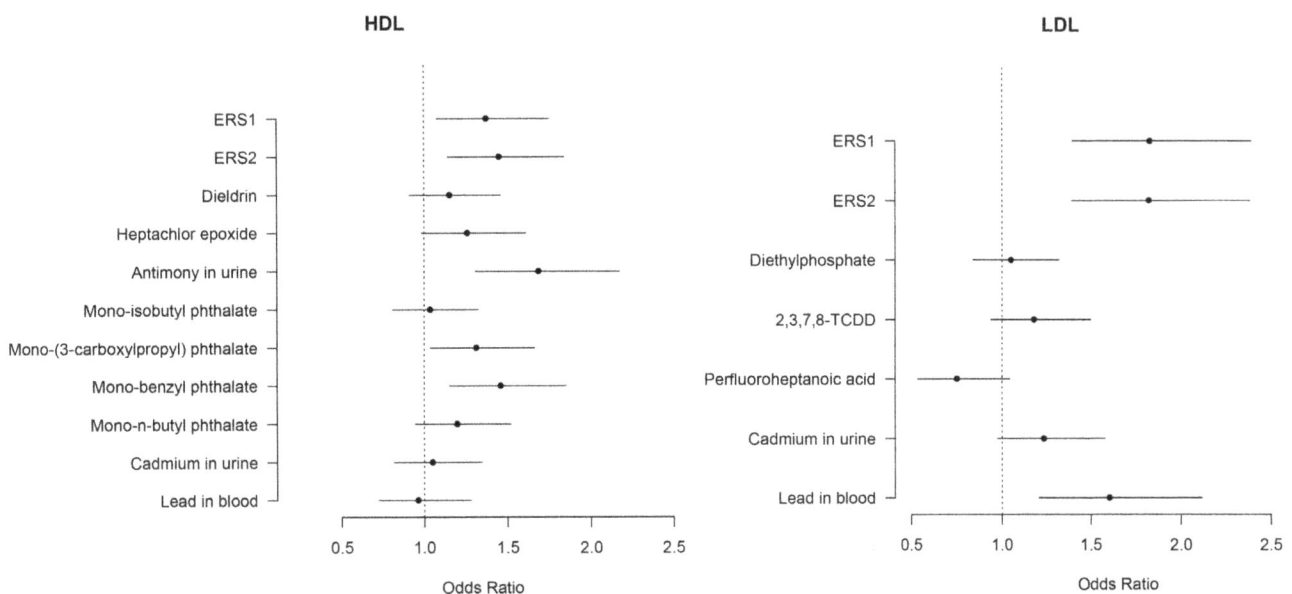

Figure 2. Odds ratios (95% confidence intervals) of having adverse levels of HDL (40 mg/dL for men and 50 mg/dL for women) and LDL (130 mg/dL) comparing the highest vs. the lowest quintiles of ERS and individual pollutants that compose the ERS. Models were adjusted for age, gender, race/ethnicity, education, BMI, and phenotype-specific micronutrients.

Although we include many environmental pollutants that are widespread and available in NHANES, we were not able to account for *all* environmental pollutants as it is unrealistic, the data were not available and not all environmental pollutants have been identified as yet. Also, we limited our analysis to chemical environmental pollutants in constructing the ERS. Recently, a new concept of the exposome, that is, the totality of exposures over the course of a lifetime, has been proposed [60–65] and the need for more complete *non-genetic* exposure assessment in epidemiologic research has been emerging, as emphasized in the strategic themes defined by the National Institute of Environmental Health Sciences (NIEHS) (http://www.niehs.nih.gov/about/strategicplan/). Our proposed approach will be useful to identify important individual factors and to combine their risks, which eventually will advance our understanding of health responses to the complex nature of multi-pollutant exposures.

Each individual pollutant has different degrees of measurement error. Exposure measurement errors are generally non-differential when the errors are independent of each other and the disease status [66]. Therefore, it is expected that environmental pollutants measured with less non-differential measurement error such as those with lower temporal variability are more likely to be detected (e.g., PCBs vs. phthalates). However, differential measurement errors may occur when exposure measurement errors are not independent because some of the effects of more poorly measured exposures may be transferred to the effect estimates of better-measured exposures [67]. In addition, most of the pollutant variables used in our study are subject to a limit of detection (LOD). Several *ad hoc* substitution methods, such as substitution of LOD/2 or LOD/$\sqrt{2}$ for values below LOD, are widely used (NHANES used LOD/$\sqrt{2}$). These *ad hoc* methods, however, can lead to bias especially when the proportion of values below LOD is high [68]. Maximum likelihood estimation based on a parametric joint distribution assumption for all the exposures, for example, multivariate normal distribution, may reduce potential bias if the parametric distribution assumption is correct [69].

Exposure data were collected cross-sectionally at one point in time, yet exposures are subject to temporal variation. This issue becomes particularly important when examining health effects of non-persistent short-lived environmental pollutants, such as BPA and phthalates. A recent study of urinary BPA and type-2 diabetes using three NHANES cycles found a significant association which was confirmed in one cycle (2003–2004) but not in the other two cycles. This finding indicates possible exposure misclassification due to a single urine sample [70]. Reliable exposure biomarker data assessed based on repeatedly collected samples is warranted to reduce exposure misclassification.

We did not consider differential risk prediction in different subpopulations. Emerging evidence suggests that certain subgroups may be more responsive to environmental pollutant exposure. Women are known to take up more divalent metals such as lead and cadmium due to iron depletion [71]. Stronger associations between lead and hypertension have been found in some racial/ethnic populations [72,73]. Sex- or race/ethnic group-specific biological differences, such as differences in body iron and estrogen levels between men and women, or socially determined gender- or race/ethnic group differences, such as different psychosocial stress levels, may confer susceptibility to health responses to pollutant exposures [74,75]. Sex-specific or race/ethnic group-specific ERS's may provide better risk prediction as well as risk assessment.

Our results may be biased due to residual confounding. Urinary creatinine adjustment has been recommended for urinary biomarkers to correct for dilutions of pollutant concentrations in

spot urine samples [76]. The main purpose of the present study is to introduce a novel ERS approach as a proof of concept illustration rather than to identify potential environmental factors related to health outcomes and estimate the associations as done in previous EWAS. Variance may be somewhat underestimated and the observed findings may not be generalizable to the US population.

Because not all environmental pollutants were measured in the entire population, we imputed unmeasured or missing pollutant data to maximize the power. We used a single imputation because our main goal was to introduce the approach of ERS, but multiple imputations after taking the uncertainty in imputed values into account would be a more appropriate approach. Imputation may be necessary for meta-analyses of multiple ERS studies in the future because it is unlikely that every cohort has a uniform set of pollutants measured. Careful data harmonization and imputation may increase the power of the analysis if correlated exposures and covariates are observed in one cohort that are predictive of exposures in another cohort where those exposures are missing. However, the imputation issue will merit a complete paper in its own right, as imputation with high dimensional data is still very much an evolving topic in statistical research [77]. In summary, the present study suggests ERS is a promising tool for integrating disease risks from multi-pollutant mixture exposures. The ERS is a simplest form of data reduction, characterizing the summary exposure burden like a polygenic risk score in genetics [27]. This new approach supports the need for moving from a single-pollutant to a multi-pollutant framework for new discoveries and better risk stratification. Combining information from ERS along with known predictors can improve disease prediction. Also, the ERS along with genetic risk score can potentially provide a way to reduce dimension and increase the power in studies of gene-environment interaction. More generally, ERS can be taken as a measure of summary/background burden of environmental exposure and it will be interesting to explore whether the effect of a certain gene, behavioral factors (diet, physical activity, smoking) or another pollutant is larger if individuals are in the highest quartile of ERS. The contribution of ERS to risk prediction and classification warrants further studies.

Data Sharing: The data and codes used for illustration of our approach are available at http://www-personal.umich.edu/bhramar/software/.

Supporting Information

Figure S1 Manhattan plots representing the P value distributions of the individual environmental pollutants examined using the stage 1 samples. Y-axis indicates −log10(p-value) of the regression coefficient for each of the environmental pollutants, adjusted for age, gender, race/ethnicity, education, body mass index and phenotype-specific micronutrients. The horizontal dotted line represents the p-value of 0.01. X-axis indicates 13 classes of environmental pollutants: 1) heavy metals; 2) phthalates; 3) environmental phenols; 4) polycyclic aromatic hydrocarbons (PAHs); 5) volatile organic compounds (VOCs); 6) perfluorinated compounds (PFCs); 7) dioxins and furans; 8) dioxin-like polychlorinated biphenyls (PCBs); 9) non-dioxin-like PCBs; 10) organochlorine pesticides; 11) organophosphate dialkyl metabolites; 12) herbicides; and 13) pesticides phenols. Each color represents one class.

Figure S2 Receiver operating characteristic (ROC) curves for four phenotypes. The dotted line denotes the null curve. The black curve is for the model with only covariates. The blue curve is for the model with both covariates and phenotype-specific micronutrients. The red curve is for the model with environmental risk score (ERS), covariates and phenotype-specific micronutrients.

Figure S3 Odds ratios (95% confidence intervals) of having adverse levels of total cholesterol (CHOL: 200 mg/dL) and triglyceride (TRIG: 150 mg/dL) comparing the highest vs. the lowest quintiles of ERS and individual pollutants that compose the ERS. Models were adjusted for age, gender, race/ethnicity, education, BMI, and phenotype-specific micronutrients.

Table S1 Environmental pollutants evaluated in the present study (n = 134).

Table S2 Spearman correlation coefficients between four phenotypes.

Table S3 Micronutrients selected for each phenotype using Bayesian model averaging (BMA).

Table S4 Distributions of Environmental Risk Scores (ERS) (n = 3847).

Table S5 Risk prediction by continuous environmental risk score (ERS) using multi-phenotype approach[a] (n = 3847).

Table S6 Regression outputs for each lipid outcome in relation to ERS1.

File S1 Diagnostic Analysis for the Imputation.

Author Contributions

Conceived and designed the experiments: SKP BM. Analyzed the data: YT SKP BM. Wrote the paper: SKP YT JDM SDH BM.

References

1. Mauderly JL, Samet JM (2009) Is there evidence for synergy among air pollutants in causing health effects? Environmental Health Perspectives 117: 1–6.
2. Guallar E, Sanz-Gallardo MI, van't Veer P, Bode P, Aro A, et al. (2002) Mercury, fish oils, and the risk of myocardial infarction. The New England journal of medicine 347: 1747–1754.
3. Stern AH, Korn LR (2011) An approach for quantitatively balancing methylmercury risk and omega-3 benefit in fish consumption advisories. Environmental health perspectives 119: 1043–1046.
4. Porta M, Pumarega J, Gasull M (2012) Number of persistent organic pollutants detected at high concentrations in a general population. Environment international 44: 106–111.
5. Patel CJ, Bhattacharya J, Butte AJ (2010) An Environment-Wide Association Study (EWAS) on type 2 diabetes mellitus. PLoS One 5: e10746.
6. Patel CJ, Cullen MR, Ioannidis JP, Butte AJ (2012) Systematic evaluation of environmental factors: persistent pollutants and nutrients correlated with serum lipid levels. International journal of epidemiology 41: 828–843.
7. Tzoulaki I, Patel CJ, Okamura T, Chan Q, Brown IJ, et al. (2012) A nutrient-wide association study on blood pressure. Circulation 126: 2456–2464.
8. Patel CJ, Rehkopf DH, Leppert JT, Bortz WM, Cullen MR, et al. (2013) Systematic evaluation of environmental and behavioural factors associated with all-cause mortality in the United States National Health and Nutrition Examination Survey. International journal of epidemiology 42: 1795–1810.
9. Lind PM, Riserus U, Salihovic S, Bavel B, Lind L (2013) An environmental wide association study (EWAS) approach to the metabolic syndrome. Environment international 55: 1–8.
10. Sun Z, Tao Y, Li S, Ferguson KK, Meeker JD, et al. (2013) Statistical strategies for constructing health risk models with multiple pollutants and their interactions: possible choices and comparisons. Environmental health: a global access science source 12: 85.
11. Tibshirani R (1996) Regression Shrinkage and Selection via the Lasso. Journal of the Royal Statistical Society Series B (Methodological) 58: 267–288.
12. Madigan D, Raftery AE (1994) Model selection and accounting for model uncertainty in graphical models using Occam's window. Journal of the American Statistical Association 89: 1535–1546.
13. Bair E, Hastie T, Paul D, Tibshirani R (2006) Prediction by supervised principal components. Journal of the American Statistical Association 101: 119–137.
14. Zou H (2006) The Adaptive Lasso and Its Oracle Properties. Journal of the American Statistical Association 101: 1418–1429.
15. Zou H, Zhang HH (2009) On the Adaptive Elastic-Net with a Diverging Number of Parameters. Annals of statistics 37: 1733–1751.
16. Billionnet C, Sherrill D, Annesi-Maesano I (2012) Estimating the health effects of exposure to multi-pollutant mixture. Annals of epidemiology 22: 126–141.
17. Bobb JF, Dominici F, Peng RD (2013) Reduced hierarchical models with application to estimating health effects of simultaneous exposure to multiple pollutants. Journal of the Royal Statistical Society Series C, Applied statistics 62.
18. Dominici F, Peng RD, Barr CD, Bell ML (2010) Protecting human health from air pollution: shifting from a single-pollutant to a multipollutant approach. Epidemiology 21: 187–194.
19. Park SK, O'Neill MS, Stunder BJ, Vokonas PS, Sparrow D, et al. (2007) Source location of air pollution and cardiac autonomic function: trajectory cluster analysis for exposure assessment. Journal of exposure science & environmental epidemiology 17: 488–497.
20. Sarnat SE, Suh HH, Coull BA, Schwartz J, Stone PH, et al. (2006) Ambient particulate air pollution and cardiac arrhythmia in a panel of older adults in Steubenville, Ohio. Occupational and environmental medicine 63: 700–706.
21. Laden F, Neas LM, Dockery DW, Schwartz J (2000) Association of fine particulate matter from different sources with daily mortality in six U.S. cities. Environmental health perspectives 108: 941–947.
22. Ostro B, Tobias A, Querol X, Alastuey A, Amato F, et al. (2011) The effects of particulate matter sources on daily mortality: a case-crossover study of Barcelona, Spain. Environmental health perspectives 119: 1781–1787.
23. Wilson PW, D'Agostino RB, Levy D, Belanger AM, Silbershatz H, et al. (1998) Prediction of coronary heart disease using risk factor categories. Circulation 97: 1837–1847.
24. Janssens AC, Ioannidis JP, Bedrosian S, Boffetta P, Dolan SM, et al. (2011) Strengthening the reporting of genetic risk prediction studies (GRIPS): explanation and elaboration. European journal of human genetics: EJHG 19: 18 p preceding 494.
25. Janssens AC, Ioannidis JP, van Duijn CM, Little J, Khoury MJ (2011) Strengthening the reporting of Genetic RIsk Prediction Studies: the GRIPS Statement. PLoS medicine 8: e1000420.
26. Willems SM, Mihaescu R, Sijbrands EJ, van Duijn CM, Janssens AC (2011) A methodological perspective on genetic risk prediction studies in type 2 diabetes: recommendations for future research. Current diabetes reports 11: 511–518.
27. Garcia-Closas M, Rothman N, Figueroa JD, Prokunina-Olsson L, Han SS, et al. (2013) Common genetic polymorphisms modify the effect of smoking on absolute risk of bladder cancer. Cancer research 73: 2211–2220.
28. Mondul AM, Shui IM, Yu K, Travis RC, Stevens VL, et al. (2013) Genetic variation in the vitamin d pathway in relation to risk of prostate cancer–results from the breast and prostate cancer cohort consortium. Cancer epidemiology, biomarkers & prevention: a publication of the American Association for Cancer Research, cosponsored by the American Society of Preventive Oncology 22: 688–696.
29. van Meurs JB, Pare G, Schwartz SM, Hazra A, Tanaka T, et al. (2013) Common genetic loci influencing plasma homocysteine concentrations and their effect on risk of coronary artery disease. The American journal of clinical nutrition 98: 668–676.
30. Kim S, Sohn KA, Xing EP (2009) A multivariate regression approach to association analysis of a quantitative trait network. Bioinformatics 25: i204–212.
31. O'Reilly PF, Hoggart CJ, Pomyen Y, Calboli FC, Elliott P, et al. (2012) MultiPhen: joint model of multiple phenotypes can increase discovery in GWAS. PLoS One 7: e34861.
32. Stephens M (2013) A unified framework for association analysis with multiple related phenotypes. PLoS One 8: e65245.
33. Satagopan JM, Verbel DA, Venkatraman ES, Offit KE, Begg CB (2002) Two-stage designs for gene-disease association studies. Biometrics 58: 163–170.
34. Raghunathan TE, Solenberger PW, Van Hoewyk J (2002) IVEware: Imputation and variance estimation software. Ann Arbor, MI: Survey Research Center, Institute for Social Research, University of Michigan.
35. Raghunathan TE, Lepkowski JM, Van Hoewyk J, Solenberger P (2001) A multivariate technique for multiply imputing missing values using a sequence of regression models. Survey Methodology 27: 85–95.
36. Skol AD, Scott LJ, Abecasis GR, Boehnke M (2006) Joint analysis is more efficient than replication-based analysis for two-stage genome-wide association studies. Nature genetics 38: 209–213.

37. Yang J, Ferreira T, Morris AP, Medland SE, Madden PA, et al. (2012) Conditional and joint multiple-SNP analysis of GWAS summary statistics identifies additional variants influencing complex traits. Nature genetics 44: 369–375, S361–363.

38. National Institutes of Health, National Heart Lung, and Blood Institute (2001) Third Report of the National Cholesterol Education Program (NCEP) Expert Panel on Detection, Evaluation, and Treatment of High Blood Cholesterol in Adults (Adult Treatment Panel III). U.S. Department of Health and Human Services, National Institutes of Health, National Heart, Lung, and Blood Institute.

39. Carpenter J, Bithell J (2000) Bootstrap confidence intervals: when, which, what? A practical guide for medical statisticians. Statistics in medicine 19: 1141–1164.

40. Robin X, Turck N, Hainard A, Tiberti N, Lisacek F, et al. (2011) pROC: an open-source package for R and S+ to analyze and compare ROC curves. BMC bioinformatics 12: 77.

41. Gibson G (2010) Hints of hidden heritability in GWAS. Nature genetics 42: 558–560.

42. Johns DO, Stanek LW, Walker K, Benromdhane S, Hubbell B, et al. (2012) Practical advancement of multipollutant scientific and risk assessment approaches for ambient air pollution. Environmental health perspectives 120: 1238–1242.

43. Billionnet C, Gay E, Kirchner S, Leynaert B, Annesi-Maesano I (2011) Quantitative assessments of indoor air pollution and respiratory health in a population-based sample of French dwellings. Environmental research 111: 425–434.

44. Qian Z, Zhang J, Korn LR, Wei F, Chapman RS (2004) Factor analysis of household factors: are they associated with respiratory conditions in Chinese children? International journal of epidemiology 33: 582–588.

45. Roberts S, Martin M (2005) A critical assessment of shrinkage-based regression approaches for estimating the adverse health effects of multiple air pollutants. Atmospheric Environment 39: 6223–6230.

46. Roberts S, Martin MA (2006) Using supervised principal components analysis to assess multiple pollutant effects. Environmental health perspectives 114: 1877–1882.

47. Hong YC, Leem JH, Ha EH, Christiani DC (1999) PM(10) exposure, gaseous pollutants, and daily mortality in Inchon, South Korea. Environmental health perspectives 107: 873–878.

48. Pepe MS, Janes H, Longton G, Leisenring W, Newcomb P (2004) Limitations of the odds ratio in gauging the performance of a diagnostic, prognostic, or screening marker. American journal of epidemiology 159: 882–890.

49. Brautbar A, Pompeii LA, Dehghan A, Ngwa JS, Nambi V, et al. (2012) A genetic risk score based on direct associations with coronary heart disease improves coronary heart disease risk prediction in the Atherosclerosis Risk in Communities (ARIC), but not in the Rotterdam and Framingham Offspring, Studies. Atherosclerosis 223: 421–426.

50. Rhomberg LR, Goodman JE (2012) Low-dose effects and nonmonotonic dose-responses of endocrine disrupting chemicals: has the case been made? Regulatory toxicology and pharmacology: RTP 64: 130–133.

51. Vandenberg LN, Colborn T, Hayes TB, Heindel JJ, Jacobs DR, Jr., et al. (2012) Hormones and endocrine-disrupting chemicals: low-dose effects and nonmonotonic dose responses. Endocrine reviews 33: 378–455.

52. Hammond EC, Selikoff IJ, Seidman H (1979) ASBESTOS EXPOSURE, CIGARETTE SMOKING AND DEATH RATES*. Annals of the New York Academy of Sciences 330: 473–790.

53. Saracci R (1977) Asbestos and lung cancer: An analysis of the epidemiological evidence on the asbestos–smoking interaction. International Journal of Cancer 20: 323–331.

54. Park SK, O'Neill MS, Vokonas PS, Sparrow D, Spiro A, 3rd, et al. (2008) Traffic-related particles are associated with elevated homocysteine: the VA normative aging study. American journal of respiratory and critical care medicine 178: 283–289.

55. Samet JM, Hatch GE, Horstman D, Steck-Scott S, Arab L, et al. (2001) Effect of antioxidant supplementation on ozone-induced lung injury in human subjects. American journal of respiratory and critical care medicine 164: 819–825.

56. Tong H, Rappold AG, Diaz-Sanchez D, Steck SE, Berntsen J, et al. (2012) Omega-3 fatty acid supplementation appears to attenuate particulate air pollution-induced cardiac effects and lipid changes in healthy middle-aged adults. Environmental health perspectives 120: 952–957.

57. Morrison AC, Bare LA, Chambless LE, Ellis SG, Malloy M, et al. (2007) Prediction of coronary heart disease risk using a genetic risk score: the Atherosclerosis Risk in Communities Study. American journal of epidemiology 166: 28–35.

58. Derks EM, Vorstman JA, Ripke S, Kahn RS, Ophoff RA (2012) Investigation of the genetic association between quantitative measures of psychosis and schizophrenia: a polygenic risk score analysis. PLoS One 7: e37852.

59. Chatterjee N, Wheeler B, Sampson J, Hartge P, Chanock SJ, et al. (2013) Projecting the performance of risk prediction based on polygenic analyses of genome-wide association studies. Nature genetics 45: 400–405, 405e401–403.

60. Brunekreef B (2013) Exposure science, the exposome, and public health. Environmental and molecular mutagenesis 54: 596–598.

61. Buck Louis GM, Sundaram R (2012) Exposome: time for transformative research. Statistics in medicine 31: 2569–2575.

62. Rappaport SM (2011) Implications of the exposome for exposure science. Journal of exposure science & environmental epidemiology 21: 5–9.

63. Rappaport SM, Smith MT (2010) Epidemiology. Environment and disease risks. Science 330: 460–461.

64. Wild CP (2005) Complementing the genome with an "exposome": the outstanding challenge of environmental exposure measurement in molecular epidemiology. Cancer epidemiology, biomarkers & prevention: a publication of the American Association for Cancer Research, cosponsored by the American Society of Preventive Oncology 14: 1847–1850.

65. Wild CP (2012) The exposome: from concept to utility. International journal of epidemiology 41: 24–32.

66. Rothman KJ, Greenland S (1998) Precision and validity in epidemiologic studies. In: Rothman KJ, Greenland S, editors. Modern Epidemiology. 2nd ed. Philadelphia, PA: Lippincott-Raven. pp. 115–134.

67. Zeger SL, Thomas D, Dominici F, Samet JM, Schwartz J, et al. (2000) Exposure measurement error in time-series studies of air pollution: concepts and consequences. Environmental health perspectives 108: 419–426.

68. Cole SR, Chu H, Nie L, Schisterman EF (2009) Estimating the odds ratio when exposure has a limit of detection. International journal of epidemiology 38: 1674–1680.

69. Nie L, Chu H, Liu C, Cole SR, Vexler A, et al. (2010) Linear regression with an independent variable subject to a detection limit. Epidemiology 21 Suppl 4: S17–24.

70. Silver MK, O'Neill MS, Sowers MR, Park SK (2011) Urinary bisphenol A and type-2 diabetes in U.S. adults: data from NHANES 2003–2008. PLoS One 6: e26868.

71. Vahter M, Akesson A, Liden C, Ceccatelli S, Berglund M (2007) Gender differences in the disposition and toxicity of metals. Environmental research 104: 85–95.

72. Scinicariello F, Abadin HG, Murray HE (2011) Association of low-level blood lead and blood pressure in NHANES 1999–2006. Environmental research 111: 1249–1257.

73. Vupputuri S, He J, Muntner P, Bazzano LA, Whelton PK, et al. (2003) Blood lead level is associated with elevated blood pressure in blacks. Hypertension 41: 463–468.

74. Clougherty JE (2010) A growing role for gender analysis in air pollution epidemiology. Environmental health perspectives 118: 167–176.

75. Hicken MT, Gee GC, Connell C, Snow RC, Morenoff J, et al. (2013) Black-white blood pressure disparities: depressive symptoms and differential vulnerability to blood lead. Environmental health perspectives 121: 205–209.

76. Barr DB, Wilder LC, Caudill SP, Gonzalez AJ, Needham LL, et al. (2005) Urinary creatinine concentrations in the U.S. population: implications for urinary biologic monitoring measurements. Environmental health perspectives 113: 192–200.

77. Boonstra PS, Taylor JM, Mukherjee B (2013) Incorporating auxiliary information for improved prediction in high-dimensional datasets: an ensemble of shrinkage approaches. Biostatistics 14: 259–272.

2

Proteomic Strategy for the Analysis of the Polychlorobiphenyl-Degrading Cyanobacterium *Anabaena* PD-1 Exposed to Aroclor 1254

Hangjun Zhang*, Xiaojun Jiang, Wenfeng Xiao, Liping Lu

College of Life and Environmental Sciences, Hangzhou Normal University, Hangzhou, Zhejiang, China

Abstract

The cyanobacterium *Anabaena* PD-1, which was originally isolated from polychlorobiphenyl (PCB)-contaminated paddy soils, has capabilities for dechlorinatin and for degrading the commercial PCB mixture Aroclor 1254. In this study, 25 upregulated proteins were identified using 2D electrophoresis (2-DE) coupled with matrix-assisted laser desorption/ionization time of flight mass spectrometry (MALDI-TOF MS). These proteins were involved in (i) PCB degradation (i.e., 3-chlorobenzoate-3,4-dioxygenase); (ii) transport processes [e.g., ATP-binding cassette (ABC) transporter substrate-binding protein, amino acid ABC transporter substrate-binding protein, peptide ABC transporter substrate-binding protein, putrescine-binding protein, periplasmic solute-binding protein, branched-chain amino acid uptake periplasmic solute-binding protein, periplasmic phosphate-binding protein, phosphonate ABC transporter substrate-binding protein, and xylose ABC transporter substrate-binding protein]; (iii) energetic metabolism (e.g., methanol/ethanol family pyrroloquinoline quinone (PQQ)-dependent dehydrogenase, malate-CoA ligase subunit beta, enolase, ATP synthase β subunit, F_OF_1 ATP synthase subunit beta, ATP synthase α subunit, and IMP cyclohydrolase); (iv) electron transport (cytochrome b_6f complex Fe-S protein); (v) general stress response (e.g., molecular chaperone DnaK, elongation factor G, and translation elongation factor thermostable); (vi) carbon metabolism (methanol dehydrogenase and malate-CoA ligase subunit beta); and (vii) nitrogen reductase (nitrous oxide reductase). The results of real-time polymerase chain reaction showed that the genes encoding for dioxygenase, ABC transporters, transmembrane proteins, electron transporter, and energetic metabolism proteins were significantly upregulated during PCB degradation. These genes upregulated by 1.26- to 8.98-fold. These findings reveal the resistance and adaptation of cyanobacterium to the presence of PCBs, shedding light on the complexity of PCB catabolism by *Anabaena* PD-1.

Editor: Sompop Bencharit, University of North Carolina at Chapel Hill, United States of America

Funding: This work was supported by the Natural Science Foundation of China (21077030), the Program for Excellent Young Teachers in Hangzhou Normal University (JTAS 2011-01-012), the Program for talents in Hangzhou City, and the Undergraduates innovation ability promotion project in Hangzhou Normal University (CX2013078). The funders had no role in study design, data collection and analysis, decision to publish, or preparation of the manuscript.

Competing Interests: The authors have declared that no competing interests exist.

* E-mail: zhanghangjun@gmail.com

Introduction

Persistent organic pollutants (POPs), such as polychlorinated biphenyls (PCBs) and organochlorine pesticides, are ubiquitous chloroorganic chemicals in the environment. PCBs were first manufactured in the United States in 1929 [1]. They are a complex class of hydrophobic, lipophilic chemicals that often slowly decompose and metabolize in natural systems [2]. Although PCB production has been banned in 1979, they still pose environmental and human risks in areas of hotspot contamination because of their stable physicochemical properties, hydrophobic properties, and high toxicity [3,4]. Therefore, cleaning residual PCB-contaminated environments has elicited significant research attention in the last few decades.

Several physical, chemical, and biological methods are available for PCB degradation [5]. Biological methods for PCB degradation have been extensively studied and considered crucial for the biodegradation of PCBs because of their low environmental impact and economic advantage compared with physicochemical methods [6]. PCB degradation is exhibited by several bacteria and

fungi, such as *Pseudomonas pseudoalcaligenes* KF707 [7], *Burkholderia cepacia* LB400 [7], *Sinorhizobium meliloti* [8], *Hydrogenophaga* sp. strain IA3-A [9], *Pleurotus ostreatus* [10], and *Ceriporia* sp. ZLY-2010 [11].

Microalgae and cyanobacteria are common species in the natural environment. They can act as distinctive biological agents for organic pollutant degradation [12] and can be used to degrade organic pollutants. The use of microalgae and cyanobacteria has become a new method for POP degradation in recent years. Kotzabasis et al. [13,14] have reported that *Scenedesmus obliquus* can biodegrade dichlorophenols. *Chlorella fusca* var. *vacuolata* can remove 23% of 2,4-dichlorophenol after 4 d [15]. The microalga *Cyclotella caspia* can degrade the aromatic pollutant nonylphenol [16]. The *Anabaena flos-aquae* strain 4054 can decompose endocrine-disrupting pollutants, such as phthalate esters [17]. *Anabaena azotica*, another common cyanobacterium, can effectively degrade the organochlorine pesticide γ-hexachlorocyclohexane (lindane) [18]. Therefore, cyanobacterial species with degradation functions may be a potential choice for PCB degradation.

The survival of wild-type microorganisms with degradation function may be limited by adverse environmental conditions, leading to reduced degradation efficiency [19]. Fortunately, genetically engineered microbes can enhance degradation efficiency by enhancing the activity of key enzymes via genetic engineering [20]. The genes and proteins in microbes have important functions in organic pollutant degradation [21–23]. The molecular mechanism by which microorganisms degrade PCB has also been explored by utilizing proteomic technologies, including 2D electrophoresis (2-DE) and matrix-assisted laser desorption/ionization time-of-flight mass spectrometry (MALDI-TOF MS) [24]. The enzymes in the PCB degradation pathway include biphenyl 2,3-dioxygenases, cis-2,3-dihydro-2,3-dihydroxybiphenyl dehydrogenase, 2,3-dihydroxybiphenyl 1,2-dioxygenases, 2-hydroxy-6-phenyl-6-oxohexa-2,4-dienoate hydrolases, 2-hydroxy-penta-2,4-dienoate hydratase, 4-hydroxy-2-oxovalerate aldolase, and acetaldehyde dehydrogenase [25]. Exposure to aromatic compounds stimulates metabolic enzymes and other polypeptides in microorganisms. Proteins involved in energy metabolism and substrate transport are upregulated during the degradation of aromatic pollutants and organochlorine pesticides by various microorganisms [26–28].

Our previous study produced encouraging results in PCB biodegradation by cyanobacteria. The cyanobacterium *Anabaena* PD-1 was originally isolated from PCB-contaminated paddy soil and exhibited strong ability to degrade PCB congeners (data not shown). However, proteomic analyses of cyanobacterial responses to stressors have mainly focused on salt [29], acid [30], and arsenic [31]. Limited information is available on the cyanobacterial catabolism of PCBs and the responses of cyanobacteria to PCBs [12]. Therefore, the key enzymes involved in PCB degradation need to be identified and the degradation mechanism should be explored to gain important information on the construction of genetically engineered cyanobacteria. Such genetically engineered cyanobacteria may achieve the same or higher PCB degradation efficiency compared with laboratory conditions.

In this study, we separated differentially expressed proteins through 2-DE and identified polypeptides through MALDI-TOF tandem mass spectrometry (MS/MS). Protein information was obtained from the NCBInr database through its Mascot search engine. Real-time PCR was utilized to analyze the genes encoding for highly expressed key proteins during PCB degradation in *Anabaena* PD-1 cells. The present contribution can provide new insights into the biodegradation of PCBs by *Anabaena* PD-1 and into the construction of genetically engineered PCB-degrading cyanobacterial species.

Materials and Methods

Strain *Anabaena* PD-1 and culture conditions

Anabaena PD-1, a PCB-tolerant strain, was isolated from PCB-contaminated paddy soil in Taizhou, Zhejiang, China (No specific permissions were required for the sampling locations and activities. The field studies did not involve endangered and protected species. The sampling site in the study is located at Latitude 28°32′N Longitude 121°27′E.). *Anabaena* PD-1 cells were grown at 25°C, 998 lux, in BG-11 medium [32], under discontinuous illumination (light : dark = 12 h : 12 h). Cyanobacterial cells in their exponential phase were cultivated with and without Aroclor 1254 (2 mg/L) for 30 d for the PCB-degrading experiment.

Preparation of cellular proteins

The cells were cultured for 30 d and ground to powder with liquid nitrogen. Subsequently, 10 mL of cooled acetone containing 10% trichloroacetic acid and 0.07% DTT was added to 1 g of sample powder at −20°C for 1 h. The deposit was collected after centrifugation at 15000 g for 15 min at 4°C. Cooled acetone containing 0.07% DTT was then added to the deposit at −20°C for 1 h. After another centrifugation at 15000 g for 15 min at 4°C, the deposit was collected and dried with a vacuum freeze dryer. The powder was dissolved in a lysis solution [9 M urea, 4% CHAPS, 1% DTT, 1% IPG buffer (GE Healthcare)] at 50 μL: 1 mg, dissolved at 30°C for 1 h, and then centrifuged again at 15000 g for 15 min at room temperature. The concentrations of the protein extracts were determined with the Bradford method [33]. The extracts were then stored at −80°C for isoelectric focusing electrophoresis (I FE).

2-DE

IFE. Samples containing 200 μg of proteins were mixed with a fresh rehydration buffer [9 M urea, 4% CHAPS, 1% DTT, 1% IPG buffer (GE Healthcare), trace amount of bromophenol blue] to form a 450 μL mixture. DryStrip (GE Healthcare, 24 cm, pH 3 to 10, NL) was obtained from a −20°C freezer, placed at room temperature for 10 min, added to the protein sample in the strip holder, and then subjected to IEF according to the following protocol: rehydration at 50 V (12 h), 500 V (1 h), 1000 V (1 h), 10000 V (1 h), and 10000 V (10 h). All steps were controlled at 50 μA/gel at 20°C.

Equilibration and SDS-PAGE

After IEF, the strip removed from the strip holder was incubated in equilibration buffer 1 [6 M urea, 30% glycerol, 2% SDS, 50 mM Tris-HCl (pH 8.8), 1% DTT, and trace amount of bromophenol blue] for 15 min and then in equilibration buffer 2 [6 M urea, 30% glycerol, 2% SDS, 50 mM Tris-HCl (pH 8.8), 2.5% iodoacetamide, and trace amount of bromophenol blue] for 15 min. After the strip was rinsed with SDS-PAGE buffer for 10 s, a sealing solution was added to the surface of the SDS-PAGE gel. The gel was then moved to the electrophoresis apparatus for electrophoresis at the following parameters: 100 V, 15°C, 45 min, followed by 200 V for 6 h to 8 h (Ettan DALTsix system). The gels were then stained with silver nitrate according to the method described by Shevchenk et al. [34].

Gel visualization, scanning, and analysis

The gels were visualized by silver staining for analysis. The stained gels were scanned by an image scanner (GE Healthcare, USA) at a resolution of 300 dots per inch. All gel images were processed by spot detection, volumetric quantification, and matching with PDQuest 8.0 software. The differences in protein content between the treatment and control groups were calculated as fold ratios. A fold change ≥2.0 or ≤0.5 was utilized to differentially select protein spots.

In-gel digestion and MS analysis

The proteins were digested by 50% ceric ammonium nitrate for 5 min, followed by 100% ACN for 5 min, and then rehydrated in 2 μL to 4 μL trypsin (Promega, Madison, USA) solution (20 μg/mL in 25 mM NH$_4$HCO$_3$) for 30 min. A 20 μL cover solution (25 mM NH$_4$HCO$_3$) was then added for 16 h of digestion at 37°C. Afterward, the supernatants were transferred into another tube, and the gels were extracted once with a 50 μL extraction buffer (67% ACN and 5% TFA). The peptide extracts and supernatant of the gel spot were combined and completely dried. The samples were analyzed with an ABI 4800 MALDI-TOF/TOF Plus mass spectrometer (Applied Biosystems, Foster City, USA). Data were obtained with a positive MS reflector. CalMix5 standard was utilized to calibrate the instrument (ABI4800 calibration mixture). MS and MS/MS data were integrated and processed with GPS Explorer V3.6 (Applied Biosystems, USA) with default parameters. According to combined MS and MS/MS spectra, the proteins were successfully identified at 95% or higher confidence interval of their scores in the MASCOT V2.1 search engine (Matrix Science, London, UK). The following search parameters were used: NCBInr database; trypsin as the digestion enzyme, one missed cleavage site, fixed modifications of carbamidomethyl (C), partial modifications of acetyl (protein N-term), and oxidization (M); 200 ppm for precursor ion tolerance and 0.5 Da for fragment ion tolerance.

RNA preparation and real time-PCR

Total RNA was extracted from 50 mg to 80 mg of cyanobacterial cell pellets ($OD_{680} = 0.38$) using TRIzol Reagent (Invitrogen, Carlsbad, CA, USA). RNA was then purified by removal of genomic DNA contaminants using an RNase-free DNase I kit (Invitrogen) and verified by determination of 260/280 nm ratios and 1% agarose–formaldehyde gel electrophoresis with ethidium bromide staining. Total RNA was subjected to cDNA synthesis using a NuGEN OvationW Prokaryotic RNA-Seq System according to the manufacturer's instructions (Haoji Biotechonlogy Hangzhou, China). RT-PCR was carried out with a multiplex real-time PCR detector (BioRad, USA). The reaction mixture included a Power Master Mix (Invitrogen), 0.5 mM of the primers, MilliQ water, and 1 mL of cDNA. The thermal cycling program was as follows: predegeneration at 95°C for 1 min, followed by 40 cycles of denaturation at 95°C for 10 s, and then annealing at 62°C for 25 s. Primer sequences for 3-chlorobenzoate-3,4-dioxygenase, cytochrome b_6f complex Fe-S protein, transporter proteins, and energetic metabolism proteins are shown in Table 1.

Statistical analysis

The statistical differences of the experimental data were determined using one-way ANOVA followed by two-sided Dunnett's t-test. Statistical tests were conducted using SPSS11.0, and the statistical significance values were defined as *$P<0.05$ and **$P<0.01$. All data were expressed as mean±standard deviation (S.D.).

Results

Protein expression patterns of *Anabaena* PD-1 exposed to Aroclor 1254

Total proteins extracted from the PCB-treated and reference cyanobacterial samples were separated through 2-DE. Most proteins were located between pH 4.0 and 6.8 and then weighed 17 kDa to 96 kDa (Figure 1 and Figure 2). Exposure to Aroclor 1254 diversified the overall protein expression patterns of *Anabaena* PD-1. A total of 25 protein spots were up-regulated (multiple changes were twice greater; Student's t-test, $P<0.05$). Twenty-six differentially expressed proteins were identified through MALDI-TOF MS/MS. Detailed information on these proteins is summarized in Table 2. The proteins involved in PCB-degradation, transport processes, energy metabolism, and electron transport are shown in Figure 3.

Real time-PCR analysis of upregulated protein-encoding genes in *Anabaena* PD-1 cells

The expression levels of eight genes in the PCB-degradation groups were significantly upregulated compared with those in the control groups ($P<0.01$). The results are presented in Figure 4. Genes encoding for dioxygenase were upregulated by 1.26-fold. The cytochrome b_6f complex Fe-S protein-encoding gene upregulated by 2.64 fold. The ABC transporter substrate-binding protein gene expression level in the PCB-treated groups was 9.98-fold of that in the control groups. Transmembrane protein-encoding genes (i.e., heterocyst to vegetative cell connection protein and porin genes) were upregulated by 2.66- and 3.19-fold, respectively. Enolase gene expression level was upregulated by 2.88-fold compared with the control groups. The upregulation levels of the malate-CoA ligase subunit beta gene and methanol dehydrogenase gene were 3.40 and 5.22, respectively.

Discussion

Anabaena PD-1 isolated from PCB-contaminated paddy soil in South Zhejiang, China, efficiently degrades PCBs. Rodrigues et al. [36] suggested that *Burkholderia xenovorans* LB400, one of the most extensively studied PCB-degrading bacteria, can degrade 57% of

Table 1. Primers used for the quantitative real-time polymerase chain reaction in this study.

Gene name	Primer Sequence (5'-3')	GenBank Accession #[a]	Size (bp)
3-chlorobenzoate-3,4-dioxygenase gene	F: GCCCCAAATCAGAAACTACCA R: CCATCACCGGGAAATAACCAA	-	89
Cytochrome b6f complex Fe-S subunit gene	F: TTAAATGCCCTTGCCACGGTTCTC R: AGCGTGACTCAAAGCCAGAGACTT	NC_007413	89
ABC transporter substrate-binding protein gene	F: GCTGCATCGCAACCAATCAAA R: GGTATATCTGCCAGCCGGAACA	NC_007413	120
Enolase gene	F: TTGCCTGTGCCTTTAATGAACGT R: AAGCCTTTGTCATGCAGCACTTC	-	170
Porin gene	F: CCACAACAAAGCTGCAAGGACA R: TGAACAAGGTATCTCGCCCAGTAAA	NC_007413	159
fraH gene	F: ATGTTGATGTTTCCGGCTTTGC R: GGTCTGAGGCGGTGTCTATTGC	NC_007413	163
methanol dehydrogenase gene	F: TTAGCAGAGGTGGCAGAATTACGA R: CCCGTGGACTGACACCGAGA	NC_007413	133
malate-CoA ligase subunit beta gene	F: TTTGCGTAATTGGCATACCAGATAA R: TGGGGGTTACGGGGTAAGGTATT	NC_007413	175

[a]Primers with accession numbers belong to a gene cluster. Primers without accession numbers were designed according to the reference [35].

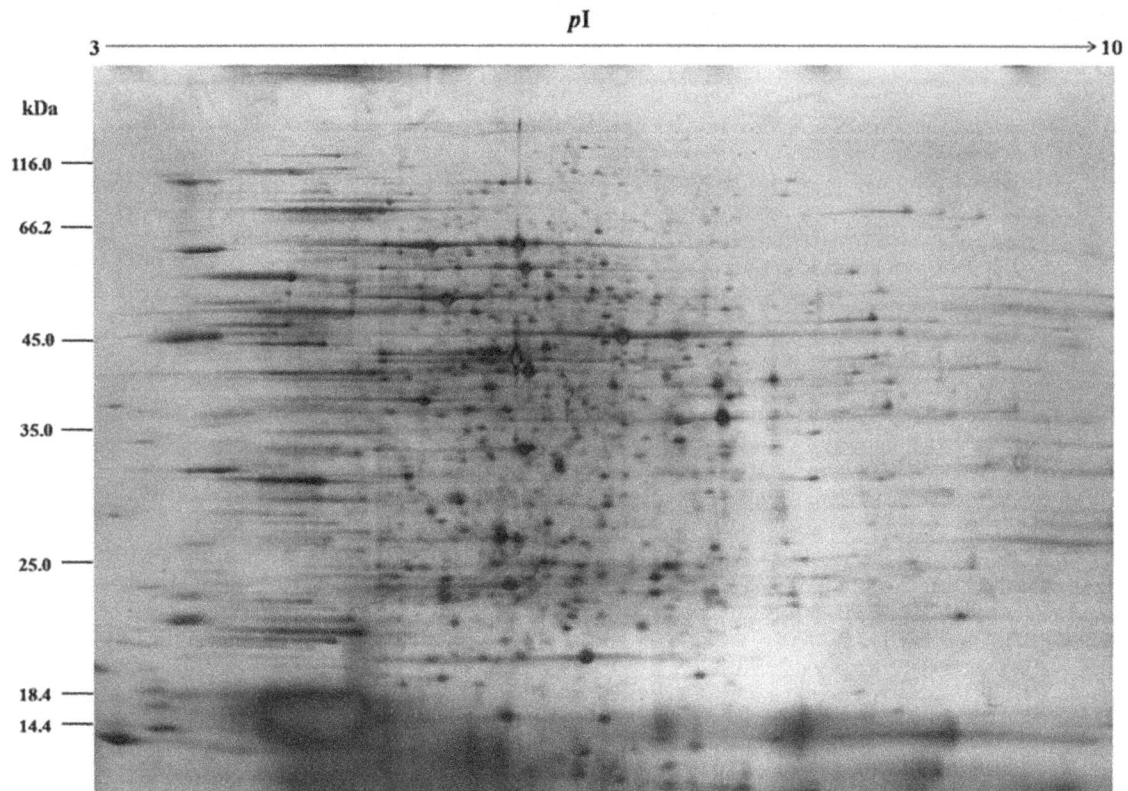

Figure 1. 2D gel maps for differentially expressed proteins in *Anabaena* **PD-1 cells in the control group.**

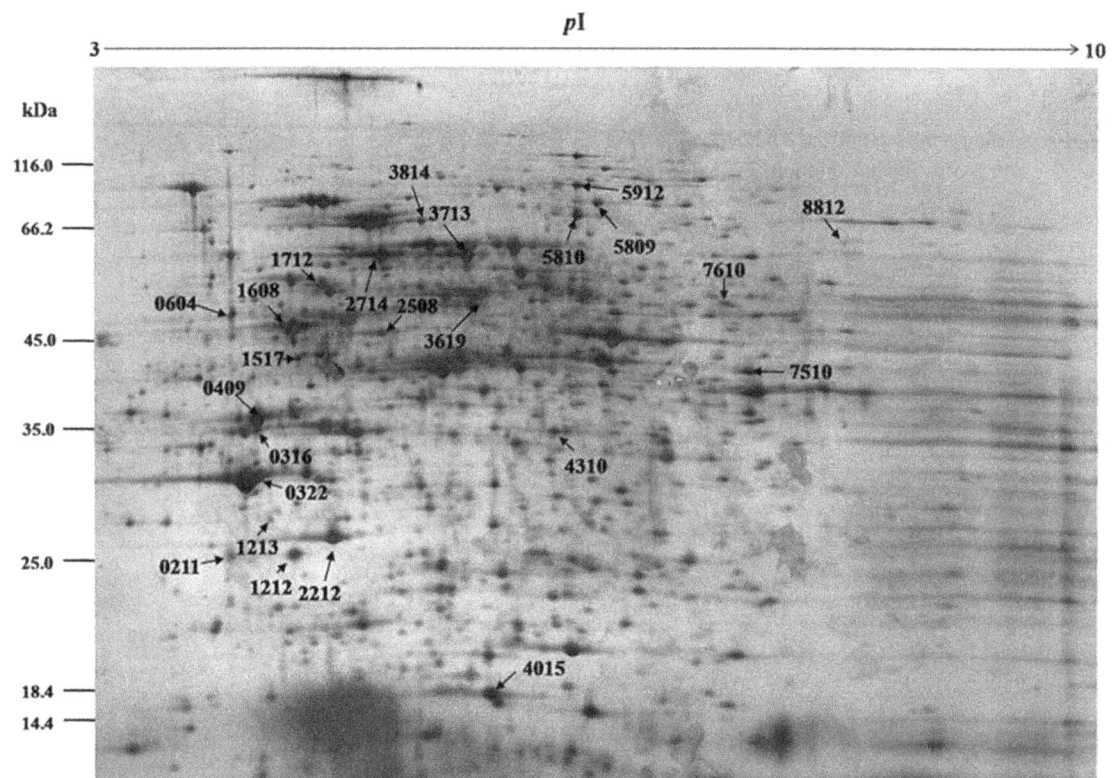

Figure 2. 2D gel maps for differentially expressed proteins in *Anabaena* **PD-1 cells in the Aroclor 1254 degradation group.** Arrow-directed spots are upregulated proteins in PCB degradation by *Anabaena* PD-1. The detailed information of upregulated proteins are listed in Table 2.

Table 2. Upregulated proteins in *Anabaena* PD-1 cells during PCB degradation.

Protein function	Spot No.	Identified protein	Accession No.	Amino acid sequences	Sequence Coverage	Score	pI Expected	pI Observed	Molecular Mass (kDa) Expected	Molecular Mass (kDa) Observed	Fold change
transportation	0409	ABC transporter substrate-bind ing protein	gi|222149178	GNTGPQAPDVIDVGLSFGPAAK	6%	183	5.67	4.51	38.9	36.5	734.5
	1212	amino acid ABC transporter substrate-binding protein	gi|222149574	TPYGQGLADETKK	11%	283	6.04	4.68	39.0	27.3	623.4
	1712	peptide ABC transporter substrate-binding protein	gi|328542580	QAIDNLVFAITPDAAVR	3%	123	5.11	4.80	59.0	58.8	381.6
	0316	putrescine ABC transporter periplasmic putrescine-binding protein	gi|117617823	QIKAGVFQK	2%	61	6.04	4.50	40.3	35.7	218.2
	1213	phosphonate ABC transporter substrate-binding protein	gi|493775881	GFTEVNVDFYKPIIEAR	13%	187	4.92	4.65	32.2	29.0	15.7
	1517	urea short-chain amide or branched-chain amino acid uptake ABC transporter periplasmic solute-binding protein	gi|39936731	ELNSILFYPVQYEGEESER	10%	239	7.63	4.72	48.2	45.3	12.6
	0322	phosphonate ABC transporter substrate-binding protein	gi|497516437	FCEGVGKNTIDIANASR	4%	135	4.24	4.54	37.4	32.6	9.8
	0211	xylose ABC transporter substrate-binding protein	gi|517198395	AQNEGIPVVGYDR	3%	91	5.48	4.28	36.3	28.5	7.3
Energetic metabolism	2508	enolase	gi|493227560	VNQIGSLTETLDAVETAHK	16%	445	5.11	4.99	45.3	48.6	239.9
	7610	bifunctional phosphoribosylaminoimidazolecarboxamide formyltransferase/IMP cyclohydrolase	gi|17230585	TAAAAGISAIVQPGGSLR	17%	401	5.56	6.24	54.3	54.8	23.0
	8812	ATP synthase beta subunit	gi|9909749	AHGGYSVFAGVGER	16%	67	6.51	6.72	15.6	68.1	10.4
	3619	FoF1 ATP synthase subunit beta	gi|154251918	AHGGYSVFAGVGER	17%	539	4.95	5.37	51.3	55.4	7.6
	3713	ATP synthase subunit alpha	gi|17134983	EAYPGDVFYIHSR	13%	290	5.11	5.31	54.4	63.4	5.01
Electron transport	4015	cytochrome b₆f complex iron-sulfur subunit	gi|17229945	CPCHGSQYDATGK	21%	246	5.31	5.42	19.2	20.0	11.5
	4310	Rieske-FeS protein	gi|14272374	FLESHNVGDR	5%	63	5.31	5.64	19.2	35.5	9.56
Carbon metabolism	2714	methanol/ethanol family PQQ-dependent dehydrogenase	gi|170743819	QDPNVIPVMCCDTVNR	4%	129	6.23	5.00	65.5	64.0	499.2
	7510	malate-CoA ligase subunit beta	gi|227823162	GGLAYSPEQAAYR	3%	107	5.16	6.31	43.0	43.1	449.1
Nitrogen metabolism	2212	ureidoglycolate lyase	gi|518240147	YIDESNALDHVAGYCVINDVSER	8%	111	4.89	4.88	30.6	29.4	172.8
	5810	nitrous oxide reductase	gi|226346680	ILTEGLLPETR	6%	83	5.64	5.72	37.4	74.0	9.7
Transmembrane protein	1608	porin; major outer membrane protein	gi|17130179	VNNADIVDTNTTLGVR	15%	255	4.61	4.69	54.4	50.1	379.9
	0604	putative heterocyst to vegetative cell connection protein	gi|556608	IPPDVDVSGFANSEIVSR	31%	473	4.46	4.38	29.9	52.5	192.9
Other proteins	5912	elongation factor G	gi|516958835	LNIIDTPGHVDFTIEVER	7%	227	5.08	5.72	76.1	83.4	8.1
	5809	100RNP protein	gi|1588265	AHGSALFTR	1%	68	5.89	5.79	85.2	78.1	7.4
	1216	hypothetical protein	gi|518291170	NLGLVDPNSTSGNNVPR	8%	217	9.27	4.82	33.9	28.3	4.9
	3814	molecular chaperone DnaK	gi|519030894	DAGLSAGQIDEVLVGGMTR	7%	227	4.92	5.15	67.9	72.4	4.7

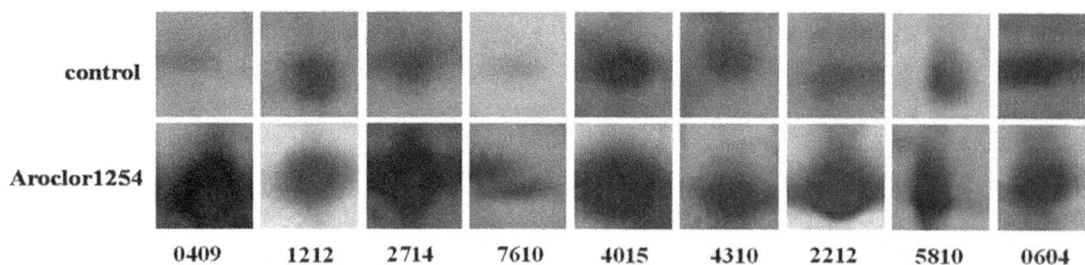

Figure 3. Upregulated proteins of *Anabaena* PD-1 exposed to Aroclor 1254 for 30 d. 0409-ABC transporter substrate-binding protein, 1212-amino acid ABC transporter substrate-binding protein, 2714-methanol/ethanol family PQQ-dependent dehydrogenase, 7610-bifunctional phosphoribosylaminoimidazolecarboxamide formyltransferase/IMP cyclohydrolase, 4015-cytochrome b6-f complex iron-sulfur subunit, 4310-Rieske FeS protein, 2212-ureidoglycolate lyase, 5810-nitrous oxide reductase, and 0604-putative heterocyst to vegetative cell connection protein.

Aroclor 1242 in 30 d. Singer et al. [35] combined *Arthrobacter* sp. strain B1B with *Ralstonia eutrophua* H850 to degrade PCB mixtures and achieved a maximum degradation rate of 59% in over 18 weeks. These bacterial species are typical PCB degraders. However, the application of these species to the bioremediation of PCB-contaminated paddy soils is limited because of their specific living conditions. By contrast, *Anabaena* PD-1, an associated cyanobacterial species in paddy soils, adapts to the condition in paddy soils very well. This condition is one of the essential requirements for this functional species to degrade PCBs in contaminated paddy soils. Thus, *Anabaena* PD-1 may be an excellent choice for the remediation of PCB-contaminated paddy soils.

This study is the first to report on the proteome files for *Anabeana* PD-1 in the PCB degradation and control groups and the RT-PCR gene data of upregulated proteins during PCB degradation. We synthesized the information from the protein and gene levels of the PCB-degrading cyanobacterial species *Anabaena* PD-1 and proposed a putative scheme to demonstrate the possible biodegradation mechanism of PCBs by *Anabaena* PD-1 (Figure 5). We assumed that 3-chlorobenzate-3,4-dioxygenase, transporter proteins, electron transport proteins, transmembrane proteins, and energetic metabolism proteins and genes encoding for the said proteins have important functions during PCB dechlorination by *Anabaena* PD-1. As a complex bioreaction, PCB degradation is involved with the transport system, energy system, and photosynthetic system in *Anabaena* PD-1 cells.

Organic pollutant stressors can generally change the cell membrane structure and function [36] and exert general stress on cells. The different (chlorinated) aromatic compounds induced the overexpression of different proteins (Table 3). Stressors with chlorinated structures can enhance the expression of stress protein DnaK, ABC transporter, and enolase, whereas pollutants without chlorinated structures can increase the expression of branched-chain amino acid uptake ABC transporter periplasmic solute-binding proteins and elongation factor G. This difference in protein expression may be attributed to the various structures of the organic stressors. The upregulated proteins with different functions can provide new insights into the adaptation of *Anabaena* PD-1 to the presence of PCBs.

In this study, 3-chlorobenzoate-3,4-dioxygenase, a new enzyme that may be involved in PCB degradation by *Anabaena* PD-1, belongs to the family of Rieske protein family (http://pfam.janelia.org/family/PF00355.). Thus, this dioxygenase is closely related to cytochrome b6f complex Fe-S protein, another typical Rieske protein [44]. The gene encoding for 3-chlorobenzoate-3,4-dioxygenase is named *cbaA* [45]. This enzyme has an important function in the degradation of pollutants belonging to the toluene/

biphenyl family [46]. Thus, we believe that 3-chlorobenzoate-3,4-dioxygenase may be closely related to the direct biodegradation of Aroclor 1254 by *Anabaena* PD-1. Nevertheless, the mechanism by which 3-chlorobenzoate-3,4-dioxygenase participates in PCB dechlorination in *Anabaena* PD-1 cells remains intriguing.

The PCB-treated and reference gels indicated that protein spots (4015, 4310) are significantly upregulated. These spots are *petC* gene products or PetC proteins that were first discovered and isolated by Rieske et al. [47]. Thus, the proteins are also called Rieske proteins. PetC protein is a subunit of cytochrome bc1 and cytochrome b6f complexes. Two to four genes of the *petC* gene family are found in the cyanobacterial cells of *Nostoc* sp. PCC 7120 [48]. PetC protein is an essential protein for the functioning of the cytochrome b6f complex. Cytochrome b6f complex has an important function in the aerobic photosynthetic electron transport chain reaction center [49]. The reaction center is involved in energy metabolism and electron transfer. Organic pollutants, such as PCBs, usually act as electron acceptors. Thus, as a key enzyme-mediating electron transporter, PetC protein upregulation may influence the biodegradation of PCBs in cyanobacterial cells. The changes in the expression levels of cytochrome b6f complex Fe-S protein confirmed this inference (Figure 4b). Comparing Figure 1 and Figure 2, the expression of proteins 0409, 1212, 1712, 0316, 1213, 1517, 2508, 0322, and 0211 were significantly enhanced. These proteins belong to the ABC transporter family and are ABC transporter substrate-binding proteins (0409, 1212, and 1712), ABC transporter periplasmic putrescine-binding protein (0316), phosphonate ABC transporter substrate-binding protein (1213, 0322), urea short-chain amide or branched-chain amino acid uptake periplasmic solute-binding protein (1517), and xylose ABC transporter substrate-binding protein (0211). The ABC transporter family comprises proteins that can transport ions, saccharides, lipids, and heavy metals across membranes [42–43,50]. The upregulation of these transport proteins indicates that they are likely to participate in PCB uptake or metabolite efflux. The upregulation of ABC transport proteins in *Pseudomonas putida* P8 suggests that the increased transport of amino acids is a cellular response to external stress [27]. The degradation of PCBs by the nitrogen-fixing species *Anabaena* PD-1 consumes energy. Thus, cyanobacterial cells bind to synthesize large amounts of ATP, thereby increasing the substrates required for ATP synthesis. The upregulated expression of transport proteins in algal cells may be due to the increased need for transporting ATP synthesis substrates. Several other studies have detected the efflux of organic solvents induced by transporters in *Pseudomonas putida* in response to aromatic and aliphatic solvents and alcohols [51].

The upregulation of enzymes involved in energetic metabolism have also been observed and found to be consistent with the

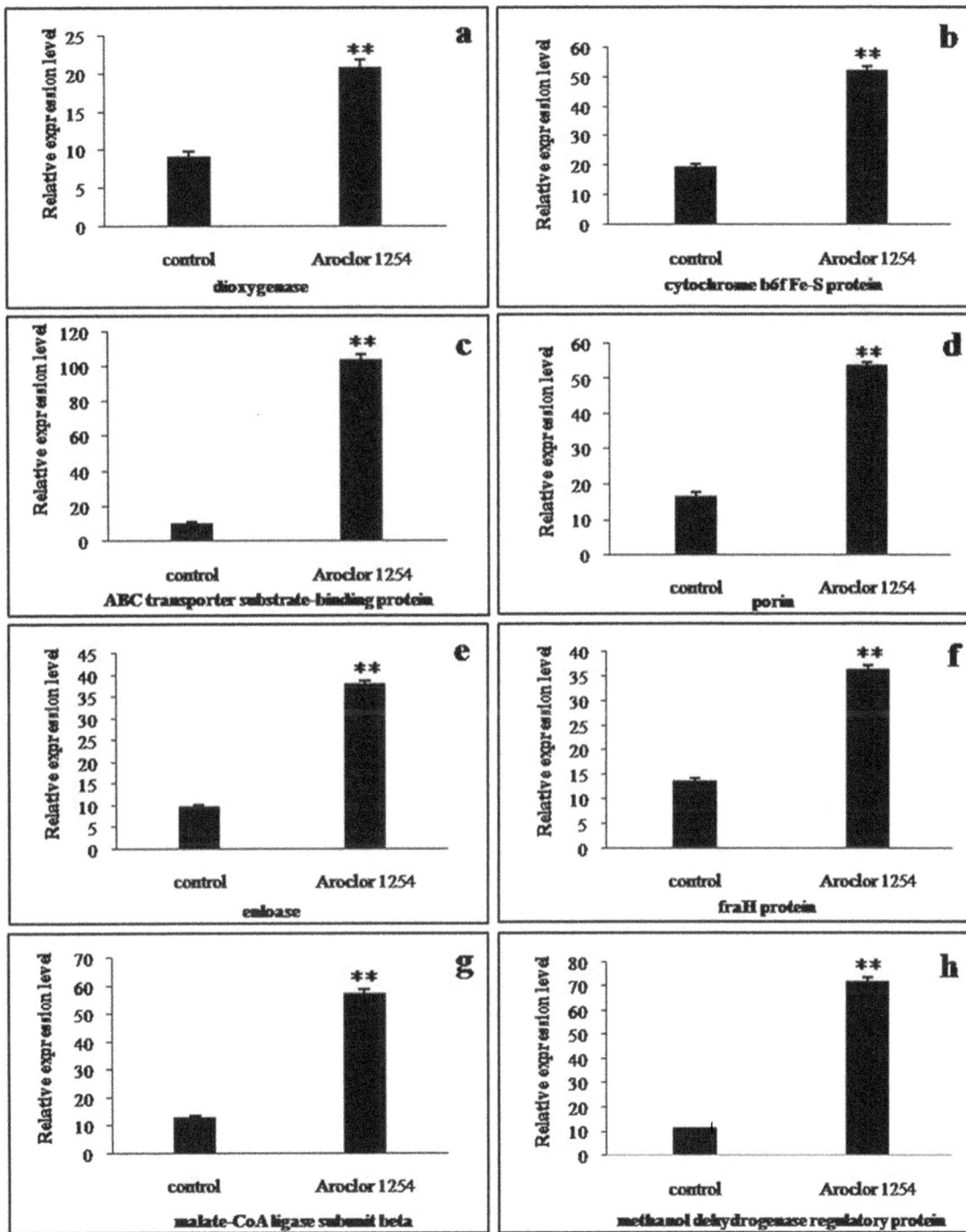

Figure 4. Gene expression levels in control and PCB-treated groups. Expression levels of genes encoding for (a) dioxygenase, (b) cytochrome b6f Fe-S protein, (c) ABC transporter substrate-binding protein, (d) porin, (e) enloase, (f) fraH, (g) malate-CoA ligase, and (h) methanol dehydrogenase. Data represented the mean $\pm$ SD, and significant difference from the control group was determined by $*p<0.05$ and $**p<0.01$.

extra-energetic requirements of cells that trigger several energetically expensive short-term adaptation mechanisms to survive and adapt to the toxicity of PCBs. Enolase (2508) and ATP synthase β subunit (8812) are enzymes involved in energy synthesis and metabolism. ATP synthase β subunit is involved in ATP synthesis under the conditions of a transmembrane proton gradient. The upregulation of this enzyme benefits the cells exposed to toxic organic compounds that can cause membrane lesions [52].

Enolase participates in glycolysis, which converts 2-phosphoglycerate to phosphoenolpyruvate [26]. Enolase consumes energy for cells to actively transport organic compounds to the intracellular space or transport such compounds from the intracellular space to the extracellular space [53]. Pérez-Pantoja [54] reported that the enolase superfamily does not participate in aromatic pollutant degradation by *Mycobacterium smegmatis*. Nevertheless, we still believe that enolase may be indirectly involved in PCB degrada-

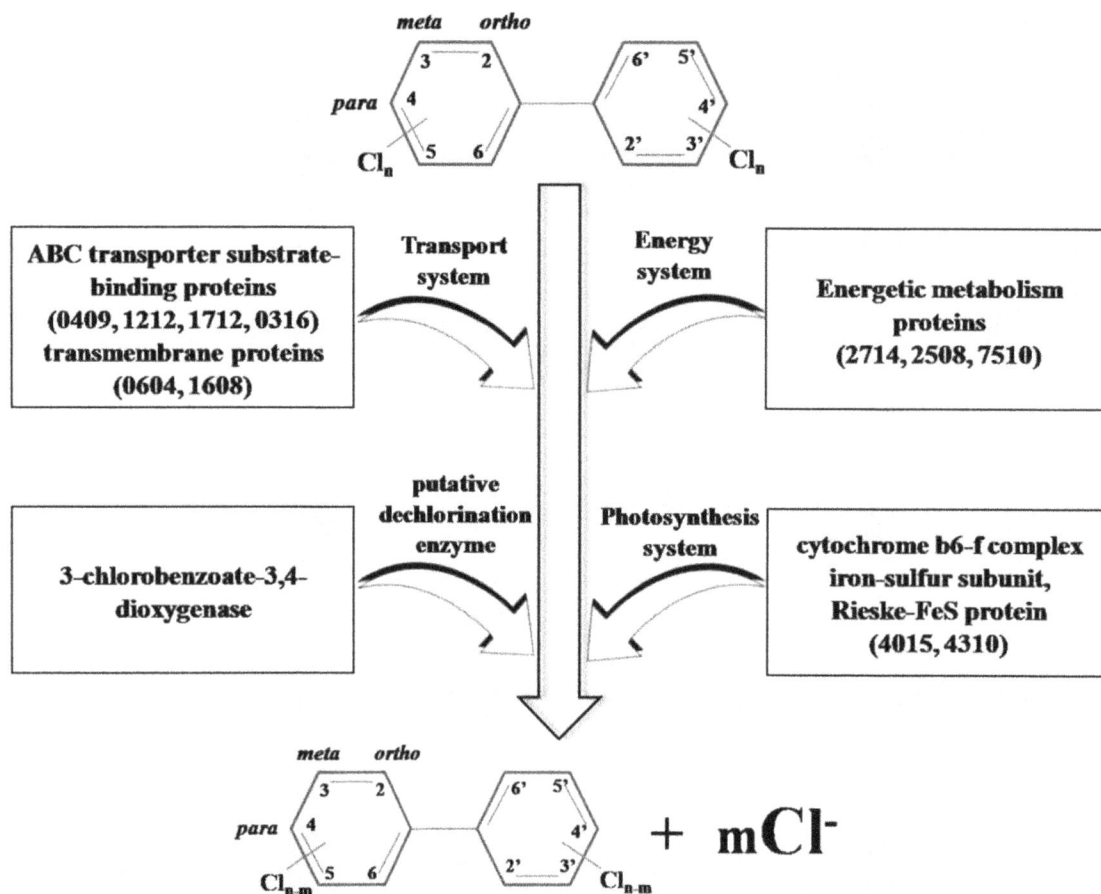

Figure 5. Hypothetic scheme of PCB degradation by Anabaena PD-1. Numbers in brackets present the different proteins in Table 2.

tion by *Anabaena* PD-1 because of the significant upregulation in both protein (Figure 3) and gene levels (Figure 4). The upregulation of ATP synthase β subunit and F_OF_1-ATP synthase β subunit (3619) indicates that the ATP consumption of nitrogen-fixing cyanobacterial cells exposed to Aroclor 1254 significantly

increases because cyanobacterial cells actively transport PCB molecules to the extracellular space to protect themselves from the toxicity of PCBs. Another possibility is that cyanobacterial cells consume energy to transport PCB molecules to the intracellular space for degradation. Stress proteins are consistently upregulated

Table 3. Comparison of induced proteins by (chlorinated) aromatic compounds in *Anabaena* PD-1 and other microorganisms.

Induced proteins in this study	Organic pollutant stressors	Microorganisms	Approaches	References
molecular chaperone DnaK	quinclorac	*Burkholderia cepacia* WZ1	2-DE, MALDI-TOF MS/MS	[37]
	3,4-dichloroaniline	*Variovorax* sp. WDL1	2-DE, MALDI-MS(/MS)	[38]
	4-chlorophenol	*Pseudomonas putida*	2-DE, MALDI-TOF MS	[39]
	4-chloronitrobenzene	*Comamonas* sp. strain CNB-1	2-DE, MALDI-TOF MS	[40]
	benzoate	*Pseudomonas putida* P8	2-DE, MALDI-TOF MS	[28]
ABC transporter substrate binding protein	2,4-dichlorophenoxy acetic acid	*Corynebacterium glutamicum*	2-DE, MALDI-TOF MS/MS	[41]
	phenol	*Pseudomonas putida* KT2440	2-DE, MALDI-TOF MS	[42]
	benzoate	*Pseudomonas putida* P8	2-DE, MALDI-TOF MS	[28]
	benzoate & succinate	*Pseudomonas putida* KT2440	iTRAQ, 1-DEMudPIT	[43]
translation elongation factor TS	4-chloronitrobenzene	*Comamonas* sp. strain CNB-1	2-DE, MALDI-TOF MS	[40]
enolase	4-chloronitrobenzene	*Comamonas* sp. strain CNB-1	2-DE, MALDI-TOF MS	[40]
branched-chain amino acid uptake ABC transporter periplasmic solute-binding protein	benzoate	*Pseudomonas putida* P8	2-DE, MALDI-TOF MS	[28]

to protect cells from organic pollutant stressors. Therefore, the upregulation of several proteins involved in polypeptide folding and synthesis is expected [55]. Molecular chaperone DnaK (3814) is a member of the HSP 70 (heat shock protein weighing 70 kDa) family. HSPs are highly conversed proteins that can assist in the refolding or hydrolysis of abnormal proteins [56]. The expression of these proteins in cells is usually upregulated under several stress conditions to protect cells [57–58]. Aromatic compounds can disrupt the synthesis of cell membranes, produce unfolded membrane proteins, and activate the stress response of membrane proteins [59]. Thus, the presence of PCBs causes stress to nitrogen-fixing cyanobacteria. The upregulated expression of chaperone proteins indicates that *Anabaena* PD-1 cells produce several stress responses to PCBs probably to protect themselves from the toxicity of PCBs.

In summary, changes in the proteome of *Anabaena* PD-1 cells during PCB degradation, gene encoding, and upregulation of proteins were observed for the first time. Twenty-five upregulated proteins were successfully identified. The *cbaA* gene encoding for 3-chlorobenzoate-3,4-dioxygenase was upregulated. Electron transport protein *petC* gene product was upregulated as well. The largest group of proteins enhanced by PCBs consists of the substrate-binding proteins of ABC transporters and proteins involved in energy metabolism. Although stress protein DnaK and several other proteins, such as elongation factors, compose a small portion of the upregulated proteins, they may still have important functions in the adaptation of *Anabaena* PD-1 to PCB and the degradation of PCBs by *Anabaena* PD-1. Thus, more studies should be conducted to identify the metabolites of PCBs and explore the PCB degradation pathway by *Anabaena* PD-1.

Acknowledgments

We are grateful to Dr Liu Jun for suggestions and encouragement.

Author Contributions

Conceived and designed the experiments: HJZ. Performed the experiments: HJZ XJJ WFX. Analyzed the data: HJZ LPL. Contributed reagents/materials/analysis tools: HJZ XJJ WFX LPL. Wrote the paper: HJZ XJJ.

References

1. Waid JS (1986) PCBs and the environment. Florida: CRC Press.
2. Ren NQ, Que MX, Li YF, Liu Y, Wan XN, et al. (2007) Polychlorinated biphenyls in Chinese surface soils. Environ Sci Technol 41: 3871–3876.
3. Hassine SB, Ameur WB, Gandoura N, Driss MR (2012) Determination of chlorinated pesticides, polychlorinated biphenyls, and polybrominated diphenyl enthers in human milk from Bizerte (Tunisia) in 2010. Chemosphere 89: 369–377.
4. Jones K, Voogt P (1999) Persistent organic pollutants (POPs): state of the science. Environ Pollut 100: 209–221.
5. Robinson GK (1998) (Bio)remediation of polychlorinated biphenyls (PCBs): problems, perspectives and solutions. Biochem Soc T 26: 686–690.
6. Field JA, Sierra-Alvarez R (2008) Microbial transformation and degradation of polychlorinated biphenyls. Environ Pollut 155: 1–12.
7. Kumamaru T, Suenaga H, Mitsuoka M, Watanabe T, Furukawa K (1998) Enhanced degradation of polychlorinated biphenyls by directed evolution of biphenyl dioxygenase. Nat Biotechnol 16: 663–666.
8. Tu C, Teng Y, Luo YM, Li XH, Sun XH, et al. (2011) Potential for biodegradation of polychlorinated biphenyls (PCBs) by *Sinorhizobium meliloti*. J Hazard Mater 186: 1438–1444.
9. Lambo AJ, Patel TR (2007) Biodegradation of polychlorinated biphenyls in Aroclor 1232 and production of metabolism from 2,4,4′-trichlorobiphenyl at low temperature by psychrotolerant *Hydrogenophaga* sp. strain IA3-A. J Appl Microbiol 102: 1318–1329.
10. Monika C, Zdena K, Alena F (2012) Biodegradation of PCBs by ligninolytic fungi and characterization of the degradation products. Chemosphere 88: 1317–1323.
11. Hong CY, Gwak KS, Lee SY. Kim SH, Lee SM (2012) Biodegradation of PCB congeners by white rot fungus, *Ceriporia* ZLY-2010, and analysis of metabolites. J Environ Sci Heal A 47: 1878–1888.
12. Subashchandrabose SR, Ramakrishnan B, Megharaj M, Venkateswarlu K, Naidu R (2013) Mixotrophic cyanobacteria and microalgae as distinctive biological agents for organic pollutant degradation. Environ Int 51: 59–72.
13. Papazi A, Kotzabasis K (2013) "Rational" management of dichlorophenols biodegradation by the microalga *Scenedesmus obliquus*. PLoS ONE 8: e61682.
14. Papazi A, Andronis E, Ioannidis NE, Chaniotakis N, Kotzabasis K (2012) High yields of hydrogen production induced by meta-substituted dichlorophenols biodegradation from the green alga *Scenedesmus obliquus*. PLoS ONE 7: e51852.
15. Naoki T, Takashi H, Hiroyasu N, Kazumasa H, Kazuhisa M (2003) Photosynthesis-dependent removal of 2,4-dichlorophenol by *Chlorella fusca* var. *vacuolata*. Biotechnol Lett 25: 241–244.
16. Liu Y, Dai XK, Wei J (2013) Toxicity of the xenoestrogen nonylphenol and its biodegradation by the alga *Cyclotella caspia*. J Environ Sci-China 25: 1662–1671.
17. Babu B, Wu JT (2010) Biodegradation of phthalate esters by cyanobacteria. J Phycol 46: 1106–1113.
18. Zhang HJ, Hu CM, Jia XY, Xu Y, Wu CJ, et al. (2012) Characteristics of γ-hexachlorocyclohexane biodegradation by a nitrogen-fixing cyanobacterium, *Anabaena azotica*. J Appl Phycol 24: 221–225.
19. Samanta SK, Singh OV, Jain RK (2002) Polycyclic aromatic hydrocarbons: environmental pollution and bioremediation. TRENDS Biotechnol 20: 243–248.
20. Furukawa K (2000) Engineering dioxygenases for efficient degradation of environmental pollutants. Curr Opin Biotech 11: 244–249.
21. Colbert CL, Agar NYR, Kumar P, Chakko MN, Sinha SC, et al. (2013) Structural characterization of *Pandoraea pnomenusa* B-356 biphenyl dioxygenase reveals features of potent polychlorinated biphenyl-degrading enzyme. PLoS ONE 8: e52550.
22. Nam JW, Nojiri H, Yoshida T, Habe H, Yamane H, et al. (2001) New classification system for oxygenase components involved in ring-hydroxylating oxygenations. Biosci Biotech Bioch 65: 254–263.
23. Barriault D, Simard C, Chatel H, Sylvestre M (2001) Characterization of hybrid biphenyl dioxygenases obtained by recombining *Burkholderia* sp. strain LB400 *bphA* with the homologous gene of *Comamonas testosteroni* B-356. Can J Microbiol 47: 1025–1032.
24. Kim SJ, Kweon O, Cerniglia CE (2009) Proteomic applications to elucidate bacterial aromatic hydrocarbon metabolic pathways. Curr Opin Microbiol 12: 301–309.
25. Ohtsubo Y, Kudo T, Tsuda M, Nagata Y (2004) Strategies for bioremediation of polychlorinated biphenyls. Appl Microbiol Biot 65: 250–258.
26. Agulló L, Cámara B, Martínez P, Latorre V, Seeger M (2007) Response to (chloro)biphenyls of the polychlorobiphenyl-degrader *Burkholderia xenovorans* LB400 involves stress proteins also induced by heat shock and oxidative stress. FEMS Microbiol Lett 267: 167–175.
27. Cao B, Loh KC (2008) Catabolic pathways and cellular responses of *Pseudomonas putida* P8 during growth on benzoate with a proteomics approach. Biotechnol Bioeng 101: 1297–1312.
28. Endo R, Ohtsubo Y, Tsuda M, Nagata Y (2007) Identification and characterization of genes encoding a putative ABC-type transporter essential for utilization of γ-hexachlorocyclohexane in *Sphingobium japonicum* UT26. J Bacteriol 189: 3712–3720.
29. Huang F, Fulda S, Hagemann M, Norling B (2006) Proteomic screening of salt-stress-induced changes in plasma membranes of *Synechocystis* sp. strain PCC 6803. Proteomics 6: 910–920.
30. Kurian D, Phadwal K, Mäenpää P (2006) Proteomic characterization of acid stress response in *Synechocystis* sp. PCC 6803. Proteomics 6: 3614–3624.
31. Pandey S, Rai R, Rai L C (2012) Proteomics combines morphological, physiological and biochemical attributes to unravel the survival strategy of *Anabaena* sp. PCC 7120 under arsenic stress. J Proteomics 75: 921–937.
32. Rippka R, Deiuelles J, Waterbury JB, Herdman M, Stanier RY (1979) Generic assignments, strain histories and properties of pure cultures of cyanobacteria. Microbiology 111: 1–61.
33. Bradford MM (1976) A rapid and sensitive method for the quantitation of microgram quantities of protein utilizing the principle of protein-dye binding. Anal Biochem 72: 248–254.
34. Shevchenko A, Wilm M, Vorm O, Mann M (1996) Mass spectrometric sequencing of proteins from silver-stained polyacrylamide gels. Anal Biochem 68: 850–858.
35. Kaneko T, Nakamura Y, Wolk CP, Kuritz T, Sasamoto S, et al. (2001) Complete genomic sequence of the filamentous nitrogen-fixing cyanobacterium *Anabaena* sp. strain PCC 7120. DNA Res 8: 205-213.
36. Rodrigues JLM, Kachel CA, Aiello MR, Quensen JF. Olga V, et al. (2006) Degradation of aroclor 1242 dechlorination products in sediments by *Burkholderia xenovorans* LB400 (*ohb*) and *Rhodococcus* sp. strain RHA1 (*fcb*). Appl Environ Microb 72: 2476–2482.
37. Singer AC, Gilbert ES, Luepromchi E, Crowley DE (2000) Bioremediation of polychlorinated biphenyl-contaminated soil using carvone and surfactant-grown bacteria. Appl Microbiol Biot 54: 838–843.

38. Li ZM, Shao TJ, Min H, Lu ZM, Xu XY (2009) Stress response of *Burkholderia cepacia* WZ1 exposed to quinclorac and the biodegradation of quinclorac. Soil Biol Biochem 41: 984–990.

39. Breugelmans P, Leroy B, Bers K, Dejonghe W, Wattiez R, et al. (2010) Proteomic study of linuron and 3,4-dichloroaniline degradation by *Variovorax* sp. WDL1: evidence for the involvement of an aniline dioxygenase-related multicomponent protein. Res Microbiol 161: 208–218.

40. Cao B, Loh KC (2009) Physiological comparison of *Pseudomonas putida* between two growth phases during cometabolism of 4-chlorophenol in presence of phenol and glutamate: a proteomics approach. J Chem Technol Biot 84: 1178–1185.

41. Zhang Y, Wu JF, Zeyer J, Meng B, Liu L (2009) Proteomic and molecular investigation on the physiological adaptation of *Comamonas* sp. strain CNB-1 growing on 4-chloronitrobenzene. Biodegradation 20: 55–66.

42. Fanous A, Weiland F, Lück C, Görg A, Friess A, et al. (2007) A proteome analysis of *Corynebacterium glutamicum* after exposure to the herbicide 2,4-dichlorophenoxy acetic acid (2,4-D). Chemosphere 69: 25–31.

43. Santos PM., Benndorf D, Sá-Correia I(2004) Insights into *Pseudomonas putida* KT2440 response to phenol-induced stress by quantitative proteomics. Proteomics 4: 2640–2652.

44. Yun SH, Park GW, Kim JY, Kwon SO. Choi CW, et al. (2011) Proteomic characterization of the *Pseudomonas putida* KT2440 global response to a monocyclic aromatic compound by iTRAQ analysis and 1DE-MudPIT. J Proteomics 74: 620–628.

45. Balka J, Lobréaux S (2005) Biogenesis of iron-sulfur proteins in plants. Trends Plant Sci 10: 324–331.

46. Nakatsu CH, Straus NA, Wyndham RC (1995) The nucleotide sequence of the Tn5271 3-chlorobenzoate 3,4-dioxygenase genes (*cbaAB*) unites the class IA oxygenases in a single lineage. Microbiology 141: 485–495.

47. David T Gibson, Rebecca E Parales. Aromatic hydrocarbon dioxygenases in environmental biotechnology. Current Opinion in Biotechnology, 2000, 11(3): 236–243.

48. Rieske JS, Maclennan DH, Coleman R (1964) Isolation and properties of an iron-protein from the (reduced coenzyme Q)-cytochrome C reductase complex of the respiratory chain. Biochem Bioph Res Co 15: 338–344.

49. Schultze M, Forberich B, Rexroth S, Dyczmons NG, Roegner M, et al. (2009) Localization of cytochrome b6f complexes implies an incomplete respiratory chain in cytoplasmic membranes of the cyanobacterium *Synechocystis* sp. PCC 6803. BBA-Biomembranes 1787: 1479–1485.

50. Baniulis D, Yamashita E, Whitelegge JP, Zatsman AI, Hendrich MP, et al. (2009) Structure-function, stability, and chemical modification of the cyanobacterial cytochrome b6f complex from *Nostoc* sp. PCC 7120. J Biol Chem 284: 9861–9869.

51. Cuthbertson L, Kimber MS, Whitfield C (2007) Substrate binding by a bacterial ABC transporter involved in polysaccharide export. P Natl Acad Sci USA 104: 19529–19534.[51] Rojas A, Duque E, Mosqueda G, Golden G, Hurtado A, et al. (2010) Three efflux pumps are required to provide efficient tolerance to toluene in *Pseudomonas putida* DOT-T1E. J Bacteriol 183: 3967–3973.

52. Eixarch H, Constantí M (2010) Biodegradation of MTBE by *Achromobacter xylosoxidans* MCM1/1 induces synthesis of proteins that may be related to cell survival. Process Biochem 45: 794–798.

53. Segura A, Duque E, Mosqueda G, Ramos JL, Junker F (1999) Multiple responses of gram-negative bacteria to organic solvents. Environ Microbiol 1: 191–198.

54. Pérez-Pantoja D, Donoso R, Junca H, Gonzalez B, Pieper DH (2009a) Phylogenomics of aerobic bacterial degradation of aromatics. In Handbook of Hydrocarbon and Lipid Microbiology. Timmis, K.N. (ed.). Berlin Heidelberg, Germany: Springer-Verlag, 1356–1397.

55. Domínguez-Cuevas P, González-Pastor JE, Marqués S, Ramos JL, de Lorenzo V (2006) Transcriptional tradeoff between metabolic and stress-response programs in *Pseudomonas putida* KT2440 cells exposed to toluene. J Biol Chem 281: 11981–11991.

56. Lund PA (2001) Microbial molecular chaperones. Adv Microb Physiol 44: 93-140.

57. Giuffrid MG, Pessione E, Mazzoli R, Dellavalle G, Barello C, et al. (2001) Media containing aromatic compounds induce peculiar proteins in *Acinetobacter radioresistens*, as revealed by proteome analysis. Electrophoresis 22: 1705–1711.

58. Park SH, Oh KH, Kim CK (2001) Adaptive and cross-protective responses of *Pseudomonas* sp. DJ-12 to several aromatics and other stress shocks. Curr Microbiol 43: 176–181.

59. Alba B, Gross C (2004) Regulation of the *Escherichia coli* sigma-dependent envelope stress response. Mol Microbiol 52: 613–619.

3

Exposure to Bisphenol A Correlates with Early-Onset Prostate Cancer and Promotes Centrosome Amplification and Anchorage-Independent Growth *In Vitro*

Pheruza Tarapore[1,2,3], Jun Ying[1,2Ꙩ], Bin Ouyang[1,2,3Ꙩ], Barbara Burke[4], Bruce Bracken[4], Shuk-Mei Ho[1,2,3,5]*

1 Department of Environmental Health, University of Cincinnati College of Medicine, Cincinnati, Ohio, United States of America, 2 Center for Environmental Genetics, University of Cincinnati College of Medicine, Cincinnati, Ohio, United States of America, 3 Cincinnati Cancer Center, University of Cincinnati College of Medicine, Cincinnati, Ohio, United States of America, 4 Department of Surgery, University of Cincinnati College of Medicine, Cincinnati, Ohio, United States of America, 5 Cincinnati Veteran Affairs Hospital Medical Center, Cincinnati, Ohio, United States of America

Abstract

Human exposure to bisphenol A (BPA) is ubiquitous. Animal studies found that BPA contributes to development of prostate cancer, but human data are scarce. Our study examined the association between urinary BPA levels and Prostate cancer and assessed the effects of BPA on induction of centrosome abnormalities as an underlying mechanism promoting prostate carcinogenesis. The study, involving 60 urology patients, found higher levels of urinary BPA (creatinine-adjusted) in Prostate cancer patients (5.74 μg/g [95% CI; 2.63, 12.51]) than in non-Prostate cancer patients (1.43 μg/g [95% CI; 0.70, 2.88]) ($p = 0.012$). The difference was even more significant in patients <65 years old. A trend toward a negative association between urinary BPA and serum PSA was observed in Prostate cancer patients but not in non-Prostate cancer patients. *In vitro* studies examined centrosomal abnormalities, microtubule nucleation, and anchorage-independent growth in four Prostate cancer cell lines (LNCaP, C4-2, 22Rv1, PC-3) and two immortalized normal prostate epithelial cell lines (NPrEC and RWPE-1). Exposure to low doses (0.01–100 nM) of BPA increased the percentage of cells with centrosome amplification two- to eight-fold. Dose responses either peaked or reached the plateaus with 0.1 nM BPA exposure. This low dose also promoted microtubule nucleation and regrowth at centrosomes in RWPE-1 and enhanced anchorage-independent growth in C4-2. These findings suggest that urinary BPA level is an independent prognostic marker in Prostate cancer and that BPA exposure may lower serum PSA levels in Prostate cancer patients. Moreover, disruption of the centrosome duplication cycle by low-dose BPA may contribute to neoplastic transformation of the prostate.

Editor: Natasha Kyprianou, University of Kentucky College of Medicine, United States of America

Funding: This work was supported in part by grants from the National Institutes of Health (P30-ES006096, U01-ES019480, U01-ES020988), a Veterans Administration Merit Award (I01-BX000675), an internal funding source from the University of Cincinnati to SMH and PT, and a Congressionally Directed Medical Research Program Department of Defense Award (PC094619) to PT. The funders had no role in study design, data collection and analysis, decision to publish, or preparation of the manuscript.

Competing Interests: The authors have declared that no competing interests exist.

* E-mail: shuk-mei.ho@uc.edu

Ꙩ These authors contributed equally to this work.

Introduction

Prostate cancer (PCa) is the second most common malignancy among men in North America. Aging is a well-established risk factor for PCa [1]. One in six men will develop PCa over their lifetime; however, the cancer is rarely diagnosed in men <40 years old, with almost two-thirds cases reported [2], [3] in men at age 65. From 2006 to 2010, the median age at diagnosis was 66 years according to the statistics from National Cancer Institute's Surveillance Epidemiology and End Results Studies (2013) [4]. Major contributing factors other than age are race and family history [1], whereas little is known about the impact of endocrine disruptors on PCa.

Bisphenol A (BPA) is an organic compound with the chemical formula $(CH_3)_2C(C_6H_4OH)_2$. BPA is used to make polycarbonate plastic and epoxy resins, which are present in thousands of consumer products [5], [6]. In the United States, exposure to BPA

is widespread, exceeding 90% in the general population [7]. Dermal absorption, inhalation, and ingestion from contaminated food and water are the major routes of exposure [8]. As an endocrine disruptor that mimics estrogen and thyroid hormone, BPA also acts as a metabolic and immune disruptor. Thus, the adverse health effects of BPA are extensive [9], [10], and higher levels of BPA exposure correlate with increased risk of cardiovascular disease, obesity, diabetes, immune disorders, and a host of reproductive dysfunctions [11], [12], [13]. Moreover, *in vitro* and animal studies have shown that BPA exposure can increase the risk of mammary gland, brain, and prostate cancers [9]. However, human studies linking BPA exposure to heightened cancer risk are scarce. One such study in China showed that the incidence of meningioma was 1.6 times higher in adults with higher concentrations of BPA in urine than in those with lower

Table 1. Summary of baseline characteristics (n = 60).

Variable	Category or unit	PCa (n = 27) Descriptive statistics*	Non-PCa (n = 33) Descriptive statistics*	p value†
Age	Year	69.67±10.29	62.76±7.15	0.003
Ln(serum PSA)	ng/mL	1.73±0.87	1.42±0.65	0.121
Gleason score	6 = 3+3	15 (71.4%)		
	7 = 3+4	4 (19.0%)		
	7 = 4+3	2 (9.5%)		
Treatment	Watchful waiting	14 (51.9%)		
	Prostatectomy	13 (48.1%)		
Rising PSA	No	23 (85.2%)		
	Yes	4 (14.8%)#		
Other cancer	No	25 (92.6%)		
	Yes	2 (7.4%)		
Recurrence	No	13 (85%)		
	Yes	2 (15%)		

*Numerical variables are summarized using mean ± standard deviation (SD). Categorical variables are summarized using frequency (in %).
†p values are calculated from t-tests.
#Serum PSA significantly rose during follow-up.

concentrations [14]. Similar studies for PCa have not been available until now.

A centrosome comprises a pair of cylindrical structures called *centrioles* surrounded by pericentriolar material. Centrosomes are involved in organizing the interphase microtubule cytoskeleton, mitotic spindles, and cilia. Centrosome dysfunction (number and integrity), a hallmark of many cancers, is believed to initiate neoplastic transformation and promote disease progression [15], [16]. An abnormal number of centrosomes can result in mono- or multipolar mitosis, leading to increased aneuploidy [15], [16]. Another feature of centrosomal disruption is abnormalities in microtubule (MT) nucleation and anchoring. Such abnormalities were more frequently observed in breast cancer cells than in normal breast epithelial cells [15], [16]. Also, a significant number of genes associated with increased PCa risk are in pathways leading to centrosome dysfunction [17], [18]. These observations have prompted us to examine, in cell-based models, the adverse effects of BPA on the centrosome cycle as a mechanism contributing to prostate carcinogenesis.

We used a cross-sectional clinical study to examine the association between BPA exposure and PCa. We hypothesized that BPA plays a role in prostate carcinogenesis. We found that patients with PCa are more likely than those without PCa to have higher levels of BPA in their urine. We observed a trend toward a negative correlation between urinary BPA and serum PSA levels in PCa patients. We performed *in vitro* studies to assess the effects of BPA on centrosome number, the formation of MT asters, and colonization in soft agar in two immortalized normal prostate epithelial cell lines (RWPE-1 and NPrEC) and four PCa cell lines (LNCaP, C4-2, 22Rv1, PC-3). We found that the percentage of cells with centrosome amplification (CA) increased in response to low-dose BPA exposure and that the relationship was non-monotonic for most cell lines. Moreover, exposure to low-dose BPA promoted MT aster organization in the non-cancerous RWPE-1 and increased anchorage-independent growth in the androgen-independent C4-2 PCa cell line. In aggregate, these findings reveal a previously unknown relationship between BPA

exposure and PCa and suggest a mechanism underlying the role of BPA in neoplastic transformation and disease progression.

Materials and Methods

Patients and the collection of urine samples

Patients were recruited from the urologic clinic at the University of Cincinnati Medical Center under a protocol approved by the University of Cincinnati Institutional Review Board. Table 1 lists patient characteristics and diagnostic information. After signing an informed consent form, patients underwent a digital rectal examination and were asked to provide a 20- to 50-ml urine specimen before their scheduled ultrasound-guided prostate biopsy. All procedures in this study were approved by the University of Cincinnati Institutional Review Board. Urine samples were centrifuged, the sediments were collected for a PCa biomarker study [19], and the supernatants were stored in aliquots at −80°C for BPA analysis. Among the 60 samples used for this study, 27 were from patients with PCa (PCa) and 33 were from patients without PCa (non-PCa).

Measurement of BPA in urine samples

BPA levels in samples were determined in the Laboratory of Organic Analytical Chemistry of Wadsworth Center, New York State Department of Health, (Albany, NY). High-performance liquid chromatography (HPLC) coupled with electrospray triple-quadrupole mass spectrometry (ESI-MS/MS) was used to quantify BPA, a technique similar to that described earlier, with some modifications [20], [21]. In brief, 500 μl of each urine sample was mixed with 1 ml of glucuronidase (2 μl/ml) for digestion and extraction. For quality control, 5 ng of $^{13}C_{12}$-BPA was added to each mixture. Extracts were applied to an Agilent 1100 series HPLC interfaced with an Applied Biosystems API 2000 electrospray MS/MS (Applied Biosystems, Foster City, CA) for quantitative of BPA. Data were acquired using multiple-reaction monitoring for the transitions of 227>212 for BPA, and 239>224 for $^{13}C_{12}$-BPA. The minimum detection limit (MDL) of BPA in this protocol was 0.05 ng/ml. For concentrations below the MDL,

Table 2. Summary of LnBPA (log-transformed BPA) values and cancer-related characteristics (n = 27) for PCa patients.

Factor	Category	n	Mean ± SD	p value
Gleason score	6 = 3+3	15	1.49±0.44	0.595
	7 = 3+4	4	0.54±0.85	
	7 = 4+3	2	0.94±1.20	
Treatment	Watchful waiting	14	1.68±0.54	0.848
	Prostatectomy	13	1.83±0.56	
Rising PSA	No	23	1.89±0.41	0.367
	Yes	4	0.91±0.99	
Other cancer	No	25	1.66±0.40	0.435
	Yes	2	2.82±1.41	
Recurrence	No	25	1.81±0.40	0.558
	Yes	2	0.94±1.41	

a value equal to the MDL divided by the square root of 2 was used in statistical analyses [22]. Reported concentrations were corrected for the recoveries of surrogate standard (isotopic dilution method). The BPA standard spiked to selected sample matrices and passed through the entire analytical procedure yielded a recovery of 88%±8% (mean ± SD). An external calibration curve was

prepared by injecting 10 μl of 0.05, 0.1, 0.2, 0.5, 1, 2, 5, 10, 50, and 100 ng/ml standards, and the regression coefficient was 0.99.

Normalization of urine BPA

Urinary creatinine levels were used to adjust for variability in dilution and to determine the validity of a spot urine sample for assessing chemical exposure [23]. A creatinine (urinary) assay kit from Cayman Chemical Company (Ann Arbor, MI) was used according to the manufacturer's protocol to measure urinary creatinine levels. The creatinine levels were used to adjust the urinary concentrations of BPA measured by the HPLC-ESI-MS/MS to obtain the "creatinine-adjusted" BPA levels (BPA levels) in μg/g.

Cells

The PCa cell lines PC-3, LNCaP, C4-2, and 22Rv1 were obtained from the American Type Culture Collection (ATCC, Manassas, VA) and cultured under standard, recommended, conditions. A description of the origin of the immortalized normal prostate epithelial NPrEC cell line has been published [24]; the other immortalized normal prostate epithelial cell line, RWPE-1, was purchased from ATCC (Manassas, VA) and was grown in Defined Keratinocyte-SFM medium (Invitrogen, Carlsbad, CA) with growth-promoting supplement. Cell cultures were maintained at 37°C in a humidified incubator with a 5% CO_2 atmosphere.

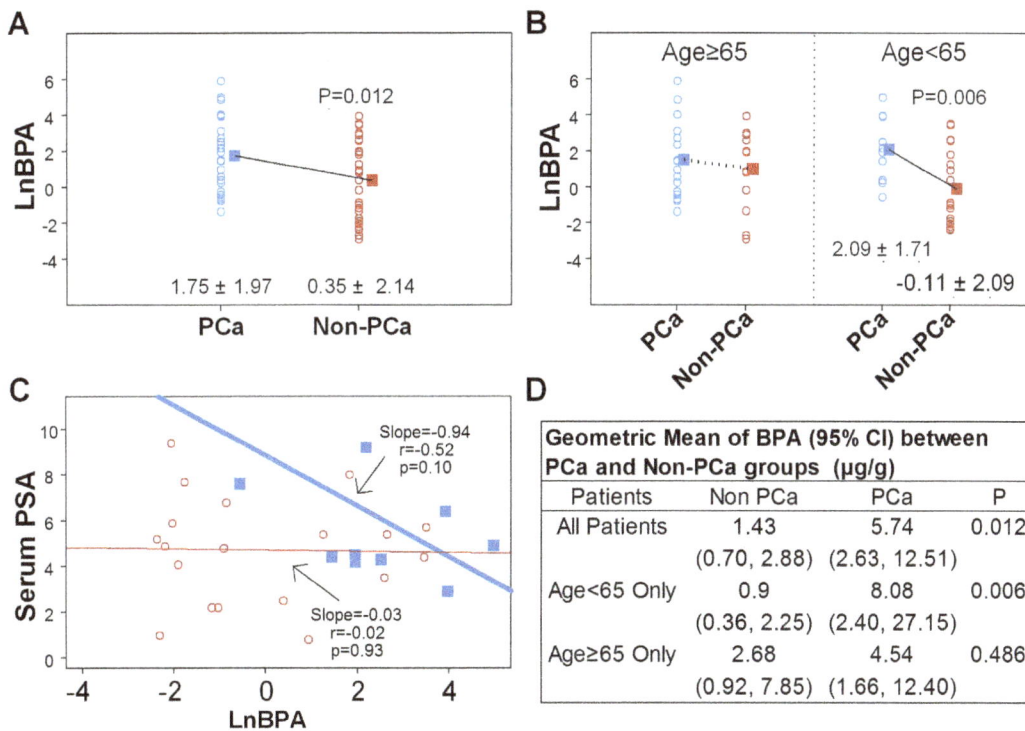

Figure 1. Scatter plots of LnBPA. Urine BPA levels are associated with PCa. The log-transformed BPA is referred to as LnBPA. Values in graph are mean ± SD of LnBPA. (A) Urine BPA levels are higher in PCa patients than in non-PCa patients. Means of LnBPA = 1.75±1.97 in PCa (blue, n = 27) vs. 0.35±2.14 in non-PCa (red, n = 33), p = 0.012. (B) LnBPA in PCa vs. LnBPA in non-PCa, stratified by age = 65. Urine BPA levels are significantly higher in young PCa patients than in the respective non-PCa patients only in the age group <65 years old; p = 0.006. (C) Linear regression analyses of Serum PSA vs. LnBPA in patients <65 years old only (n = 30). Blue solid squares represent PCa patients; red inverse-circles represent non-PCa patients. Blue and red solid lines represent their regression lines, respectively. (D) Comparison of the geometric mean of BPA in PCa and non-PCa groups. The geometric mean (Geo) is defined as the exponential of the mean of LnBPA. Values are geometric means (95% CI) of BPA in unit of μg/g creatinine.

BPA treatments

Cells from each cell line were seeded into six-well plates with glass cover slips at 25,000 cells/well. After 24 h, the medium was changed to phenol red–free media with 10% charcoal-stripped serum for another 24 h, at which time BPA was added to achieve a final concentration of 0, 0.01 nM, 0.1 nM, 1 nM, 10 nM, or 100 nM. The experiment was repeated five times to generate a total of five samples per cell line per BPA concentration.

Indirect immunofluorescence

For immunostaining of centrosomes, cells were fixed with methanol for 5 min at $-20°C$ and then processed for γ-tubulin (clone GTU88 antibody, Sigma Immunochemicals), α-tubulin (clone DM1A, Sigma Immunochemicals), and centrin (sc-50452, Santacruz Biotechnology) staining as previously described [25]. In brief, cells were extracted in 1% NP-40 in PBS for 10 min. Cells were probed with primary antibodies, and the antibody-antigen complexes were detected with Alexa fluor 488- or 594-conjugated antibodies (Molecular Probes). Cells were also stained for DNA with 4′, 6-diamidino-2-phenylindole (DAPI, Invitrogen). Immunostained cells were examined by fluorescence microscopy.

Microtubule (MT) aster formation assay

The effect of BPA on microtubule dynamics was determined by an assay of MT aster formation described previously [25]. In brief, cells were treated with nocodazole (1.5 μg/ml) for 40 min on ice to depolymerize interphase MTs, washed with PBS to remove the nocodazole, and incubated in fresh warm medium for 10 min at 37°C to allow for MT regrowth.

Measurements

The number of centrosomes per cell was scored by fluorescence microscopy. At least 150 cells were examined per treatment, and the percentage of cells with an abnormal number of centrosomes calculated from the total number of cells examined was used as the outcome measure for the analysis. A major abnormality in CA was defined as a cell with more than two centrosomes.

Anchorage-independent growth assay

Cells were assayed for anchorage-independent growth by measuring the efficiency of colony formation in semisolid medium as described [26]. In brief, cells were cultured under conditions described above, in the presence or absence of 0.1 nM BPA, for ~10 passages. We chose 0.1 nM because this concentration induced the highest percentage of cells with CA for most cell lines

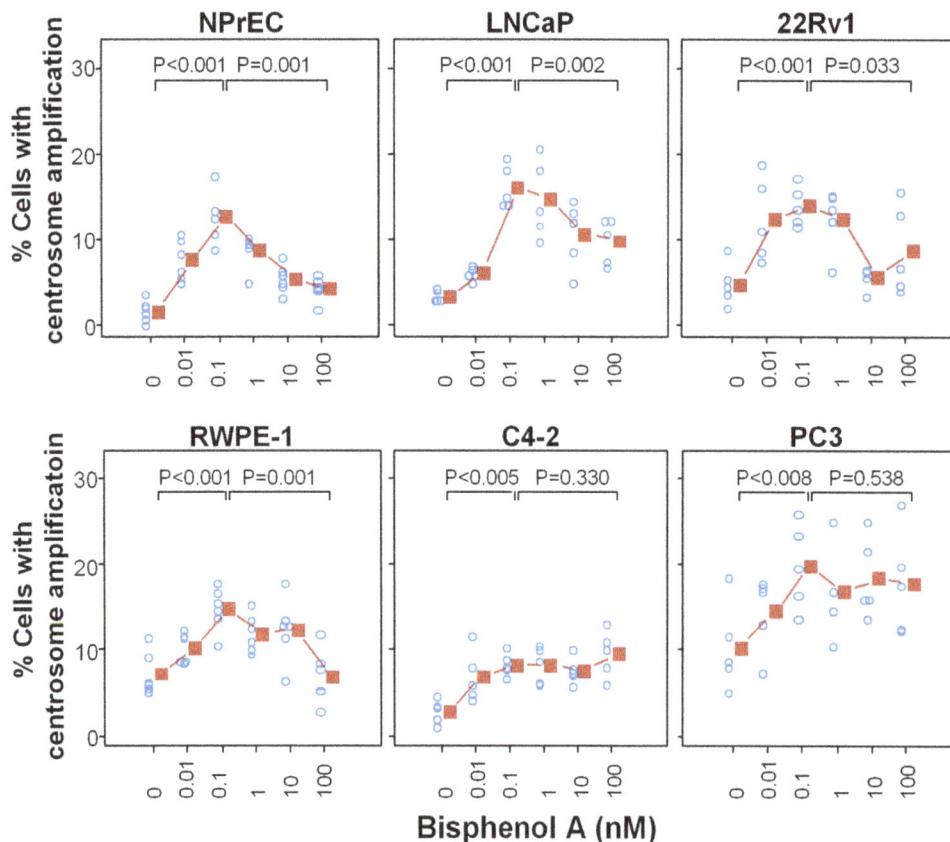

Figure 2. Low doses of BPA have an adverse effect on centrosome numbers in prostate cancer cells. The cell lines NPrEC, RWPE1, LNCaP, C4-2, 22Rv1, and PC3 were treated with medium containing 10% CSS plus 0, 0.01 nM, 0.1 nM, 1 nM, 10 nM and 100 nM BPA for 72 h. Cells were fixed with 100% cold methanol and immunostained for centrosomes and nuclei. The number of centrosomes per cell was scored by fluorescence microscopy. The results are shown as an average determined from five separate experiments. The scatter plot was generated of the percentage of cells with an abnormal number of centrosomes in response to BPA. Analyses was performed using a fixed effect model for each cell line. *Post hoc* comparisons of means were adjusted using Bonferroni's tests. The fold change is the percentage of cells with abnormal centrosomes at 0.1 nM BPA/ the percentage of cells with abnormal centrosomes at 0 nM BPA.

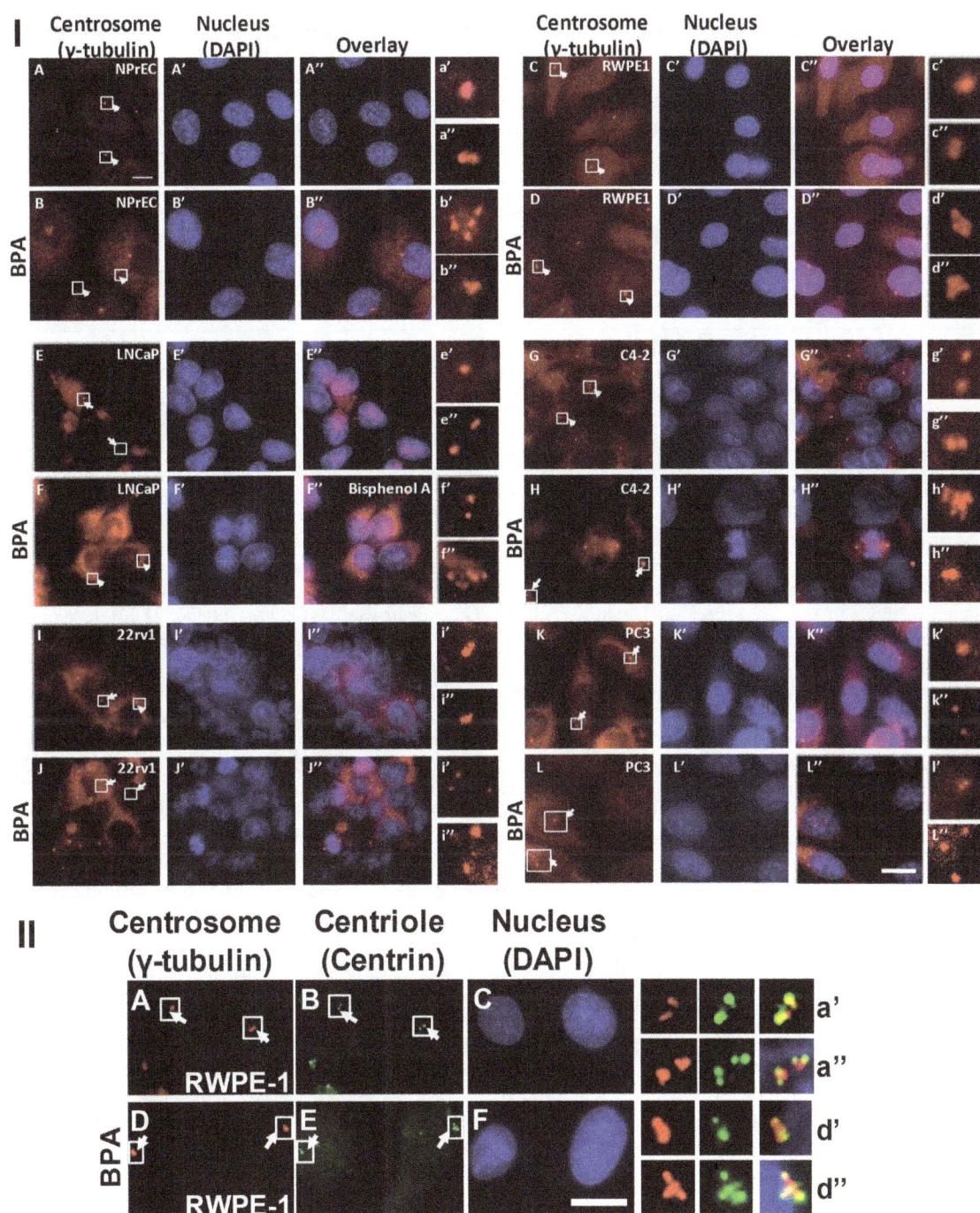

Figure 3. An increase in centrosome numbers is seen in prostate cancer cells exposed to BPA. (I) An increase in centrosome numbers. The cell lines NPrEC, RWPE1, LNCaP, C4-2, 22Rv1 and PC3 were treated with medium containing 10% CSS plus 0 or 0.1 nM BPA for 72 h. Cells were fixed with 100% cold methanol and immunostained for centrosomes (anti-γ-tubulin, red) and nucleus (DAPI, blue). The cells were examined by fluorescence microscopy. Arrows point to the positions of centrosomes, and panels on the right show magnified images of the indicated areas. Scale bar, 10 μm. (II) Centrosome amplification in the presence of BPA is not due to centriole separation. RWPE-1 cells were treated with 0.1 nM BPA for 3 days. Cells were fixed and immunostained for centrosomes (anti-γ-tubulin, red), centrioles (anti-centrin, green), and nucleus (DAPI, blue). Arrows point to the positions of centrosomes. Panels on right show magnified images of the indicated areas. Scale bar, 10 μm.

(see Results). Approximately 2,500 cells/35-mm well were embedded in soft agar. Cells were fed twice a week with fresh medium with and without BPA. After 2–3 weeks, colonies were counted under a microscope. Experiments were performed in triplicate and repeated twice. Colony-forming efficiency is the number of colonies obtained divided by the total number of cells plated, multiplied by 100.

Statistical analysis

The primary measure in the clinical analysis was a continuous variable of urinary BPA level after normalization or adjustment for

Table 3. Fold change in the percentage of cells with centrosomal amplification in presence of 100 pM BPA in indicated cell lines compared with untreated controls.

Cell line	Mean fold change ± SE (from BPA = 0 to BPA = 0.1 nM)*	p value of indicated cell lines vs. NPrEC-1†
NPrEC-1	8.1±2.4	-
LNCaP	4.7±0.5	0.047
C4-2	3.8±1.1	0.013
22RV1	3.6±0.8	0.009
RWPE-1	2.2±0.3	0.001
PC3	2.1±0.3	0.000

*Fold change is defined as % of cells with abnormal centrosomes at 0.1 nM BPA/% cells with abnormal centrosomes in untreated cells.
†Post hoc comparisons were performed under a fixed effect model and adjusted using Bonferroni's methods. Only the p-values of comparing NPrEC-1 to other cell lines are presented. Other comparisons between the cell lines were not statistically different.

urinary creatinine level. Initial inspection of the distribution showed that this variable was highly skewed to the right. Hence, its log-transformed variable (LnBPA) was used as the dependent variable in the statistical models. The principal statistical model was a fixed-effect model to assess the association between the LnBPA and PCa status (1 = yes; 0 = no). We applied both unadjusted and adjusted methods to our fixed-effect model. In the unadjusted method, the PCa status was the only independent variable. In the adjusted method, we included age (stratified as age ≥65 vs. <65 years) and serum PSA levels as controlling covariates. We performed *post hoc* comparisons of means between PCa and non-PCa patients and a similar comparisons in subsets of patients stratified by age. Wilcoxon rank sum tests were used to validate the findings from the fixed-effect models to ensure that all their findings were robust (data not shown). For urinary BPA and other numeric independent variables such as serum PSA levels, the relationships were assessed with linear regression models and/or correlation coefficients.

In the *in vitro* analyses for each cell line, we used the fixed effect model to assess the association of the percentage of cells with CA to the BPA concentration used to treat the cells and *post hoc* analyses adjusted for multiple comparisons using a Bonferroni's test. The anchorage-independent growth assay data were analyzed by two-sample *t*-tests. All statistical tests were performed with an SAS 9.3 software (SAS, Cary, NC) package. *P*-values <0.05 were considered statistically significant.

Results

Urinary BPA level is associated with PCa and may have prognostic value

We studied 60 urology patients, 27 with PCa and 33 without PCa. The mean age (± standard deviation [SD]) of PCa patients was 69.7±10.3 yr (min, 56 yr; max. 87 yr); they were older than non-PCa patients, who were 62.8±7.15 yr (min. 46 yr; max. 77 yr; *p* = 0.003). Serum PSA levels of PCa and non-PCa patients were not different. The Gleason score of 71% of the PCa patients was 6, and 7 in the others. Their baseline characteristics are summarized in Table 1.

In all subjects (PCa and non-PCa), levels of urinary BPA were not associated with age and serum PSA and did not correlate with Gleason score of the cancer and cancer-related characteristics in PCa subjects (Table 2). However, patients with PCa had higher levels of urinary BPA (creatine adjusted), with a geometric mean of 5.74 [95% CI; 2.63, 12.51] μg/g (mean ± SD of LnBPA of 1.75±1.97), whereas the urinary BPA levels of non-PCa patients

had a geometric mean of 1.43 [95% CI; 0.70, 2.88] μg/g (mean ± SD of LnBPA of 0.35±2.14; *p* = 0.012, Fig. 1A & 1D). Stratified analyses showed that the positive association was significant only among the 30 urologic patients younger than 65 (mean and median age = 58 yr, minimum age = 46 yr). In the younger patients (<65 yr), the geometric mean of urinary BPA levels among PCa patients was 8.08 [95% CI; 2.40, 27.15] μg/g (mean ± SD of LnBPA of 2.09±1.71) vs. a geometric mean of 0.90 [95% CI; 0.36, 2.25] μg/g (mean ± SD of LnBPA of −0.11±2.09) among non-PCa patients (*p* = 0.006; Fig. 1B & 1D). Moreover, linear regression analyses of this younger group revealed a trend toward a negative association between urinary BPA levels and serum PSA concentrations in the PCa patients (n = 10, r = −0.52, *p* = 0.10) but no such trend in non-PCa patients (Fig. 1C). The correlation did not reach significance at the 5% level because of the small sample size.

Low doses of BPA promoted centrosome amplification (CA)

CA is commonly observed in human tumors and is a major factor contributing to chromosome instability [15], [27]. Depending on whether the cell is in the G1 or S/G2/M phase of the cell cycle, normal cells show one or two centrosomes, respectively. We determined whether treating cells with BPA changed the number of centrosomes, by treating cell cultures with increasing concentrations of BPA (0.01–100 nM) (Figs. 2 and 3). Untreated cells that served as controls showed the expected normal centrosome profile, in which most of the cells (>90%) contained either one or two centrosomes (Fig. 3-I, panels A, C, E, G, I, K). The untreated NPrEC had the fewest cells with centrosomal aberrations (1.7%), followed by C4-2 (2.9%), LNCaP (3.5%), 22Rv1 (4.9%), RWPE-1 (7.3%), and PC-3 (10.4%) (Fig.2). In contrast, all cell lines treated with BPA showed an increase (two- to eight-fold, Table 3) in the number of cells with three or more centrosomes (Fig. 2, Fig. 3-I panels B, D, F, H, J, L). The dose-response curves of the two non-cancerous cell lines, NPrEC and RWPE-1, and two PCa cell lines, LNCaP and 22Rv1, reveal a non-monotonic (biphasic) response relationship, with the maximal response with 0.1 nM BPA (Fig. 2). On the other hand, the two other PCa lines, C4-2 and PC-3, displayed an increasing dose-response curve that plateaus at the same low concentration of BPA (0.1 nM) (Fig. 2). The immortalized non-cancerous prostate epithelial cell line NPrEC-1, showed the highest fold change (mean ± SD, 8.1±2.4) in centrosome profile (Table 3), suggesting that its centrosome duplication cycle may be most sensitive to the effects of low-dose BPA on the promotion of CA.

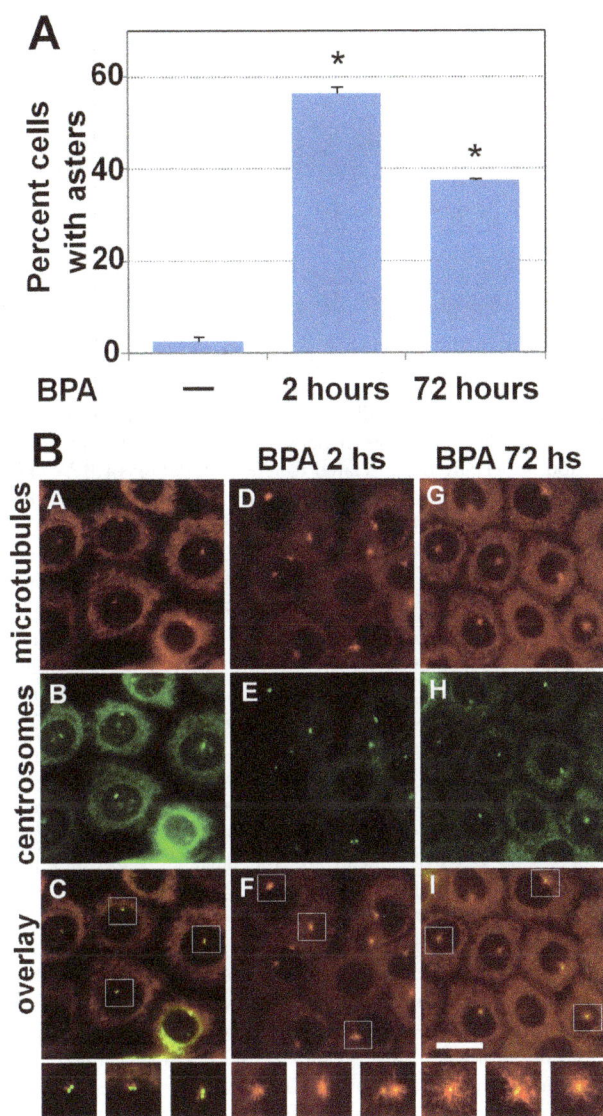

Figure 4. BPA enhances centrosomal aster formation. The microtubule aster formation assay was performed 2 h after and 3 days after treatment with 0.1 nM BPA. For the assay, microtubules were depolymerized by treatment with nocodazole on ice, followed by the addition of fresh warm medium for 10 min to allow for microtubule regrowth, and co-immunostained for centrosomes (anti-γ-tubulin, red) and MTs (anti-α-tubulin, green). The centrosomal aster formation was assessed as positive if centrosomes had an MT aster with more than 15 MTs. The results shown in (A) are the average ± standard error (SE) from three experiments. For each experiment, >200 cells were examined. Significance was calculated using Student's t-test vs. 0 pM. *$p \leq 0.00002$.

Low-dose BPA did not affect centriole splitting

Structurally, the centrosome consists of a pair of cylindrical structures called centrioles that act as the duplicating units. To verify the integrity of the centrosomes, we immunostained cells for centrin, a major constituent of the centriole cylinder, allowing visualization of the centriole pair within the centrosome. Fig. 3 shows representative images for RWPE-1 cells. Each dot detected by antibody to γ-tubulin (Fig. 3-II; panels A and D) was resolved to a pair of dots (representing a centriole pair) revealed by antibody to centrin at a higher magnification (Fig. 3-II; panels B and E, panels a, a″; d′, d″). These data thus indicate that the centrosomes

are intact, containing a pair of centrioles. The centrosome profiles determined by counting the centrin signal were similar to those determined by counting the γ-tubulin signal (Fig. 2). Results for LNCaP, C4-2, 22Rv1, and NPrEC cells were similar. Hence, BPA had no effects on centrosome separation or centriole splitting.

Low-dose BPA enhanced MT aster formation

The anchoring of MTs and their subsequent elongation to form radial MT arrays (asters) are critical events during interphase and also lead to the formation of the mitotic spindle associated with normal centrosome function [28]. RWPE-1 prostate cells assayed for MT aster formation (Fig. 4). Cells were first treated with nocodazole on ice to completely depolymerize interphase MTs; nocodazole was then removed, and cells were incubated in fresh warm medium for MT regrowth. The ability of the centrosomes to nucleate, anchor, and elongate MTs was determined by co-immunostaining for centrosomes (anti-γ-tubulin) and MTs (anti-α-tubulin). The MT aster forming activity of centrosomes was assessed according to the previously established protocol [25]. Untreated RWPE-1 cells showed negligible aster formation. After acute 2-h treatment with 0.1 nM BPA, short asters were seen 56% of cells. Three days post-treatment with BPA (chronic exposure), ~37% cells showed asters (Fig. 4A, 4B panels g–i). Our data thus indicate that BPA enhances MT aster formation.

Chronic BPA exposure promotes anchorage-independent growth in C4-2 cells

The ability of chronic BPA exposure to transform or promote malignant growth of NPrEC, RWPE-1, LNCaP, and C4-2 cells was determined by a soft-agar colony-formation assay. The cells were grown in medium with or without 0.1 nM BPA for 10–14 passages before they were seeded on soft agar. Colony formation for NPrEC, RWPE-1, and LNCaP was <2%, and exposure to 0.1 nM BPA did not change the efficiency of colony formation. However, BPA-exposed C4-2 cells produced substantially more, larger, faster-growing soft-agar colonies (Table 4, Fig. 5). The percent efficiency of colony formation (mean ± SD) increased to 19.25±7.05% with BPA treatment compared with 2.03±0.40% in unexposed controls ($p<0.001$). The colony diameter was 50–400 μm in controls vs. 100–1,200 μm in BPA-treated C4-2 cells.

Discussion

Evidence that BPA exposure contributes to PCa was derived from animal studies [29], [30], [31], [32] or cell-based [33], [34], [35], [36] models. To the best of our knowledge, this is the first study that provides preliminary evidence of an association of BPA exposure with PCa in a clinical setting. Our findings in 60 urologic patients show that urinary BPA level is an independent prognostic biomarker of PCa, as higher urinary BPA levels were detected in the 27 PCa patients (geometric mean, 5.74 [95% CI; 2.63, 12.51] μg/g creatinine) as compared with those in the 33 non-PCa patients (geometric mean, 1.43 [95% CI; 0.70, 2.88] μg/g creatinine) ($p = 0.012$). The detection limit for this study was 0.05 ng/ml. Several population studies have now established BPA as a ubiquitous environmental contaminant detectable in the urine of most individuals in US populations. In the first large-scale cross-sectional study in the US involving 2,517 participants of the 2003–2004 National Health and Nutrition Examination Survey (NHANES) [7], BPA was detected in 93% of the population at a geometric mean and a 95th percentile concentration of 2.6 μg/g and 11.2 μg/g, respectively; the limit of detection was 0.4 ng/ml vs. 0.05 ng/ml in our study. A later study of 2,747 adult participants in the 2003–2006 NHANES [37] reported a

Table 4. Anchorage- independent growth in the presence and absence of BPA.

| Cell type | Mean ± SD of anchorage- independent growth (% colonies) | | p value* |
	Control (n = 6 replicates)	Exposed to BPA (n = 6 replicates)	
RWPE1	0.10±0.05	0.13±0.07	0.397
NPrEC	0.15±0.08	0.21±0.18	0.504
LNCaP	0.64±0.29	0.60±0.25	0.836
C4-2	2.03±0.40	19.25±7.05	<0.001

*p values were computed using t-tests.

geometric mean of 2.05 µg/g creatinine (25th percentile: 1:18, 75th percentile: 3.33); the lower limit of detection was 0.36 ng/ml. Thus, the geometric mean of urinary BPA levels in the PCa patients in the present study was ~2–2.5 times higher than the geometric means of those in large US cross-sectional studies. In contrast, the geometric mean in the non-PCa patients in this study was ~50% lower than the geometric means of these two population studies.

A strength of this study is our use of the method recommended by the Centers for Disease Control and Prevention involving solid-phase extraction coupled with isotope dilution-HPLC-MS/MS to measure total urinary BPA in a reference laboratory. Furthermore, we corrected for variations caused by factors that affect urinary dilution by expressing our data relative to urinary creatinine concentrations. Finally, all patients had biopsy-confirmed, rather than self-reported, PCa. A potential limitation of our study was that total urinary BPA in our patients was measured only once. However, according to current literature, total urinary BPA concentrations (free plus conjugated) in spot samples (one-time measurement) is a reliable method of evaluating baseline exposure from all sources across time when the sample size is sufficiently large [38]. Although toxicokinetic studies have shown that BPA and its major metabolite, BPA-glucuronide, have rather short half-lives (~2.5 h) in the bloodstream and that they are rapidly excreted with urine [39], [40], cross-sectional population studies

have suggested substantially longer half-lives due to nonfood exposure, bioaccumulation in body tissues such as fat, and liver function, especially those related to glucuronidation of BPA [12], [39]. The presence of high BPA concentrations in urine may suggest that the lifestyle habits of these patients may sustain higher levels of exposure. In this regard, in one clinical study, BPA levels in urine samples collected on the same day from male and female partners correlated [41], supporting the premise that similar lifestyle choices may determine the level of BPA exposure. Moreover, a recent study showed higher within-person variability (over 1–3 years) in BPA levels as compared with the total variability in 80 women [42]. Collectively, these studies highlight the significance of our finding that a one-time sampling of urinary BPA correlates with PCa.

Stratified analyses showed that the association between urinary BPA levels and PCa is highly significant ($p = 0.006$) among the 30 patients <65 years old (mean and median age = 58, minimum age = 46) but that this association does not reach significance among the half of patients >65 years (Fig. 1). These findings are intriguing, but perplexing. Taken at face value, they suggest that higher BPA exposure is associated with earlier onset of PCa. However, on the basis of the theory of developmental reprogramming of cancer risk [43], our findings raise the possibility of early-life reprogramming of PCa in humans. In rat studies, neonates fed environmentally relevant levels of BPA had an increased risk of developing prostate neoplasms [29], [44]. According to this reasoning, one should note that the younger PCa patients were either just born or young children when BPA was introduced for commercial use in the US in 1957. For example, the patient aged 64 years old would have been around 11 years old (prepuberty) when first exposed to BPA and those younger might have been exposed *in utero*.

Further analyses of the age groups <65 years old revealed that BPA levels negatively correlated with PSA levels in the younger patients but not the non-PCa patients While this observation needs to be validated in a larger clinical study to reach significance, this has crucial repercussions for young patients who take PSA tests for PCa screening. If exposure to high levels of BPA suppresses their serum levels of PSA, this may result in a misdiagnosis. This problem is similar to the under-detection of PCa in hypogonadal men because of the androgen dependency of PSA [45], [46], [47]. The inhibitory effect of BPA may be indirect, acting through the hypothalamic pituitary testicular axis [48]. Alternatively, it might be a direct inhibition on the cancer cells, similar to a report of direct suppression by genistein of PSA production [49].

BPA is not a recognized carcinogen. The question thus arises as to the mechanism behind the positive correlation of BPA exposure with PCa. Several studies have shown that centrosome amplification is a major contributing factor to aneuploidy in human tumors [15], [16]. We hence examined the centrosome profile of PCa cells

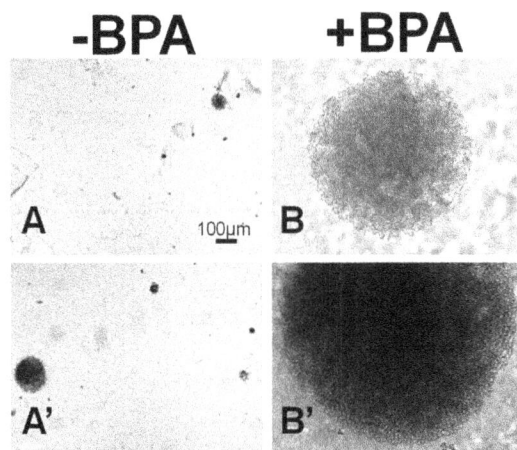

Figure 5. Cells grown in the absence and presence of 0.1 nM BPA were assayed for anchorage-independent growth. Representative pictures of colonies after 2 weeks of incubation in agar. C4-2 cells in the presence of 0.1 nM BPA formed larger colonies (B, B', 100–1200 µm diameter) compared with those grown in the absence of BPA (A, A', 50–400 µm diameter).

treated with BPA and found that treatment with BPA increased the number of cells with abnormal centrosomes. One can speculate that BPA may be affecting the cell-cycle machinery involved in centrosome duplication or the structural components required for centrosome duplication and maturation [50], [51]. Perturbations in these events have the potential to induce CA and increase genomic instability. Moreover, the estrogenic action of BPA may affect the expression of genes regulating centrosome cycle. For example, while *AurkA* is not a specific direct target of estrogen *in vitro*, *AurkA* is implicated in estrogen-induced oncogenesis, with long-term treatment of rats with estrogen having been shown to upregulate its expression [52]. Thus, the mechanism by which BPA deregulates the centrosome cycle and induces CA needs further clarification.

An interesting finding was of the greatest sensitivity of the immortalized normal prostate epithelial cell line to the effects of low-dose BPA (Table 3), suggesting that BPA might perturb the centrosome cycle in normal cells and contribute towards aneuploidy. This result is similar to that of previously published studies indicating that a BPA-related increase of DNA adducts was more pronounced in a non-tumorigenic epithelial cell line (PNT1) than in PC3 metastatic carcinoma cells [34]. On the whole, these experimental findings support the hypothesis that BPA plays a role in prostate carcinogenesis, in addition to promoting disease progression.

Another intriguing observation was the non-monotonic response observed in immortalized normal epithelial cells (NPrEC, RWPE-1) and androgen-dependent PCa (LNCaP) cells, suggesting that low concentrations of BPA elicit CA, with the greatest effect at 0.1 nM. This concentration is at least 10- to100-fold lower than most studies reporting a low-dose effect of BPA *in vitro* [53], [54]. At higher BPA concentrations, the detrimental effects on centrosomes appear to disappear. This observation could be explained by findings in the literature that BPA differentially interacts with various receptors such as estrogen receptors α and β, GPR30, or ERRγ, depending on the cell context [55], [56], [57], [58], [59], [60]. Alternatively, it may be a result of checkpoint mechanisms activated, blocking CA at higher BPA doses, causing either cell-cycle arrest or death of cells with dysregulated centrosome duplication. Future studies needs to address the underlying cause of non-monotonic dose-responses in these cell lines.

We found increased MT aster formation in RWPE-1 cells in the presence of BPA. The interphase MT dynamics tightly regulates mitosis. It also maintains normal subcellular localization of organelles, vesicular transport, cell migration, and the overall directionality of cells within the milieu of tissue architecture. In this context, androgen receptor (AR) nuclear localization has been shown to be dependent on the MTs [61], [62]. Since AR nuclear localization is essential for its transcriptional activity [63], it would be interesting to determine whether BPA induced perturbations in MT dynamics impacts AR trafficking and nuclear translocation, and hence alters AR functionality. Moreover, both AR and BPA

directly interact with tubulin [62], [64], [65]. One can thus speculate that BPA and AR may compete for tubulin, thus affecting the function of AR. Alternatively, the effects of BPA on MT-dynamics may increase the translocation of AR to the nucleus. Thus, studies on AR trafficking in response to BPA need to be performed, especially in light of reports on the adverse effects of MT-disrupting chemotherapeutic drugs on AR accumulation in nucleus [62]. Hence it is possible that in the non-tumorigenic cells, BPA may initiate or promote PCa progression by interfering with AR function. A previous report has shown that treatment with BPA stimulates human PCa cell migration [33] and affects MT dynamics [66]. Moreover, a change in MT dynamics could be linked to our observation that BPA increased cloning efficiencies of C4-2 cells in soft agar, which could be indicative of enhanced tumorigenicity and/or aggressiveness for these cells *in vivo*. This latter finding supports the notion that BPA may promote PCa progression in addition to its speculative role in neoplastic transformation.

The centrosome is emerging as a potential therapeutic target of drugs in castration resistant PCa (CRPC). Targeted inhibitory compounds are available for inhibition of kinases such as Polo-like kinases, Cyclin-dependent kinases, Aurora kinases, as well as molecular motor proteins [67], some of which have progressed to early clinical trials [68], [69]. Recently, histone deacetylases HDAC1, HDAC5 and SIRT1 have been identified to suppress centrosome duplication and amplification [70], suggesting that HDAC activation could be an important therapeutic avenue in CRPC. Aryl hydrocarbon receptor agonists such as indirubins also reduced centriole overduplication, implying involvement of aryl hydrocarbon receptor signaling in the centrosome cycle [71]. Additionally, the MT-disrupting agents are first line treatments for CRPC [72]. However, because of the ubiquitous presence of BPA, the possible adverse interactions of BPA with these centrosome and MT targeting drugs necessitate evaluation for CRPC.

In short, our findings provide the first evidence that urinary BPA level may have prognostic value for PCa and that disruption of the centrosome duplication cycle by low-dose BPA is a previously unknown mechanism underlying neoplastic transformation and cancer progression in the prostate.

Acknowledgments

We thank Paulina Haight for her assistance in the centrosome studies as a Summer Undergraduate Research Fellowship student at the University of Cincinnati, Ms. Hong Xiao for her assistance in measuring urine creatinine, and Ms. Nancy K. Voynow for her excellent editing of this manuscript.

Author Contributions

Conceived and designed the experiments: PT BO SMH. Performed the experiments: PT BO SMH. Analyzed the data: PT JY BO SMH. Contributed reagents/materials/analysis tools: BB BB PT SMH. Wrote the paper: PT BO JY SMH.

References

1. Bostwick DG, Burke HB, Djakiew D, Euling S, Ho SM, et al. (2004) Human prostate cancer risk factors. Cancer 101: 2371–2490. doi: 10.1002/cncr.20408.
2. Siegel R, Naishadham D, Jemal A (2012) Cancer statistics, 2012. CA Cancer J Clin 62: 10–29. doi: 10.3322/caac.20138.
3. American Cancer Society (2013) Cancer Facts & Figures. Atlanta: American Cancer Society.
4. SEER Cancer Statistics Review (2013) 1975–2010, National Cancer Institute. Bethesda, MD. based on November 2012 SEER data submission, posted to the SEER web site.
5. Erickson BE (2008) Bisphenol A under Scrutiny. Chemical & Engineering News 86: 36–39. doi: 10.1021/cen-v086n022.p036.
6. Fiege H, Voges H-W, Hamamoto T, Umemura S, Iwata T, et al. (2002) Phenol derivatives in Ullmann's encyclopedia of industrial chemistry. Wiley-VCH, Weinheim.
7. Calafat AM, Ye X, Wong LY, Reidy JA, Needham LL (2008) Exposure of the U.S. population to bisphenol A and 4-tertiary-octylphenol: 2003–2004. Environ Health Perspect 116: 39–44. doi: 10.1289/ehp.10753.
8. Geens T, Aerts D, Berthot C, Bourguignon JP, Goeyens L, et al. (2012) A review of dietary and non-dietary exposure to bisphenol-A. Food Chem Toxicol 50: 3725–3740. doi: S0278-6915(12)00537-6;10.1016/j.fct.2012.07.059.

9. Keri RA, Ho SM, Hunt PA, Knudsen KE, Soto AM, et al. (2007) An evaluation of evidence for the carcinogenic activity of bisphenol A. Reprod Toxicol 24: 240–252. doi: S0890-6238(07)00195-5;10.1016/j.reprotox.2007.06.008.

10. Rogers JA, Metz L, Yong VW (2013) Review: Endocrine disrupting chemicals and immune responses: a focus on bisphenol-A and its potential mechanisms. Mol Immunol 53: 421–430. doi: S0161-5890(12)00414-2;10.1016/j.molimm.2012.09.013.

11. Ehrlich S, Williams PL, Missmer SA, Flaws JA, Ye X, et al. (2012) Urinary bisphenol A concentrations and early reproductive health outcomes among women undergoing IVF. Hum Reprod 27: 3583–3592. doi: des328;10.1093/humrep/des328.

12. Lang IA, Galloway TS, Scarlett A, Henley WE, Depledge M, et al. (2008) Association of urinary bisphenol A concentration with medical disorders and laboratory abnormalities in adults. JAMA 300: 1303–1310. doi: 300.11.1303;10.1001/jama.300.11.1303.

13. Donohue KM, Miller RL, Perzanowski MS, Just AC, Hoepner LA, et al. (2013) Prenatal and postnatal bisphenol A exposure and asthma development among inner-city children. J Allergy Clin Immunol 131: 736–742. doi: S0091-6749(12)00060-2;10.1016/j.jaci.2012.12.1573.

14. Duan B, Hu X, Zhao H, Qin J, Luo J (2013) The relationship between urinary bisphenol A levels and meningioma in Chinese adults. Int J Clin Oncol 18: 492–497. doi: 10.1007/s10147-012-0408-6.

15. Lingle WL, Lukasiewicz K, Salisbury JL (2005) Deregulation of the centrosome cycle and the origin of chromosomal instability in cancer. Adv Exp Med Biol 570: 393–421. doi: 10.1007/1-4020-3764-3_14.

16. Fukasawa K (2005) Centrosome amplification, chromosome instability and cancer development. Cancer Lett 230: 6–19. doi: S0304-3835(04)00988-7;10.1016/j.canlet.2004.12.028.

17. Gorlov IP, Sircar K, Zhao H, Maity SN, Navone NM, et al. (2010) Prioritizing genes associated with prostate cancer development. BMC Cancer 10: 599. doi: 1471-2407-10-599;10.1186/1471-2407-10-599.

18. Lam YW, Yuan Y, Isaac J, Babu CV, Meller J, et al. (2010) Comprehensive identification and modified-site mapping of S-nitrosylated targets in prostate epithelial cells. PLoS One 5: e9075. doi: 10.1371/journal.pone.0009075.

19. Ouyang B, Bracken B, Burke B, Chung E, Liang J, et al. (2009) A duplex quantitative polymerase chain reaction assay based on quantification of alpha-methylacyl-CoA racemase transcripts and prostate cancer antigen 3 in urine sediments improved diagnostic accuracy for prostate cancer. J Urol 181: 2508–2513. doi: S0022-5347(09)00280-8;10.1016/j.juro.2009.01.110.

20. Padmanabhan V, Siefert K, Ransom S, Johnson T, Pinkerton J, et al. (2008) Maternal bisphenol-A levels at delivery: a looming problem? J Perinatol 28: 258–263. doi: 7211913;10.1038/sj.jp.7211913.

21. Zhang Z, Alomirah H, Cho HS, Li YF, Liao C, et al. (2011) Urinary bisphenol A concentrations and their implications for human exposure in several Asian countries. Environ Sci Technol 45: 7044–7050. doi: 10.1021/es200976k.

22. Hornung RW, Reed LD (1990) Estimation of Average Concentration in the Presence of Nondetectable Values. Applied Occupational and Environmental Hygiene 5: 46–51.

23. WHO (1996) Biological Monitoring of Chemical Exposure in the Workplace. Occupational Health for All, Geneva: World Health Organization vol 1.

24. Mobley JA, Leav I, Zielie P, Wotkowitz C, Evans J, et al. (2003) Branched fatty acids in dairy and beef products markedly enhance alpha-methylacyl-CoA racemase expression in prostate cancer cells in vitro. Cancer Epidemiol Biomarkers Prev 12: 775–783.

25. Tarapore P, Hanashiro K, Fukasawa K (2012) Analysis of centrosome localization of BRCA1 and its activity in suppressing centrosomal aster formation. Cell Cycle 11: 2931–2946. doi: 21396;10.4161/cc.21396.

26. Stavnezer E, Gerhard DS, Binari RC, Balazs I (1981) Generation of transforming viruses in cultures of chicken fibroblasts infected with an avian leukosis virus. J Virol 39: 920–934.

27. Carroll PE, Okuda M, Horn HF, Biddinger P, Stambrook PJ, et al. (1999) Centrosome hyperamplification in human cancer: chromosome instability induced by p53 mutation and/or Mdm2 overexpression. Oncogene 18: 1935–1944. doi: 10.1038/sj.onc.1202515.

28. Delgehyr N, Sillibourne J, Bornens M (2005) Microtubule nucleation and anchoring at the centrosome are independent processes linked by ninein function. J Cell Sci 118: 1565–1575. doi: jcs.02302;10.1242/jcs.02302.

29. Ho SM, Tang WY, Belmonte de FJ, Prins GS (2006) Developmental exposure to estradiol and bisphenol A increases susceptibility to prostate carcinogenesis and epigenetically regulates phosphodiesterase type 4 variant 4. Cancer Res 66: 5624–5632. doi: 66/11/5624;10.1158/0008-5472.CAN-06-0516.

30. Prins GS, Ye SH, Birch L, Ho SM, Kannan K (2011) Serum bisphenol A pharmacokinetics and prostate neoplastic responses following oral and subcutaneous exposures in neonatal Sprague-Dawley rats. Reprod Toxicol 31: 1–9. doi: S0890-6238(10)00306-0;10.1016/j.reprotox.2010.09.009.

31. Tang WY, Morey LM, Cheung YY, Birch L, Prins GS, et al. (2012) Neonatal exposure to estradiol/bisphenol A alters promoter methylation and expression of Nsbp1 and Hpcal1 genes and transcriptional programs of Dnmt3a/b and Mbd2/4 in the rat prostate gland throughout life. Endocrinology 153: 42–55. doi: en.2011-1308;10.1210/en.2011-1308.

32. Jenkins S, Wang J, Eltoum I, Desmond R, Lamartiniere CA (2011) Chronic oral exposure to bisphenol A results in a nonmonotonic dose response in mammary carcinogenesis and metastasis in MMTV-erbB2 mice. Environ Health Perspect 119: 1604–1609. doi: 10.1289/ehp.1103850.

33. Derouiche S, Warnier M, Mariot P, Gosset P, Mauroy B, et al. (2013) Bisphenol A stimulates human prostate cancer cell migration remodelling of calcium signalling. Springerplus 2: 54. doi: 10.1186/2193-1801-2-54;93.

34. De FS, Micale RT, La MS, Izzotti A, D'Agostini F, et al. (2011) Upregulation of clusterin in prostate and DNA damage in spermatozoa from bisphenol A-treated rats and formation of DNA adducts in cultured human prostatic cells. Toxicol Sci 122: 45–51. doi: kfr096;10.1093/toxsci/kfr096.

35. Nomura H, Kawashima H, Masaki S, Hosono TY, Matsumura K, et al. (2009) Effect of selective estrogen receptor modulators on cell proliferation and estrogen receptor activities in normal human prostate stromal and epithelial cells. Prostate Cancer Prostatic Dis 12: 375–381. doi: pcan200920;10.1038/pcan.2009.20.

36. Hess-Wilson JK, Webb SL, Daly HK, Leung YK, Boldison J, et al. (2007) Unique bisphenol A transcriptome in prostate cancer: novel effects on ERbeta expression that correspond to androgen receptor mutation status. Environ Health Perspect 115: 1646–1653. doi: 10.1289/ehp.10283.

37. Carwile JL, Michels KB (2011) Urinary bisphenol A and obesity: NHANES 2003-2006. Environ Res 111: 825–830. doi: S0013-9351(11)00143-5;10.1016/j.envres.2011.05.014.

38. Ye X, Wong LY, Bishop AM, Calafat AM (2011) Variability of urinary concentrations of bisphenol A in spot samples, first morning voids, and 24-hour collections. Environ Health Perspect 119: 983–988. doi: 10.1289/ehp.1002701.

39. Volkel W, Colnot T, Csanady GA, Filser JG, Dekant W (2002) Metabolism and kinetics of bisphenol a in humans at low doses following oral administration. Chem Res Toxicol 15: 1281–1287. doi: tx025548t.

40. Dekant W, Volkel W (2008) Human exposure to bisphenol A by biomonitoring: methods, results and assessment of environmental exposures. Toxicol Appl Pharmacol 228: 114–134. doi: S0041-008X(07)00561-3;10.1016/j.taap.2007.12.008.

41. Mahalingaiah S, Meeker JD, Pearson KR, Calafat AM, Ye X, et al. (2008) Temporal variability and predictors of urinary bisphenol A concentrations in men and women. Environ Health Perspect 116: 173-178. doi: 10.1289/ehp.10605.

42. Townsend MK, Franke AA, Li X, Hu FB, Eliassen AH (2013) Within-person reproducibility of urinary bisphenol A and phthalate metabolites over a 1 to 3 year period among women in the Nurses' Health Studies: a prospective cohort study. Environ Health 12: 80. doi: 1476-069X-12-80;10.1186/1476-069X-12-80.

43. CL, Ho SM (2012) Developmental reprogramming of cancer susceptibility. Nat Rev Cancer 12: 479–486. doi: nrc3220;10.1038/nrc3220.

44. Prins GS, Tang WY, Belmonte J, Ho SM (2008) Developmental exposure to bisphenol A increases prostate cancer susceptibility in adult rats: epigenetic mode of action is implicated. Fertil Steril 89: e41. doi: S0015-0282(07)04305-1;10.1016/j.fertnstert.2007.12.023.

45. Morgentaler A, Rhoden EL (2006) Prevalence of prostate cancer among hypogonadal men with prostate-specific antigen levels of 4.0 ng/mL or less. Urology 68: 1263–1267. doi: S0090-4295(06)01962-5;10.1016/j.urology.2006.08.1058.

46. Rhoden EL, Riedner CE, Morgentaler A (2008) The ratio of serum testosterone-to-prostate specific antigen predicts prostate cancer in hypogonadal men. J Urol 179: 1741–1744. doi: S0022-5347(08)00051-7;10.1016/j.juro.2008.01.045.

47. Guay AT, Perez JB, Fitaihi WA, Vereb M (2000) Testosterone treatment in hypogonadal men: prostate-specific antigen level and risk of prostate cancer. Endocr Pract 6: 132–138. doi: ep99055.or.

48. Xi W, Lee CK, Yeung WS, Giesy JP, Wong MH, et al. (2011) Effect of perinatal and postnatal bisphenol A exposure to the regulatory circuits at the hypothalamus-pituitary-gonadal axis of CD-1 mice. Reprod Toxicol 31: 409–417. doi: S0890-6238(10)00353-9;10.1016/j.reprotox.2010.12.002.

49. Davis JN, Muqim N, Bhuiyan M, Kucuk O, Pienta KJ, et al. (2000) Inhibition of prostate specific antigen expression by genistein in prostate cancer cells. Int J Oncol 16: 1091–1097.

50. Brownlee CW, Rogers GC (2013) Show me your license, please: deregulation of centriole duplication mechanisms that promote amplification. Cell Mol Life Sci 70: 1021-1034. doi: 10.1007/s00018-012-1102-6.

51. Avidor-Reiss T, Gopalakrishnan J (2013) Building a centriole. Curr Opin Cell Biol 25: 72–77. doi: S0955-0674(12)00180-9;10.1016/j.ceb.2012.10.016.

52. Lee HH, Zhu Y, Govindasamy KM, Gopalan G (2008) Downregulation of Aurora-A overrides estrogen-mediated growth and chemoresistance in breast cancer cells. Endocr Relat Cancer 15: 765–775. doi: ERC-07-0213;10.1677/ERC-07-0213.

53. Sheng ZG, Tang Y, Liu YX, Yuan Y, Zhao BQ, et al. (2012) Low concentrations of bisphenol a suppress thyroid hormone receptor transcription through a nongenomic mechanism. Toxicol Appl Pharmacol 259: 133–142. doi: S0041-008X(11)00475-3;10.1016/j.taap.2011.12.018.

54. Qin XY, Kojima Y, Mizuno K, Ueoka K, Muroya K, et al. (2012) Identification of novel low-dose bisphenol a targets in human foreskin fibroblast cells derived from hypospadias patients. PLoS One 7: e36711. doi: 10.1371/journal.pone.0036711;PONE-D-12-06650.

55. Watson CS, Bulayeva NN, Wozniak AL, Finnerty CC (2005) Signaling from the membrane via membrane estrogen receptor-alpha: estrogens, xenoestrogens, and phytoestrogens. Steroids 70: 364–371. doi: S0039-128X(05)00075-9;10.1016/j.steroids.2005.03.002.

56. Safe SH, Pallaroni L, Yoon K, Gaido K, Ross S, et al. (2002) Problems for risk assessment of endocrine-active estrogenic compounds. Environ Health Perspect 110 Suppl 6: 925-929. doi: sc271_5_1835.

57. Thomas P, Dong J (2006) Binding and activation of the seven-transmembrane estrogen receptor GPR30 by environmental estrogens: a potential novel mechanism of endocrine disruption. J Steroid Biochem Mol Biol 102: 175–179. doi: S0960-0760(06)00262-7;10.1016/j.jsbmb.2006.09.017.

58. Matsushima A, Teramoto T, Okada H, Liu X, Tokunaga T, et al. (2008) ERRgamma tethers strongly bisphenol A and 4-alpha-cumylphenol in an induced-fit manner. Biochem Biophys Res Commun 373: 408-413. doi: S0006-291X(08)01180-7;10.1016/j.bbrc.2008.06.050.

59. Okada H, Tokunaga T, Liu X, Takayanagi S, Matsushima A, et al. (2008) Direct evidence revealing structural elements essential for the high binding ability of bisphenol A to human estrogen-related receptor-gamma. Environ Health Perspect 116: 32–38. doi: 10.1289/ehp.10587.

60. De CS, van LN (2012) Endocrine-disrupting chemicals: associated disorders and mechanisms of action. J Environ Public Health 2012: 713696. doi: 10.1155/2012/713696.

61. Darshan MS, Loftus MS, Thadani-Mulero M, Levy BP, Escuin D, et al. (2011) Taxane-induced blockade to nuclear accumulation of the androgen receptor predicts clinical responses in metastatic prostate cancer. Cancer Res 71: 6019–6029. doi: 0008-5472.CAN-11-1417;10.1158/0008-5472.CAN-11-1417.

62. Zhu ML, Horbinski CM, Garzotto M, Qian DZ, Beer TM, et al. (2010) Tubulin-targeting chemotherapy impairs androgen receptor activity in prostate cancer. Cancer Res 70: 7992-8002. doi: 0008-5472.CAN-10-0585;10.1158/0008-5472.CAN-10-0585.

63. Balk SP, Knudsen KE (2008) AR, the cell cycle, and prostate cancer. Nucl Recept Signal 6: e001. doi: 10.1621/nrs.06001.

64. Lehmann L, Metzler M (2004) Bisphenol A and its methylated congeners inhibit growth and interfere with microtubules in human fibroblasts in vitro. Chem Biol Interact 147: 273–285. doi: 10.1016/j.cbi.2004.01.005;S0009279704000079.

65. George O, Bryant BK, Chinnasamy R, Corona C, Arterburn JB, et al. (2008) Bisphenol A directly targets tubulin to disrupt spindle organization in embryonic and somatic cells. ACS Chem Biol 3: 167–179. doi: 10.1021/cb700210u.

66. Pfeiffer E, Rosenberg B, Deuschel S, Metzler M (1997) Interference with microtubules and induction of micronuclei in vitro by various bisphenols. Mutat Res 390: 21–31. doi: S0165-1218(96)00161-9.

67. Korzeniewski N, Hohenfellner M, Duensing S (2013) The centrosome as potential target for cancer therapy and prevention. Expert Opin Ther Targets 17: 43–52. doi: 10.1517/14728222.2013.731396.

68. Cheung CH, Coumar MS, Chang JY, Hsieh HP (2011) Aurora kinase inhibitor patents and agents in clinical testing: an update (2009-10). Expert Opin Ther Pat 21: 857–884. doi: 10.1517/13543776.2011.574614.

69. Schoffski P (2009) Polo-like kinase (PLK) inhibitors in preclinical and early clinical development in oncology. Oncologist 14: 559–570. doi: theoncologist.2009-0010;10.1634/theoncologist.2009-0010.

70. Ling H, Peng L, Seto E, Fukasawa K (2012) Suppression of centrosome duplication and amplification by deacetylases. Cell Cycle 11: 3779–3791. doi: 21985;10.4161/cc.21985.

71. Chan JY (2011) A clinical overview of centrosome amplification in human cancers. Int J Biol Sci 7: 1122–1144.

72. Heidenreich A, Bastian PJ, Bellmunt J, Bolla M, Joniau S, et al. (2014) EAU Guidelines on Prostate Cancer. Part II: Treatment of Advanced, Relapsing, and Castration-Resistant Prostate Cancer. Eur Urol 65: 467–479. doi: S0302-2838(13)01199-8;10.1016/j.eururo.2013.11.002.

4

Motor Oil Classification Based on Time-Resolved Fluorescence

Taotao Mu, Siying Chen*, Yinchao Zhang, Pan Guo, He Chen, Fandong Meng

School of Optoelectronics, Beijing Institute of Technology, Beijing, China

Abstract

A time-resolved fluorescence (TRF) technique is presented for classifying motor oils. The system is constructed with a third harmonic Nd:YAG laser, a spectrometer, and an intensified charge coupled device (ICCD) camera. Steady-state and time-resolved fluorescence (TRF) measurements are reported for several motor oils. It is found that steady-state fluorescence is insufficient to distinguish the motor oil samples. Then contour diagrams of TRF intensities (CDTRFIs) are acquired to serve as unique fingerprints to identify motor oils by using the distinct TRF of motor oils. CDTRFIs are preferable to steady-state fluorescence spectra for classifying different motor oils, making CDTRFIs a particularly choice for the development of fluorescence-based methods for the discrimination and characterization of motor oils. The two-dimensional fluorescence contour diagrams contain more information, not only the changing shapes of the LIF spectra but also the relative intensity. The results indicate that motor oils can be differentiated based on the new proposed method, which provides reliable methods for analyzing and classifying motor oils.

Editor: Sabato D'Auria, CNR, Italy

Funding: These authors have no support or funding to report.

Competing Interests: The authors have declared that no competing interests exist.

* Email: csy@bit.edu.cn

Introduction

Motor oils, as products of petroleum refinery, are irreplaceable in car industry. They are used to protect the engine from many physically and chemically related malfunctions like heating, corrosion and contamination [1,2]. Due to the large variations in price of motor oils, cheap substituent products instead of more expensive petroleum products have been used by some profit-driven businesses to maximize the interest. It arisen the question about quality authentication [1].

There are many studies about the petroleum [3–6]. However the studies concentrating on motor oils are rare. Recently motor quality authentication is gaining increasing attention. As to Roman M. Balabin, motor oil classification is an important task for quality control and identification of oil adulteration [7]. Over the past several years many new technologies have surged in motor oils identification. Due to the usage for exploratory analysis and classification, chemometric techniques are often used in oils analysis [1]. However, sample preparation is time consuming in the way described above. In addition, it is destructive analysis methods.

Liquid chromatography has been widely used in many fields including petroleum industry. [8–10] Dielectric spectroscopy is used in classification of engine oil by Guan, L. [11]. As a non-destructive measurement technique, near infrared (NIR) spectroscopy has been widely used in petroleum industry [12–16]. Near infrared (NIR) spectroscopy is used to classify motor oils by base stock and viscosity in Ref. [17]. The spectral range is from 780 nm to 1400 nm. A commercial IR spectrometer combined with multivariate data analysis is employed. The probabilistic neural network (PNN) method and classifiers have shown good results. But NIR measurements are scarcely selective, so it should be used together with chemometric techniques. And the low sensitivity is another disadvantage of IR [18].

In some recent years, fluorescence has been widely used in food [19,20], medicine [21–23], petroleum industry [24–26]. The application of time-resolved fluorescence (TRF) has a development in petroleum industry [27–30]. TRF has also been used to characterize of crude oil in Ref. [31,32]. The variation in the spectral profile of the emitted fluorescence bands as a function of time is measured in the experiment. Nine oil samples are successfully distinguished by this approach.

Experiments demonstrate that the TRF shapes of motor oil are distinct from each other. In this paper, Contour diagrams of normalized TRF intensities (CDTRFIs) are constructed to serve as fingerprints for classifying motor oils. The oils can be distinguished by the new presented methods. Superior to ordinary fluorescence spectra, the two-dimensional fluorescence contour diagrams of motor oils contain more information, not only the changing shapes of the LIF spectra but also the relative intensity [33]. Due to the non-destructive and high sensitive characteristic, CDTRFIs make identification more accurate and reliable.

Materials and Methods

Oil samples are obtained from local market and stored in a dark room during the period of analysis. Nine motor oil samples of five

Table 1. Motor oils used in the study.

No.	API service	SAE grade	company
1	SM	5W-40	Shell
2	SF	15W-40	Shell
3	SN	5W-40	Mobil
4	SN	5W-30	Mobil
5	SN	0W-40	Mobil
6	SF	20W-40	Mobil
7	SN/CF	5W-40	Castrol
8	SM	5W-40	Prestone
9	SJ	10W-50	GreatWall

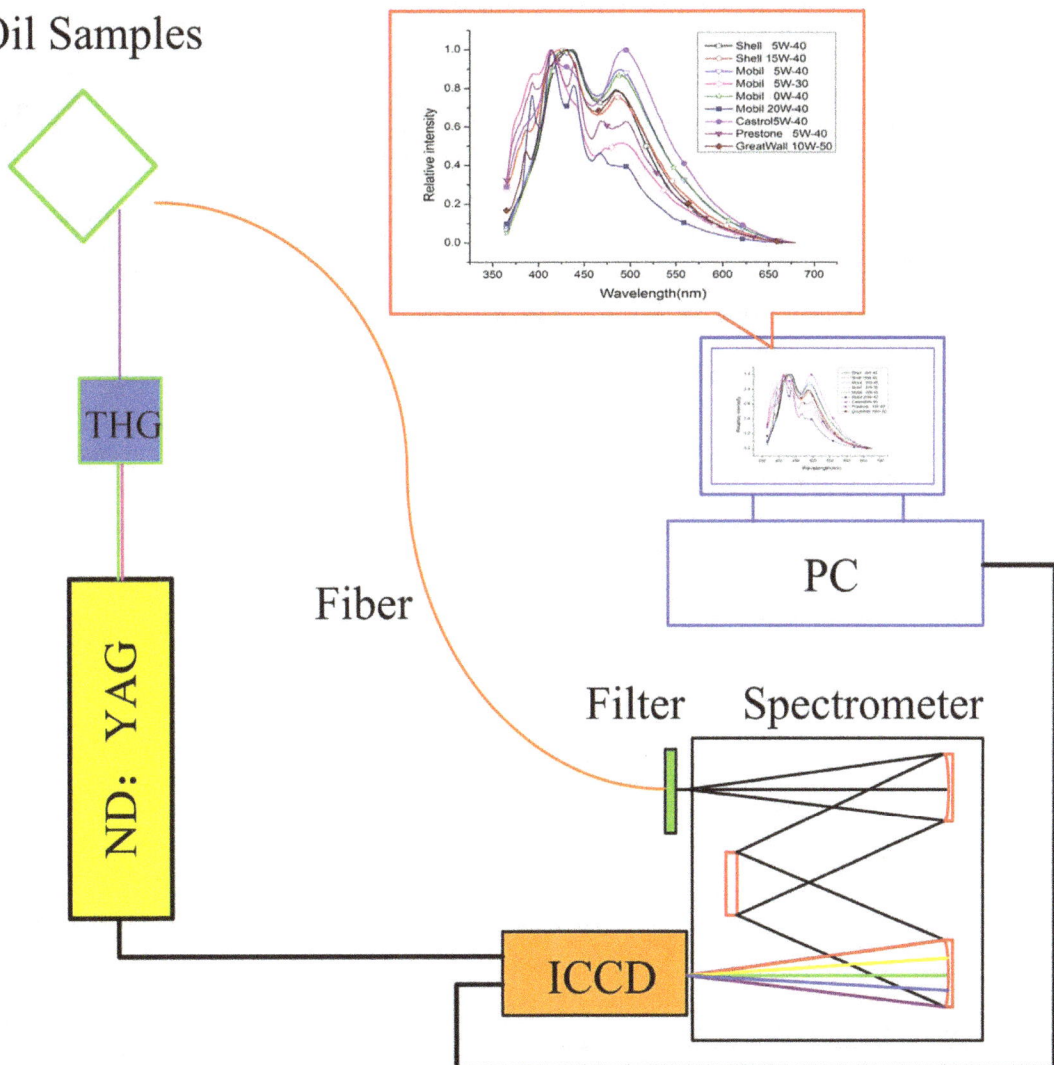

Figure 1. Schematic of the experimental set-up. PC: personal computer; Filter: 355 nm long-pass filter; ICCD: intensified charge-coupled device; THG: third harmonic generator output at 355 nm.

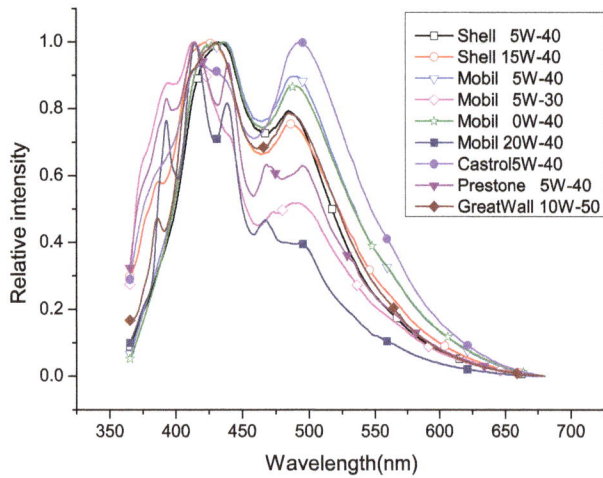

Figure 2. Normalized steady-state fluorescence spectra of motor oils, including, Shell 5W-40, Shell 15W-40, Mobil 5W-30, Mobil 5W-40, Mobil 0W-40, Mobil 20W-40, Castrol 5W-40, Prestone 5W-40, GreatWall 10W-50, under 355 nm laser pulse excitation.

popular brands (Table 1) and different SAE grades are used for classification in this paper.

Steady-state and time-resolved fluorescence spectra are collected and analyzed in the study. CDTRFIs are based on measuring the variations in the shapes of the TRF spectra at specific gate delay time (GDT) within the laser-pulse time profiles [33]. GDT should be set to 0 ns and the gate-width of ICCD should be set to 75 ns when steady-state fluorescence is collected. Experiments demonstrated that fluorescence intensity is too weaker to detect when GDT is larger than 75 ns. Before CDTRFIs constructed, GDT is increased from 0 ns to 75 ns by intervals of 5 ns. Thus, 16 independent fluorescence spectra, used to construct CDTRFIs, are collected for each oil sample. Background and noises are removed from all the fluorescence spectra at emission wavelengths (λ_{em}) between 360 and 675 nm. In order to avoid the geometrical effects on fluorescence measurement, we normalize the maximum fluorescence intensities to 1. All the measurements are repeated three times to ensure the repeatability of this approach.

Experimental setup

Figure 1 presents the configuration of experimental setup. a third-harmonic Nd:YAG laser (laser repetition rate 10 Hz, pulse width 3 ns, and pulse energy 40 mJ at 355 nm.) is employed. Motor oil samples are placed in a 10 mm fused-quartz cuvette. Front face illumination is used in the study to decrease inner effect, which is caused by high optical densities. A quartz optical fiber (NA = 0.22) are used to collect fluorescence directly. To block the strong elastic light from reaching the fluorescence detector, a 355 nm long-pass filter (Semrock BLP01-355R; cutoff wavelength of 361 nm) is inserted before the entrance slit (0.1 mm) of spectrometer. An intensified charge-coupled device (ICCD) camera (LI2CAM, Lambert Instruments) is placed behind the spectrometer to detect the dispersed fluorescence. The light is finally focused through the entrance slit and dispersed by a spectrometer (spectral range from 360 nm to 675 nm). On one hand, ICCD shows high sensitivity of a single photon level combined with fast gating of less than 3 ns. On the other hand, ICCD contain an embedded digital delay/pulse generator, which

can synchronize laser and ICCD itself without any other equipment. A computer is employed to sample and digitize the acquired fluorescence and finally complete experiment data processing and results analysis.

Results and Discussion

0.1 Steady-state Fluorescence Spectra of Motor Oil

The normalized steady-state LIF spectra of nine motor oils (including Shell 5W-40, Shell 15W-40, Mobil 5W-30, Mobil 5W-40, Mobil 0W-40, Mobil 20W-40, Castrol 5W-40, Prestone 5W-40, and GreatWall 10W-50) are shown in Figure 2. The excitation wavelength (λ_{ex}) is 355 nm. The GW is set to 75 ns, approximately including all the fluorescence emitted by the samples. However, some of the normalized fluorescence spectra (Figure 2) are similar at one excitation wavelength only (355 nm). Four main characteristic peaks (centered at 380 nm, 410 nm, 435 nm, and 490 nm) could be found in Figure 2.

0.2 Influence of GDT on Fluorescence Spectra

The normalized LIF spectra of different GDT (5 ns, 20 ns, 35 ns, 50 ns, 65 ns, and 75 ns) of oil samples are presented in Figure 3 (The data of oils can be found in Data S1). The GW is set to 5 ns. AS GDT is increased from 0 ns to 75 ns, the shapes of fluorescence of motor oils change a lot. Some new characteristics peaks appear while some old peaks disappear. And the intensity of each peak varies as the GDT increasing. As to E. HEGAZI, because not all the excited aromatic compounds have the same lifetime, the fluorescence spectrum of any oil will have different shapes when measured at different time windows [31]. Then time-character will reflect in the fluorescence when GW is narrower that fluorescence lifetime. The fluorescence spectrum should be correlated with GDTs. Equation 1, 2 can be derived [33]:

$$S(t, GW) = \int_t^{t+GW} I(t')dt' + \int_0^t I(t-t')\varepsilon(t)dt' \quad (1)$$

$$I(t) = \frac{L(t)\lambda_0}{hc} \Delta R \xi \eta(\lambda) \int_0^{\Delta R} e^{-(\alpha_{\lambda_0} + \alpha_\lambda)r} dr \sum_k N_k \sigma_k \tau_k(t) \quad (2)$$

Where:

$S(t, GW)$ is the fluorescence photon numbers collected by detector at GDT $= t$ with gate width $=$ GW, Including excited directly by the laser and other relaxation of energy transfer processes.

k refers to any compounds capable of fluorescing at $\lambda_{em} = \lambda$.

$\tau_k(t)$ is the time distribution function of fluorescence of compounds. t is detection time (s).

$I(t')$ is the fluorescence excited directly, which occurs within the 5 ns GW.

$I(t-t')$ is the fluorescence excited at earlier time, which will encounter a decay according to some exponential function $\varepsilon(t)$.

$\frac{\lambda_0}{hc}\tau_k(t)I(t')\varepsilon(t)$ is the reciprocal of laser photo energy ($photo/J$).

N_k is the concentration of each compound (g/cm^3).

σ_k is fluorescence cross section of different compounds ($cm^2/sr/mg$).

ΔR is optical path in the sample (cm).

$\eta(\lambda)$ is total receiver efficiency at $\lambda_{em} = \lambda$.

ξ is the geometric overlap factor.

$\alpha_{\lambda_0}, \alpha_\lambda$ are extinction coefficient at $\lambda_{ex} = \lambda_0$ and $\lambda_{em} = \lambda$.

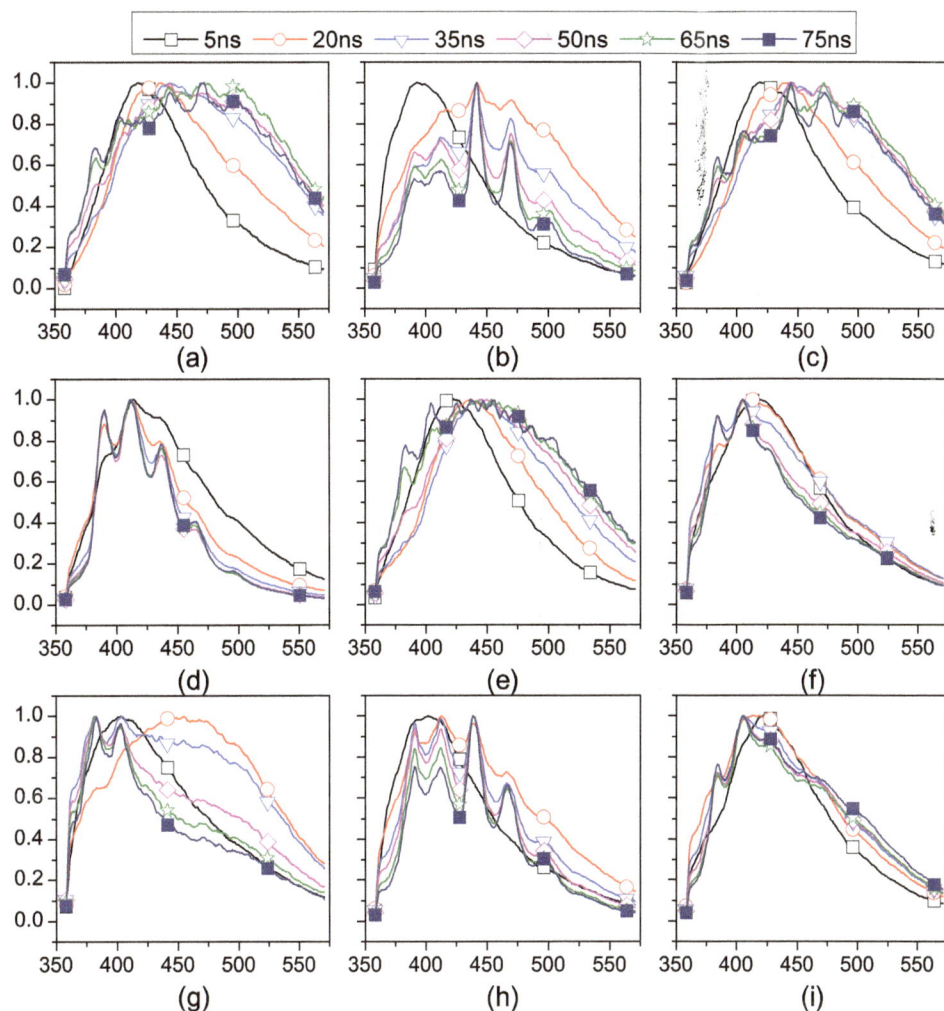

Figure 3. LIF spectra at specific time gates (including 5 ns, 20 ns, 35 ns, 50 ns, 65 ns, 75 ns). (a) Mobil 0W-40, (b) Mobil 5W-30, (c) Mobil 5W-40, (d) Mobil 20W-40, (e) Shell 5W-40, (f) Shell 15W-40, (g) Castrol 5W-40, (h) Prestone 5W-40, and (i) GreatWall 10W-50. The normalized fluorescence intensity (y-axis) and the wavelength (x-axis) are used as axes. The wavelength range is 360 nm to 575 nm. The integration time is set to 5 ns.

These variations in the spectra will be different for different oils, as we will see shortly, and can be conveniently taken advantage of to characterize these oils.

0.3 Contour Diagrams of TRF Intensities of Motor Oils

CDTRFIs are constructed based on the time-character of LIF of motor oils (Figure 4, The data of oils can be found in Data S1). Wavelength and GDTs are used as axis of CDTRFI to show the variation of fluorescence spectra shapes over time. It should be pointed out that the zero point of the GDT ($x-axis$) is set when the sample illuminated by laser. GDTs increase from 0 ns to 75 ns with a 5 ns sampling interval. The spectra ($y-axis$) range from 360 nm to 675 nm at only one excitation wavelength (355 nm). Contour diagrams are constructed in this way when all the spectra are normalized. The most important parameter of CDTRFIs is the increment between the contour lines. The smaller the increment, the more contour features will be revealed and the higher the resolution of the fingerprints will become. However, the limit of the increment will depend on the uncertainty of the TRF spectra [31]. The increment is set to 0.02 in this study.

Then fluorescence should be measured at 16 different GDTs for each oil to construct intact CDTRFI. To eliminate the influence of energy fluctuation, this process should be repeated at least three times. It is found that CDTRFIs, showing not only the variation in the shape of TRF spectra over time (along $x-axis$) but also a series of normalized LIF spectra of different GDTs (along $y-axis$), can serve as unique fingerprints for motor oils. The CDTRFIs of motor oils are quite different from each other, suggesting that the effective of TRF-based method. Outperform to the steady-state LIF, the distinction among the CDTRFIs of different oils is more significant. The Motor oil samples can be easily distinguished by this method, while it is hard to classify all the oil samples just by steady-state LIF fluorescence spectra. All the measurements are repeated three times to ensure the repeatability of this method.

In this paper, the contour diagrams are constructed at only one excitation wavelength 355 nm. The CDTRFIs of motor oils providing additional information of motor oils to enhance the recognition rate.

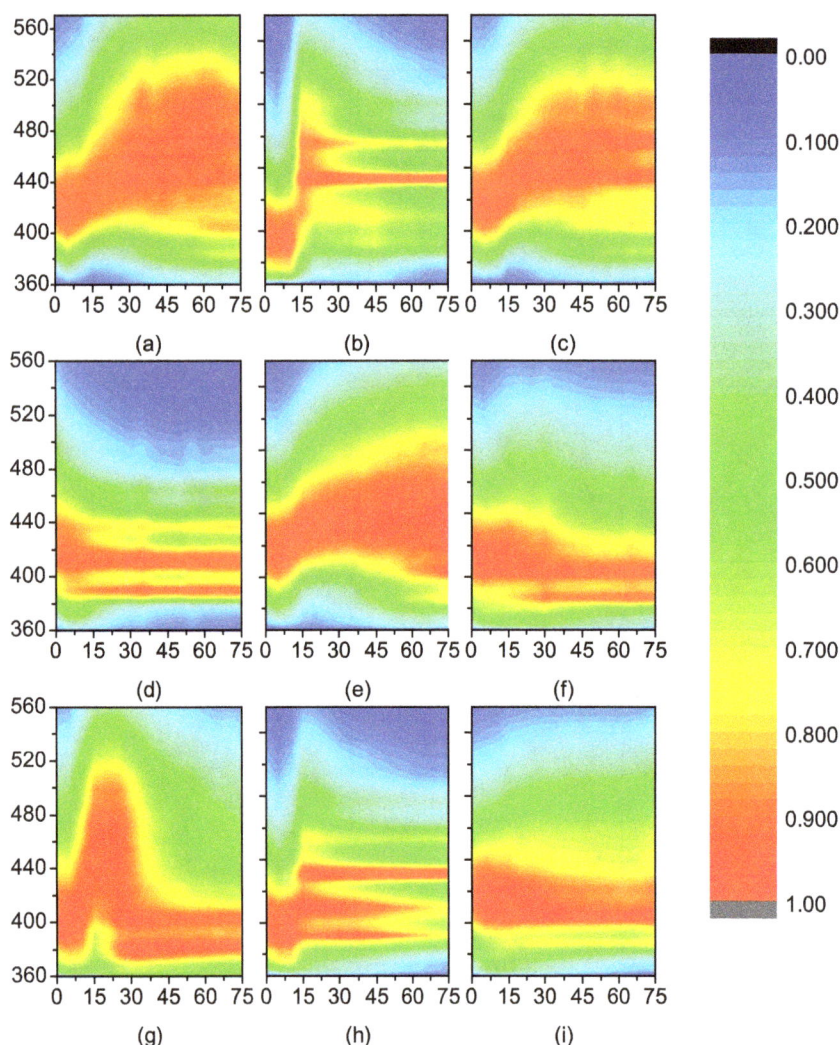

Figure 4. CDTRFIs of (a) Mobil 0W-40, (b) Mobil 5W-30, (c) Mobil 5W-40, (d) Mobil 20W-40, (e) Shell 5W-40, (f) Shell 15W-40, (g) Castrol 5W-40, (h) Prestone 5W-40, and (i) GreatWall 10W-50. The fluorescence wavelength (y-axis) and the detection time (x-axis) are used as axes. The wavelength range is 360 nm to 560 nm, and the time range is 75 ns with an excitation wavelength of 355 nm. The gate width of the ICCD is set to 5 ns.

Conclusions

A new method based on TRF is proposed for motor oils classification. The LIF spectra of nine kinds of motor oils are measured at 16 different GDTs in the study. Then CDTRFIs are constructed by using these spectra measured at different GDTs. Superior to steady-state LIF, this approach presents not only a series of normalized fluorescence spectra (along $y-axis$) but also the variation in the spectra shapes (along $x-axis$) as the GDTs increasing. The proposed method is shown successful in classification of nine motor oils, and to be capable in distinguishing between more similar oils. The method may be further improved by reducing the GW of the ICCD and by using smaller increments in the counter. Considering the high sensitive and non-destructive ability of CDTRFIs, the presented method can be used as unique fingerprints of motor oils, providing researchers with a reliable means of characterizing and distinguishing motor oils.

Supporting Information

Data S1 The time-resolved fluorescence data of motor oils. (Including Mobil 0W-40, Mobil 5W-30, (c) Mobil 5W-40, Mobil 20W-40, Shell 5W-40, Shell 15W-40, Castrol 5W-40, Prestone 5W-40, and GreatWall 10W-50). The wavelength range is 360 nm to 560 nm, and the time range is 75 ns with an excitation wavelength of 355 nm. The gate width of the ICCD is set to 5 ns.

Author Contributions

Conceived and designed the experiments: TTM SYC. Performed the experiments: TTM YCZ PG HC FDM. Analyzed the data: TTM. Contributed reagents/materials/analysis tools: FDM. Contributed to the writing of the manuscript: TTM SYC. Designed the software used in analysis (MATLAB): PG HC.

References

1. Bassbasi M, Hafid A, Platikanov S, Tauler R, Oussama A (2013) Study of motor oil adulteration by infrared spectroscopy and chemometrics methods. Fuel 104: 798–804.

2. Al-Ghouti MA, Al-Degs YS, Amer M (2010) Application of chemometrics and FTIR for determination of viscosity index and base number of motor oils. Talanta 81: 1096–1101.

3. Wang CY, Wang XS, Wang YH, Gao JW, Zheng RE (2006) Fluorescence analysis of crude oil samples with different spectral approaches. Spectroscopy and Spectral Analysis 26: 728–732.

4. Chen ZW, Song SH, Dunphy J (2012) Direct measurement of trace metals in crude oils and related upstream stocks using high definition X-ray fluorescence. Abstracts of Papers of the American Chemical Society 244.

5. Wan W, Wang S (2013) Determination of Residual Oil Saturation and Connectivity between Injector and Producer using Interwell Tracer Tests. Journal of Petroleum Engineering & Technology. 3: 18–24.

6. Zhao Y, Zhang SY, Ling P, Li SY, Huang KS, et al. (2009) Discussion of the Characteristic Index in Using Three-Dimensional Fluorescence Method to Identify Different Crude Oils. Spectroscopy and Spectral Analysis 29: 3335–3338.

7. Balabin RM, Safieva RZ, Lomakina EI (2011) Near-infrared (NIR) spectroscopy for motor oil classification: From discriminant analysis to support vector machines. Microchemical Journal 98: 121–128.

8. Wu SB, Meyer RS, Whitaker BD, Litt A, Kennelly EJ (2013) A new liquid chromatography-mass spectrometry-based strategy to integrate chemistry, morphology, and evolution of eggplant (Solanum) species. Journal of Chromatography A 1314: 154–172.

9. Giruts MV, Gordadze GN (2013) Differentiation of crude oils and condensates by distribution of saturated hydrocarbons: 1. Oil types determinable by gas-liquid chromatography. Petroleum Chemistry 53: 209–219.

10. Kudasheva FK, Nabiullina ER (1989) Study of Crude-Oil Composition by the High-Performance Liquid-Chromatography Method. Izvestiya Vysshikh Ucheb-nykh Zavedenii Khimiya I Khimicheskaya Tekhnologiya 32: 63–66.

11. Guan L, FengXL, Xiong G (2008) Engine lubricating oil classification by SAE grade and source based on dielectric spectroscopy data. Analytica Chimica Acta 628: 117–120.

12. De Peinder P, Petrauskas DD, Singelenberg F, Salvatori F, Visser T, et al. (2008) Prediction of long and short residue properties of crude oils from their infrared and near-infrared spectra. Applied Spectroscopy 62: 414–422.

13. Khanmohammadi M, Garmarudi AB, Ghasemi K, de la Guardia M (2013) Quality based classification of gasoline samples by ATR-FTIR spectrometry using spectral feature selection with quadratic discriminant analysis. Fuel 111: 96–102.

14. Cooper JB, Larkin CM, Schmitigal J, Morris RE, Abdelkader MF (2011) Rapid Analysis of Jet Fuel Using a Handheld Near-Infrared (NIR) Analyzer. Appl Spectrosc. 65: 187–92.

15. Chung H, Ku MS (2003) Near-infrared spectroscopy for on-line monitoring of lube base oil processes. Appl Spectrosc. 57: 545–50.

16. Cramer JA, Morris RE, Hammond MH, Rose-Pehrsson SL (2009) Ultra-low Sulfur Diesel Classification with Near-Infrared Spectroscopy and Partial Least Squares. Energ Fuel. 23: 1132–3.

17. Balabin RM, Safieva RZ (2008) Motor oil classification by base stock and viscosity based on near infrared (NIR) spectroscopy data. Fuel. 87: 2745-52.

18. Blanco M, Villarroya I (2002) NIR spectroscopy: a rapid-response analytical tool. Trac-Trend Anal Chem. 21: 240–50.

19. Poulli KI, Mousdis GA, Georgiou CA (2006) Synchronous fluorescence spectroscopy for quantitative determination of virgin olive oil adulteration with sunflower oil. Anal Bioanal Chem. 386: 1571–5.

20. Poulli KI, Mousdis GA, Georgiou CA (2007) Rapid synchronous fluorescence method for virgin olive oil adulteration assessment. Food Chem. 105: 369–75.

21. Yuvaraj M, Udayakumar K, Jayanth V, Prakasa Rao A, Bharanidharan G, et al. (2014) Fluorescence spectroscopic characterization of salivary metabolites of oral cancer patients. Journal of Photochemistry and Photobiology B: Biology. 130: 153–60.

22. Gharekhan AH, Biswal NC, Gupta S, Panigrahi PK, Pradhan A (2012) Characteristic Spectral Features of the Polarized Fluorescence of Human Breast Cancer in the Wavelet Domain. Appl Spectrosc. 66: 820–7.

23. Ionita I (2009) Early diagnosis of tooth decay using fluorescence and polarized Raman spectroscopy. Optoelectron Adv Mat. 3: 1122–6.

24. Guedes CLB, Di Mauro E, De Campos A, Mazzochin LF, Bragagnolo GM, et al. (2006) EPR and fluorescence spectroscopy in the photodegradation study of Arabian and Colombian crude oils. Int J Photoenergy 1–6.

25. Camagni P, Colombo G, Koechler C, Pedrini A, Omenetto N, et al. (1988) Diagnostics of Oil Pollution by Laser-Induced Fluorescence. Ieee T Geosci Remote. 26: 22–6.

26. Ralston CY, Wu X, Mullins OC (1996) Quantum yields of crude oils. Appl Spectrosc. 50: 1563–8.

27. Wang X, Mullins OC (1994) Fluorescence Lifetime Studies of Crude Oils. Appl Spectrosc. 48: 977–84.

28. Hegazi E, Hamdan A (2002) Estimation of crude oil grade using time-resolved fluorescence spectra. Talanta. 56: 989–95.

29. Ryder AG (2004) Time-resolved fluorescence spectroscopic study of crude petroleum oils: Influence of chemical composition. Appl Spectrosc. 58: 613–23.

30. Pantoja PA, Lopez-Gejo J, Le Roux GAC, Quina FH, Nascimento CAO (2011) Prediction of Crude Oil Properties and Chemical Composition by Means of Steady-State and Time-Resolved Fluorescence. Energ Fuel. 25: 3598–604.

31. Hegazi E, Hamdan A, Mastromarino J (2001) New approach for spectral characterization of crude oil using time-resolved fluorescence spectra. Appl Spectrosc. 55: 202–7.

32. Ryder AG, Glynn TJ, Feely M, Barwise AJG (2002) Characterization of crude oils using fluorescence lifetime data. Spectrochim Acta A. 58: 1025–37.

33. Mu TT, Chen SY, Zhang YC, Chen H, Guo P (2014) Characterization of edible oils using time-resolved fluorescence. Anal Methods-Uk. 6: 940–3.

5

Nitrophenol Chemi-Sensor and Active Solar Photocatalyst Based on Spinel Hetaerolite Nanoparticles

Sher Bahadar Khan[1,2]*, Mohammed M. Rahman[1,2], Kalsoom Akhtar[3], Abdullah M. Asiri[1,2], Malik Abdul Rub[1,2]

1 Center of Excellence for Advanced Materials Research (CEAMR), King Abdulaziz University, Jeddah, Saudi Arabia, **2** Chemistry Department, Faculty of Science, King Abdulaziz University, Jeddah, Saudi Arabia, **3** Division of Nano Sciences and Department of Chemistry, Ewha Womans University, Seoul, Korea

Abstract

In this contribution, a significant catalyst based on spinel $ZnMn_2O_4$ composite nanoparticles has been developed for electro-catalysis of nitrophenol and photo-catalysis of brilliant cresyl blue. $ZnMn_2O_4$ composite (hetaerolite) nanoparticles were prepared by easy low temperature hydrothermal procedure and structurally characterized by X-ray powder diffraction (XRD), field emission scanning electron microscopy (FESEM), X-ray photoelectron spectroscopy (XPS), Fourier transform infrared (FTIR) and UV-visible spectroscopy which illustrate that the prepared material is optical active and composed of well crystalline body-centered tetragonal nanoparticles with average size of $\sim38\pm10$ nm. Hetaerolite nanoparticles were applied for the advancement of a nitrophenol sensor which exhibited high sensitivity ($1.500~\mu Acm^{-2}~mM^{-1}$), stability, repeatability and lower limit of detection ($20.0~\mu M$) in short response time (10 sec). Moreover, hetaerolite nanoparticles executed high solar photo-catalytic degradation when applied to brilliant cresyl blue under visible light.

Editor: Andrew C. Marr, Queen's University Belfast, United Kingdom

Funding: This work was funded by the Deanship of Scientific Research (DSR), King Abdulaziz University, Jeddah, under grant No. (130-031-D1433). The authors, therefore, acknowledge with thanks DSR technical and financial support. The funders had no role in study design, data collection and analysis, decision to publish, or preparation of the manuscript.

Competing Interests: The authors have declared that no competing interests exist.

* E-mail: sbkhan@kau.edu.sa

Introduction

Environmental pollution has received extensive interest because of the uncertain consequences on human health and living organisms [1,2]. Industrial activity causes many environmental problems by discharging vast quantities of toxic compounds into the environment. Many hazardous waste sites have been created by gathering pollutants in soil and water for many years [3,4]. Therefore inspection of health risky pollutants in water and environment is an urgent demand for pollution controlling option. Thus easy, efficient and low cost processes for the recognition and detoxification of organic pollutants in aqueous solutions are needed to check and protect water resources and food supplies.

Several instrumental techniques exist for the recognition of organic pollutants, but are of less significance on the basis of efficiency and expense. However, sensor technology plays an important role in environmental safety and water treatment [5,6]. Thus for environmental and health safety, it is essential to produce simple, reliable, and inexpensive sensors for the detection of p-nitrophenol (p-NP) in water, because nitrophenols are hazardous and toxic pollutants with inhibitory and biorefractory nature and have adverse effect on living organisms. Nitrophenols have been widely used in the manufacture of pesticides, dyes and pharmaceuticals. p-NP is a toxic derivative of the parathion insecticide and is carcinogenic, hazardous, mutagenic and toxic (cytotoxic and embryo-toxic) to mammals [7]. Due to high solubility and stability of p-NP in water, it has been found in freshwater, marine environments and has been detected in industrial wastewaters.

Thus detection and monitoring of nitrophenols are crucial for environmental pollution control and industrial applications. Various chromatographic and spectroscopic techniques have been utilized for the detection and finding of hazardous solvents but their use is limited due to complication and sluggishness. Electrochemical sensors have achieved immense interest in the recognition and quantification of unsafe compounds since it is uncomplicated and rapid operation, response and recognition [8]. The sensitivity and selectivity of electrochemical sensor strongly dependent on the size, structure and properties of electrode materials and thus semiconductor nanostructured materials have received much importance and has extensively been utilized as a redox mediator in various sensors [9].

Photo-catalysis is also one of the low cost processes for detoxification of lethal organic compounds but the lack of active catalysts excluded this process from a wide range of applications. Among different photo-catalysts, TiO_2 and ZnO have proven to act as dynamic photocatalysts and play pivotal roles in the detoxification of lethal organic compounds [10,11]. However these photo-catalysts only encourage photo-catalysis upon irradiation by UV light because they absorb in the UV region. For solar photo-catalysis, a photo-catalyst must promote photo-catalysis by irradiation with visible light because solar spectrum consists 46% of visible light while the UV light is only 5–7% of the solar spectrum. Thus it is an urgent demand to develop an active photo-catalyst which can promote photo-catalysis in the visible region. Manganese oxides with various crystalline structures have been explored for various applications such as catalysis and electrode materials for lithium cathodes [12,13]. However, to enhance the various properties of manganese oxide to meet the increasing needs for different applications, the features of manganese oxide must be modified. One of the main noteworthy methods to amend the characteristics of these nanomaterials is the introduction of doped materials in the parent system because recently doped metal oxides have shown excellent properties in various applications [14].

Therefore in this study, spinel (AB_2O_4) like zinc doped manganese oxide (spinel hetaerolite) has been synthesized by a simple low-temperature hydrothermal process and characterized by XRD, FESEM, XPS, FTIR, XPS and UV-vis. spectroscopy. From an application point of view, spinel hetaerolite was investigated as a sensor substance for the discovery of p-NP and solar photo-catalyst for the degradation of brilliant cresyl blue under visible light. To the best of our knowledge, this is the first detail of a spinel heterolite as a nitrophenol sensor and solar photo-catalyst for brilliant cresyl blue.

Methods

1. Synthesis of hetaerolite nanoparticles

An equi-molar aqueous solution was prepared by dissolving 1.36 g of $ZnCl_2$ and $MnCl_2.4H_2O$ in 100 ml distilled water and further titrated by NH_4OH solution to increase the pH above 10.0. This high basic solution was then shifted to a Teflon autoclave and kept at 150.0°C for 16 hours in an oven. Finally a black product was obtained by washing the precipitate. The precipitate was dried and calcined at 400.0°C for 5 hours.

2. Possible growth of hetaerolite nano-particles

In the case of manganese oxides formation, it has been reported that manganese gets reduced to Mn^{3+} or Mn^{2+} states and generates lamellae structure which then appear in the form of rod like structure to reduce the surface energy [15–17]. After hydrothermal process it crystallizes to form MnO-OH nanorods. The calcination of MnO-OH gives Mn_3O_4 due to dehydration and oxygen out diffusion at high temperature [18]. However the incorporation of a Zn precursor has great influence on the reduction of manganese. It enhances the reduction process of Mn by first reducing to Mn^{+3} state and further reduce Mn^{+3} to Mn^{+2} state which result in the formation of $ZnMn_2O_4$ nanoparticles. Schematically probable growth process of $ZnMn_2O_4$ nanoparticles is given in Figure 1.

3. Characterization

JEOL Scanning Electron Microscope (JSM-7600F, Japan) was used to analyze the morphology of the prepared material while the crystallography was studied by X'Pert Explorer, PANalytical X-ray diffractometer. The XPS spectrum was recorded in the range of 0 to 1350 eV by using a Thermo Scientific K-Alpha KA1066

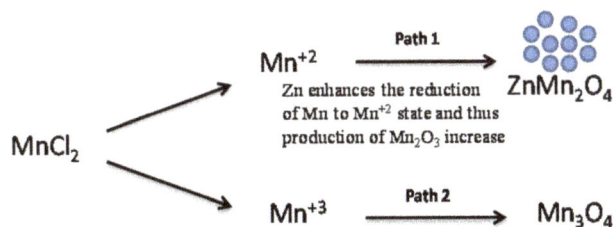

Figure 1. Possible growth mchanism of hetaerolite nanoparticles.

spectrometer (Germany). For compositional, optical and degradation studies of the nanomaterial, FT-IR and UV spectrum were record by a PerkinElmer (spectrum 100) FT-IR spectrometer and PerkinElmer (Lambda 950) UV-visible spectrometer. For sensing study, I–V measurements were carried out using an electrometer (Kethley, USA).

4. Fabrication of chemical sensor

Hetaerolite nanoparticles were coated as a thin film on the surface of a glassy carbon electrode (GCE, surface area 0.0314 cm^2) and dried at 60.0°C for 12 hours. Time delaying and response time were 5.0 sec and 10.0 sec, respectively. The sensing ability of the hetaerolite nanoparticle modified GCE was evaluated using nitrophenol in a similar way as we published earlier [19,20].

5. Solar photo-catalytic degradation

Solar photo-catalytic activity of hetaerolite nanoparticles was evaluated through degradation of brilliant cresyl blue under sun light. The dye is stable under visible light irradiation in the absence of a photo-catalyst. In photo-catalytic degradation, two different 100.0 mL, 1×10^{-4} M of brilliant cresyl blue solutions were taken in different beakers and adjusted the pH to 5, 8 and 10, by drop wise addition of 0.2 M NaOH solution under vigorous stirring. 0.1193 g and 0.1132 g catalyst were then added into the dye reaction solutions. The solution was then irradiated under sunlight at constant stirring and 4–5 mL of dye solutions were pipetted out at regular interval and measured the absorbance at $\lambda_{max} = 595.0$ nm using UV visible spectrophotometer. Dye degradation without catalyst was also studied under visible light to check any degradation of dye.

Results and Discussion

1. Physiochemical characterization of hetaerolite nanoparticles

FESEM was utilized to explore the morphology and size of hetaerolite. The morphology of hetaerolite was illustrated by FESEM which are shown in Figure 2 (a–c). It is evident from the FESEM images that the synthesized material is composed of particles having average diameter of $\sim 38 \pm 10$ nm. These nanoparticles were prepared in large quantity possessing a spherical shape.

The crystallographic information of the hetaerolite nanoparticles were corroborated by X-ray diffraction (Figure 2 (d)). All the characteristic diffraction peaks coincided with those for well-crystalline distinct body-centered tetragonally structured of $ZnMn_2O_4$. X-ray diffraction peaks of the samples are in outstanding agreement and excellent accordance with the JCPDS card no. 077-0470 [21]. According to the JCPDS card, the synthesized product is a body-centered tetragonal phase $ZnMn_2O_4$ with cell parameters of a = 5.72 Å and c = 9.24 Å and space group of 141/amd(141). All diffraction peaks were only related to $ZnMn_2O_4$ without any impurity peaks and thus the synthesized product therefore consist of pure $ZnMn_2O_4$ crystals. The size of hetaerolite nano-particles (38 nm) suggested by FESEM was also verified and supported by Scherrer formula.

$$D = 0.9\lambda / \beta \cos\theta$$

Where λ is the wavelength of X-ray radiation, β is the full width at half maximum (FWHM) of the peaks at the diffracting angle θ.

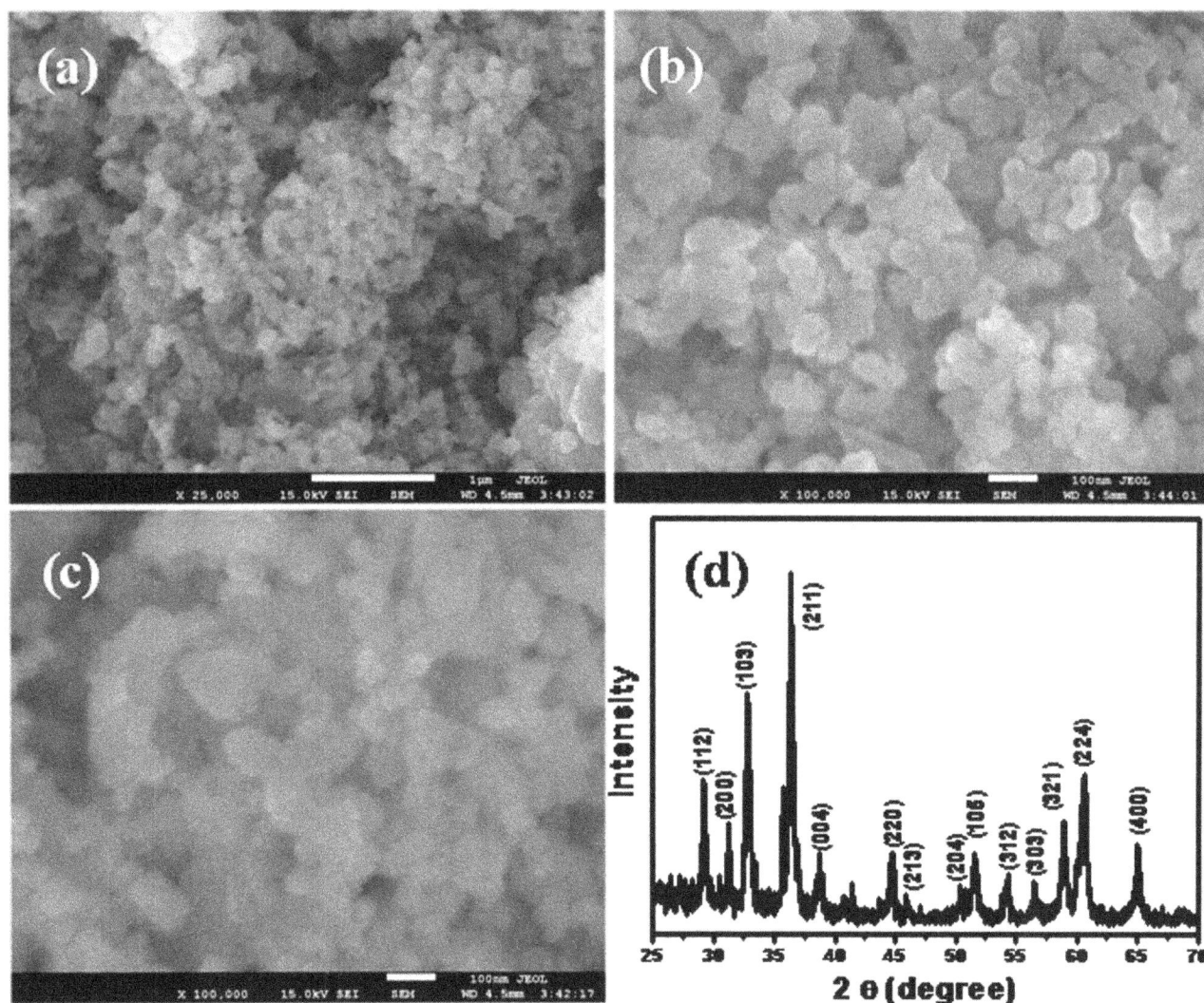

Figure 2. Typical low and high-resolution FESEM images (a–c) and XRD pattern (d) of the synthesized hetaerolite nano-particles.

XPS was analyzed to study the composition of hetaerolite nano-particles and the graphs of hetaerolite nano-particles are depicted in Figure 3. Hetaerolite nano-particles XPS spectrum exhibited peaks at 530.6, 642.6, 654.2, 1022.1 and 1045.1 eV which are responsible for O 1 s, Mn $2p_{3/2}$, Mn $2p_{1/2}$, Zn $2p_{3/2}$ and Zn $2p_{1/2}$ peaks, respectively. The spin–orbit splitting between Mn $2p_{3/2}$ and Mn $2p_{1/2}$ is 11.6 eV while 23.0 eV is observed among Zn $2p_{3/2}$ and Zn $2p_{1/2}$. These results are comparable to the values report earliar [18,22]. The above results are consisted with XRD which confirms the synthesis of $ZnMn_2O_4$.

The chemical structure of the hetaerolite nano-particles was also examined by FTIR analysis which was recorded in the range of $400{\sim}4000$ cm^{-1} and shown in Figure 4 (a). The very intense bands observed at 510 and 621 cm^{-1} were attributed to M–O (M = Zn, Mn) and M-O-M bonds, respectively. Supplementary peaks centered at 3417 and 1625 cm^{-1} were assigned to H_2O absorbed from the environment [23].

The photocatalytic performance of a photocatalyst can be evaluated by its optical properties, one of the essential requirements for a photocatalyst. Thus optical properties of hetaerolite nano-particles were scrutinized by using a UV-Vis. spectrophotometer and the spectrum is shown in Figure 4 (b). In UV/visible

absorption method, energy band gap of nanomaterials can be acquired by analyzing their optical absorption. UV–vis absorption spectra exhibited a broad absorption peak and showed band gap energy equal to 2.91 eV which is calculated by Tauc's formula. The relation between absorption coefficient (α) and the incident photon energy (hυ) is given by the equation [24].

$$\alpha h\upsilon = A(h\upsilon - E_g)^n$$

where A is a constant, E_g is the band gap of the material and the exponent n depends on the type of transition, n = 1/2, 2, 3/2 and 3 corresponding to allowed direct, allowed indirect, forbidden direct and forbidden indirect, respectively. Taking n = 1/2, we have calculated the direct energy band gap from the $(\alpha h\upsilon)^{1/n}$ vs. hυ plots (Figure 4 (c).

2. Chemical sensing properties

The hetaerolite nano-particles were utilized for the recognition of p-nitrophenol in aqueous media in order to study their chemical sensing properties toward p-nitrophenol [25,26]. I–V technique was used to measure the electrical response hetaerolite nano-

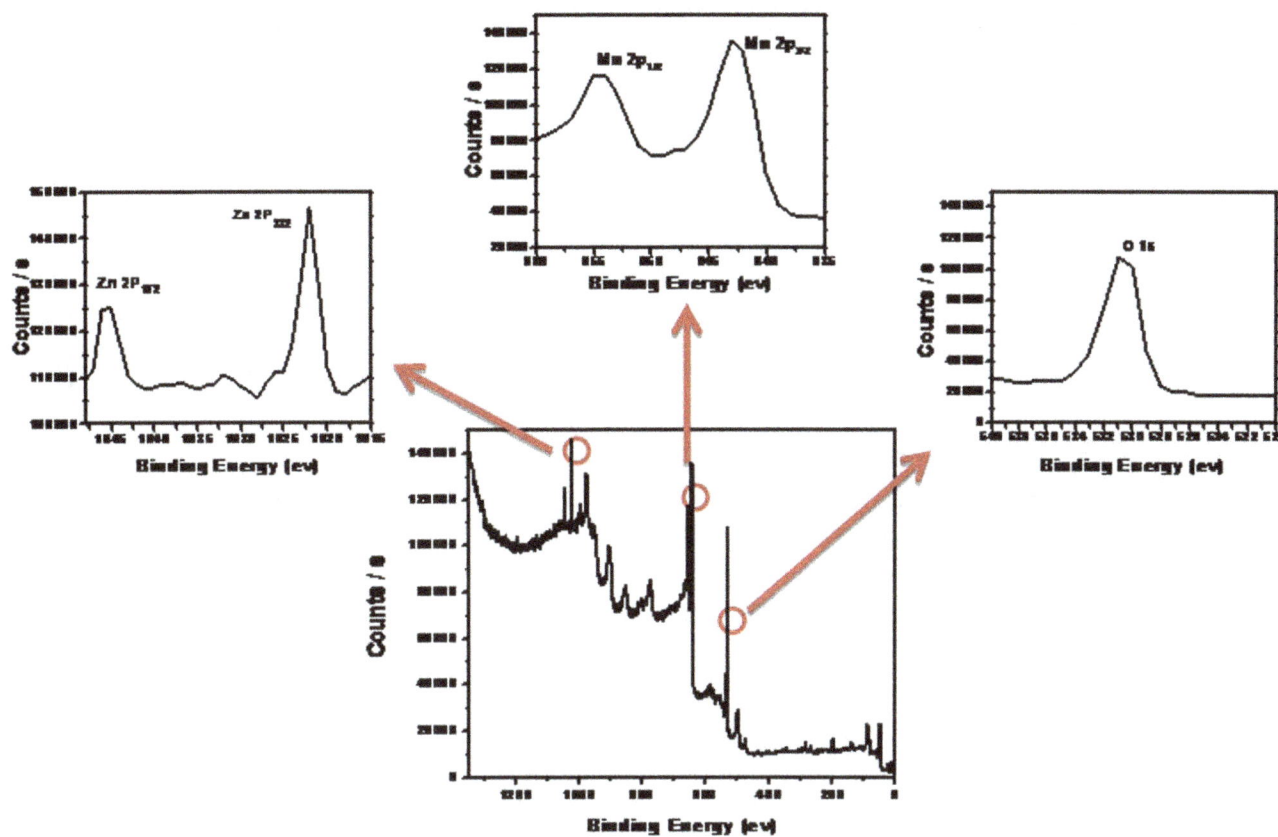

Figure 3. Typical XPS spectrum of the synthesized hetaerolite nano-particles.

particles sensor for p-nitrophenol which is shown in Figure 5 and Figure 6.

I–V curves for the glassy carbon electrode without coating (without hetaerolite nano-particles) and after coating (with hetaerolite nano-particles) were measured and shown in Figure 6 (a). The hetaerolite coated glassy carbon electrode (black square) illustrates not as much current response in contrast to the naked glassy carbon electrode (gray square), which might be attributed to resistance originated by hetaerolite nano-particles along with binders coated on the electrode surface [27].

Figure 4 (b) illustrates electrical reactions of the hetaerolite nano-particles without p-nitrophenol (gray-dotted line) and with 100.0 µL p-nitrophenol (dark-dotted line) in 0.1 M phosphate buffer solution (pH = 7.0). It is observed from the Figure 6 (b) that by adding the target chemical, hetaerolite nano-particles demonstrate a noteworthy enhancement in electrical current that reveals the sensitivity of hetaerolite nano-particles to p-nitrophenol. Consequently by injection of analyte, the augmentation in electrical response shows that hetaerolite nano-particles exhibit a fast and sensitive reply to the target chemical which might be due to speedy redox reaction (electron exchange) and fine electro-catalytic oxidation properties of the hetaerolite sensor. p-Nitrosophenol generally undergo reduction and produces p-hydroxylaminophenol. In second step, oxidation of p-hydroxyla-minophenol takes place which give rise to 4-nitrosophenol and the subsequent reversible reduction.

The influence of p-nitrophenol concentration on the electrical reaction of hetaerolite nano-particles was examined by consecutive addition of p-nitrophenol in the range of 50.0 µM to 1.0 M into 0.1 M PBS solution (pH = 7.0) and the graph is portrayed in

Figure 6 (c). Enhancement of the electrical current with rising p-nitrophenol concentration is observed which designates that the hetaerolite nano-particles conductivity was enhanced by the increase in the concentration of the target chemical [28]. The mechanism of p-nitrophenol sensing is graphically shown in Figure 5.

Calibration curve (Figure 6 (d)) was plotted from the difference in target concentration. The calibration curve portrays two sensitivity areas; the region at inferior concentrations (physisorp-tion process) is linear up to 0.005 M with correlation coefficient (R) of 0.7599. The sensitivity is determined from the slope of the lesser concentration section of calibration curve, which is 1.500 µA.cm^{-2}.mM^{-1}. The linear dynamic range reveals from 50.0 µM to 0.05 M and the detection limit was estimated, based on signal to noise ratio (S/N), to be 20.0 µM. Above 0.005 M concentration the sensor become saturated due to chemisorption method which could be due to the lack of free hetaerolite nano-particles sites for p-nitrophenol adsorption [28].

3. Photocatalysis

3.1. Solar photocatalytic performance of hetaerolite nanoparticles. The solar photo-catalytic performance of he-taerolite nanoparticles was evaluated by degrading brilliant cresyl blue under solar light [29]. In this study two sets of photo-catalytic reaction were carried out utilizing hetaerolite nanoparticles. Irradiation of brilliant cresyl blue under visible light degraded a little amount of dye at pH 7 without the catalyst (hetaerolite nano-particles) indicating a photolysis reaction. Photo-catalytical degra-dation of brilliant cresyl blue solution was carried out in the presence of hetaerolite nano-particles under visible light irradia-

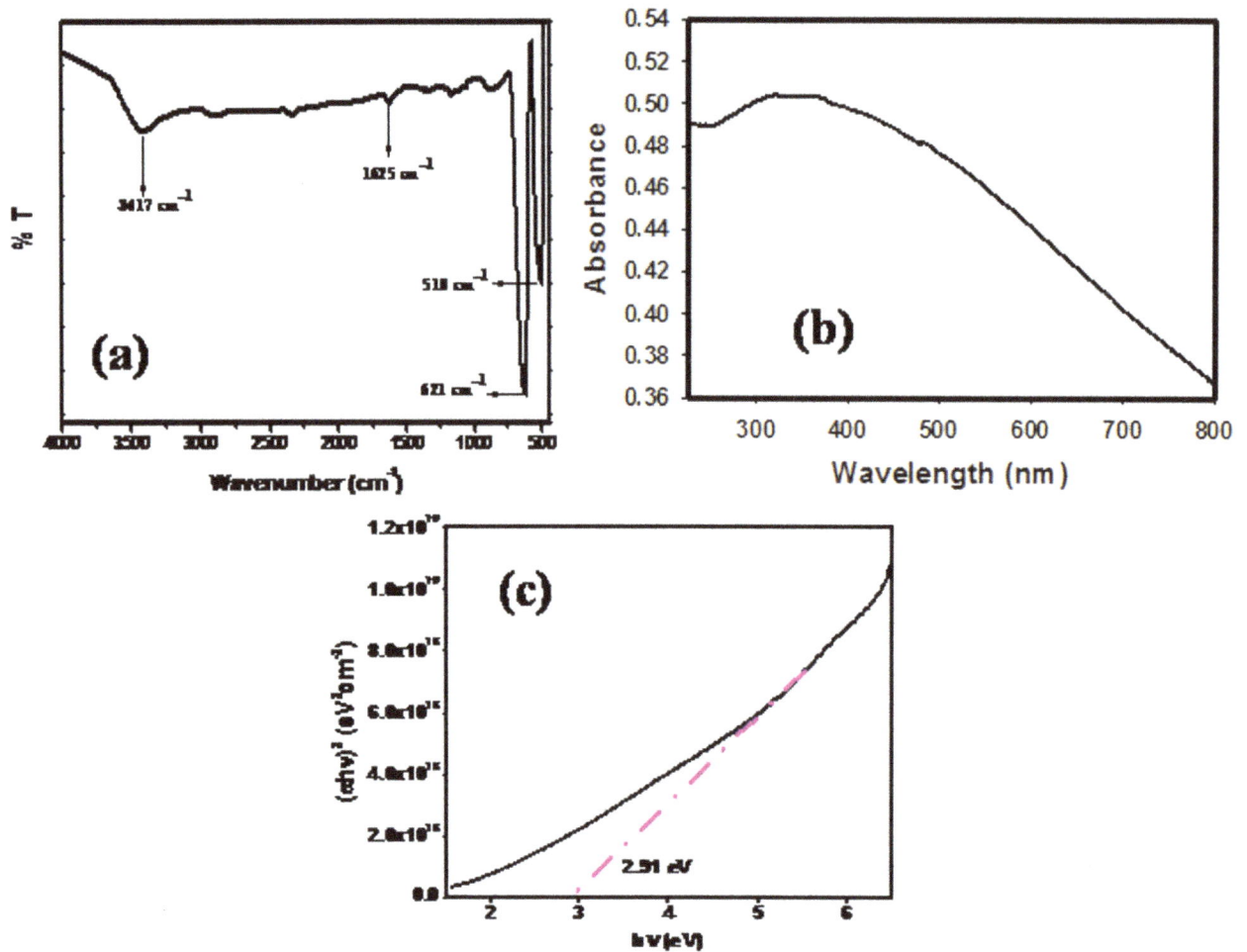

Figure 4. Typical FT-IR (a), UV-vis (b) spectrum and (ahυ)^2 vs hυ graph (c) of the synthesized hetaerolite nano-particles.

tion at different pHs and the effects of pH on the photo-catalytic degradation of brilliant cresyl blue studied under solar light irradiation [6]. Hetaerolite nanoparticles showed efficient activity for degradation of brilliant cresyl blue at different pH under solar light irradiation.

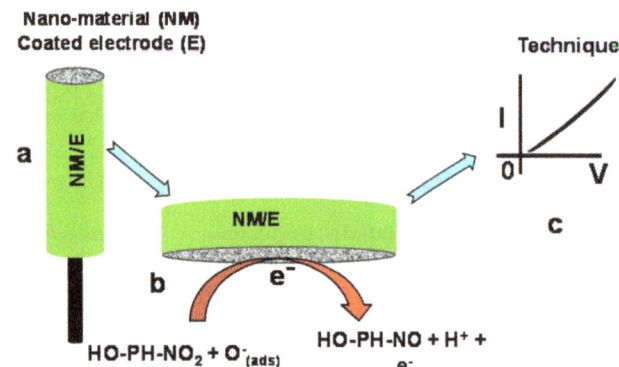

Figure 5. Schematic views of (a) fabricated AgE with hetaerolite nanoparticles and conducting binders, (b) reaction occurred at fabricated AgE (c) I–V detection technique.

An aqueous suspension of brilliant cresyl blue was irradiated with solar light in the presence of hetaerolite nanoparticles and this led to the alteration of absorbance with irradiation time [8]. Figure 7 (a) display transformation in absorption for the photo-catalytic degradation of brilliant cresyl blue at different time intervals which showed decline in absorption strength. It is clear that the absorbance at 595 nm steadily reduces with increase in irradiation time. Figure 7 (b) illustrates the transformation in absorbance with change in irradiation time for the brilliant cresyl blue in the presence and absence of hetaerolite nano-particles. Irradiation of brilliant cresyl blue solution in the presence of hetaerolite nano-particles leads to decline in absorption intensity. Figure 7 (c) demonstrates % degradation of brilliant cresyl blue in the presence and absence of hetaerolite nano-particles with respect to irradiation time (min). Degradation (%) graph shows that 27%, 52% and 83% of brilliant cresyl blue is degraded at pH 5, 8 and 10, correspondingly in the presence of hetaerolite nano-particles after 120 minutes of irradiation time while in the absence of hetaerolite nano-particles, no apparent loss of brilliant cresyl blue might be seen.

Hetaerolite nanoparticles were further compared with TiO$_2$ which is a dynamic photo-catalyst and displayed an imperative role in the detoxification of polutants [9]. However TiO$_2$ only promotes photo-catalysis upon irradiation by UV light because it absorbs only in the UV region. Hetaerolite nanoparticles promote

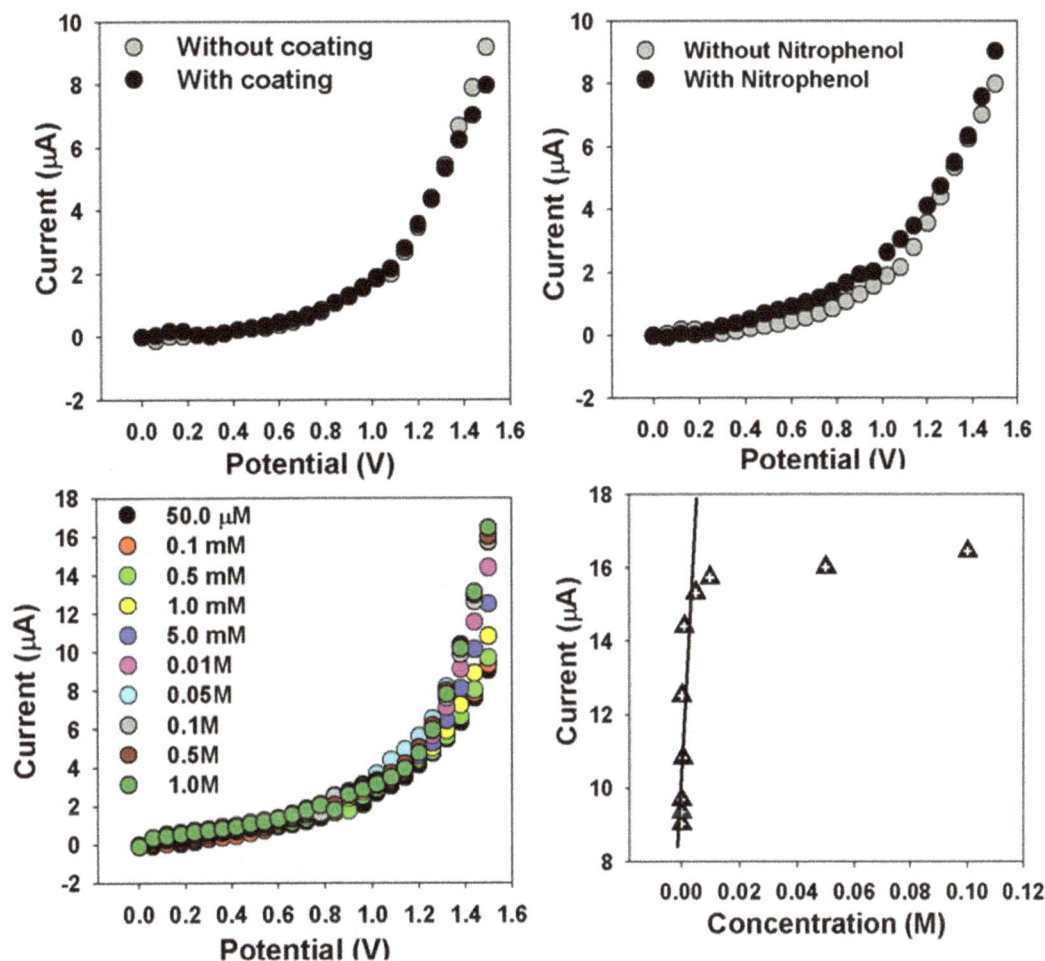

Figure 6. I–V characterization of hetaerolite nano-particles (a) Comparison of with and without hetaerolite nano-particles coating on GCE, (b) Comparison of with and without *p*-nitrophenol injection, (c) Concentration variation of *p*-nitrophenol, and (d) calibration plot.

photo-catalysis by irradiation with visible light because the solar spectrum consists of 46% visible light while the UV light is only 5–7% in the solar spectrum. Thus it is vital to grow an active photo-catalyst which can promote photo-catalysis in the visible region.

3.2. Effect of pH. The influence of pH on the visible light photocatalytic degradation of brilliant cresyl blue was examined in pH range 5–10 and compared with the brilliant cresyl blue degradation at various pH in the presence of hetaerolite nano-particles (Figure 7). It exhibited the same trend of degradation at all pH but exhibited high photo-catalytic degradation at pH 10 as compared to pH 5 and 8. The results showed that the rate of decomposition of brilliant cresyl blue increases with increase in pH and at pH 10, brilliant cresyl blue was degraded 83% in the presence of hetaerolite nano-particles. The photocatalytic perfor-mance of hetaerolite nano-particles was attributed to the surface electrical properties, which facilitate the brilliant cresyl blue adsorption. It is beneficial for the promotion of a visible light generated charge carrier i.e. electron to the surface which leads to formation of hydroxide radical. Thus high pH makes the surface of hetaerolite nano-particles negatively charged and also supports the formation of OH^- radicals [30]. Both these factors favor the attraction and degradation of brilliant cresyl blue. It is clear from the results that pH has a substantial influence on the photocat-alytic degradation of brilliant cresyl blue and hetaerolite nano-

particles synthesized by a very straightforward synthesis process exhibit considerable solar photo-catalytic activity towards brilliant cresyl blue degradation. This demonstrates the applicability of the solar photo-catalyst towards organic pollutants and the potential to extend the applications to environmental pollutants.

3.3. Reaction kinetics of photo-degradation. In order to realize the degradation behavior, we studied the degradation pattern of brilliant cresyl blue by Langmuir–Hinshelwood (L–H) model. Langmuir–Hinshelwood (L–H) model well defines the relationship among the rate of degradation and the preliminary concentration of brilliant cresyl blue in photo-catalytic reaction [31]. The rate of photo-degradation was calculated by using Equation 1:

$$r = -dC/dt = K_r KC = K_{app} C \qquad (1)$$

Where r represents the degradation rate of brilliant cresyl blue, K_r is the reaction rate constant, K is the equilibrium constant, C is the reactant concentration. When C is very diminutive, then KC is insignificant and Equation 1 turns into first order kinetic. Consider preliminary conditions (t = 0, C = C_0), Equation 1 turn into Equation 2.

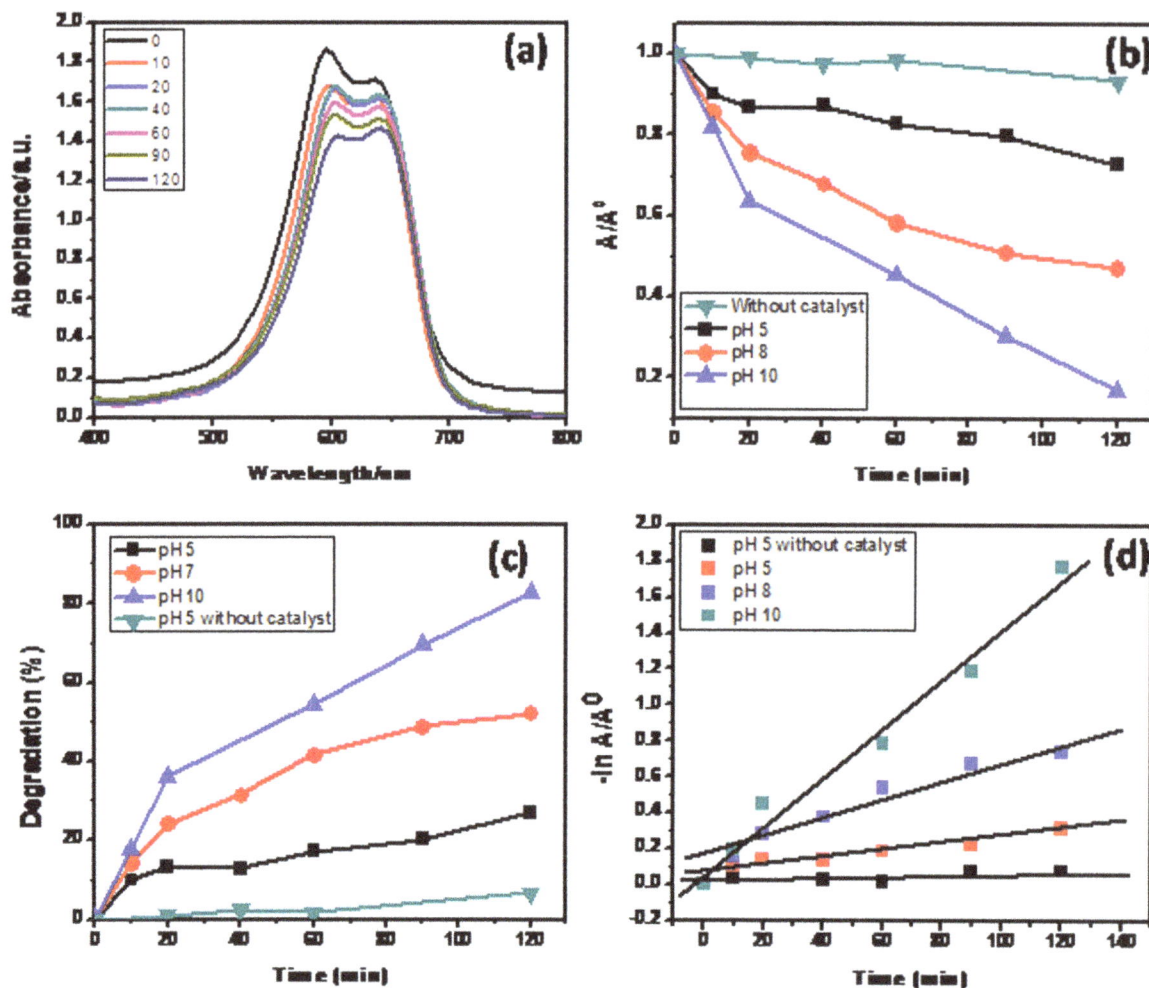

Figure 7. Typical plot for (a) Change in the absorption spectrum of brilliant cresyl blue (b) Change in absorbance vs irradiation time for brilliant cresyl blue at different pH, (c) % degradation vs irradiation time and (d) reaction kinetic for brilliant cresyl blue at different pH in the presence of hetaerolite nanoparticles.

$$-\ln C/C_0 = kt \qquad (2)$$

Half-life, $t_{1/2}$ (in min) is

$$t_{1/2} = 0.693/k \qquad (3)$$

Figure 7 (d) demonstrated degradation of brilliant cresyl blue which pursued first-order kinetics (plots of $\ln(C/C_0)$ vs time showed linear relationship). First-order rate constants, calculated from the slopes of the $\ln(C/C_0)$ vs time plots and the half-life of the degraded brilliant cresyl blue can simply determined by Equation 3 [30,31]. Rate constant for hetaerolite nano-particles were found to be 0.000526 min^{-1} ($t_{1/2} = 1317.5$ min), 0.00212 min^{-1} ($t_{1/2} = 326.9$ min), 0.00596 min^{-1} ($t_{1/2} = 116.3$ min) and 0.0136 min^{-1} ($t_{1/2} = 51.0$ min). Thus the kinetic study revealed that hetaerolite nano-particles is a proficient photo-catalyst for degradation of organic noxious wastes.

3.4. Mechanism of photo-degradation. In the current study, a heterogeneous photo-catalysis method was employed for the degradation of brilliant cresyl blue. Briefly, when hetaerolite nano-particles were exposed to light having energy the same or superior to the band gap of nanoparticles, the development of an electron and hole pair occurs on the surface of hetaerolite nano-particles. If charge partition is sustained, subsequently the electron hole pair reacts with brilliant cresyl blue in the presence of oxygen. Hydroxyl radicals (OH$^{\bullet}$) and superoxide radical anions (O$_2^{\bullet-}$) are assumed to be the major degrading mediators (oxidizing species)

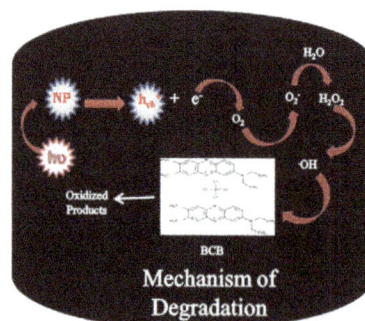

Figure 8. Mechanism of photodegradation of brilliant cresyl blue in the presence of hetaerolite nano-particles.

and oxidative reactions result in the degrading (oxidation) of the brilliant cresyl blue. The whole mechanism of photo-activity of hetaerolite nano-particles is depicted in Figure 8 [30,31].

Conclusion

Well-crystalline body-centered tetragonal hetaerolite nanoparticles based p-nitrophenol electrochemical sensor has been fabricated. Hetaerolite nano-particles were produced by a hydrothermal process. The featured structural characterizations proved that the manufactured hetaerolite is optically active, well-crystalline, body-centered tetragonal nanoparticles. Sensing toward p-nitrophenol was executed which showed excellent sensitivity and limit of detection with quick response time. Moreover the solar photo-catalytic property of hetaerolite nanoparticles was utilized in the degradation of brilliant cresyl blue. Thus it is concluded that the hetaerolite nanoparticles are an attractive sensor material and active photo-catalyst for accomplishing a proficient chemical sensor and photo-catalyst for application within water resources and health monitoring.

Acknowledgments

This work was funded by the Deanship of Scientific Research (DSR), King Abdulaziz University, Jeddah, under grant No. (130-031-D1433).

Author Contributions

Conceived and designed the experiments: SBK MMR KA MAA. Performed the experiments: SBK MMR KA MAA MAR. Analyzed the data: SBK MMR KA MAA. Contributed reagents/materials/analysis tools: SBK MMR KA MAA. Wrote the paper: SBK MMR KA MAA MAR.

References

1. Jamal A, Rahman MM, Khan SB, Faisal M, Akhtar K, et al. (2012) Cobalt doped antimony oxide nano-particles based chemical sensor and photo-catalyst for environmental pollutants. App Surf Sci 261: 52–58.
2. Stanca SE, Popescu IC, Oniciu L (2003) Biosensors for phenol derivatives using biochemical signal amplification. Talanta 61: 501–507.
3. Khan SB, Akhtar K, Rahman MM, Asiri AM, Seo J, et al. (2012) Thermally and mechanically stable green environmental composite for chemical sensor applications. New J Chem 36: 2368–2375.
4. Jain RK, Kapur M, Labana S, Lal B, Sharma PM, et al. (2005) Microbial diversity: Application of microorganisms for the biodegradation of xenobiotics. Curr Sci 89: 101–112.
5. Rahman MM, Jamal A, Khan SB, Faisal M (2011) Characterization and applications of as-grown b-Fe₂O₃ nanoparticles prepared by hydrothermal method. J Nanoparticle Res 13: 3789–3799.
6. Faisal M, Khan SB, Rahman MM, Jamal A (2011) Synthesis, characterizations, photocatalytic and sensing studies of ZnO nanocapsules. Appl Surf Sci 258: 672–677.
7. Banik RM, Prakash MR, Upadhyay SN (2008) Microbial biosensor based on whole cell of Pseudomonas sp. for online measurement of p-Nitrophenol. Sens Actuat B 131: 295–300.
8. Khan SB, Rahman MM, Akhtar K, Asiri AM, Seo J, et al. (2012) Novel and sensitive ethanol chemi-sensor based on nanohybrid materials. Int J Electrochem Sci 7: 4030–4038.
9. Rahman MM, Khan SB, Jamal A, Faisal M, Asiri AM (2012) Fabrication of a methanol chemical sensor based on hydrothermally prepared α-Fe₂O₃ codoped SnO₂ nanocubes. Talanta 95: 18–24.
10. Khan SB, Faisal M, Rahman MM, Jamal A (2011) Low-temperature growth of ZnO nanoparticles: Photocatalyst and acetone sensor. Talanta 85: 943–949.
11. Faisal M, Khan SB, Rahman MM, Jamal A (2011) Smart chemical sensor and active photo-catalyst for environmental pollutants. Chem Engineer J 173: 178–184.
12. Qamar MM, Lofland SE, Ramanujachary KV, Ganguli AK (2009) Magnetic and photocatalytic properties of nanocrystalline ZnMn₂O₄. Bull Mater Sci 32: 231–237.
13. Jamal A, Rahman MM, Khan SB, Faisal M, Asiri AM, et al. (2013) Hydrothermally preparation and characterization of un-doped manganese oxide nanostructures: Efficient photocatalysis and chemical sensing applications. Micro and Nanosystems 5: 22–28.
14. Rahman MM, Jamal A, Khan SB, Faisal M (2011) Fabrication of highly sensitive ethanol chemical sensor based on Sm-doped Co₃O₄ nanokernels by a hydrothermal method. J Phys Chem C 115: 9503–9510.
15. Jha A, Thapa R, Chattopadhyay KK (2012) Structural transformation from Mn3O4 nanorods to nanoparticles and band gap tuning via Zn doping. Mater Res Bull 47: 813–819.
16. Du J, Gao Y, Chai L, Zou G, Li Y, et al. (2006) Hausmannite Mn~3O~4 nanorods: synthesis, characterization and magnetic properties. Nanotech 17: 4923–4928.
17. Wang X, Li Y (2003) Synthesis and formation mechanism of manganese dioxide Nanowires/Nanorods. Chem Eur J 9: 300–306.
18. Yang J, Zeng JH, Yu SH, Yang L, Zhou GE, et al. (2000) Formation process of CdS nanorods via solvothermal route. Chem Mater 12: 3259–3263.
19. Khan SB, Rahman MM, Akhtar K, Asiri AM, Alamry KA, et al. (2012) Copper oxide based polymer nanohybrid for chemical sensor applications. Int J Electrochem Sci 7: 10965–10975.
20. Rahman MM, Khan SB, Faisal M, Asiri AM, Tariq MA (2012) Detection of aprepitant drug based on low-dimensional un-doped iron oxide nanoparticles prepared by a solution method. Electrochimica Acta 75: 164–170.
21. Zhang P, Li X, Zhao Q, Liu S (2011) Synthesis and optical property of one-dimensional spinel ZnMn₂O₄ nanorods. Nanoscal Res Lett 6: 323–331.
22. Apte SK, Naik SD, Sonawane RS, Kale BB, Pavaskar N, et al. (2006) Nanosize Mn3O4 (Hausmannite) by microwave irradiation method. Mater Res Bull 41: 647–654.
23. Faisal M, Khan SB, Rahman MM, Jamal A, Umar A (2011) Ethanol chemi-sensor: Evaluation of structural, optical and sensing properties of CuO nanosheets. Mater Lett 65: 1400–1403.
24. Rahman MM, Jamal A, Khan SB, Faisal M (2011) Highly sensitive ethanol chemical sensor based on Ni-doped SnO₂ nanostructure materials. Biosens Bioelectron 28: 127–134.
25. Ansari SG, Ansari ZA, Seo HK, Kim GS, Kim YS, et al. (2008) Urea sensor based on tin oxide thin films prepared by modified plasma enhanced CVD. Sensors Actuators B 132: 265–271.
26. Ansari SG, Ansari ZA, Wahab R, Kim YS, Khang G, et al. (2008) Glucose sensor based on nano-baskets of tin oxide templated in porous alumina by plasma enhanced CVD. Biosens Bioelectron 23: 1838–1842.
27. Ansari SG, Wahab R, Ansari ZA, Kim YS, Khang G, et al. (2009) Effect of nanostructure on the urea sensing properties of sol-gel synthesized ZnO. Sensors Actuators B 137: 566–73.
28. Khan SB, Rahman MM, Jang ES, Akhtar K, Han H (2011) Special susceptive aqueous ammonia chemi-sensor: Extended applications of novel UV-curable polyurethane-clay nanohybrid. Talanta 84: 1005–1010.
29. Khan SB, Faisal M, Rahman MM, Jamal A (2011) Exploration of CeO₂ nanoparticles as a chemi-sensor and photo-catalyst for environmental applications. Sci Tot Environ 409: 2987–2992.
30. Mohapatra L, Parida KM (2012) Zn-Cr layered double hydroxide: Visible light responsive photocatalyst for photocatalytic degradation of organic pollutants Separat Purif Technol 91: 73–80.
31. Parida KM, Mohapatra L (2012) Characterization of inverted pyramidal hollow cathode microplasma devices operating in reactive gases for maskless scanning plasma etching. Chem Engineer J 179: 131–139.

6

The Combined Toxic and Genotoxic Effects of Cd and As to Plant Bioindicator *Trifolium repens* L

Alessandra Ghiani, Pietro Fumagalli, Tho Nguyen Van, Rodolfo Gentili, Sandra Citterio*

Department of Earth and Environmental Sciences, University of Milano-Bicocca, Milan, Italy

Abstract

This study was undertaken to investigate combined toxic and genotoxic effects of cadmium (Cd) and arsenic (As) on white clover, a pollutant sensitive plant frequently used as environmental bioindicator. Plants were exposed to soil spiked with increasing concentrations of cadmium sulfate (20, 40 and 60 mg Kg^{-1}) or sodium arsenite (5, 10 and 20 mg Kg^{-1}) as well as with their combinations. Metal(loid) bioavailability was assessed after soil contamination, whereas plant growth, metal(loid) concentration in plant organs and DNA damage were measured at the end of plant exposition. Results showed that individual and joint toxicity and genotoxicity were related to the concentration of Cd and As measured in plant organs, and that As concentration was the most relevant variable. Joint effects on plant growth were additive or synergistic, whereas joint genotoxic effects were additive or antagonistic. The interaction between Cd and As occurred at both soil and plant level. In soil the presence of As limited the bioavailability of Cd, whereas the presence of Cd increased the bioavailability of As. Nevertheless only As biovailability determined the amount of As absorbed by plants. The amount of Cd absorbed by plant was not linearly correlated with the fraction of bioavailable Cd in soil suggesting the involvement of additional factors, such as plant uptake mechanisms. These results reveal that the simultaneous presence in soil of Cd and As, although producing an additive or synergistic toxic effect on *Trifolium repens* L. growth, generates a lower DNA damage.

Editor: Massimo Labra, University of Milano Bicocca, Italy

Funding: This work was supported by the Ministry of University, Research, Science and Technology and Italian Lombardy Region (Project: Soil mapping). The funders had no role in study design, data collection and analysis, decision to publish, or preparation of the manuscript.

Competing Interests: The authors have declared that no competing interests exist.

* E-mail: sandra.citterio@unimib.it

Introduction

The use of efficient early warning bioindication systems represents a powerful approach for assessing and interpreting the impact of natural or anthropogenic perturbations in soil ecosystems preventing environmental alteration and human disease.

Living organisms provide information on the cumulative effects of environmental stressors and as such bioindication is complementary to direct physical and chemical measurements [1]. *T. repens* is a pollutant-sensitive plant, suitable for biomonitoring campaigns. Specifically, its environmental exposition followed by a DNA analysis with molecular markers allows the detection of sublethal levels of genotoxic compounds in the environment [2]. However, given the limited information available on the joint-genotoxic-effect of chemicals, the interpretation of biomonitoring results is often difficult. In addition, most environmental risk assessments of contaminated lands are currently based on guideline values derived from the ecotoxicological properties of specific chemicals, whereas it is well known that environmental pollutants interact producing additive, antagonistic or synergistic effects on exposed organisms [3–7]; it is then evident that there is a clear need to improve the knowledge about the combined effects of stressors on bioindicators.

In this work we considered the combined toxic and genotoxic effects of Cd and As, two of the most dangerous pollutants for both environment and human health. They are ranked among the top ten priority hazardous compounds by the Agency for Toxic Substances and Disease Registry [8]. Over the last century, their natural presence on the earth's crust along with improper industrial and agricultural practices has led to the pollution of many areas, which are now of public concern. In fact, since Cd and As cannot be degraded by living organism, after their natural or anthropogenic release, they persist in soil and sediments where they can accumulate up to harmful levels. As consequence, detrimental effects on life in the environment and on human health are exerted by these two elements directly or indirectly, through the food chain [9].

Among the damages induced by As and Cd to living organisms, genome alteration is one of the most dangerous. Genomic instability is particularly related to cancer induction and progression in animals and to inhibition of growth and even to death in plants. In wild and agro ecosystems the presence of genotoxic compounds significantly reduces the number of species and decreases the yield and quality of crops [4,10].

Many in vitro and in vivo studies demonstrated Cd and As induction of micronuclei, chromosomal aberrations, DNA strand breaks and oxidative DNA base damage [11].

Moreover the two elements are classified as Group 1 carcinogens by the International Agency for Research on Cancer (IARC). The human exposure to Cd or As has been associated with various cancers, principally kidney and lung for Cd [12] and skin, lung, liver and bladder for As [11]. The underlying cellular pathways leading to cancer are similar for the two metals and include mechanisms inducing DNA damage, such as oxidative stress and DNA repair inhibition. Thus the available literature clear shows that Cd and As are genotoxic compounds and that they exert their effect even at low concentrations. However most of this information comes from studies on single chemical whereas no data are available on their genotoxic joint action. In our experiment the bioindication system, based on the use of *T. repens*, as bioindicator, and molecular markers, as suitable tool to detect DNA changes, were used to investigate the combined toxic and genotoxic effect of increasing concentrations of Cd and As in soil. Effects were also correlated to the uptake and distribution of Cd and As in plant organs.

Materials and Methods

Plant bioindicator and experimental design

T. repens seeds cv. Ladino (Ingegnoli, Milan, Italy) were surface sterilized and directly grown in soil containing 3.0% organic matter for 30 days.

For the exposition to Cd or As, the nearly 5 cm high plantlets obtained after 30 days from germination, were transferred to pots filled with 2.0 kg of soil contaminated with or without (control) increasing concentrations of cadmium sulfate (20, 40 and 60 mg Kg^{-1} Sigma, St. Louis, MO) or sodium arsenite (5, 10 and 20 mg Kg^{-1}; Sigma). Concentrations were defined on the basis of the results obtained by preliminary trials by which the sensitivity of *T. repens* plantlets to cadmium sulfate and sodium arsenite was tested.

For the joint-exposition to Cd and As, the following 9 soils were prepared by combining the 3 concentrations of cadmium sulfate and sodium arsenite, used to prepare soils for the single-expositions: (1) Cd 20 and As 5 mg Kg^{-1}, (2) Cd 20 and As 10 mg Kg^{-1}, (3) Cd 20 and As 20 mg Kg^{-1}, (4) Cd 40 and As 5 mg Kg^{-1}, (5) Cd 40 and As 10 mg Kg^{-1}, (6) Cd 40 and As 20 mg Kg^{-1}, (7) Cd 60 and As 5 mg Kg^{-1}, (8) Cd 60 and As 10 mg Kg^{-1}, (9) Cd 60 and As 20 mg Kg^{-1}.

For each treatment, three different pots with 15 plantlets were prepared and placed in a growth room for 2 weeks (25°C; 10 h dark/14 h light, 150 µmol $m^{-2} s^{-1}$).

At the end of exposition, plant survival and growth were determined and RAPD analysis was carried out to evaluate DNA damage. Three independent repetitions of the entire experiment were performed.

Plant survival and growth measurements

The assessment of plant survival and plant growth was performed after exposition. Plant growth was evaluated by measuring plant organ dry weight (DW): for each treatment the shoots and roots from 30 plants were placed in a drying cabinet at 40 °C until a constant weight was reached. Statistical analyses were performed using the GraphPad Prism software for Windows (version 4.0 GraphPad Software Inc., San Diego CA): ANOVA and Dunnet or Tukey test were applied to the data when normality and homogeneity of variance were satisfied. Data which did not conform to the assumptions were alternatively transformed into logarithms, or were analysed by Kruskal Wallis non-parametric procedures.

Analytical methods for metal(loid) quantification

Before plant exposition, the bioavailable fraction of Cd and As was quantify in control and contaminated soils following the protocol of Lindsay and Norwell [13]. Briefly, 5 g of soil were extracted with 10 ml of 5 mM DTPA (Sigma-Aldrich), 0.1 M trietanolamine (Sigma-Aldrich) and 0.01 M $CaCl_2$ (Sigma-Aldrich), for 2 h at 20°C under stirring. Samples were then filtered and metal concentrations were determined by graphite furnace atomic absorption spectroscopy (GFAAS; AAnalyst600, Perkin-Elmer). Nine soil samples for each treatment were processed.

Cd and As were also quantified in plant organs applying the USEPA 3051a protocol. The harvested plants were carefully washed with tap water and then with distilled water to remove soil debries before analysis. All the samples were dried at 100 °C overnight. For each sample 10 mL of HNO3 and 2 mL of HClO3 were added to 0.2 g of dry plant matter. The samples were digested by using the ETHOS HPR 100/10 microwave lab station (FKV, Bergamo, Italy) reaching the 180°C temperature. After their complete mineralization, they were opportunely diluted and analysed by graphite furnace atomic absorption spectroscopy (GFAAS; AAnalyst600, Perkin-Elmer). Standards (from ENEA Research Centre, Roma, Italy) and blanks were run with all sample series for quality control.

Random amplified polymorphic DNA (RAPD)

RAPD technique was used to quantify DNA sequence changes in test-plants. DNA was isolated separately from shoots and roots by using DNeasy isolation and purification kit (Qiagen, Italy). The kit was used to obtain high quality DNA, free of polysaccharides or other metabolites which might interfere with DNA amplification. At least three independent extractions were performed for each treatment.

Extracted genomic DNA and twelve different 10-bp-long random primers were used in RAPD-Polymerase chain reaction (RAPD-PCR); the sequences of the primers are reported in Table S1. DNA amplification was performed in a 20 µl final reaction volume containing 15 ng genomic DNA, 1 µM primer and 1X Taq PCR Master Mix (Qiagen). The RAPD amplification protocol consisted of an initial denaturing step of 5 min at 95°C, followed by 45 cycles at 95°C for 30 sec (denaturation), 35°C for 30 sec (annealing) and 72°C for 30 sec (extension), with an additional extension period of 8 min at 72°C. DNA amplification products were separated on a ethidium bromide-stained 2.0% high resolution agarose gel (Sigma-Aldrich), using a Tris-borate-EDTA (TBE) buffer (90 mM Tris base, 90 mM boric acid and 2 mM EDTA). Three independent replicates were performed for each sample.

Visual inspection under UV light of the resulting gels, allowed for scoring of polymorphic bands. For statistical analysis, each RAPD band detected after electrophoresis of the amplification DNA products was scored as a binary character for its absence (0) or presence (1). The percentage of polymorphism (P%), that represents the ratio between the number of polymorphic bands and total detected bands X 100, was determined for each sample and data were statistically analysed using the software program GraphPad Prism for Windows version 4.0 (GraphPad Software Inc., San Diego CA USA). The statistical significance between the treated samples and the control were obtained by applying ANOVA and Dunnet test (P<0,05).

Statistical determination of Cd and As interaction type

The interaction type existing between Cd and As in each treatment and concerning their joint effect on plant growth and DNA sequence change were evaluated by applying the statistical method reported by Ince et al. [14]. The method was based on testing the null hypothesis of "additive effect" at 95% confidence level.

Specifically, the interaction of Cd and As in each treatment was assessed by comparing the observed toxicity at the ith test level and at the concentration (x+y)i (where x and y were the concentrations of the first and second element, respectively) with the value of the null hypothesis at that level, defined as "the sum of the toxicity indices of the two elements, tested previously at x and y".

For the joint effect on plant growth, evaluation of the null hypothesis was based on multiplication of plant dry weigh (PDW) of each element as percentage of control, whereas for the joint effect on DNA sequence changes the null hypothesis was evaluated by the addition of plant damage induced by each element, defined as PP = polymorphism percentage. Thus, toxic and genotoxic interactions at each binary test level were assessed by statistical testing of the two null hypotheses PDW_H and PP_H, defined by Equation 1 and Equation 2 for growth and DNA damage data, respectively:

$$H0 \text{ Plant growth}: PDW_H(x+y)i = (PDWx)i*(PDWy)i/100 \qquad (1)$$

$$H0 \text{ Plant sequence changes}: PP_H(x+y)i = (PPx)i + (PPy)i \quad (2)$$

where $(x+y)_i$ was the ith combination of Cd and As concentrations in soil, $(PDWx)i$ and $(PDWy)i$ the plant dry weight (as%) for each metal ion, recorded at the xi^{th} and yi^{th} singular concentrations, and $(PPx)i$, $(PPy)i$ the percentage of polymorphism induced by each element, recorded at the xi^{th} and yi^{th} singular concentrations.

The compound interactions were called "antagonistic," "additive," or "synergistic" according to the statistical significance (t student) and the sign of the difference between the tested hypothesis and the value of the observed effect.

Regression and Redundance statistical analyses (RDA) were also applied to investigate the relationships between variables and their relevance to the joint-effects of Cd and As.

Results

Bioavailability of Cd and As in soil

The bioavailable amount of Cd and As in artificially contaminated soils was assessed just before *T. repens* exposition. The measured concentrations of DTPA-extractable Cd and As are reported in Table 1. In keeping with literature [15–18] Cd was much more bioavailable than As: the percentage of bioavailable As and Cd ranged from 0.016 to 0.055 and from 0.43 to 0.79, respectively. In soils contaminated with single compounds the bioavailable amounts of both Cd and As increased in parallel with the increase of metal concentration added to soil ($r^2_{Cd} = 0.99$ $r^2_{As} = 0.97$). A different bioavailability trend was instead observed in soil simultaneously contaminated with the two elements: the presence of As reduced the amounts of bioavailable Cd, whereas the presence of Cd increased the amounts of bioavailable As.

Effect of Cd and As contaminated soils on plant survival and growth

Single and joint effects of Cd and As on plant survival and plant development were assessed after 15 days of exposition. Plant development was evaluated by measuring plant organ dry weight (DW). As expected on the basis of preliminary trials, none of the used single Cd or As concentrations negatively affected plant survival and plant DW (Fig.1). Plant survival was not affected also by all the combined treatments. On the contrary, the combination of As5 with the higher Cd concentration (Cd 60) and the combination of As10 with Cd 40 or Cd 60 and of As 20 with all the tested Cd concentrations statistically reduced the shoot

development (Fig.1). Concerning the effect of these combined concentrations on roots, although a growth reduction trend was observed, the results obtained were not statistically significant, given the root very low DW and the consequent difficulty in assessment (Fig.1).

A statistiscal analysis, according to Ince et al. [14], was applied to evaluate the type of interaction existing between As and Cd, responsible for the joint effect on plant growth observed in each treatment. Table 2 shows the results of the analysis. A synergistic effect leading to plant growth reduction was found when the higher tested Cd concentration (Cd 60) was combined with As 5 or As 10 or As 20. An additive effect was instead determined for all the other soil binary mixture.

Accumulation of Cd and As in plant organs

The total amount of Cd and As accumulated in plant organs at the end of the experiment, was calculated by multiplying the element concentration, determined by AAS in root and shoot (Fig.S1), with the correspondent organ DW (Fig.1). The obtained results are reported in Fig.2.

On average, plants grown in soil contaminated with As accumulated an amount of metalloid proportional to the concentration of As added to the soil, which was also related to the amount of bioavailable element ($r^2_{bioav-As} = 0.97$, P<0.05). Differently, plants grown in soil contaminated with Cd accumulated a mean amount of metal not proportional to its soil bioavailable concentration; although plants tended to accumulate higher Cd amounts in presence of higher concentrations of Cd in soil, the differences were not statistically significant (Fig.2A). Moreover, with respect to the available amounts of Cd and As, plants accumulated a greater relative amount of As than Cd. Indeed, considering that the available amounts of Cd in each pot containing 2kg of soil were much higher (ranging from about 32 to74 mg) than those of As (ranging from about 0.16 to 1.6 mg), the relative mean amounts of Cd accumulated per plant (ranging from about 0.4 to 0.7 μg) were proportionally lower than those of As (ranging from 0.05 to 0.2 μg;), suggesting different plant absorbtion mechanisms for the two metal(loid)s.

Similarly, in soils contaminated with both metal(loid)s, the mean total amounts of As accumulated in plants were related to element bioavailability (multiple $r^2 = 0.90$, P<0.05) and, since the presence of both metal(loid)s in soil increased As biovailability, its amount, accumulated in plants grown in the presence of both elements, was higher than that measured in the plants grown in presence of As alone. On the contrary, Cd accumulation was not proportional to the bioavailable amount in soil and was lowered by the presence of As (Fig.2A).

Regarding the distribution of Cd and As in plant organs, most of them were accumulated in root (Fig.2B) and the very low amounts translocated to shoot (Fig.2C) were proportional to the amounts accumulated in root ($r^2_{Cd} = 0.51$, $r^2_{As} = 0.69$, P<0.05).

The same trend of Cd and As accumulation and distribution was also observed analyzing the mean metal(loid) concentration measured in plant organs (Fig.S1). However it can be observed that, due to the different reduction in plant growth, induced by the different metal(loid) treatments, the mean total amount of Cd and As (calculated multiplying metal concentration for DW), did not always reflect the mean concentration of elements in plant organs. For instance, the mean concentration of Cd measured in roots of plants grown in As20+Cd60 soil was statistically higher than that found in root of plants grown in As20+Cd40 soil whereas the mean total amount of Cd was not statistically different between the two treatments, due to the higher growth reduction of plants grown in As20+Cd 60 soil. Thus, in our data elaboration, the total

Table 1. Bioavailable content ($\mu g\ g^{-1}$) of Cd and As in soil before plant exposition.

Soil Sample	pH	Bioavailabe As (μg g-1)	Bioavailabe Cd (μg g-1)
CTR	7.9	BDL	BDL
As 5	7.8	0.08±0.01	BDL
As 10	8	0.25±0.04	BDL
As 20	7.8	0.80±0.06	BDL
Cd 20	7.8	BDL	15.76±2.72
Cd 40	7.8	BDL	26.81±4.32
Cd 60	7.9	BDL	36.79±5.91
As 5+Cd 20	8	0.13±0.02	9.87±1.59
As 5+Cd 40	8	0.12±0.03	18.65±2.96
As 5+Cd 60	7.8	0.14±0.02	32.57±5.41
As 10+Cd 20	7.9	0.33±0.03	8.91±1.49
As 10+Cd 40	7.9	0.32±0.04	17.41±2.72
As 10+Cd 60	7.9	0.37±0.03	31.99±5.18
As 20+Cd 20	7.9	1.11±0.05	9.58±1.55
As 20+Cd 40	7.9	1.04±0.04	19.83±3.20
As 20+Cd 60	7.9	0.93±0.05	30.70±4.95

Data are the mean+SD of 9 soil samples (3 from each pot).
BDL: below instrument detection limit.

amount of metal(loid)s was calculated to properly correlate the amount of element absorbed by plant with its bioavailable soil quantity, whereas the concentration of elements in plant organs was also taken in to account to better evaluate the observed toxic and genotoxic effects of metal(loid)s.

Single and joint genotoxic effects of Cd and As

DNA sequence changes were evaluated by means of RAPD analysis, a technique which detects mutations at the primer annealing sites and also within the amplified DNA fragments (*i.e.* deletions or insertions). Twelve single primers were applied for the shoot and root analysis revealing a total of 130 and of 152 reproducible bands, respectively. Of these bands, 3.52% and 4.62% were polymorphic among the shoot and root controls, respectively. These values were considered as a basal polymorphic level among *T. repens* plants (*i.e.* intra-species variability).

Taking into account all the independent repetitions, DNA sequence damage, induced by Cd and As, was calculated as the percentage of polymorphism (P%) of the treated samples compared to that of the control plants and reported in Fig. 3.

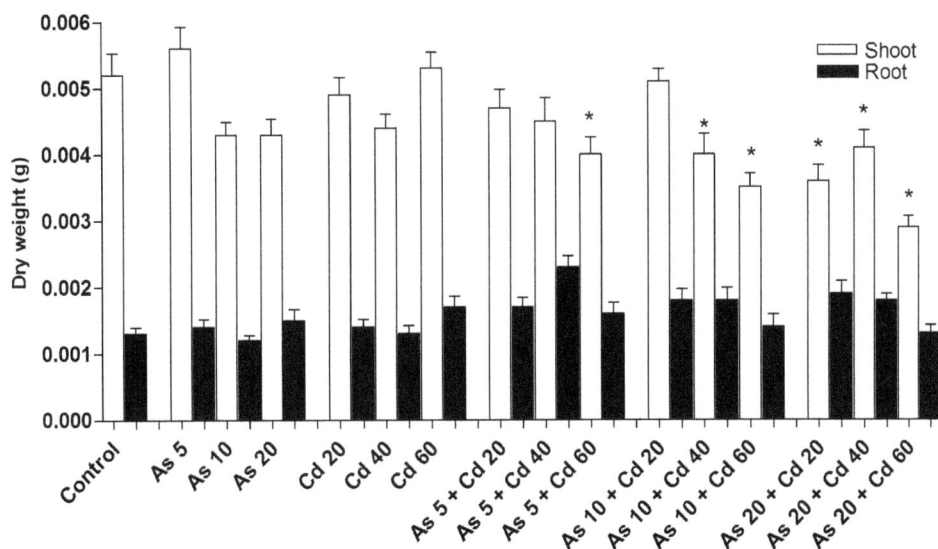

Figure 1. Effect of metal(loid) stress on *T. repens* **growth, measured as dry weight (DW).** Data are the mean of 30 measurements from single plants per each treatment. The asterisk (*) indicates statistically significant differences with respect to the control (ANOVA and Dunnet test; $P<0.05$).

Table 2. Observed and calculated toxic effects at binary test combinations x:y, and single metal concentrations x, y, respectively (predicted interaction types).

	x (μg g⁻¹) As	y (μg g⁻¹) Cd	Observed toxicity PDW$_{obs}$	Calculated toxicity PDW$_{calc}$ (PDW$_x$ * PDW$_y$/100)	Difference (PDW$_{obs}$ - PDW$_{calc}$)	Difference standard error	t Student (df=34)	Significance (P<0.05)	Interactive Effect
SHOOT									
	5	20	84.3±5.1	101.3±8.1	−17.0	9.6	−1.8	I	additive
	5	40	80.0±6.2	91.6±5.9	−11.6	8.5	−1.4	I	additive
	5	60	70.1±4.7	110.7±7.8	−40.6	9.1	−4.5	S	synergistic
	10	20	90.6±7.6	77.1±6.2	13.5	9.8	1.4	I	additive
	10	40	71.4±5.4	69.7±4.5	1.7	7.0	0.2	I	additive
	10	60	69.0±3.9	84.2±5.9	−15.2	7.1	−2.1	S	synergistic
	20	20	64.7±4.1	76.9±6.2	−12.2	7.4	−1.7	I	additive
	20	40	73.2±4.8	69.5±4.5	3.7	6.5	0.6	I	additive
	20	60	51.6±3.0	84.0±5.9	−32.4	6.6	−4.9	S	synergistic
ROOT									
	5	20	92.9±7.7	120.8±13.5	−27.9	15.5	−1.8	I	additive
	5	40	123.5±8.7	96.8±11.1	−6.8	14.1	−0.5	I	additive
	5	60	83.6±8.4	142.8±26.6	−59.2	27.9	−2.1	S	synergistic
	10	20	95.2±8.4	81.0±9.0	14.2	13.1	1.1	I	additive
	10	40	98.3±10.4	64.9±7.4	9.1	12.8	0.7	I	additive
	10	60	73.2±9.7	95.8±17.9	−.8	20.3	−2.0	S	synergistic
	20	20	103.6±10.1	97.8±10.9	5.8	14.8	0.4	I	additive
	20	40	97.6±5.6	78.4±9.0	19.3	10.6	1.8	I	additive
	20	60	69.1±6.0	115.6±21.6	−44.2	22.4	−2.0	S	synergistic

PDW: plant dry weight; S: statistically significant; I: statistically insignificant; df: degrees of freedom.

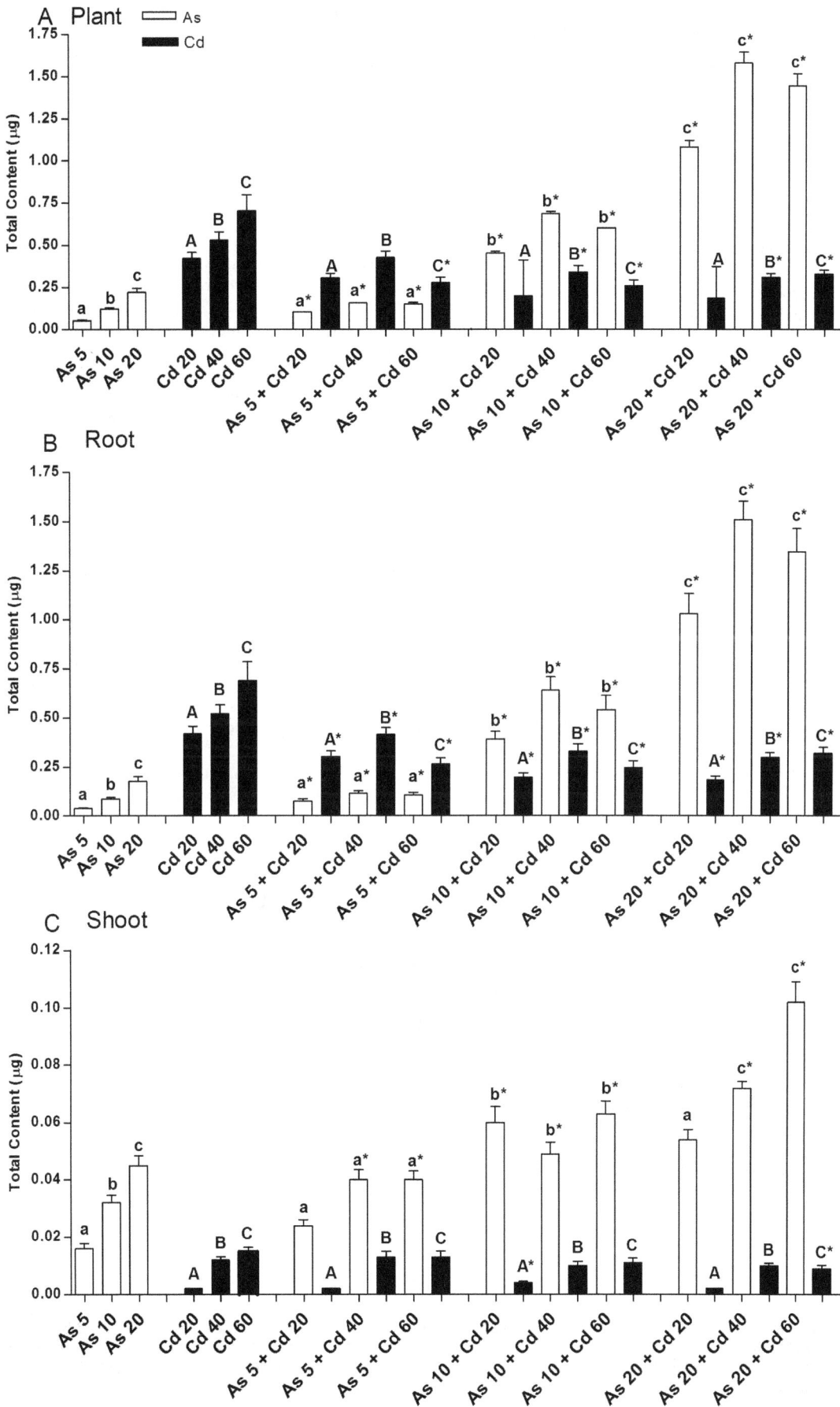

Figure 2. Metal(loid) total content (μg) in *T. repens* plants after exposition. Mean total amount of Cd and As accumulated in plant organs during exposition, was calculated for each treatment by multiplying the metal(loid) concentration, determined by AAS in root and shoot, with the correspondent organ dry weight. Uppercase letters represent significant differences with the correspondent concentration of Cd control (P<0.05); Lowercase letters represent significant differences with the correspondent concentration of As control (P<0.05).

All tested As and Cd concentrations (alone or in combination) determined a statistically higher percentage of polymorphisms in the shoots and in the roots compared to the control plants. For both Cd and As, induced plant damage was approximately two-three fold higher in the roots than in the shoots, according to the low amounts of both metal(oid)s translocated to shoot. Moreover, DNA damage was related to the concentration of Cd and As accumulated in shoot and in root. Finally, As was more genotoxic than Cd: 5 μg g^{-1} of As induced a double amount of DNA polymorphisms (14%) than 5 μg g^{-1} of Cd (6%), and 20 μg g^{-1} of As induced a significant higher amount of DNA polymorphism (32%) than 20 μg g^{-1} of Cd (25%).

The interactions between Cd and As, responsible for the joint genotoxic effects observed in Fig. 3, were defined applying the Ince et al. [14] statistical analysis. The results are reported in Table 3. Differently from the interactions responsible for the joint effects on plant development, an antagonistic interaction, leading to a DNA damage reduction, was observed in roots of plants exposed to all the combined concentrations tested. In shoots the interaction was additive except for soils contaminated with the lower Cd concentration (Cd 20) combined with As 5, or As 10, or As 20, which was antagonistic.

RDA analysis

In order to better understand the correlation among the soil metal(loid) concentrations, their accumulation in plant organs and their effects on plant growth and DNA sequence, a RDA statistical analysis was carried out. Fig.4 shows that 4 of the 6 variables considered (Cd and As bioavailability, Cd and As concentrations in plant organs) were significant (P<0.05) in determining the toxic and genotoxic effects and that the concentration of As found in plant organs was the most relevant factor (Fig.4 inset).

Discussion

Cd and As are two of the main environmental contaminants, often occurring simultaneously in polluted sites. Although, their individual toxicity and genotoxicity is well proved, few data are available about their joint effects and in particular no information is available about their joint genotoxic action. In our study we investigated the effect of combined concentrations of Cd and As on the growth and DNA damage of *T. repens*, a sensitive plant to metals, widely used in biomonitoring campaigns. Plants are efficient bioindicators to get information on cumulative effects of environmental pollutants. They are used as early warning systems for preventing environment alterations and human diseases. However, given the complexity of the mechanisms causing the final effects, the results obtained through bioindication systems should be better interpreted if the knowledge about the interaction of pollutants had improved.

Individual and joint effects of soil inorganic pollutants on bioindicators depend on different factors. First of all, at soil level, the mobility of chemicals influences the amount of compounds which can be absorbed by test-plant. Nevertheless, the uptake is not only dependent on pollutant bioavailability but it is also dependent on plant uptake mechanisms, which are compound-specific. In addition plant possess detoxification strategies, such as metal exclusion, which influence the final concentration of compounds inside the cells [18,19]. Finally when two or more compounds are simultaneously present in soil, the toxic final effects depend also on the interaction among pollutants which can occur at all levels.

In our experiment we found that all the concentrations of Cd and As, supplemented alone to the soils did not induce any relevant effect on plant survival and growth, whereas they induced a DNA damage related to the metal(loid) concentration measured

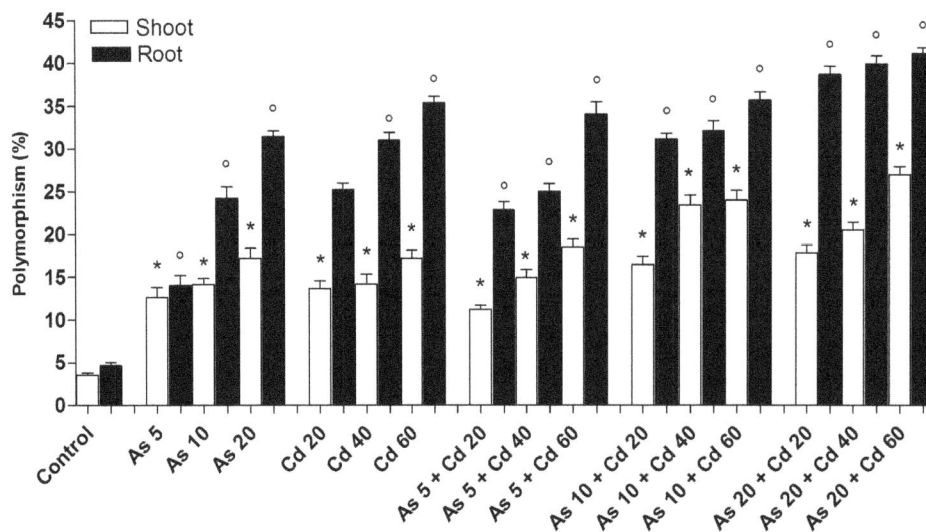

Figure 3. Analysis of the percentage of polymorphism (P% = number of polymorphic loci/number of total loci) detected by RAPD in DNA from *T. repens* plants exposed to increasing concentrations of Cd. Root and Shoot mean percentages ± SD for each treatment are reported. The asterisk and circle show statistically significant differences with respect to the control (ANOVA and Dunnet test; P<0.05).

Table 3. Observed and calculated genotoxic effects at binary test combinations x:y, and single metal concentrations x, y, respectively (predicted interaction types).

x (μg g⁻¹) As	y (μg g⁻¹) Cd	Observed genotoxicity PP_{obs}	Calculated genotoxicity PP_{calc} ($PP_x + PP_y$)	Difference ($PP_{obs} - PP_{calc}$)	Difference standard error	t Student (df = 4)	Significance (P<0.05)	Interactive Effect
SHOOT								
5	20	11.3±1.2	26.2±3.7	−15.0	3.9	−.9	S	antagonistic
5	40	14.9±2.3	26.8±4.1	−11.9	4.7	−2.5	I	addidive
5	60	18.5±2.3	29.8±3.7	−11.3	4.4	−2.6	I	additive
10	20	16.5±2.3	27.8±2.9	−11.3	3.7	−3.1	S	antagonistic
10	40	23.5±2.9	28.3±3.4	−4.8	4.4	−1.1	I	additive
10	60	24.0±2.9	31.3±2.9	−7.3	4.1	−1.8	I	additive
20	20	17.9±2.3	30.8±3.7	−13.0	4.4	−3.0	S	antagonistic
20	40	20.5±2.3	31.3±4.1	−10.9	4.7	−2.3	I	additive
20	60	27.0±2.3	34.4±3.7	−7.4	4.4	−1.7	I	additive
ROOT								
5	20	22.8±2.3	39.3±3.4	−16.4	4.1	−4.0	S	antagonistic
5	40	25.0±2.3	45.0±3.7	−20.0	4.4	−4.6	S	antagonistic
5	60	34.1±3.5	49.4±3.4	−15.3	4.8	−3.2	S	antagonistic
10	20	31.1±1.7	49.5±3.9	−18.4	4.2	−4.3	S	antagonistic
10	40	32.1±2.9	55.2±4.2	−23.1	5.1	−4.6	S	antagonistic
10	60	35.7±2.3	59.6±3.9	−23.9	4.5	−5.3	S	antagonistic
20	20	38.7±2.3	56.7±2.4	−18.0	3.4	−5.3	S	antagonistic
20	40	39.9±2.3	62.4±2.9	−22.5	3.7	−6.1	S	antagonistic
20	60	41.1±1.7	66.8±2.4	−25.6	3.0	−8.5	S	antagonistic

PP: percentage of polymorphism; S: statistically significant; I: statistically insignificant; df: degrees of freedom.

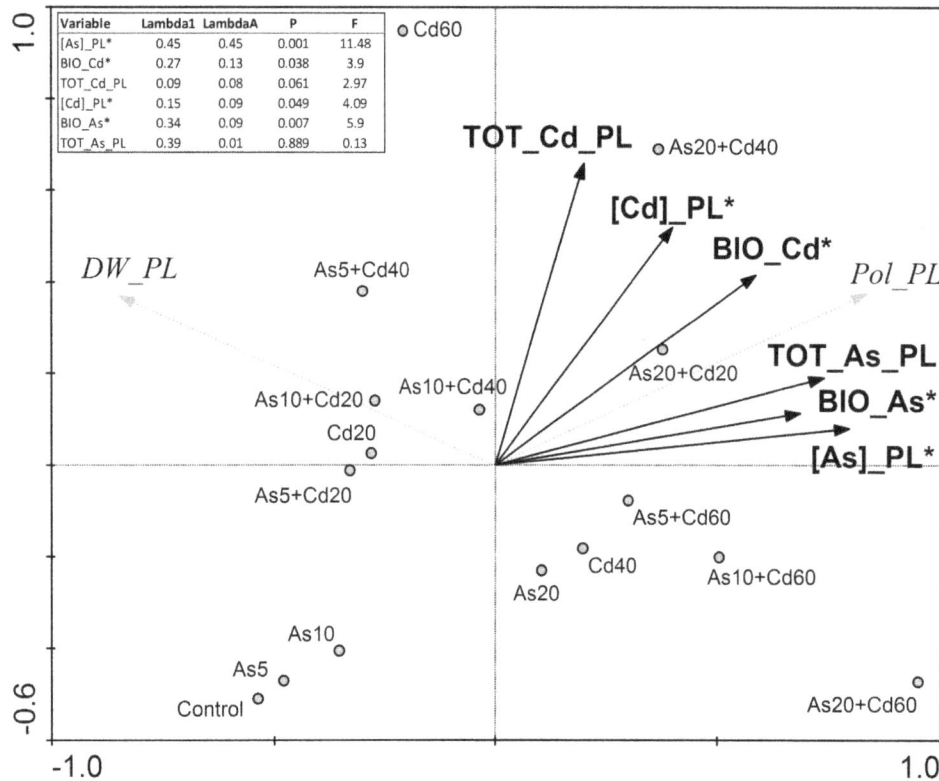

Figure 4. RDA analysis showing the relationship between the metal(loid) effects on plant growth (DW_PL) and DNA sequence (Pol_PL) and the following variables: total content of metal(loid)s in plant (TOT_Cd_PL and TOT_As_PL), concentration of metal(loid)s in plant ([Cd]_PL and [As]_PL). * Statistically different (P<0.05).

in the different plant organs. Moreover, we found that some of the concentrations of Cd and As, supplemented as a mixture to the soils, produced a synergistic effect on plant growth and an antagonistic effect on DNA, suggesting an interaction between the two compounds.

In order to understand the main factors which determined our results, we measured the soil bioavailability of Cd and As and the total amounts, and their concentrations accumulated in the different plant organs.

Concerning soil Cd and As bioavailability, in keeping with literature, Cd was much more bioavailable than As [15–18]. The very low availability of As that we measured can be ascribed to the form of As that we used to contaminate the soil (arsenite) along with an alkaline soil pH. In fact, Smith et al. [15] observed that the proportion of arsenite sorbed by soil increased with increasing pH. Specifically they observed that sorption by the soil ranged from approximately 0.80 of added As(III) at low pH, to approximately 0.95 of added As(III) at pH 6 to 7. In addition the low availability of As that we recorded should be related to the DTPA-based method that we used. This method was applied because, according to several studies, it provides the prediction of trace elements uptake by plants from soils. In particular, Karak et al. [20] showed a very high correlation between DTPA-extractable As and the labile pool of As suggesting that the latter is the portion of As most hazardous for human health, due to the possibility of entering the food chain.

Interestingly, for both the metal(loid)s, bioavailability increased with increasing metal concentration in the soils only when the two compounds were used separately, whereas, when they were used as a mixture to contaminate soil, the presence of Cd increased the

amount of bioavailable As and, on the contrary, the presence of As reduced Cd bioavailability. The reduction of Cd bioavailability in the presence of As was also observed by Sun and collaborators [17]. This type of result suggests a sort of competition between the two metal(loid)s for binding with soil constituents (clays, Al or Fe or Mn oxides, organic matter, etc.). Generally, both Cd and As retention in soil is due to their primary association to organic matter and amorphous Fe and Mn oxides [9,20,21]. It is then likely that in our experiment the interaction between Cd and As, involved these soil constituents. Anyway, given the different characteristic of As and Cd, it is very difficult to shade light on the mechanisms determining the bioavailability changes that we observed when the two compounds were simultaneously present in a soil, therefore further works, beyond the aim of the present study, are needed to clarify the Cd and As sorption-desorption processes.

In our experiment, bioavailability was a very important factor for As accumulation, given the linear correlation found between the total As in plant and the soil As bioavailability.

The result was consistent with previous works [16,17,22] showing a significant (p<0.01) correlation between As uptake by plants in various treatments and total soil As. On the contrary, regression analysis indicated that Cd accumulation was not linearly correlated to soil bioavailability. This is also in agreement with previous studies, which showed that the uptake of Cd by plant increases proportionally to increasing soil Cd only up to about 20 mg kg^{-1} above which the trend becomes curvilinear [23]. The different behavior of the two metal(loid)s could be explained by considering their absorption mechanisms. The uptake of Cd from the soil occurs mainly via Ca^{2+}, Fe^{2+}, Mn^{2+} and Zn^{2+} transporters

[24], whereas that of arsenite [As(III)] occurs mainly by diffusion across membrane through members of the NIP (nodulin 26-like intrinsic protein) subfamily of aquaporins [25,26]. Thus we can assume that in our conditions, the main factor determining As accumulation in *T. repens* was bioavailability, whereas the limiting factor for Cd accumulation was related to the uptake system. Moreover, the possible combination of that fraction of arsenate [As (V)], likely formed in soil from [As (III)], with Cd (Cd^{2+} + AsO_4^{3-} $Cd_3(AsO_4)_2$) could have decreased the ion exchange on the root surfaces, playing a role in the reduction of Cd uptake, as demonstrated by Liu et al. [4], and explaining the reduction of Cd accumulation we observed in plants grown in presence of both the metal(loid)s.

Interestingly, as shown by RDA analysis, in our study the accumulated total amounts of Cd and As in the plant organs were not statistically significant to explain the observed toxic and genotoxic effects. This because some treatments caused a plant organ reduction, so that the effects were related to the concentration of metal(loid)s measured in plant organs and not to the total absorbed amounts. Specifically As concentration was the most important variable due to both its intrinsic toxicity, that was higher than that of Cd at equal concentration (in agreement with Luan and collaborators [16]), and to its chemical characteristics allowing a plant uptake proportional to soil bioavailability, which was also increased by the presence of Cd in the soil. Moreover, although the concentration of Cd was also important in determining the observed effects, we should consider that, differently from arsenite, which is chemically neutral, a fraction of the total amount of Cd^{2+}, accumulated in the different plant organs, was likely stored in cell walls, since the negative charges of the cell wall bind and retain heavy metals [27,28]. It is one of the several mechanisms evolved by plants to cope with Cd^{2+} toxicity, limiting cellular internalization and its associated toxicity [24,29].

Concerning the observed toxic effect, a reduction of plant growth was induced by most of the combined concentrations of Cd and As tested. The type of interaction between the two metal(loid)s was additive except for the combinations of the higher Cd concentration (Cd 60) with any As concentration, which were synergistic. Joint plant Cd and As toxicity was previously investigated with contrasting results: Luan and collaborators [16] reported a synergistic effect on soybean plants whilst, Liu et al. [4] and Sun et al. [17] observed an antagonistic effect on wheat and rice biomass production. These opposing results are probably due to the different experimental conditions and to the plant response mechanisms to metal stress, which are species-specific and even development stage and organ specific [7]. *T. repens* is a pollutants-sensitive plant and lack of effective tolerance mechanisms, therefore it is not able to tolerate high metal(loid)s concentrations, whose effect might be exacerbated whenever they act simultaneously. Accordingly, in our experimentation a synergistic effect on plant growth reduction was observed in those plants showing a higher total concentration of metal(loid)s. Likely a consistent inhibition of enzyme activity due to the high Cd and As reactivity to sulfhydryl groups along with oxidative stress and deregulation of homeostasis of essential element or their displacement from protein, primarily due to Cd chemical similarity to Zn, Cu and Fe, led to the inhibition of cellular functions and growth.

In addition, the observed plant growth reduction could be due to an arrest of cell cycle specifically caused by plant in response to high DNA damage induced by high concentrations of metal(loid)s. The temporary inhibition of cell cycle progression and DNA synthesis would provide a longer time for DNA repair and for the production of free radical scavengers. In support of this hypothesis we found an antagonistic genotoxic effect in most of the combined treatments. The antagonism could be also related to the similar genotoxic mechanisms of Cd and As involving the induction of ROS and the inhibition of DNA repair enzymes, which could be reach a maximum in the presence of a defined concentration of metal(loid)s beyond which it does not increase. Anyway, further investigations are needed to clarify the cellular molecular mechanisms involved in the interaction between Cd and As.

In conclusion, our experiment showed that Cd and As can interact at different levels producing additive, synergistic or antagonistic effects. In our experimental condition the presence of Cd increased As soil bioavailability, whereas As presence reduced that of Cd. Nevertheless bioavailability determined the absorption of As but not that of Cd, which was likely limited by its uptake mechanisms. Toxicity and genotoxicity were related to the total concentration of Cd and As in plant organs and As concentration was the most significant variable. Joint effects on plant growth were additive or synergistic, whereas joint genotoxic effects were additive or antagonists. We have supposed that growth reduction was due to both toxic effects of Cd and As and to plant response to high DNA damage, which led to a temporary arrest of cell cycle providing a longer time for DNA repair and for free radical scavenger production. This hypothesis is consistent with the antagonistic genotoxic effect observed in most of the combined treatments. Nevertheless the antagonistic interaction of Cd and As could be also associated to the similar genotoxic mechanisms own of the two metal(loid)s.

Supporting Information

Figure S1 Metal(loid) concentration ($\mu g\ g^{-1}$ dry matter) in white clover plants after exposition. The mean concentration obtained by AAS ± standard deviation for each plant organ and for each soil is shown. Uppercase letters represent significant differences with the correspondent concentration of Cd control (P<0.05); lowercase letters represent significant differences with the correspondent concentration of As control (P<0.05).

Table S1 Sequences of primers used for RAPD analysis.

Acknowledgments

The authors wish to thank Maria Tringali and Fabio Moia for technical help.

Author Contributions

Conceived and designed the experiments: SC. Performed the experiments: AG TNV. Analyzed the data: RG PF. Contributed to the writing of the manuscript: AG.

References

1. Heger TJ, Imfeld G, Mitchell EAD (2012) Special issue on "Bioindication in soil ecosystems": Editorial note. Eur J Soil Biol 49: 1–4. Doi: 10.1016/j.ejsobi.2012.02.001.

2. Piraino F, Aina R, Palin L, Prato N, Sgorbati S, et al. (2006) Air quality biomonitoring: Assessment of air pollution genotoxicity in the Province of Novara (North Italy) by using *Trifolium repens* L. and molecular markers. Sci Total Environ 372: 350–359. Doi:10.1016/j.scitotenv.2006.09.009.

3. Zhou QX, Zhang QR, Liang JD (2006) Toxic effects of acetochlor and methamidops on earthworm *Eisenia fetida* in Phaeozem, Northeast China. J Environ Sci 18: 741–745.

4. Liu XL, Zhang SZ. (2007) Intraspecific differences in effects of co-contamination of cadmium and arsenate on early seedling growth and metal uptake by wheat. J E S 19 (10): 1221–1227. Doi:10.1016/S1001-0742(07)60199-5.

5. Wang G, Fowler BA (2008) Roles of biomarkers in evaluating interactions among mixtures of lead, cadmium and arsenic, Toxicol Appl Pharm 233: 92–99. Doi:10.1016/j.taap.2008.01.017.

6. Huang M, Choi SJ, Kim DW, Kim NY, Park CH, et al. (2009) Risk assessment of low-level cadmium and arsenic on the kidney. J Toxicol Environ Health A 72(21-22): 1493-8. Doi: 10.1080/15287390903213095.

7. Tkalec M, Stefanić PP, Cvjetko P, Sikić S, Pavlica M, et al. (2014) The effects of cadmium-zinc interactions on biochemical responses in tobacco seedlings and adult plants. PLoS One 27: 9(1):e87582. doi: 10.1371/journal.pone.0087582. eCollection 2014.

8. Agency for Toxic Substances and Disease Registry (ATSDR), CERCLA Priority List of Hazardous Substances, GA, US Department of Health and Human Services (www.atsdr.cdc.gov), 2005.

9. Keil DE, Berger-Ritchie J, McMillin GA. (2011) Testing for toxic elements: a focus on arsenic, cadmium, lead, and mercury. Lab Medicine 42: 735–742. Doi:10.1309/LMYKGU05BEPE7IAW.

10. Nagajyoti PC, Lee KD, Sreekanth TVM (2010) Heavy metals, occurrence and toxicity for plants: a review. Environ Chem Lett 8(3): 199–216. Doi: 10.1007/s10311-010-0297-8.

11. Beyersmann D, Hartwig A (2008) Carcinogenic metal compounds: recent insight into molecular and cellular mechanisms. Arch toxicol 82(8): 493–512. Doi: 10.1007/s00204-008-0313-y.

12. Hartwig A (2010) Mechanisms in cadmium-induced carcinogenicity:recent insights. Biometals 23: 951–960. Doi 10.1007/s10534-010-9330-4.

13. Lindsay WL, Norwell WA (1969) Development of a DTPA micronutrient soil test. Agronomy Abstracts 69: 87.

14. Ince NH, Dirilgen N, Apikyan IG, Tezcanli G, Üstün B (1999) Assessment of toxic interactions of heavy metals in binary mixtures: a statistical approach. Arch Environ Contam Toxicol 36(4): 365–372.

15. Smith E, Naidu R, Alston AM (1999) Chemistry of arsenic in soils: I. Sorption of arsenate and arsenite by four Australian soils. J Environ Qual 28: 1719–1726.

16. Luan ZQ, Cao HC, Yan BX (2008) Individual and combined phytotoxic effects of cadmium, lead and arsenic on soybean in Phaeozem. Plant Soil Environ 54: 403–411.

17. Sun Y, Li Z, Guo B, Chu G, Wei C, et al. (2008) Arsenic mitigates cadmium toxicity in rice seedlings. Environ Exp Bot 64: 264–270. Doi:10.1016/j.envexpbot.2008.05.009.

18. Verbruggen N, Hermans C, Schat H (2009) Mechanisms to cope with arsenic or cadmium excess in plants. Curr Opin Plant Biol 12: 364–372. Doi:10.1016/j.pbi.2009.05.001.

19. Hossain MA, Piyatida P, da Silva JAT, Fujita M (2012) Molecular mechanism of heavy metal toxicity and tolerance in plants: central role of glutathione in detoxification of reactive oxygen species and methylglyoxal and in heavy metal chelation. J Botany Article ID 872875. Doi:10.1155/2012/872875.

20. Karak T, Abollino O, Bhattacharyya P, Das KK, Paul RK (2011) Fractionation and speciation of arsenic in three tea gardens soil profiles and distribution of As in different parts of tea plant (*Camellia sinensis* L.). Chemosphere 85: 948–960. Doi:10.1016/j.chemosphere.2011.06.061.

21. Gonzaga MIS, Santos JAG, Ma LQ (2008) Phytoextraction by arsenic hyperaccumulator *Pteris vittata* L. from six arsenic-contaminated soils: repeated harvests and arsenic redistribution. Environ Pollut 154: 212–218. Doi:10.1016/j.envpol.2007.10.011.

22. Fayiga AO, Ma LQ (2005) Using phosphate rock to immobilize metals in soil and increase arsenic uptake by hyperaccumulator *Pteris vittata*. Sci Total Environ 359: 17–25. Doi:10.1016/j.scitotenv.2005.06.001.

23. Smolders E (2001) Cadmium uptake by plants. Int J Occup Med Environ Health 14 (2): 177–183. PubMed ID: 11548068.

24. Clemens S (2006) Toxic metal accumulation, responses to exposure and mechanisms of tolerance in plants. Biochimie 88: 1707–1719. Doi:10.1016/j.biochi.2006.07.003.

25. Bienert GP, Thorsen M, Schussler MD, Nilsson HR, Wagner A, at al. (2008) A subgroup of plant aquaporins facilitate the bi-directional diffusion of As(OH) 3 and Sb(OH) 3 across membranes. BMC Biol 6: 26. Doi:10.1186/1741-7007-6-26.

26. Isayenkov SV, Maathuis FJ (2008) The *Arabidopsis thaliana* aquaglyceroporin AtNIP7; 1 is a pathway for arsenite uptake. FEBS Lett 582: 1625–1628. Doi:10.1016/j.febslet.2008.04.022.

27. Polle A, Schützendübel A (2003) Heavy metal signalling in plants: linking cellular and organismic responses. In: Hirt H, Shinozaki K, editors. Plant Responses to Abiotic Stress. Springer-Verlag, ISBN 3540200371, Germany: Berlin-Heidelberg. pp.187–215.

28. Lux A, Martinka M, Vaculík M, White PJ (2011) Root responses to cadmium in the rhizosphere: a review. J Exp Bot 62(1): 21–37. Doi:10.1093/jxb/erq281.

29. Zhu XF, Wang ZW, Dong F, Lei GJ, Shi YZ, et al. (2013) Exogenous auxin alleviates cadmium toxicity in *Arabidopsis thaliana* by stimulating synthesis of hemicellulose 1 and increasing the cadmium fixation capacity of root cell walls. J Hazard Mater 263: 398–403. http://dx.doi.org/10.1016/j.jhazmat.2013.09.018.

7

Phthalic Acid Esters in Soils from Vegetable Greenhouses in Shandong Peninsula, East China

Chao Chai[1], Hongzhen Cheng[1], Wei Ge[2], Dong Ma[1], Yanxi Shi[1]*

1 College of Resources and Environment, Qingdao Agricultural University, Qingdao, China, **2** College of Life Sciences, Qingdao Agricultural University, Qingdao, China

Abstract

Soils at depths of 0 cm to 10 cm, 10 cm to 20 cm, and 20 cm to 40 cm from 37 vegetable greenhouses in Shandong Peninsula, East China, were collected, and 16 phthalic acid esters (PAEs) were detected using gas chromatography-mass spectrometry (GC-MS). All 16 PAEs could be detected in soils from vegetable greenhouses. The total of 16 PAEs (Σ_{16}PAEs) ranged from 1.939 mg/kg to 35.442 mg/kg, with an average of 6.748 mg/kg. Among four areas, including Qingdao, Weihai, Weifang, and Yantai, the average and maximum concentrations of Σ_{16}PAEs in soils at depths of 0 cm to 10 cm appeared in Weifang, which has a long history of vegetable production and is famous for extensive greenhouse cultivation. Despite the different concentrations of Σ_{16}PAEs, the PAE compositions were comparable. Among the 16 PAEs, di(2-ethylhexyl) phthalate (DEHP), di-n-octyl phthalate (DnOP), di-n-butyl phthalate (DnBP), and diisobutyl phthalate (DiBP) were the most abundant. Compared with the results on agricultural soils in China, soils that are being used or were used for vegetable greenhouses had higher PAE concentrations. Among PAEs, dimethyl phthalate (DMP), diethyl phthalate (DEP) and DnBP exceeded soil allowable concentrations (in US) in more than 90% of the samples, and DnOP in more than 20%. Shandong Peninsula has the highest PAE contents, which suggests that this area is severely contaminated by PAEs.

Editor: Raffaella Balestrini, Institute for Plant Protection (IPP), CNR, Italy

Funding: This work was supported by the "Science and Technology Plan Projects of Qingdao (No. 12-1-3-64-nsh), the Two Districts" Foundation of Shandong Province, China (No. 2011-Yellow-19) and the Talent Foundation of Qingdao Agricultural University (No. 630642). The funders had no role in study design, data collection and analysis, decision to publish, or preparation of the manuscript.

Competing Interests: The authors have declared that no competing interests exist.

* E-mail: yanxiyy@126.com

Introduction

Phthalic acid esters (PAEs) are used extensively as plasticizers of plastic products, such as polyvinyl chloride, and as nonplasticizers in consumer products, including medical devices, building materials, paints, pesticides, fertilizes, food packaging, and so on [1]. The large-scale production and application of 6.0 million tons/yr [2] of PAEs have made these materials ubiquitous environment pollutants [3–8]. Some PAEs have endocrine disruptive effects [9], and six PAEs are categorized as priority environmental pollutants by the United States Environmental Protection Agency [10].

Greenhouse cultivation has expanded dramatically in China since the 1980s, reaching up to 3.5 million ha by 2011 [11]. Greenhouse cultivation is mainly for vegetable production in China, and plastic greenhouses account for more than 99% of greenhouse cultivation relative to glass greenhouses [12–13]. Several studies detected PAEs in soils of vegetable greenhouses in Nanjing and Hangzhou [14–15], as well as in other agricultural soils, such as vegetable soils in Guangzhou and paddy soils in Leizhou Peninsula in China [16–17]. The buildup of PAEs in agricultural soils may contaminate agricultural products, and further raise the human health risk [18].

Shandong Peninsula is the largest Peninsula in China with rapid urbanization and high population density of 550 people/km². The Peninsula includes the cities of Qingdao, Yantai, Weifang, and Weihai. Shandong Peninsula has a long history of vegetable greenhouse cultivation and is a main vegetable-producing region,

with its greenhouse coverage accounting for approximately 50% of that of China. The vegetable greenhouses in this peninsula are close to the highly populated urban areas, and plastic film is widely used. Plastic film of 30000 tons/yr is estimated to be used only in one county, i.e., Shouguang in Weifang of Shandong Peninsula [15]. PAEs account for 10 wt% to 60 wt% of plastic products [9,19], thus giving rise to concerns about the potential risk of PAEs in recent years. However, few studies focused on the characteristics of PAEs in soils of vegetable greenhouses in Shandong Peninsula.

This study provides information on the concentrations, compositions, and distributions of 16 PAEs in soils from vegetable greenhouses in Shandong Peninsula and discusses possible sources, influence factors, and potential environment risk.

Materials and Methods

Chemicals and materials

Mixed standard solutions of 16 PAEs containing dimethyl phthalate (DMP), diethyl phthalate (DEP), diisobutyl phthalate (DiBP), di-n-butyl phthalate (DnBP), dimethylglycol phthalate (DMGP), di(4-methyl-2-pentyl) phthalate (DMPP), di(2-ethylhexyl) phthalate (DEHP), di(2-ethoxyethyl) phthalate (DEEP), dipentyl phthalate (DPP), di-n-hexyl phthalate (DHXP), butylbenzyl phthalate (BBP), di(2-n-butoxyethyl) phthalate (DBEP), dicyclohexyl phthalate (DCHP), di-n-octyl phthalate (DnOP), diphenyl phthalate (DPhP), and di-n-nonyl phthalate (DNP) were supplied by O2SI, Inc. (USA). The concentration of each PAE in this mixture solution was 1000 mg/L. Glassware was steeped with

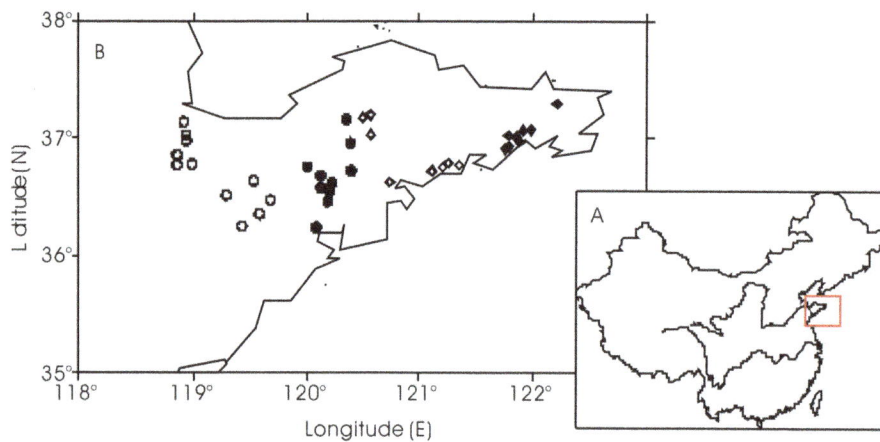

Figure 1. Schematic map showing the geographical location of (A) Shandong Peninsula and (B) the vegetable soil sampling sites in 4 regions in the Shandong Peninsula (solid round: Qingdao; solid diamond: Weihai; circle: Weifang; diamond: Yantai).

K_2CrO_7/H_2SO_4 solution for 12 h, washed with redistilled water, and then baked at 300°C for 4 h. Acetone, petroleum ether, and diethyl ether were of analytical grade and re-distilled before use to avoid PAEs contamination. Hexane was of HPLC grade and purchased from Anpel Company Inc. Florisil (60 mesh to 80 mesh) was activated at 650°C, and anhydrous sodium sulfate was baked at 420°C for 4 h.

Sampling

No specific permissions were required for sampling locations/activities. The field studies did not involve endangered or protected species. A total of 111 soil samples were collected from 37 vegetable greenhouses in Qingdao (number of samples: 30), Weihai (number of samples: 24), Weifang (number of samples: 33), and Yantai (number of samples: 24) in Shandong Peninsula in from 28 to 30 May 2012. The sampling locations are shown in Fig. 1.

Each sampling site consisted of five sub-samples (0.2 kg each) in the middle and four corners at depths of 0 cm to 10 cm, 10 cm to 20 cm, and 20 cm to 40 cm. The five sub-samples were mixed

immediately after sampling, and then the soils were collected using aluminum foil envelopes through a pre-cleaned stainless steel auger and transported to laboratory in an ice box. Soils were stored in glass bottles at −20°C until analysis after being freeze-dried, ground, and homogenized with a stainless steel sieve (60 mesh). PAE contamination was avoided during sampling and further processing.

Soil physical and chemical analyses

Soil pH was measured using a pH meter with a soil/water ratio of 1:2.5. Soil cation exchange capacity (CEC) was analyzed using the Ba^{2+} compulsive exchange method [20]. Particle-size fraction was determined using the pipette method, and the soil texture was classified according to the Soil Survey Division Staff [21]. Total organic carbon (TOC) was determined using the wet oxidation method with chromate [22] and total nitrogen (TN) using micro-Kjeldahl digestion method [23].

Table 1. The main characteristics of the soils from vegetable greenhouses in Shandong Peninsula.

Area	Soil depth (cm)	pH	TOC (g/kg)	TN (g/kg)	C/N	CEC (mol/kg)	Sand (%)	Silt (%)	Clay (%)
Qingdao	0~10	6.62±0.64	31.7±9.8	1.3±0.4	26.44±0.90	0.14±0.05	54.1±3.8	27.6±6.6	17.8±3.7
	10~20	6.52±0.56	29.6±13.3	1.1±0.3	26.02±0.36	0.15±0.06	53.2±5.4	25.9±4.6	15.5±4.6
	20~40	6.64±0.42	25.1±7.1	0.8±0.3	33.49±9.12	0.11±0.05	46.7±4.6	24.6±7.8	15.9±4.2
Weihai	0~10	6.31±0.56	30.0±3.8	1.4±0.6	23.89±6.93	0.07±0.03	50.7±3.4	22.1±0.7	17.1±2.7
	10~20	6.10±0.43	27.7±4.1	1.0±0.1	27.33±2.72	0.09±0.01	55.3±6.2	29.6±9.9	15.6±1.4
	20~40	5.90±0.55	19.3±7.2	0.7±0.2	27.32±9.13	0.08±0.04	52.6±4.5	33.2±1.6	16.0±2.5
Weifang	0~10	6.88±0.46	32.2±6.0	1.7±3.1	40.49±5.39	0.13±0.09	61.3±0.2	37.8±7.7	19.9±6.0
	10~20	6.96±0.36	27.9±12.1	0.9±0.9	41.81±8.63	0.13±0.06	57.0±9.8	33.6±2.6	17.9±5.6
	20~40	6.99±0.49	23.4±4.3	1.2±2.3	42.55±5.78	0.14±0.08	59.2±2.1	28.1±5.9	21.1±7.4
Yantai	0~10	6.46±0.66	24.0±6.4	1.3±0.4	18.95±5.92	0.10±0.02	57.8±3.4	38.7±9.5	18.9±4.6
	10~20	6.56±0.96	23.7±6.3	1.1±0.4	22.99±5.90	0.10±0.03	57.4±4.2	38.6±4.4	18.0±3.0
	20~40	7.10±0.99	20.5±5.0	0.7±0.1	29.22±9.46	0.11±0.03	55.1±6.1	33.6±6.6	16.3±1.7

Table 2. The detection rate and concentration of PAEs in all soil samples from vegetable greenhouses in Shandong Peninsula (n = 111).

PAEs	Detection rate (%)	Mean (mg/kg)	SD (mg/kg)	Minimum (mg/kg)	Maximum (mg/kg)
DMP	99.1	0.364	0.276	ND	1.245
DEP	100	0.108	0.169	0.002	1.051
DiBP	96.4	1.118	1.928	ND	11.434
DnBP	100	1.471	2.715	0.016	15.722
DMGP	23.4	0.015	0.031	ND	0.170
DMPP	48.6	0.246	0.405	ND	1.971
DEHP	100	1.465	1.207	0.073	5.323
DEEP	23.4	0.041	0.243	ND	2.556
DPP	58.6	0.088	0.098	ND	0.516
DHXP	64.9	0.084	0.157	ND	1.448
BBP	86.5	0.194	0.557	ND	5.691
DBEP	18.9	0.015	0.038	ND	0.267
DCHP	44.1	0.035	0.048	ND	0.204
DnOP	97.3	1.239	1.796	ND	14.397
DPhP	82.0	0.240	0.290	ND	2.371
DNP	19.8	0.026	0.060	ND	0.251
Σ_{16}PAEs		6.748	5.716	1.939	35.442

ND: not detected. The data labeled as "ND" were treated as zero in further statistical treatment.

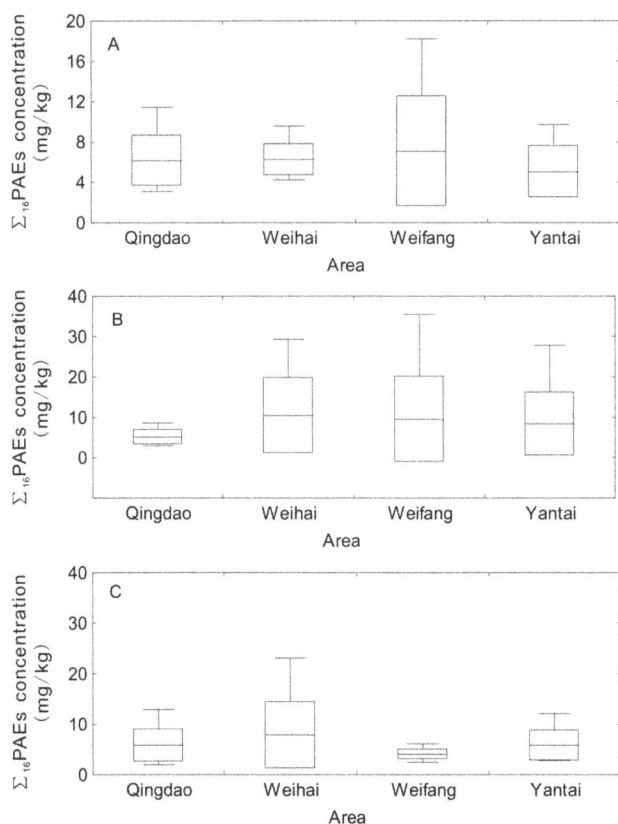

Figure 2. The concentrations of Σ_{16}PAEs in (A) soils of 0–10 cm, (B) soils of 10–20 cm and (C) soils of 20–40 cm from vegetable greenhouses in Shandong Peninsula.

Sample extraction of PAEs and instrumental analysis

PAEs extraction was conducted according to Wang's methods [24]. 5.0 g soil was spiked with surrogate standard (benzyl benzoate) and extracted through ultrasonication for 15 min thrice with 90 mL of acetone/petroleum ether (1:3, v:v). The extracts were combined, filtered, and concentrated to approximately 1 mL. The extracts were cleaned with anhydrous sodium sulfate (3 g), florisil (6 g), and anhydrous sodium sulfate (3 g) on a glass column (1 cm i.d.). The column was washed with 10 mL of petroleum ether/diethyl ether (10:0.4, v:v), and then PAEs were eluted with 90 mL of petroleum ether/diethyl ether (10:3, v:v). The extracts were reduced to 1.0 mL in hexane, and internal standard (diisophenyl phthalate) was added before instrumental analysis.

Instrumental analysis was performed on an Agilent 6890 GC-5973 MSD gas chromatography-mass spectrometry system (GC-MS) in electron impact and selective ion monitoring modes according to Zeng et al. [25]. The GC column used was DB-5MS capillary column (30 m×0.25 mm i.d. ×0.25 mm film thickness, J&W Scientific). The column temperature program was 80°C (1 min), to 180°C (10°C/min, 1 min), to 300°C (2°C/min, 10 min). The carrier gas was helium with flow rate of 0.8 mL/min. Then, 1 μL of the extracts was injected into GC-MS in splitless injection mode, and the injector temperature was 250°C. The GC-MS transfer line was 280°C, and the post run temperature was 285°C for 2 min.

Quality control and quality assurance

Quality assurance was performed by analyzing a procedural and solvent blank, a spiked blank every 10 samples, surrogate standards for each sample, and sample duplicate. DiBP, DnBP, and DEHP were subtracted from those in the soil samples because of the small amount in procedural blanks. The surrogate recoveries were 84.1%±8.5%, and no surrogate corrections were

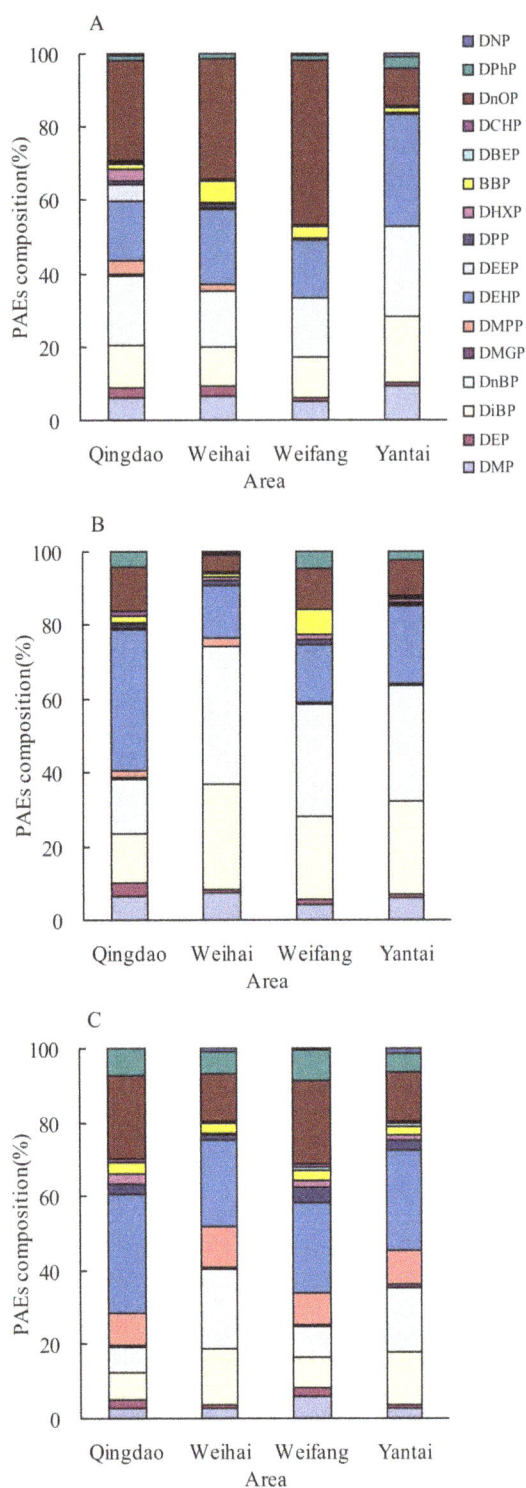

Figure 3. The compositions of PAEs in (A) soils of 0–10 cm, (B) soils of 10–20 cm and (C) soils of 20–40 cm from vegetable greenhouses in Shandong Peninsula.

ranged from 1∼9 pg, calculated by signal to noise ratio of 10. The method detection limits for PAEs were determined as mean field blanks plus three times the standard deviation of the field blanks [25], ranging from 0.002 mg/kg for DEP to 0.024 mg/kg for DEHP.

Results

Properties of the soils from vegetable greenhouses

The major characteristics of soils from vegetable greenhouses in Shandong Peninsula are presented in Table 1. The pH (H_2O) of soils was neutral in three sample areas, but was less than 6.5 in Weihai, which was moderately acidic. The TOC ranged from 19.3 g/kg to 32.2 g/kg and presented a decreasing trend with soil depth. The C/N ratio was approximately 20 to 30, except in Weifang, which had a value of more than 40 on average. This value indicated that the organic matter content was more C-rich. The C/N ratio presented an increasing trend with soil depth. The CEC followed a similar pattern as pH, with lower values in soil samples from Weihai. Most of the soils were sandy loam, and some were loam.

PAE concentrations in soils from vegetable greenhouses in Shandong Peninsula

All 16 PAEs were detected in soils from vegetable greenhouses (Table 2). Among them, three PAEs, namely, DEP, DnBP, and DEHP, were detected in all samples. The detection rates of another three PAEs (DMP, DiBP, and DnOP) were more than 90%. By contrast, the detection rates of DBEP and DNP were lower than 20%. The mean concentrations of DiBP, DnBP, DEHP, and DnOP were more than 1 mg/kg, higher than other PAEs. On the whole, the mean was almost systematically inferior to standard deviation, suggesting a very high heterogeneity between soils. The total of 16 PAEs (Σ_{16}PAEs) in Shandong Peninsula ranged from 1.939 mg/kg to 35.442 mg/kg, with an average of 6.748 mg/kg.

The concentrations of Σ_{16}PAEs in soils from vegetable greenhouses in different areas of Shandong Peninsula are presented in Fig. 2. The variability of Σ_{16}PAEs was high in Weifang in the two upper layers, but low in Qingdao (10 cm to 20 cm) and Weifang (20 cm to 40 cm). The maximum value of Σ_{16}PAEs in soils at 0 cm to 10 cm and 10 cm to 20 cm both appeared in Weifang, with values of 18.179 and 35.442 mg/kg, respectively.

PAE composition in soils of vegetable greenhouses from Shandong Peninsula

Despite the different concentration of Σ_{16}PAEs, the PAE compositions in soils from vegetable greenhouses in Shandong Peninsula were comparable (Fig. 3). DnOP had the highest proportion (27.1% to 45%) in soils at 0 cm to 10 cm in Qindao, Weihai, and Weifang, whereas DEHP had the highest proportion (30.4%) in Yantai. The proportion of DnBP and DiBP ranged from 10.6% to 24.5%, suggesting significant proportion. In soils at 10 cm to 20 cm, DEHP had the largest proportion of 38.4% in Qingdao, but DnBP had the largest and DiBP had the second largest proportion in the other three areas. In soils at 20 cm to 40 cm, DEHP was a dominant congener, ranging from 23.4% to 31.8% in four areas. DnOP in Qingdao and Weifang had the second highest proportion, whereas DnBP in Weihai and Yantai had the second highest proportion. Therefore, DEHP, DnOP, DnBP, and DiBP had the largest proportion in soils in all areas. In addition, DMP accounted for about 5∼10% in the upper two

made to the final PAE concentrations reported. The calibration curves were used with six concentration levels of a standard mixture for PAEs quantitation. The standard mixture was analyzed every 10 samples to determine instrument stability and to confirm the calibration curve. The instrumental detection limits

Table 3. Comparisons of PAEs contents in agricultural soils in China (mg/kg).

Area	DMP	DEP	DnBP	DEHP	BBP	DnOP	Σ_6PAEs	Σ_{16}PAEs	Type of soils	References
Shandong Peninsula	0.36	0.11	1.47	1.47	0.19	1.24	4.48	6.73	Soils of vegetable greenhouses	In this study
	(ND~1.24)	(ND~1.21)	(0.02~15.72)	(0.07~5.32)	(ND~5.69)	(ND~14.40)	(1.18~23.35)	(1.94~35.44)		
Nanjing	0.006	0.005	0.19	1.72	0.003	0.158	1.89		Soils of vegetable greenhouses	[14]
	(ND~0.016)	(ND~0.012)	(ND~1.41)	(0.034~9.031)	(ND~0.038)	(ND~1.739)	(0.15~9.68)			
Hangzhou	ND	0.59	0.21	1.48	0.05	0.14	2.47	2.75	Soils of vegetable greenhouses	[15]
		(0.06~1.49)	(0.14~0.35)	(0.81~2.20)	(0.03~0.16)	(0.10~0.25)		(1.90~4.36) (Σ_{11}PAEs)		
Guangzhou, Shenzhen	(ND~0.68)	(ND~1.77)	(3.75~18.45)	(2.82~25.11)	(ND~1.48)	(ND~0.92)	(3.00~45.67)		vegetable Soils	[16]
Zhangjiang,	0.02	0.09	0.23	0.15	0.05	0.03	0.56		Vegetable soil	[50]
	(ND~0.45)	(ND~1.06)	(ND~7.65)	(ND~6.38)	(ND~2.83)	(ND~0.32)	(0.01~9.30)			
Dongguan,	0.02	0.05	0.49	0.17	0.13	0.06	0.92		Paddy soil	
	(ND~0.86)	(ND~1.60)	(ND~17.51)	(ND~4.22)	(ND~5.89)	(ND~1.12)	(0.01~25.99)			
Zhongshan,	0.03	0.19	0.41	0.12	0.02	0.02	0.81		Banana soil	
	(ND~0.12)	(ND~2.50)	(ND~4.13)	(ND~2.69)	(ND~0.26)	(ND~0.08)	(0.05~5.92)			
Zhuhai,	0.03	0.06	0.30	0.07	0.01	0.02	0.49		Sugarcane soil	
	(ND~0.18)	(ND~0.44)	(ND~1.77)	(ND~0.26)	(ND~0.03)	(ND~0.07)	(ND~2.10)			
Shunde	0.02	0.06	0.20	0.11	0.01	0.02	0.43		Orchard soil	
	(ND~0.06)	(ND~0.40)	(ND~0.56)	(ND~0.99)	(ND~0.06)	(ND~0.15)	(0.04~1.38)			
Leizhou Peninsula	0.02	0.02	0.45	0.01	0.01	0.02	0.53	1.11	Sugarcane soil	[17]
								(0.02~5.45)		
	0.01	0.01	0.24	0.24	0.01	0.02	0.53	0.86	Paddy soil	
								(ND~2.78)		
	0.02	0.01	0.27	0.12	0.01	0.02	0.45	0.61	Vegetable soil	
								(0.02~1.87)		
	0.03	0.01	0.28	0.10	0.01	0.01	0.44	0.60	Orchard soil	
								(0.28~1.05)		
Huizhou	0.004	0.01	0.15	0.09	0.002	0.01	0.31	0.60	Agricultural soil	[51]
	(ND~0.03)	(ND~0.22)	(ND~0.39)	(ND~0.44)	(ND~0.04)	(ND~0.06)	(0.09~0.75)	(0.18~2.04)		
							0.28	0.59	Vegetable soil	
							(0.06~0.64)	(0.18~2.04)		
							0.24	0.65	Paddy soil	
							(0.08~0.64)	(0.43~1.21)		

Table 3. Cont.

Area	DMP	DEP	DnBP	DEHP	BBP	DnOP	Σ_6PAEs	Σ_{16}PAEs	Type of soils	References
							0.22 (0.08~0.64)	0.51 (0.38~0.63)	Orchard soil	

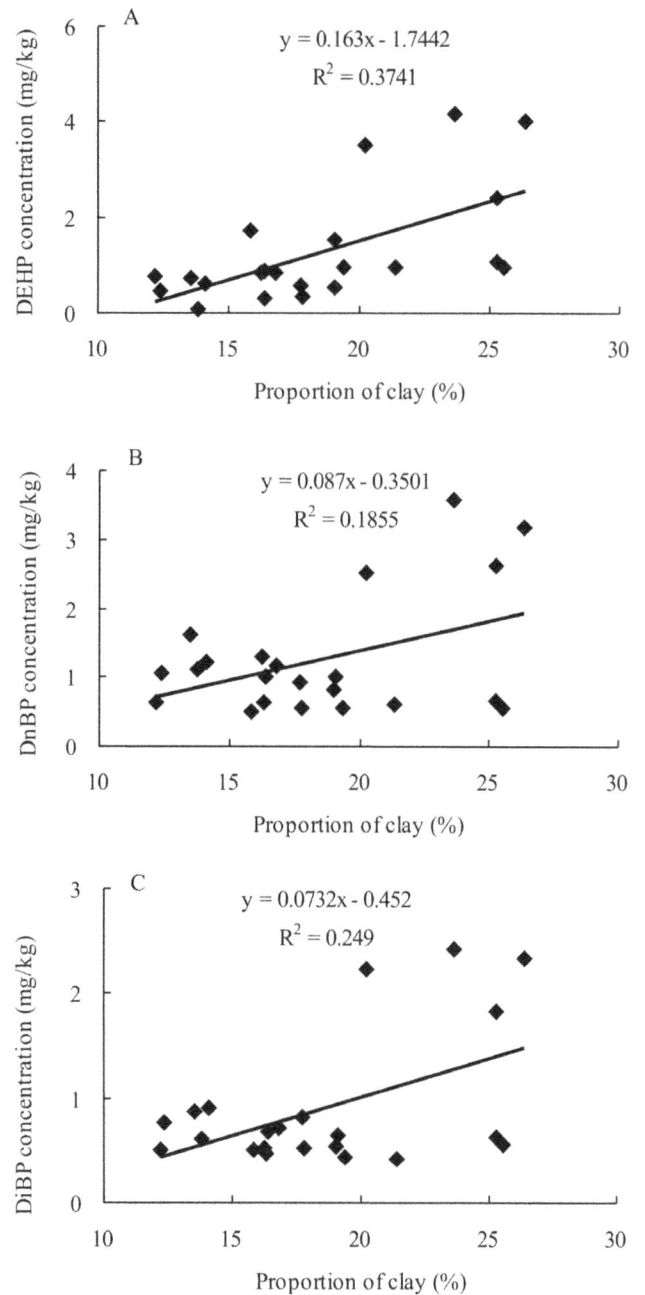

Figure 4. Correlations of the concentrations of (A) DEHP, (B) DnBP and (C) DiBP with the proportions of clay in 0–10 cm soils of vegetable greenhouses from Shandong Peninsula.

layers; by contrast, DEP, DMPP, BBP, and DPhP accounted for only approximately 1% to 5%, suggesting a small proportion.

Discussion

Potential sources of PAEs in soils from vegetable greenhouses

Soils that are being used or were used for vegetable greenhouses had higher PAE concentrations (Table 3), which suggests that PAEs are widespread in soils from vegetable greenhouses. Various PAE sources exist in soils from vegetable greenhouses. The plastic film used in vegetable greenhouses is a major source of PAEs. The

Table 4. Ratio of PAEs in samples exceeding allowable and cleanup concentrations in soils from vegetable greenhouses in Shandong Peninsula.

Soil depth (cm)	Ratio of samples exceeding allowable concentrations (%)						Ratio of samples exceeding cleanup concentrations (%)					
	DMP	DEP	DnBP	BBP	DEHP	DnOP	DMP	DEP	DnBP	BBP	DEHP	DnOP
0~10	94.6	89.2	97.3	0	2.7	45.9	0	0	0	0	0	0
10~20	100	94.6	94.6	2.7	5.4	27	0	0	13.5	0	0	0
20~30	89.2	97.3	89.2	0	0	21.6	0	0	2.7	0	0	0

maximum value of Σ_{16}PAEs in soils at 0 cm to 10 cm and 10 cm to 20 cm appeared in Weifang (Fig. 2), which has a long history of vegetable production and is famous for extensive greenhouse cultivation. Even more remarkably, the plastic film used in greenhouse cultivation in Shandong Peninsula is replaced annually, which may result in a higher concentration of PAEs in soils from vegetable greenhouses than in other soils. In addition, PAEs are found in organic fertilizers in China [26] and in compost of sewage sludge with rice straw [27]. The amount of fertilizers used for vegetable planting in greenhouses is more than that used for field crops, and the proportion of organic fertilizers has increased since 2007 [28]. Moreover, PAEs are found in the groundwater in China [29–31], and groundwater is used for irrigation in vegetable greenhouse, which may result in the buildup of PAEs in vegetable greenhouse soils. More importantly, Zeng et al. [32] found a declining trend of PAEs in agricultural soils that were far from urban centers. The highest PAE contents are found in soils close to architectural markets, where plastic materials are sold, and those close to large chemical manufacturing factories. Most vegetable greenhouses in this study are near industrialized cities with large populations. Over 300 plastic manufacturers that produced 0.3 million tons/yr of plastic existed in Shandong Peninsula by 2003. All these factors may have resulted in the high concentration of PAEs in vegetable greenhouses in Shandong Peninsula.

Among 16 PAEs, DEHP and DnBP are found to be the two most abundant PAEs in agricultural soils in Guangzhou, Shenzhen, Leizhou Peninsula, and Huizhou (Table 3). Moreover, DiBP is found to be abundant in Guangzhou agricultural soils [32], whereas DnOP is abundant in the soils of vegetable greenhouses in Nanjing [14]. Similarly, DEHP and DnOP are the two most abundant PAEs in soils of vegetable greenhouses in Shandong Peninsula, followed by DnBP and DiBP (Fig. 3). The relative contribution of PAEs in agricultural soils is consistent with that in sediment [33], air [5,34], and waters [35]. The global consumption of PAEs is about 6.0 million tones/yr, mainly as plasticizers in the plastic industry. Among plasticizers of PAEs, DEHP, DnOP, and DiBP/DnBP are widely used. It is found that DEHP and DnBP are two dominant PAE components in white and black mulch film used in vegetable production systems, ranging from 48.0~115.6 mg/kg and 2.3~3.2 mg/kg, respectively [14]. We also found that besides DEHP, DnOP and DiBP were two dominant PAEs in polyvinyl chloride (PVC) plastics mainly used in vegetable greenhouses in Shandong Peninsula, accounting for 20% and 10% of total of 16 PAEs, respectively. DMP and DEP are also detected in the plastics film, though their contents are low. Therefore, the plastics film may be a major potential source of some PAEs. Furthermore, PAEs are found in fertilizer and manure. DEHP, DnBP, DMP and DEP are the major organic pollutants in fertilizers, with contents more than

2.5 mg/kg [27]. Similarly, six PAEs (DEHP, DnBP, DnOP, DMP, DEP and BBP) are found in chicken, pig, cow and duck manure, in the range of 2.24~6.84 mg/kg [14]. These potential sources may lead to the high detection rates of DMP, DEP DnBP, DiBP, DEHP and DnOP (Table 2).

Relationship between PAEs and soil properties or age of vegetable greenhouses

Soil properties, such as pH, organic matter, texture, and redox potential, have a certain effect on the migration of hydrophobic organic compounds (HOCs) in soil [36–37]. A positive correlation between HOCs and TOC is found in several research [32,38]; however, it is not in this study. Katsoyiannis [39] reported that no correlation can be found between HOCs and TOC if continuous sources of HOCs exist in soils. In this study, several continuous inputs of PAEs in vegetable greenhouse soils, including plastic film, fertilizers, and irrigation, may hinder the achievement of equilibrium between PAEs and TOCs.

The relationship among the major PAEs, including DEHP, DnBP and DiBP, with the proportions of clay in 0 cm to 10 cm soils was analyzed (Fig. 4). DiBP, DnBP, and DEHP have a significantly positive correlation with the proportion of clay ($r = 0.431$~0.611, $p<0.05$). A similar relationship of HOCs, such as PAEs, organic chlorinated pesticides, and PAHs, with clay is also found in soils [32,40–41]. The clay of sediment or soil shows stronger capability to adsorb HOCs than sand and silt, due to small granulometry but high specific surface area [42–43]. Besides, the aging of organic matter, such as humic material, distributes around clay complexes, resulting in the formation of films of organic material [44]. These films of organic material are very difficult to remove, and so organic matter builds up and becomes a permanent part of the clay complexes. The clay-organic complexes supply rich reactive sites for the adsorption of organic pollutants [45].

A positive correlation between Σ_{16}PAE concentration and age of vegetable greenhouses was found in this study ($r = 0.294$, $p<0.01$), suggesting that PAEs in soils may be related with the cumulative use of potential PAE sources over years in greenhouse vegetable cultivation. However, the correlation coefficient is not high. Studies demonstrate the biodegradability of some PAEs in aerobic condition [46–47], though PAEs are resistant to degradation through hydrolysis and photolysis. The biodegradability of PAE congeners is different. Shanker et al found the degradation rates of DMP and DBP were greater than that of DEHP under aerobic conditions [48]. Additionally, PAEs migrates deeper in soils profiles, and the TOC of soil and volume of leaching water can affect the migration of PAEs [49]. These factors may result in the low correlation coefficient between PAE contents and age of vegetable greenhouses.

Comparison of PAE concentrations in the different soils in China

Comparisons of PAE contents in agricultural soils from China are presented in Table 3. The average Σ_6PAE (DMP, DEP, DnBP, DEHP, BBP, and DnOP) contents in the soils from vegetable greenhouses in Shandong Peninsula, Nanjing, and Hangzhou are approximately 2 mg/kg to 6 mg/kg, higher than other types of soils. High Σ_6PAE concentrations are also found in vegetable soils from Guangzhou and Shenzhen, where soils are previously used to plant greenhouse vegetables. In comparison, the Σ_6PAE concentrations in vegetables, paddy, banana, sugarcane, or orchard soil are low, ranging from 0.2 mg/kg to 1 mg/kg.

Potential risk assessment of soils of vegetable greenhouses from Shandong Peninsula

PAEs have a variety of toxic effects. Long term exposure to PAEs results in decreased fertility in females, fetal defect, altered hormone levels, uterine damage and male reproduction abnormalities such as reduced sperm production and motility, cell damage, cell tumors, etc [52–55]. According to human health based levels that correspond to excess lifetime cancer risks and human health based levels for systemic toxicant calculated from reference doses, allowable soil concentrations and cleanup levels of PAEs have been recommended in New York, USA [56]. The allowable soil concentrations are 0.02, 0.071, 0.081, 1.125, 4.35, and 1.20 mg/kg for DMP, DEP, DnBP, BBP, DEHP, and DnOP, respectively; and soil cleanup levels are 2, 7.1, 8.1, 50, 50, and 50 mg/kg, respectively [56]. PAEs exceeding allowable and cleanup concentrations may be a menace to human health. According to these criteria, the ratios of PAE concentration in this study exceeding allowable and cleanup concentrations are

presented in Table 4. The ratios of DMP, DEP, and DnBP exceeding allowable concentrations are 90% to 100% at different soil depths, suggesting high PAE pollution. Moreover, the ratios of DnOP exceeding allowable concentrations are also high, particularly in soils at 0 cm to 10 cm; however, the ratios of BBP and DEHP are low. Similarly, in agricultural soils around Guangzhou, DMP, DEP, and DnBP also exceed allowable concentrations [32]. Notably, DnBP in some samples is approximately twice to thrice higher than the recommended cleanup concentration. These soil samples are mostly from the vegetable greenhouses with ages of approximately 10 years, suggesting that long-term application of plastic film or manure in vegetable greenhouses may increase environmental and health risks.

The cultivated vegetables can uptake and accumulate PAEs, but the difference is found in accumulated amount of PAE congeners by vegetables. Compared with DEHP, more DBP in soils is accumulated in stalk and leaf of carrot, cucumber and tomato [24]. The physical and chemical properties, such as molecular weight and octanol/water partition coefficient (K_{ow}), have effects on the accumulation of PAEs by plants. Due to the smaller molecular weight and lower K_{ow}, DBP is more easily absorbed and transported by vegetables than DEHP. Furthermore, several studies report a positive correlation between accumulated PAE amount by vegetables and contents in soils [24,57]. Thus, mitigation of PAEs in soils is important to lower the risks of PAEs to human health.

Author Contributions

Conceived and designed the experiments: CC HC YS. Performed the experiments: CC HC. Analyzed the data: HC WG. Contributed reagents/materials/analysis tools: WG DM. Wrote the paper: CC HC YS.

References

1. Staples CA, Peterson DR, Parkerton TF, Adams WJ (1997) The environmental fate of phthalate esters, a literature review. Chemosphere 4: 667–749.
2. Xie Z, Ebinghaus R, Temme C, Lohmann R, Caba A, et al. (2007) Occurrence and air-sea exchange of phthalates in the Arctic. Environmental Science & Technology 41: 4555–4560.
3. Lin C, Lee C, Mao W, Nadim F (2009) Identifying the potential sources of di-(2-ethylhexyl) phthalate contamination in the sediment of the Houjing River in southern Taiwan. Journal of Hazardous Materials 161: 270–275.
4. Buszka PM, Yeskis DJ, Kolpin DW, Furlong ET, Zaugg SD, et al. (2009) Waste-indicator and pharmaceutical compounds in landfill-leachate-affected ground water near Elkhart, Indiana, 2000–2002. Bulletin of Environmental Contamination and Toxicology 82: 653–659.
5. Wang P, Wang SL, Fan CQ (2008) Atmospheric distribution of particulate- and gas-phase phthalic esters (PAEs) in a Metropolitan City, Nanjing, East China. Chemosphere 72(10):1567–1572.
6. Zhang LF, Dong L, Ren LJ, Shi SX, Zhou L, et al. (2012) Concentration and source identification of polycyclic aromatic hydrocarbons and phthalic acid esters in the surface water of the Yangtze River Delta, China. Journal of Environmental Sciences 24(2): 335–342.
7. Bauer MJ, Herrmann R (1997) Estimation of the environmental contamination by phthalic acid esters leaching from household wastes. Science of the Total Environment 208(1–2): 49–57.
8. Amir S, Hafidi M, Merlina G, Hamdi H, Jouraiphy A, et al. (2005) Fate of phthalic acid esters during composting of both lagooning and activated sludges. Process Biochemistry 40(6): 2183–2190.
9. Hens GA, Caballos AMP (2003) Social and economic interest in the control of phthalic acid esters. Trends in Analytical Chemistry 22, 847–857.
10. United States Environmental Protection Agency (USEPA) (2013) Electronic Code of Federal Regulations, Title 40-Protection of Environment, Part 423-Steam Electric Power Generating Point Source Category. Appendix A to Part 423–126, Priority Pollutants. http://www.ecfr.gov/cgi-bin/text-idx?c¼ecfr&SID¼b960051a53c9015d817718d71f1617b7&rgn¼div5&view¼text&node¼40.30.0.1.1.23&idno¼40#40:30.0.1.1.23.0.5.9.9
11. Li ZH, Wang GZ, Qi F (2012) Current Situation and Thinking of Development of Protected Agriculture in China. Chinese Agricultural Mechanization 239(1): 7–10(in Chinese with English abstract).
12. Costa JM, Heuvelink E (2004) Protected cultivation rising in China. Fruit & Vegetable Technology 4: 8–11.
13. Zou ZR (2002) Facility Horticulture Science. China agriculture press, Beijing (in Chinese).
14. Wang J, Luo YM, Teng Y, Ma WT, Christie P, et al. (2013) Soil contamination by phthalate esters in Chinese intensive vegetable production systems with different modes of use of plastic film. Environmental Pollution 180: 265–273.
15. Chen YS, Luo YM, Zhang HB, Song J (2011) Preliminary study on PAEs pollution of greenhouse soils. ACTA Pedologica Sinica 48(3): 516–523(in Chinese with English abstract).
16. Cai QY, Mo CH, Li YH, Zeng QY, Wang BG, et al. (2005) The study of PAEs in soils from typical vegetable fields in areas of Guangzhou and Shenzhen, South China. ACTA Ecologica Sinica 25(2): 283–288(in Chinese with English abstract).
17. Guan H, Wang JS, Wan HF, Li PX, Yang GY (2007) PAEs Pollution in soils from typical agriculture area of Leizhou Peninsula. Journal of Agro-Environment Science 26(2): 622–628(in Chinese with English abstract).
18. Mariko M, Mutsuko HK, Makoto E (2008) Potential adverse effects of phthalic acid esters on human health: A review of recent studies on reproduction. Regulatory Toxicology and Pharmacology 50(1): 37–49.
19. Chou K, Robert OW (2006) Phthalates in food and medical devices. Journal of Medical Toxicology 2: 126–135.
20. Bascomb CL (1964) Rapid method for the determination of cation-exchange capacity of calcareous and non-calcareous soils. Journal of the Science of Food and Agriculture 15(12): 821–823.
21. Soil Survey Division Staff. Soil Survey Manual (1993) In: Agriculture Handbook, Revised Edition, vol. 18. United States Department of Agriculture, Washington DC.
22. Schwartz V (1995) Fractionated combustion analysis of carbon in forest soils - new possibilities for the analysis and characterization of different soils. Fresenius' journal of analytical chemistry 351(7): 629–631.
23. Flowers TH, Bremner JM (1991) A rapid dichromate procedure for routine estimation of total nitrogen in soils. Communications in Soil Science and Plant Analysis 22(13–14): 1409–1416.
24. Wang ML (2007) Research on analytical method and environmental behavior of PAEs in vegetable greenhouse. Shandong Agricultural University. Doctor thesis (in Chinese with English abstract).
25. Zeng F, Cui KY, Xie ZY, Wu L, Luo DL, et al. (2009) Distribution of phthalate esters in urban soils of subtropical city, Guangzhou, China. Journal of Hazardous Materials 164: 1171–1178.

26. Cai QY, Mo CH, Wu QT, Zeng QY, Katsoyiannis A (2007) Quantitative determination of organic priority pollutants in the composts of sewage sludge with rice straw by gas chromatography coupled with mass spectrometry. Journal of Chromatography A 1143: 207–214.

27. Mo CH, Cai QY, Li YH, Zeng QY (2008) Occurrence of priority organic pollutants in the fertilizers, China. Journal of Hazardous Materials 152: 1208–1213.

28. Liu ZH, Jiang LH, Zhang WJ, Zheng FL, Wang M, et al. (2008) Evolution of fertilization rate and variation of soil nutrient contents in greenhouse vegetable cultivation in shandong. ACTA Pedologica Sinica 45 (2): 296–303 (in Chinese with English abstract).

29. Zhang D, Liu H, Liang Y, Wang C, Liang HC, et al. (2009) Distribution of phthalate esters in the groundwater of Jianghan plain, Hubei, China. Frontiers of Earth Science in China 3(1): 73–79.

30. Xiong PX, Gong X, Deng L (2008) Analysis of PAE Pollutants in Farm Soil and Water Samples in Nanchang City. Chemistry 8: 636–640 (in Chinese with English abstract).

31. Wang C, Liu H, Cai HS, Liang Y, Liang HC, et al. (2009) Source Analysis and Detection of Trace Phthalate Esters in Groundwater in Wuhan. Environmental Science & Technology 32(10): 118–123 (in Chinese with English abstract).

32. Zeng F, Cui KY, Xie ZY, Wu LN, Liu M, et al. (2008) Phthalate esters (PAEs): Emerging organic contaminants in agricultural soils in peri-urban areas around Guangzhou, China. Environmental Pollution 156: 425–434.

33. Liu H, Liang HC, Liang Y, Zhang D, Wang C, et al. (2010) Distribution of phthalate esters in alluvial sediment: A case study at JiangHan Plain, Central China. Chemosphere 78(4): 382–388.

34. Zeng F, Lin YJ, Cui KY, Wen JX, Ma YQ, et al. (2010) Atmospheric deposition of phthalate esters in a subtropical city. Atmospheric Environment 44(6): 834–840.

35. He W, Qin N, Kong XZ, Liu WX, He QS, et al. (2013) Spatio-temporal distributions and the ecological and health risks of phthalate esters (PAEs) in the surface water of a large, shallow Chinese lake. Science of The Total Environment 461–462: 672–680.

36. Hitch RK, Day HR (1992) Unusual persistence of DDT in some western USA soils. Bulletin of Environmental Contamination and Toxicology 48: 259–264. http://www.dec.ny.gov/docs/remediation_hudson_pdf/cpsoil.pdf.

37. Cousins IT, Bondi G, Jones KC (1999) Measuring and modelling the vertical distribution of semivolatile organic compounds in soils. I: PCB and PAH soil core data. Chemosphere 39: 2507–2518.

38. Jiang YF (2009) Preliminary study on composition, distribution and source indentification of persistent organix pollutants in soil of Shanghai. Shandong University. Doctor thesis (in Chinese with English abstract).

39. Katsoyiannis A (2006) Occurrence of polychlorinated biphenyls (PCBs) in the Soulou stream in the power generation area of Eordea, northwestern Greece. Chemosphere 65: 1551–1561.

40. Wang L (2013) Pollution characteristics of organochlorine pesticides in Daling River estuary. Dalian Maritime University. Master thesis (in Chinese with English abstract).

41. Chen J, Wang XJ, Tao S (2005) The Influences of soil total organic carbon and clay contents on PAHs vertical distributions in soils in Tianjin area. Research of Environmental Sciences 18(4): 79–83 (in Chinese with English abstract).

42. Amellal N, Portal JM, Berthelin J (2001) Effect of soil structure on the bioavailability of polycyclic aromatic hydrocarbons within aggregates of a contaminated soil. Applied Geochemistry 16: 1611–1619.

43. Benlahcen KT, Chaoui A, Budzinski H, Bellocq J, Garrigues Ph (1997) Distribution and sources of polycyclic aromatic hydrocarbons in some Mediterranean coastal sediments. Marine Pollution Bulletin 34(5): 298–305.

44. Gjessing ET (1976) Physical & Chemical Characteristics of Aquatic Humus. Ann Arbor Science Publishers Inc. (Ann Arbor), Mich.

45. Evans KM, Gill RA, Robotham PWJ (1990) The PAH and organic content of sediment particle size fractions. Water, Air, and Soil Pollution 51: 13–31.

46. Shelton DR, Boyd SA, Tiedje JM (1984) Anaerobic biodegradation of phthalic acid esters in sludge. Environmental Science & Technology 18: 93–97.

47. Ejlertsson J, Meyerson U, Svensson BH (1996) Anaerobic degradation of phthalic acid esters during digestion of municipal solid waste under landfilling conditions. Biodegradation 7: 345–352.

48. Shanker R, Ramakrishna C, Seth PK (1985) Degradation of some phthalic acid esters in soil. Environmental Pollution Series A, Ecological and Biological 39(1):1–7.

49. Wan TT, He GX, Zhang ZH, Zhu L (2013) Simulation on soil column leaching of oxygen nonhydrocarbon migration in soil profiles. Acta Scientiae Circumstantiae 33(10): 2795–2806(in Chinese with English abstract).

50. Yang GY, Zhang TB, Gao ST, Huo ZX, Wan HF, et al. (2007) Distribution of phthalic acid esters in agricultural soils in typical regions of Guangdong Province. Chinese Journal of Applied Ecology 18(10): 2308–2312(in Chinese with English abstract).

51. Tan Z, Li CH, Mo CH (2012) Distribution of Phthalic Acid Esters in Agricultural Soils of Huizhou City. Environmental Science and Management 37(5): 120–123(in Chinese with English abstract).

52. Biscardi D, Monarca S, De Fusco R, Senatore F, Poli P, et al. (2003) Evaluation of the migration of mutagens/carcinogens from PET bottles into mineral water by Tradescantia/micronuclei test, Comet assay on leukocytes and GC/MS. Science of the Total Environment 302:101–108.

53. Sharpe RM, Fisher JS, Millar MM, Jobling S, Sumpter JP (1995) Gestational and lactational exposure of rats to xenoestrogens results in reduced testicular size and sperm production. Environmental Health Perspectives 103:1136–1143.

54. Jones HB, Garside DA, Liu R, Roberts JC (1993) The influence of phthalate-esters on leydig-cell structure and function in-vitro and in-vivo. Experimental and Molecular Pathology 58:179.

55. Giuseppe L, Alberto V, Claudio DF (2004) Di-2-ethylhexyl phthalate and endocrine disruption: A review. Current Drug Targets-Immune, Endocrine & Metabolic Disorders 4: 37–40.

56. Department of Environmental Conservation, New York, USA (1994) Determination of soil cleanup objectives and cleanup levels (TAGM 4046). http://www.dec.ny.gov/regulations/2612.html.

57. Chiou CT, Sheng GY, Manes M (2001) A partition-limited model for the plant uptake of organic contaminants from soil and water. Environmental Science Technology 35: 1437–1444.

8

Household Ventilation May Reduce Effects of Indoor Air Pollutants for Prevention of Lung Cancer: A Case-Control Study in a Chinese Population

Zi-Yi Jin[1,2,9,¶], Ming Wu[1,9,¶], Ren-Qiang Han[1], Xiao-Feng Zhang[3], Xu-Shan Wang[3], Ai-Ming Liu[4], Jin-Yi Zhou[1], Qing-Yi Lu[5], Claire H. Kim[6], Lina Mu[7], Zuo-Feng Zhang[6,‡], Jin-Kou Zhao[1,*,‡]

1 Department of Non-communicable Chronic Disease Control, Jiangsu Provincial Center for Disease Control and Prevention, Nanjing, Jiangsu, China, 2 Jiangyin Center for Disease Control and Prevention, Jiangyin, Jiangsu, China, 3 Ganyu Center for Disease Control and Prevention, Ganyu, Jiangsu, China, 4 Dafeng Center for Disease Control and Prevention, Dafeng, Jiangsu, China, 5 Center for Human Nutrition, David Geffen School of Medicine, University of California Los Angeles (UCLA), Los Angeles, California, United States of America, 6 Department of Epidemiology, Fielding School of Public Health, University of California Los Angeles (UCLA), Los Angeles, California, United States of America, 7 Department of Social and Preventive Medicine, State University of New York at Buffalo, Buffalo, New York, United States of America

Abstract

Background: Although the International Agency for Research on Cancer (IARC) has classified various indoor air pollutants as carcinogenic to humans, few studies evaluated the role of household ventilation in reducing the impact of indoor air pollutants on lung cancer risk.

Objectives: To explore the association between household ventilation and lung cancer.

Methods: A population-based case-control study was conducted in a Chinese population from 2003 to 2010. Epidemiologic and household ventilation data were collected using a standardized questionnaire. Unconditional logistic regression was employed to estimate adjusted odds ratios (OR_{adj}) and their 95% confidence intervals (CI).

Results: Among 1,424 lung cancer cases and 4,543 healthy controls, inverse associations were observed for good ventilation in the kitchen ($OR_{adj} = 0.86$, 95% CI: 0.75, 0.98), bedroom ($OR_{adj} = 0.90$, 95% CI: 0.79, 1.03), and both kitchen and bedroom ($OR_{adj} = 0.87$, 95% CI: 0.75, 1.00). Stratified analyses showed lung cancer inversely associated with good ventilation among active smokers ($OR_{adj} = 0.85$, 95% CI: 0.72, 1.00), secondhand smokers at home ($OR_{adj} = 0.77$, 95% CI: 0.63, 0.94), and those exposed to high-temperature cooking oil fumes ($OR_{adj} = 0.82$, 95% CI: 0.68, 0.99). Additive interactions were found between household ventilation and secondhand smoke at home as well as number of household pollutant sources.

Conclusions: A protective association was observed between good ventilation of households and lung cancer, most likely through the reduction of exposure to indoor air pollutants, indicating ventilation may serve as one of the preventive measures for lung cancer, in addition to tobacco cessation.

Editor: Xiaoping Miao, MOE Key Laboratory of Environment and Health, School of Public Health, Tongji Medical College, Huazhong University of Science and Technology, China

Funding: This project was supported by Jiangsu Provincial Health Department (RC 2003090); was partially supported by the National Institutes of Health, National Institute of Environmental Health Sciences, National Cancer Institute, Department of Health and Human Services, Grants ES06718, ES01167, CA90833, CA077954, CA09142, CA96134, DA11386; and the Alper Research Center for Environmental Genomics of the University of California, Los Angeles Jonsson Comprehensive Cancer Center. The funders had no role in study design, data collection and analysis, decision to publish, or preparation of the manuscript.

Competing Interests: The authors have declared that no competing interests exist.

* Email: jinkouzhao@hotmail.com

9 These authors contributed equally to this work.

¶ Z-YJ and MW are first authors on this work.

‡ Z-FZ and J-KZ also contributed equally to this work and are senior authors on this work.

Introduction

Lung cancer is the leading cause of cancer death in males and the second leading cause of cancer death in females in the world. It accounted for 18.2% of all cancer deaths worldwide in 2008, with an age-standardized mortality rate (ASR) of 19.4 per 100,000 [1,2]. Lung cancer was also the most common cause of cancer

death in China, leading to 452,813 deaths and an ASR of 28.7 per 100,000 in 2008 [3,4].

Risk factors associated with lung cancer have been examined extensively. Sufficient evidence confirmed tobacco smoking (including involuntary or secondhand smoke) is the most critical risk factor for lung cancer [5,6]. Family history of lung cancer and occupational exposure to carcinogens such as asbestos, arsenic and crystalline silica are also established risk factors for lung cancer [7].

A number of indoor air pollutants, such as emissions from household combustion of coal, coal gasification, coke production, and radon-222 and its decay products, have been classified by the International Agency for Research on Cancer (IARC) as lung carcinogens with sufficient evidence in humans [7]. In addition, indoor emissions from household combustion of biomass fuel (primarily wood), and fumes from high-temperature cooking oil have been classified as probably carcinogenic to humans (Group 2A) by the IARC [8]. In China, indoor air pollutants are comprised of secondhand smoke, emissions from solid fuel used for heating and cooking, and cooking oil fumes [9].

Adequate ventilation in the household would reduce the exposure to indoor air pollutants, which in turn will reduce the risk of lung cancer for people who lived in the household. However, only a few studies have examined the protective role of ventilation on lung cancer. A previous retrospective cohort study reported that reduced lung cancer mortality is associated with changing from unvented stoves to portable stoves or stoves with chimneys in both male and female lifetime smoky coal users in Xuanwei, China [10,11]. Two case-control studies of Chinese population reported inverse associations between good ventilation and lung cancer in Guangzhou, Southern China and Taiyuan, Northern China [12,13]. However, those studies did not examine the potential effect modification between ventilation and indoor air pollution.

To evaluate the role of household ventilation in the kitchen and bedroom and the potential lung cancer risk reduction due to indoor air pollution, we conducted a population-based case–control study with a large sample in two counties of Jiangsu Province, China.

Materials and Methods

Study design, participants and data collection

Details of this study have been described in a previous study, which suggested protective association between consumption of raw garlic and lung cancer [14]. In brief, a population-based case-control study was conducted in Dafeng and Ganyu counties in Jiangsu Province, China, from 2003 to 2010. Both counties are economically under-developed, rural areas in the coastal region of northern Jiangsu. Dafeng and Ganyu have approximately 0.7 million and 1.1 million inhabitants respectively. Agricultural population accounts for 64.8% of total local residents in Dafeng and 55.1% in Ganyu. The two counties have very similar lung cancer mortality, with an average of 20.5 per 100,000 from 1996 to 2002 [15].

Eligible cases were newly diagnosed primary lung cancer patients identified from the local population-based tumor registries. Eligible controls were randomly selected from each county's demographic database, matched with cases on gender and age (± 5 years). A total of 1,424 cases (625 in Dafeng and 799 in Ganyu) and 4,543 controls (2,533 in Dafeng and 2,010 in Ganyu) were included in this study. Participation rates were 39.5% and 56.8% for cases and 87% and 85% for controls in Dafeng and Ganyu, respectively.

The Institutional Review Board of Jiangsu Provincial Health Department approved the protocol of this study. Participants read and signed the informed consent form prior to entering the study. Using a standardized epidemiological questionnaire [16], trained interviewers collected data on basic demographic factors, socio-economic status, tobacco smoking history, alcohol consumption, family history of cancers, dietary history, and physical activity. In addition, particular efforts were made to collect data on exposure to household indoor air pollutants, such as secondhand smoke at home, high-temperature cooking oil fumes, cooking fuels (kerosene, coal, natural gas, liquefied gas, coal gas, electricity, firewood and straw), and heating fuels (coal stove, firewood and straw, coke, ondol heating, central heating and electricity). Information regarding ventilation conditions in both the kitchen and the bedroom were also collected through the questionnaire. Ventilation quality was categorized into good, fair, or poor, based on interviewers' observation of household conditions such as the number and size of windows and participants' self-report on usual ventilation. For good, fair and poor ventilation of a house, the average number of windows were 7, 4 and 3, respectively, and the average total size of windows were 10, 6 and 4 square meters, respectively.

Statistical analysis

Firewood and straw were universally used for cooking (91.2% among cases and 93.5% among controls), so they were not considered in the analysis. The other types of cooking fuel were dichotomized into coal (including kerosene) versus others (natural gas, liquefied gas, coal gas, and electricity). Heating fuel was dichotomized into solid fuels (coal stove, firewood and straw, coke and ondol heating) versus others (central heating, electricity, and no heating). Because the proportions of individuals living in households with poorly ventilated kitchens or bedrooms were very low (7.7% and 7.3%, respectively), we combined the categories of "poor" and "fair." Thus, kitchen ventilation and bedroom ventilation were dichotomized into good versus not good. Overall ventilation quality in the household was classified into three categories: good (good ventilation in both the kitchen and the bedroom), fair (good ventilation in either the kitchen or the bedroom) and poor (good ventilation in neither the kitchen nor the bedroom). In the analysis of overall household ventilation in reducing exposure to indoor air pollutants, high-temperature cooking oil and coal used for cooking were included as pollutant sources in the kitchen, solid fuels used for heating was included as a pollutant source in the bedroom, and tobacco smoke from active smoking and secondhand smoke at home were considered pollutant sources in both places. The distributions of basic characteristics of cases and controls were compared using the Chi-squared test. Unconditional logistic regression was used to estimate odds ratios (ORs) and their 95% confidence intervals (CIs) for the associations between variables related to indoor air pollution and lung cancer. Joint effects of household ventilation and indoor air pollutants were evaluated. For multiplicative interaction, ratio of ORs (ROR) was generated by including main effect variables and their product terms in a logistic regression model. Additive interaction was examined based on three measures – the relative excess risk due to interaction (RERI), attributable proportion due to interaction (AP), and synergy index (S) [17–19]. The stratum with the lowest risk served as the reference category [20]. RERI and AP are equal to 0 and S is equal to 1 in the absence of additive interaction. Multivariate models were adjusted for potential confounders, which were selected based on prior knowledge and confounding assessment; these included age (continuous), gender, education level (illiterate/primary school/middle school/high school/college), income 10 years ago (Yuan/year, continuous), body mass index (BMI, kg/m^2, continuous), family history of lung cancer (yes/no), pack-years of smoking (continuous), ethanol consumption (ml/week, continuous) and study area (Dafeng/Ganyu). We also examined the associations after stratifying the data by county, ever smoking and gender. Epidata 3.0 (EpiData Association, Denmark) was used for data entry and SAS v9.2 (SAS Institute, Inc., Cary, NC) was used for data cleaning and analysis.

Table 1. Characteristics of cases and controls.

Variables	Male		P-value [a]	Female		P-value [a]
	Control (%) (N = 3,415)	Case (%) (N = 995)		Control (%) (N = 1,128)	Case (%) (N = 429)	
Study area						
Dafeng	1,815 (53.1)	442 (44.4)		718 (63.7)	183 (42.7)	
Ganyu	1,600 (46.9)	553 (55.6)	<0.001	410 (36.3)	246 (57.3)	<0.001
Age (year)						
<50	374 (11.0)	106 (10.7)		125 (11.1)	60 (14.0)	
50–	783 (22.9)	233 (23.4)		206 (18.3)	95 (22.1)	
60–	1,080 (31.6)	330 (33.2)		360 (31.9)	138 (32.2)	
70–	962 (28.2)	268 (26.9)		347 (30.8)	119 (27.7)	
≥80	216 (6.3)	58 (5.8)	0.837	90 (8.0)	17 (4.0)	0.012
Education level						
Illiteracy	1,441 (42.2)	370 (37.2)		868 (77.0)	336 (78.3)	
Primary	1,217 (35.6)	399 (40.1)		176 (15.6)	67 (15.6)	
Middle & above	757 (22.2)	226 (22.7)	0.011	84 (7.4)	26 (6.1)	0.631
Income 10 years ago (Yuan/year)						
<1000	734 (21.5)	202 (20.3)		225 (19.9)	82 (19.1)	
1000–	613 (18.0)	180 (18.1)		230 (20.4)	86 (20.0)	
1500–	925 (27.1)	262 (26.3)		309 (27.4)	118 (27.5)	
≥2500	1,143 (33.5)	351 (35.3)	0.699	364 (32.3)	143 (33.3)	0.971
BMI (kg/m²)[b]						
<18.5	209 (6.1)	125 (12.6)		98 (8.7)	79 (18.4)	
18.5–23.9	2,236 (65.5)	630 (63.3)		645 (57.2)	232 (54.1)	
24–27.9	800 (23.4)	196 (19.7)		303 (26.9)	89 (20.7)	
≥28	170 (5.0)	44 (4.4)	<0.001	82 (7.3)	29 (6.8)	<0.001
Family history of lung cancer						
No	3,342 (97.9)	949 (95.4)		1,091 (96.7)	407 (94.9)	
Yes	73 (2.1)	46 (4.6)	<0.001	37 (3.3)	22 (5.1)	0.088
Pack-years of smoking						
Never smoker	1,013 (29.7)	100 (10.1)		847 (75.1)	294 (68.5)	
<30 years	908 (26.6)	196 (19.7)		180 (16.0)	66 (15.4)	
≥30 years	1,494 (43.7)	699 (70.3)	<0.001	101 (9.0)	69 (16.1)	<0.001
Ethanol consumption						
Never	1,036 (30.3)	257 (25.8)		901 (79.9)	352 (82.1)	
<500 ml/week	1,035 (30.3)	270 (27.1)		193 (17.1)	55 (12.8)	
≥500 ml/week	1,344 (39.4)	468 (47.0)	<0.001	34 (3.0)	22 (5.1)	0.022

[a]Based on Chi-squared tests.
[b]Chinese recommend standard was used for the cut-off points of overweight and obesity: underweight (BMI <18.5), normal weight (BMI 18.5–23.9), overweight (BMI 24.0–27.9), obese (BMI≥28.0).

Results

Table 1 shows the characteristics of cases and controls. Among both males and females, cases were more likely than controls to be underweight (BMI <18.5 kg/m²) and to smoke cigarettes. Among males, cases were more likely than controls to have an education level of primary school or above, have a family history of lung cancer, and to consume alcohol. Among females, cases tended to be younger and less likely to consume alcohol than controls.

The distributions of factors related to indoor air pollution and their associations with lung cancer are presented in Table 2. After adjusting for potential confounding factors, an inverse association was found between lung cancer and good ventilation in the kitchen ($OR_{adj} = 0.86$, 95% CI: 0.75, 0.98), and a borderline inverse association was observed for good ventilation in the bedroom ($OR_{adj} = 0.90$, 95% CI: 0.79, 1.03). Compared with poor ventilation in both the kitchen and the bedroom, good ventilation in both was inversely associated with lung cancer ($OR_{adj} = 0.87$, 95% CI: 0.75, 1.00). On the other hand, positive associations were observed between lung cancer and exposure to secondhand smoke at home ($OR_{adj} = 1.41$, 95% CI: 1.24, 1.60), high-temperature cooking oil vapor ($OR_{adj} = 1.26$, 95% CI: 1.10, 1.43), coal used for cooking ($OR_{adj} = 1.38$, 95% CI: 1.17, 1.62) and solid fuels used for heating ($OR_{adj} = 1.27$, 95% CI: 1.10, 1.47). When treating the cumulative number of pollutant sources as a continuous variable in logistic regression model, it was positively associated with lung cancer in a dose-response manner ($OR_{adj} = 1.29$, 95% CI: 1.21, 1.38) for each additional indoor pollution source. Compared with no pollutant source, OR for three or four pollutant sources and one or two pollutant sources was 1.97 (95% CI: 1.48, 2.63) and 3.01 (95% CI: 2.22, 4.08), respectively.

Table 3 shows the joint effects of household ventilation and indoor air pollutants on the risk of lung cancer. In the presence of each type of indoor air pollutant, good ventilation was inversely associated with lung cancer for ever smoking ($OR_{adj} = 0.85$, 95% CI: 0.72, 1.00), secondhand smoke at home ($OR_{adj} = 0.77$, 95% CI: 0.63, 0.94), high-temperature cooking oil fumes ($OR_{adj} = 0.82$, 95% CI: 0.68, 0.99), coal used for cooking ($OR_{adj} = 0.89$, 95% CI: 0.72, 1.11), and solid fuels used for heating ($OR_{adj} = 0.94$, 95% CI: 0.81, 1.10). There was some indication of additive interaction between household ventilation and secondhand smoke at home ($RERI = -0.17$, 95% CI: -0.35, 0.01; $AP = -0.13$, 95% CI: -0.26, 0.00; $S = 0.67$, 95% CI: 0.47, 0.95).

Table 4 presents the joint effects of household ventilation and the number of pollutant sources. Compared with living in a household with poor ventilation and three or four sources of

indoor air pollutants, living in a household with good ventilation was associated with a decreased risk of lung cancer, in a dose-response manner relative to the number of pollutant sources ($OR_{adj} = 0.72$, 95% CI $= 0.56$, 0.94 for three or four pollutant sources; $OR_{adj} = 0.54$, 95% CI $= 0.44$, 0.67 for one or two pollutant sources; $OR_{adj} = 0.24$, 95% CI $= 0.15$, 0.38 for no pollutant source). An additive interaction was suggested between household ventilation and number of pollutant sources ($RERI = -0.11$, 95% CI: -0.24, 0.02; $AP = -0.07$, 95% CI: -0.14, 0.00); $S = 0.85$, 95% CI: 0.74, 0.98). Among individuals exposed to the same number of pollutant sources, those who lived in households with good ventilation were generally at a lower risk for lung cancer compared with those who lived in households with poor ventilation.

Discussion

In this large population-based case-control study, good ventilation in the kitchen and bedroom was inversely associated with lung cancer risk. Exposure to secondhand smoke at home, high-temperature cooking oil fumes, coal used for cooking or solid fuels

Table 2. Distribution of factors related to indoor air pollution and their associations with lung cancer risk.

Variables	Control (%) (N = 4,543)	Case (%) (N = 1,424)	Crude OR (95%CI)	Adjusted OR (95%CI)[a]
Good ventilation in kitchen				
No	2,618 (57.6)	887 (62.3)	1.00	1.00
Yes	1,925 (42.4)	537 (37.7)	0.82 (0.73,0.93)	0.86 (0.75,0.98)
Good ventilation in bedroom				
No	2,604 (57.3)	873 (61.3)	1.00	1.00
Yes	1,939 (42.7)	551 (38.7)	0.85 (0.75,0.96)	0.90 (0.79,1.03)
Household ventilation				
Poor	2,411 (53.1)	821 (57.7)	1.00	1.00
Fair	400 (8.8)	118 (8.3)	0.87 (0.70,1.08)	0.85 (0.68,1.07)
Good	1,732 (38.1)	485 (34.1)	0.82 (0.72,0.94)	0.87 (0.75,1.00)
Secondhand smoke at home				
No	2,742 (60.4)	688 (48.3)	1.00	1.00
Yes	1,801 (39.6)	736 (51.7)	1.63 (1.45,1.84)	1.41 (1.24,1.60)
High-temperature cooking oil				
No	3,194 (70.3)	911 (64.0)	1.00	1.00
Yes	1,349 (29.7)	513 (36.0)	1.33 (1.18,1.51)	1.26 (1.10,1.43)
Coal used for cooking				
No	3,317 (73.0)	868 (61.0)	1.00	1.00
Yes	1,226 (27.0)	556 (39.0)	1.73 (1.53,1.96)	1.38 (1.17,1.62)
Solid fuels used for heating				
No	1,238 (27.3)	348 (24.4)	1.00	1.00
Yes	3,305 (72.7)	1,076 (75.6)	1.16 (1.01,1.33)	1.27 (1.10,1.47)
Number of pollutant sources				
0	442 (9.7)	60 (4.2)	1.00	1.00
1 or 2	3,244 (71.4)	914 (64.2)	2.08 (1.57,2.75)	1.97 (1.48,2.63)
3 or 4	857 (18.9)	450 (31.6)	3.87 (2.89,5.18)	3.01 (2.22,4.08)
P_{trend}			<0.001	<0.001
OR (continuous)			1.43 (1.34,1.52)	1.29 (1.21,1.38)

[a]Adjusted for age (continuous), gender, education level (illiterate/primary school/middle school/high school/college), income 10 years ago (Yuan/year, continuous), body mass index (kg/m², continuous), family history of lung cancer (yes/no), pack-years of smoking (continuous), ethanol consumption (ml/week, continuous), and study area (Dafeng/Ganyu).

Table 3. Joint effects of household ventilation and indoor air pollutants on lung cancer risk.

Pollutant source	Household ventilation[a]	Control (%) (N = 4,543)	Case (%) (N = 1,424)	Crude OR (95%CI)	Adjusted OR (95%CI)[b]
Ever smoking					
Yes	Poor [c]	1,466 (32.3)	605 (42.5)	1.00	1.00
Yes	Fair	248 (5.5)	84 (5.9)	0.82 (0.63,1.07)	0.77 (0.59,1.01)
Yes	Good	969 (21.3)	341 (23.9)	0.85 (0.73,1.00)	0.85 (0.72,1.00)
No	Poor	945 (20.8)	216 (15.2)	0.55 (0.47,0.66)	0.39 (0.32,0.48)
No	Fair	152 (3.3)	34 (2.4)	0.54 (0.37,0.80)	0.37 (0.24,0.55)
No	Good [d]	763 (16.8)	144 (10.1)	0.46 (0.37,0.56)	0.33 (0.26,0.41)
Interaction [b]	Additive: RERI = −0.13 (−0.37,0.11); AP = −0.05 (−0.16,0.05);				
	S = 0.91 (0.78,1.07);				
	Multiplicative: ROR = 1.00 (0.87,1.15)				
Secondhand smoke at home					
Yes	Poor [c]	957 (21.1)	440 (30.9)	1.00	1.00
Yes	Fair	183 (4.0)	68 (4.8)	0.81 (0.60,1.09)	0.84 (0.62,1.15)
Yes	Good	661 (14.5)	228 (16.0)	0.75 (0.62,0.91)	0.77 (0.63,0.94)
No	Poor	1,454 (32.0)	381 (26.8)	0.57 (0.49,0.67)	0.66 (0.56,0.78)
No	Fair [d]	217 (4.8)	50 (3.5)	0.50 (0.36,0.70)	0.54 (0.38,0.76)
No	Good	1,071 (23.6)	257 (18.0)	0.52 (0.44,0.62)	0.64 (0.52,0.77)
Interaction [b]	Additive: RERI = −0.17 (−0.35,0.01); AP = −0.13 (−0.26,0.00);				
	S = 0.67 (0.47,0.95);				
	Multiplicative: ROR = 0.89 (0.78,1.02)				
High-temperature cooking oil					
Yes	Not good [c]	755 (16.6)	313 (22.0)	1.00	1.00
Yes	Good	594 (13.1)	200 (14.0)	0.79 (0.67,0.95)	0.82 (0.68,0.99)
No	Not good	1,863 (41.0)	574 (40.3)	0.58 (0.50,0.68)	0.78 (0.66,0.93)
No	Good [d]	1,331 (29.3)	337 (23.7)	0.54 (0.45,0.63)	0.67 (0.56,0.81)
Interaction [b]	Additive: RERI = 0.08 (−0.26,0.41); AP = 0.05 (−0.17,0.27);				
	S = 1.18 (0.54,2.62);				
	Multiplicative: ROR = 0.98 (0.75,1.28)				
Coal used for cooking					
Yes	Not good [c]	743 (16.4)	353 (24.8)	1.00	1.00
Yes	Good	483 (10.6)	203 (14.3)	0.89 (0.72,1.09)	0.89 (0.72,1.11)
No	Not good	1,875 (41.3)	534 (37.5)	0.60 (0.51,0.70)	0.74 (0.61,0.90)
No	Good [d]	1,442 (31.7)	334 (23.5)	0.49 (0.41,0.58)	0.62 (0.50,0.77)
Interaction [b]	Additive: RERI = −0.02 (−0.39,0.35); AP = −0.01 (−0.24,0.21);				
	S = 0.97 (0.54,1.73);				
	Multiplicative: ROR = 1.07 (0.82,1.39)				
Solid fuel used for heating					
Yes	Not good [c]	1,942 (42.7)	663 (46.6)	1.00	1.00
Yes	Good	1,363 (30.0)	413 (29.0)	0.89 (0.77,1.02)	0.94 (0.81,1.10)
No	Not good	662 (14.6)	210 (14.7)	0.93 (0.78,1.11)	0.85 (0.70,1.02)
No	Good [d]	576 (12.7)	138 (9.7)	0.70 (0.57,0.86)	0.68 (0.55,0.85)
Interaction [b]	Additive: RERI = −0.16 (−0.52,0.21); AP = −0.11 (−0.35,0.13);				
	S = 0.75 (0.43,1.29);				
	Multiplicative: ROR = 1.17 (0.87,1.57)				

[a]High-temperature cooking oil and coal used for cooking were included in kitchen ventilation, solid fuels used for heating was included in bedroom ventilation while ever smoking and secondhand smoke at home were included in both kitchen and bedroom ventilation.
[b]Adjusted for age (continuous), gender, education level (illiterate/primary school/middle school/high school/college), income 10 years ago (Yuan/year, continuous), body mass index (kg/m^2, continuous), family history of lung cancer (yes/no), pack-years of smoking (continuous), ethanol consumption (ml/week, continuous), and study area (Dafeng/Ganyu).
[c]The joint effects category for further estimation of additive interaction.
[d]The reference category for measures of interaction on additive scale.

Table 4. Joint effects of household ventilation and number of pollutant sources on lung cancer risk.

Number of pollutant sources	Household ventilation	Control (%) (N = 4,543)	Case (%) (N = 1,424)	Crude OR (95%CI)	Adjusted OR (95%CI) [a]
3 or 4	Poor [b]	435 (9.6)	273 (19.2)	1.00	1.00
3 or 4	Fair	109 (2.4)	40 (2.8)	0.59 (0.40,0.87)	0.60 (0.40,0.91)
3 or 4	Good	313 (6.9)	137 (9.6)	0.70 (0.54,0.90)	0.72 (0.56,0.94)
1 or 2	Poor	1,781 (39.2)	517 (36.3)	0.46 (0.39,0.55)	0.57 (0.47,0.69)
1 or 2	Fair	272 (6.0)	74 (5.2)	0.43 (0.32,0.58)	0.52 (0.38,0.71)
1 or 2	Good	1,191 (26.2)	323 (22.7)	0.43 (0.36,0.53)	0.54 (0.44,0.67)
0	Poor	195 (4.3)	31 (2.2)	0.25 (0.17,0.38)	0.32 (0.21,0.48)
0	Fair	19 (0.4)	4 (0.3)	0.34 (0.11,1.00)	0.40 (0.13,1.23)
0	Good [c]	228 (5.0)	25 (1.8)	0.18 (0.11,0.27)	0.24 (0.15,0.38)
Interaction [a]	Additive: RERI = −0.11 (−0.24,0.02); AP = −0.07 (−0.14,0.00);				
	S = 0.85 (0.74,0.98);				
	Multiplicative: ROR = 0.93 (0.82,1.05)				

[a]Adjusted for age (continuous), gender, education level (illiterate/primary school/middle school/high school/college), income 10 years ago (Yuan/year, continuous), body mass index (kg/m2, continuous), family history of lung cancer (yes/no), pack-years of smoking (continuous), ethanol consumption (ml/week, continuous), and study area (Dafeng/Ganyu).
[b]The joint effects category for further estimation of additive interaction.
[c]The reference category for measures of interaction on additive scale.

used for heating were found to be positively associated with lung cancer. A strong dose-response pattern was observed between the number of pollutant sources and the risk of lung cancer. Additive interactions were observed between household ventilation and indoor air pollutants. Thus, our results suggest that adequate ventilation in household could reduce the exposure to multiple co-existing indoor air pollutants.

Our findings of an inverse association between good ventilation and lung cancer are consistent with the results reported in two previous studies [12,13]. One hospital-based case-control study of 224 male and 92 female cases of lung cancer in Guangzhou Province, China, reported an increased lung cancer risk associated with no separate kitchen and poor air circulation in both males and females. The OR for lung cancer tended to decrease with increasing size of ventilation openings in living areas and kitchens [12]. Similarly, a population-based case-control study with 164 cases and 218 controls of female non-smokers in Taiyuan city, China, found that good ventilation conditions, such as multi-story houses and houses with more windows, separate kitchens, installed ventilators, and frequently open windows showed strong inverse associations with lung cancer risk [13]. However, these studies were not able to evaluate the potential effect modification of lung cancer risk between good ventilation and indoor air pollutants due to small sample sizes.

The indoor air pollutants included in our study were all associated with an increased lung cancer risk. However, the protective effects of household ventilation for lung cancer risk were different by the type of indoor air pollutants (Table 3). Second-hand smokers at home (OR_{adj} = 0.77) benefited most from good household ventilation, while exposure to high-temperature cooking oil fumes (OR_{adj} = 0.82) and active smokers (OR_{adj} = 0.85) also had benefited from good ventilation. Those exposed to coal used for cooking and solid fuels used for heating had borderline inverse associations with good ventilation. Smoking, especially second-hand smoke, is one of the major sources of indoor air pollutants in China [21]. Both smokers and non-smokers are often exposed to dense smoke in small and crowded kitchens and bedrooms without

proper ventilation. At least 17 carcinogens are emitted in higher levels from sidestream smoke than from mainstream smoke, of which benzo(a)pyrene diol epoxide is directly associated with lung cancer [22–26]. Adequate ventilation both in the kitchen and in the bedroom could mitigate the positive association between lung cancer and smoking, including secondhand smoke.

Chinese home cooking often involves the use of cooking oil at high-temperature [27]. Consistent with the majority of epidemiologic studies [28–32], the present study also found a positive association between exposure to cooking oil fumes and the risk of lung cancer, particularly in Chinese women (OR_{adj} = 1.21, 95% CI: 1.03, 1.42, and OR_{adj} = 1.43, 95% CI: 1.12, 1.83 for men and women, respectively). Our study also found the positive association of exposure to high-temperature cooking oil fumes with lung cancer risk can be modified by good household ventilation.

Indoor emissions from household coal combustion contain high levels of carcinogenic polycyclic aromatic hydrocarbons (PAHs). Consistent with the findings in many previous studies and pooled analyses [33–38], the association between household coal combustion and lung cancer was confirmed in our study. Moreover, we further analyzed the relationship between years of coal used for cooking and lung cancer, a clear dose–response pattern was observed (P_{trend} <0.001). However, household ventilation was not observed to reduce the impact of coal used for cooking and solid fuels used for heating on lung cancer risk in the present analysis.

This study has certain limitations. First, as with most case-control studies, selection bias and recall bias may exist in our study. To minimize selection bias, all cases were recruited directly from the tumor registries while controls were selected from the local population databases. Since household ventilation conditions in the kitchen and bedroom are not well established protective factors for lung cancer, differential recall bias would have been minimal. However, our results might be conservative due to non-differential recall or interview bias, which might lead to observed associations biased toward the null. Second, the ventilation effect was collected by interview using an epidemiological questionnaire. Direct measurements might give more quantitative data, however it may also suffer misclassification bias because it can only represent the current exposure when the cases and controls were interviewed without considering the latent time for cancer development. Third, the data were not collected on the duration the participants lived in the same households and household renovations. Further analyses on the association of the duration with the risk of lung cancer development and when to ventilate were not possible. Fourth, outdoor air pollution may also affect indoor air quality, which in term may affect lung cancer risk. However, since both study sites are located in countryside, the less air-polluted areas. The potential impact of outdoor air pollution on indoor would be limited. Fifth, most of the cases were diagnosed at an advanced stage without surgical treatment, resulting in a relatively low participation rate (46.3%) and a low proportion of pathologic diagnosis (17%). Finally, uncontrolled confounding may have affected our results. In particular, we only examined the effects of air pollutants inside the household without considering environmental and occupational exposures outside the household, particularly for male workers. However, there were no clear differences between exposure variables and lung cancer among the subgroups after stratifying by ever smoking and gender, except for differences in CIs due to smaller sample size caused by stratification. Moreover, the lack of obvious differences between the crude ORs and the adjusted ORs indicate that the impact of potential confounding factors was limited. Despite these limita-

tions, the present study is one of the largest population-based case-control studies to evaluate the association between household ventilation and lung cancer and the potential impact of good ventilation in reducing the exposure to indoor air pollutants as a preventive measure against lung cancer.

In conclusion, adequate ventilation in the household was observed to be inversely associated with lung cancer, most likely through the reduction of exposure to indoor air pollutants. The findings from this study suggest that household ventilation, in addition to tobacco cessation, should be considered as one of the public health measures for the prevention and control of lung cancer in the Chinese population.

Acknowledgments

The authors thank the subjects for their voluntary participation as well as the staff of local Health Bureaus and local CDCs in Dafeng and Ganyu County for their assistance in the field work.

Author Contributions

Conceived and designed the experiments: MW Z-FZ J-KZ. Performed the experiments: X-FZ X-SW A-ML. Analyzed the data: Z-YJ. Contributed reagents/materials/analysis tools: Z-YJ R-QH X-FZ X-SW A-ML J-YZ. Wrote the paper: Z-YJ Q-YL CHK LM Z-FZ J-KZ.

References

1. Ferlay J, Shin HR, Bray F, Forman D, Mathers C, et al. (2010) Estimates of worldwide burden of cancer in 2008: GLOBOCAN 2008. Int J Cancer 127: 2893–2917.
2. Jemal A, Bray F, Center MM, Ferlay J, Ward E, et al. (2011) Global cancer statistics. CA Cancer J Clin 61: 69–90.
3. Hao J, Zhao P, Chen WQ (2011) Chinese cancer registry annual report. Beijing: Military Medical Science Press. 62 p.
4. IARC. Cancer incidence, mortality and prevalence worldwide in 2008. Globocan 2008.
5. Dela CCS, Tanoue LT, Matthay RA (2011) Lung cancer: epidemiology, etiology, and prevention. Clin Chest Med 32: 605–644.
6. Hecht SS (1999) Tobacco smoke carcinogens and lung cancer. J Natl Cancer Inst 91: 1194–1210.
7. Travis W, Brambilla E, Muller-Hermelink H, Harris C (2004) World Health Organization Classification of Tumours. Pathology and genetics of tumours of the lung, pleura, thymus and heart. Lyon, France: IARC Press.
8. IARC Working Group on the Evaluation of Carcinogenic Risks to Humans (2010) Household use of solid fuels and high-temperature frying. IARC Monogr Eval Carcinog Risks Hum 95: 1–430.
9. Gao YT (1996). Risk factors for lung cancer among nonsmokers with emphasis on lifestyle factors. Lung cancer 14 Suppl 1:S39–45.
10. Hosgood HD 3rd, Chapman R, Shen M, Blair A, Chen E, et al. (2008) Portable stove use is associated with lower lung cancer mortality risk in lifetime smoky coal users. Br J Cancer 99: 1934–1939.
11. Lan Q, Chapman RS, Schreinemachers DM, Tian L, He X (2002) Household stove improvement and risk of lung cancer in Xuanwei, China. J Natl Cancer Inst 94: 826–835.
12. Liu Q, Sasco AJ, Riboli E, Hu MX (1993) Indoor air pollution and lung cancer in Guangzhou, People's Republic of China. Am J Epidemiol 137: 145–154.
13. Mu L, Liu L, Niu R, Zhao B, Shi J, et al. (2013) Indoor air pollution and risk of lung cancer among Chinese female non-smokers. Cancer Causes Control 24: 439–450.
14. Jin ZY, Wu M, Han RQ, Zhang XF, Wang XS, et al. (2013) Raw garlic consumption as a protective factor for lung cancer, a population-based case-control study in a Chinese population. Cancer Prev Res (Phila) 6: 711–718.
15. Zhao JK, Liu AM, Wang XS, Wu M, Sheng LG, et al. (2004) An analysis on death cause of cancer in high and low Incidence areas of Jiangsu province. Zhong guo zhong liu 13: 757–759.
16. Mu LN, Lu QY, Yu SZ, Jiang QW, Cao W, et al. (2005) Green tea drinking and multigenetic index on the risk of stomach cancer in a Chinese population. Int J Cancer 116: 972–983.
17. Hosmer DW, Lemeshow S (1992) Confidence interval estimation of interaction. Epidemiology 3: 452–456.
18. Andersson T, Alfredsson L, Kallberg H, Zdravkovic S, Ahlbom A (2005) Calculating measures of biological interaction. Eur J Epidemiol 20: 575–579.
19. Knol MJ, van der Tweel I, Grobbee DE, Numans ME, Geerlings MI (2007) Estimating interaction on an additive scale between continuous determinants in a logistic regression model. Int J Epidemiol 36: 1111–1118.
20. Knol MJ, VanderWeele TJ, Groenwold RH, Klungel OH, Rovers MM, et al. (2011) Estimating measures of interaction on an additive scale for preventive exposures. Eur J Epidemiol 26: 433–438.
21. Oberg M, Jaakkola MS, Woodward A, Peruga A, Pruss-Ustun A (2011) Worldwide burden of disease from exposure to second-hand smoke: a retrospective analysis of data from 192 countries. Lancet 377: 139–146.
22. Mohtashamipur E, Mohtashamipur A, Germann PG, Ernst H, Norpoth K, et al. (1990) Comparative carcinogenicity of cigarette mainstream and sidestream smoke condensates on the mouse skin. J Cancer Res Clin Oncol 116: 604–608.
23. Denissenko MF, Pao A, Tang M, Pfeifer GP (1996) Preferential formation of benzo[a]pyrene adducts at lung cancer mutational hotspots in P53. Science 274: 430–432.
24. Boffetta P (2002) Involuntary smoking and lung cancer. Scand J Work Environ Health 28 Suppl 2: 30–40.
25. Taylor R, Najafi F, Dobson A (2007) Meta-analysis of studies of passive smoking and lung cancer: effects of study type and continent. Int J Epidemiol 36: 1048–1059.
26. Hoh E, Hunt RN, Quintana PJ, Zakarian JM, Chatfield DA, et al. (2012) Environmental tobacco smoke as a source of polycyclic aromatic hydrocarbons in settled household dust. Environ Sci Technol 46: 4174–4183.
27. Ko YC, Cheng LS, Lee CH, Huang JJ, Huang MS, et al. (2000) Chinese food cooking and lung cancer in women nonsmokers. Am J Epidemiol 151: 140–147.
28. Yu IT, Chiu YL, Au JS, Wong TW, Tang JL (2006) Dose-response relationship between cooking fumes exposures and lung cancer among Chinese nonsmoking women. Cancer Res 66: 4961–4967.
29. Li M, Yin Z, Guan P, Li X, Cui Z, et al. (2008) XRCC1 polymorphisms, cooking oil fume and lung cancer in Chinese women nonsmokers. Lung cancer 62: 145–151.
30. Wang XR, Chiu YL, Qiu H, Au JS, Yu IT (2009) The roles of smoking and cooking emissions in lung cancer risk among Chinese women in Hong Kong. Ann Oncol 20: 746–751.
31. Lee CH, Yang SF, Peng CY, Li RN, Chen YC, et al. (2010) The precancerous effect of emitted cooking oil fumes on precursor lesions of cervical cancer. Int J Cancer 127: 932–941.
32. Lee T, Gany F (2013) Cooking oil fumes and lung cancer: a review of the literature in the context of the U.S. population. J Immigr Minor Health 15: 646–652.
33. Lan Q, He X, Costa DJ, Tian L, Rothman N, et al. (2000) Indoor coal combustion emissions, GSTM1 and GSTT1 genotypes, and lung cancer risk: a case-control study in Xuan Wei, China. Cancer Epidemiol Biomarkers Prev 9: 605–608.
34. Lissowska J, Bardin-Mikolajczak A, Fletcher T, Zaridze D, Szeszenia-Dabrowska N, et al. (2005) Lung cancer and indoor pollution from heating and cooking with solid fuels: the IARC international multicentre case-control study in Eastern/Central Europe and the United Kingdom. Am J Epidemiol 162: 326–333.
35. Zhao Y, Wang S, Aunan K, Seip HM, Hao J (2006) Air pollution and lung cancer risks in China–a meta-analysis. Sci Total Environ 366: 500–513.

36. Zhang JJ, Smith KR (2007) Household air pollution from coal and biomass fuels in China: measurements, health impacts, and interventions. Environ Health Perspect 115: 848–855.

37. Galeone C, Pelucchi C, La Vecchia C, Negri E, Bosetti C, et al. (2008) Indoor air pollution from solid fuel use, chronic lung diseases and lung cancer in Harbin, Northeast China. Eur J Cancer Prev 17: 473–478.

38. Hosgood HD 3rd, Boffetta P, Greenland S, Lee YC, McLaughlin J, et al. (2010) In-home coal and wood use and lung cancer risk: a pooled analysis of the International Lung Cancer Consortium. Environ Health Perspect 118: 1743–1747.

9

Fetal Exposure to Perfluorinated Compounds and Attention Deficit Hyperactivity Disorder in Childhood

Amanda Ode[1]*, Karin Källén[1], Peik Gustafsson[2], Lars Rylander[1], Bo A. G. Jönsson[1], Per Olofsson[3], Sten A. Ivarsson[4], Christian H. Lindh[1], Anna Rignell-Hydbom[1]

1 Division of Occupational and Environmental Medicine, Lund University, Lund, Sweden, **2** Child and Adolescent Psychiatry Unit, Department of Clinical Sciences, Lund University, Lund, Sweden, **3** Obstetrics and Gynecology Unit, Department of Clinical Sciences, Skåne University Hospital, Lund University, Malmö, Sweden, **4** Department of Clinical Sciences, Unit of Pediatric Endocrinology, Lund University/Clinical Research Centre (CRC), Malmö, Sweden

Abstract

Background: The association between exposure to perfluorinated compounds (PFCs) and attention deficit hyperactivity disorder (ADHD) diagnosis has been sparsely investigated in humans and the findings are inconsistent.

Objectives: A matched case-control study was conducted to investigate the association between fetal exposure to PFCs and ADHD diagnosis in childhood.

Methods: The study base comprised children born in Malmö, Sweden, between 1978 and 2000 that were followed up until 2005. Children with ADHD (n = 206) were identified at the Department of Child and Adolescent Psychiatry. Controls (n = 206) were selected from the study base and were matched for year of birth and maternal country of birth. PFC concentrations were measured in umbilical cord serum samples. The differences of the PFC concentrations between cases and controls were investigated using Wilcoxon's paired test. Possible threshold effects (above the upper quartile for perfluorooctane sulfonate (PFOS) and perfluorooctanoic acid (PFOA) and above limit of detection [LOD] for perfluorononanoic acid (PFNA)) were evaluated by conditional logistic regression.

Results: The median umbilical cord serum concentrations of PFOS were 6.92 ng/ml in the cases and 6.77 ng/ml in the controls. The corresponding concentrations of PFOA were 1.80 and 1.83 ng/ml. No associations between PFCs and ADHD were observed. Odds ratios adjusted for smoking status, parity, and gestational age were 0.81 (95% confidence interval [CI] 0.50 to 1.32) for PFOS, 1.07 (95% CI 0.67 to 1.7) for PFOA, and 1.1 (95% CI 0.75 to 1.7) for PFNA.

Conclusions: The current study revealed no support for an association between fetal exposure to PFOS, PFOA, or PFNA and ADHD.

Editor: Aimin Chen, University of Cincinnati, United States of America

Funding: The project was funded by the Swedish Research Council (via SIMSAM Early Life) (web page: www.vr.se/) and the Medical Faculty at Lund University (web page: www.med.lu.se). The funders had no role in study design, data collection and analysis, decision to publish, or preparation of the manuscript.

Competing Interests: The authors have declared that no competing interests exist.

* E-mail: amanda.ode@med.lu.se

Introduction

Emission of pollutants from densely populated areas and industries is a growing environmental problem. Contaminants present in the environment can have a negative impact on both human health and environment. Perfluorinated compounds (PFCs) are extremely stable and persistent man-made organic chemicals that have been identified as environmental pollutants. The unique properties of PFCs have made them highly useful in numerous industrial and consumer applications such as lubricants, firefighting foams, cleaning agents, and in surface coating for paper, food packaging, textiles, furniture, carpets and cookware [1–3].

PFCs, particularly perfluorooctane sulfonate (PFOS) and perfluorooctanic acid (PFOA) have been widely detected in the environment, wildlife, and humans [4–8]. Humans are exposed to PFCs through consumer products as well as contaminated air, water, and food [1]. In recent years, studies have revealed that PFCs cross the placenta and accumulate in the fetus [9–11]. The fetal brain is immature and is therefore susceptible to injury caused by toxic agents [12]. Animal data have indicated that PFCs accumulate in the brain both before and after the blood-brain barrier is formed [13–16].

Animal studies have shown that neonatal exposure to low doses of PFCs induced irreversible neurotoxic effects in adult mice and caused changes in behavior and habituation by altering the dopaminergic and cholinergic system [17,18]. PFCs also alter levels of neural proteins that are important for the formation and growth of the synapses [19]. Defects in the dopamine transporters and receptors have been suggested to be the most significant neurobiological problem in attention deficit hyperactivity disorder (ADHD) [20,21].

ADHD is a neurodevelopmental disorder defined by inattention, hyperactivity and impulsivity [22,23]. The disorder has its onset in childhood, and persists into adolescence and into adulthood in some cases [24,25]. The genetic factor is believed to play the major role in the development of ADHD [22,26,27]. In addition, exposure to environmental toxins, such as lead, mercury, and persistent chlorinated biphenyls, has also been related to ADHD [26,28,29].

Two cross-sectional studies based on parent-reported ADHD diagnosis have investigated the potential association between PFC levels in school-age children and ADHD [30,31]. The study by Hoffman et al. [30] found a positive relationship between ADHD and PFC levels in the blood of children between 12 and 15 years, whereas an association with only perfluorohexane sulfonate (PFHxS) was found in the study by Stein and Savitz [31]. In another cross-sectional study, PFC exposure was associated with impulsivity in children [32]. Other studies based on questionnaires investigated whether behavioral health and motor coordination as well as motor and mental developmental milestones were associated with maternal PFCs during pregnancy and found no such associations except for PFOS which was associated with delayed motor development in the first two years of life [33,34].

The frequency of children receiving an ADHD diagnosis has increased in recent years [35]. Improved diagnostic criteria might be responsible for the increased detection of ADHD cases. Increased exposure to environmental pollutants might also contribute to the high prevalence of ADHD. Since the human brain is susceptible to disturbance by environmental pollutants during the fetal period, it is of importance to investigate the association between exposure to these pollutants during the sensitive period of fetal development and ADHD.

The objective of this study is to investigate the association between fetal exposure to PFCs and ADHD diagnosis in childhood. Unlike previous studies, this case-control study is based on clinical ADHD diagnosis and PFCs are measured in umbilical cord serum samples which reflect the PFC concentrations in the fetus. The study is a part of the Fetal Environment and Neurodevelopment Disorders in Epidemiological Research project (the FENDER project).

Material and Methods

Participants

The selection procedure of the children with ADHD diagnosis has been previously described by Gustafsson and Kallen [36]. Briefly, at the Department of Child and Adolescent Psychiatry in the city of Malmö, 419 children born and living in Malmö between 1978 and 2000 with ADHD diagnosis were identified and were followed up until 2005. During the study period, the children with ADHD were diagnosed by one of ten experienced clinicians at the department using the Diagnostic and Statistical Manual of Mental Disorders (DSM). A child with suspected attention difficulties, hyperactivity and/or difficulties with impulse control is usually assessed to the child and adolescent psychiatry by a special teacher and a school psychologist or by the parents. The assessment begins with gathering information about the child's general medical health condition and the child's development from birth until the present time. The school psychologist or the psychologist at the psychiatric clinic performs a cognitive testing with the Wechsler Intelligence Scale (WISC). The parents and the teacher are asked to fill in questionnaires like SNAP-IV, Conner's questionnaire or the 5–15 questionnaire which all cover the symptoms of ADHD. Parents are usually asked to fill in the BRIEF-questionnaire concerning the child's executive functions in every-day life. Sometimes a member of the team at the clinic observes the child at school. The child's ability to concentrate is

tested with TEA-Ch or with a computerized test of attention such as QB-Tech or IVA+. The child psychiatrist performs a paediatric examination with assessment of neurological soft-signs. The child's behaviour in different test situations and at the visits at the clinic is observed and registered. When all parts of the assessment have been performed, a team consisting of doctor, psychologist and sometimes a social worker meet and discuss the findings to come to a consensus decision concerning the diagnosis using DSM criteria. The DSM criteria DSM-III-R$_{11}$ and DSM-IV$_{12}$ were used before 1994 and from 1994 and onwards, respectively. Age at the time of diagnosis varied between 5 and 17 years, with most children being diagnosed between the ages of 8 and 12 years.

Using the personal identification numbers, children with ADHD were linked to the Swedish Medical Birth Register (SMBR) which contains demographic and obstetric information on nearly all (99%) the mothers and the infants in Sweden. Umbilical cord serum samples for children with ADHD were collected from the Malmö Maternity Unit Serum Biobank (MMUSB) using the personal identification numbers. Nearly all deliveries in Malmö take place at the Malmö University Hospital Maternity Unit, where blood samples from the mother and from the umbilical cord of the newborn have been collected at the time of delivery and stored at −20°C at the MMUSB since 1969. Controls were selected into two phases. In the first phase, for each ADHD case with an available umbilical blood sample in the biobank, the next-baby-born with serum sample of the same sex was selected as a control. However, a new publication by Gustafsson and Kallen [36] revealed the impact of maternal country of birth on the diagnosis of ADHD. Thus, the benefit of matching for the maternal country of birth completely overrode that from matching for the infant's sex. Therefore, in the second phase, a pool of ten eligible controls per ADHD case were collected from the SMBR and were matched to the cases for year of birth (±12 months) and country of birth of the mother. The sample of the next-baby-born from the first phase was used if no newborn in that eligible pool of controls had an available umbilical blood sample in the biobank. The selection procedure for cases and controls is presented in Figure 1.

Ethics statements

At the Maternity Unit, the women were informed that the umbilical cord serum sample collected could be used for research purposes in the future and those who accepted gave their verbal informed consent that was documented in the medical records. During the study period only verbal informed consent was obtained. The written informed consent has been implemented in 2005 and therefore could not be considered for the current study. The data were analysed anonymously. The study protocol followed the requirements of the Declaration of Helsinki and the study, together with the consent procedure, was approved by the Research Ethics Committee at Lund University, Sweden.

Analysis of perfluorinated compounds and cotinine in umbilical cord serum

The analyses of PFHxS, PFOS, PFOA, PFNA, perfluorodeca-noic acid (PFDA), perfluoroundecanoic acid (PFUnDA), perfluor-ododecanoic acid (PFDoDA), and cotinine were performed as previously described [37]. Briefly, aliquots of 100 μL sera were added with isotopically labeled internal standards, the proteins were precipitated by acetonitrile and centrifugation, and analysis was then performed using a hybrid triple quadrupole linear ion-trap mass spectrometer (LC/MS/MS; UFLCXR, Shimadzu Corporation, Kyoto, Japan; QTRAP 5500; Sciex, Framingham, MA, USA). The limits of detections for the detected PFCs and

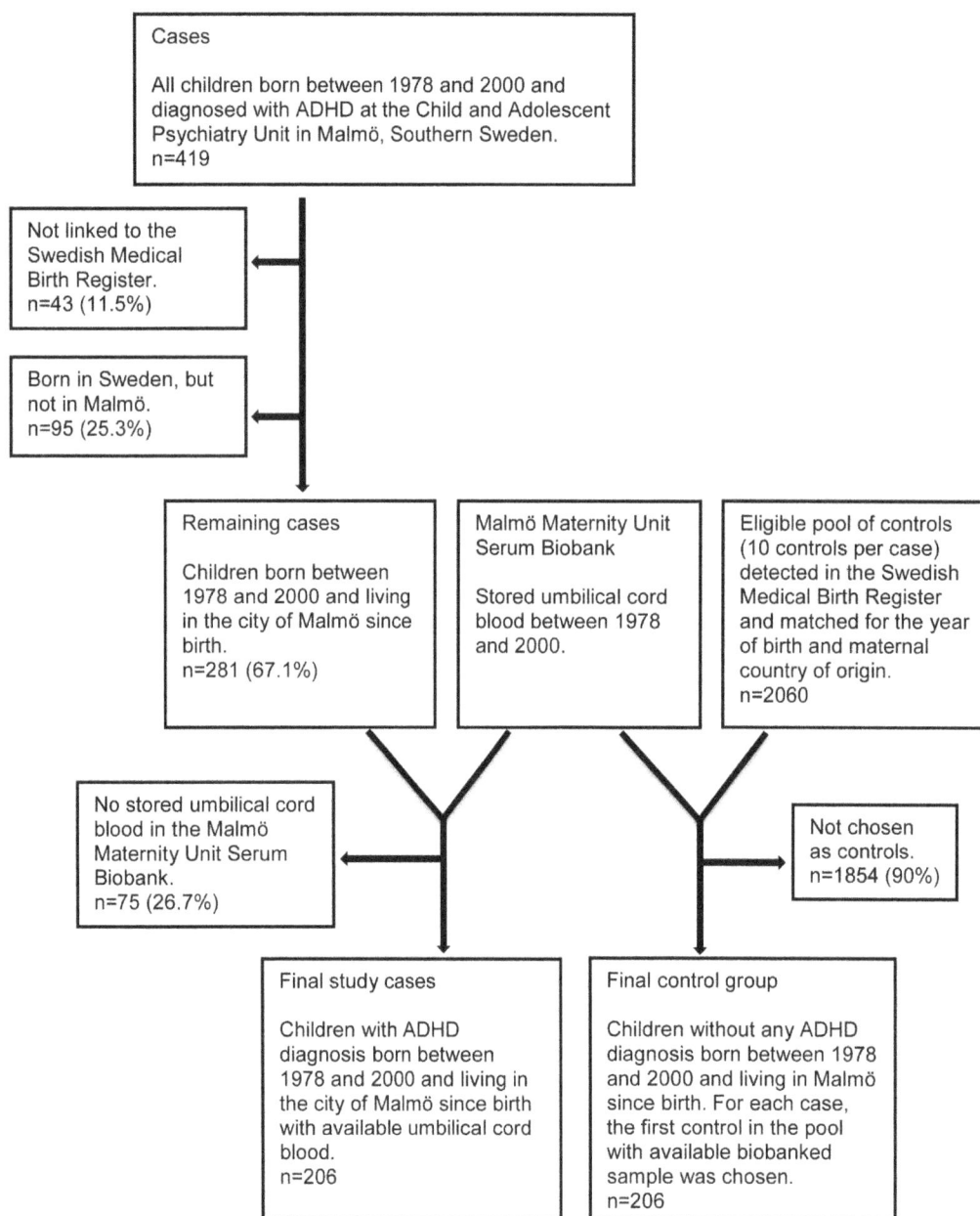

Figure 1. Flowchart for the selection procedure of the children with attention deficit hyperactivity disorder and controls.

cotinine were 0.2 ng/ml. To increase the accuracy, the result reported is the average of two measurements from the same sample worked up and analyzed on different days. In all sample batches, the quality of the measurements was controlled by analyzing chemical blanks and in-house quality control (QC) samples. The reproducibility, determined as the relative standard deviation, between measured duplicate samples was 11% for PFOS, 12% for PFOA, 12% for PFNA, and 9% for cotinine. The reproducibility in QC samples was 8% for PFOS, 11% for PFOA, 8% for PFNA, and 5% for cotinine. Usually we are able to analyze several more PFCs with the method but some factor, probably during the storage of the samples, resulted in a high background noise in the chromatograms making detection impossible. Thus, PFHxS, PFDA, PFUnDA, and PFDoDA could not be detected in the samples due to this effect. On the other hand, due to the high

correlation between PFOS and other PFCs often only PFOS and PFOA are reported in studies of PFCs. Although contamination of samples during collection is believed to be minimal, field blanks could not be provided to control for eventual contamination of the samples with PFCs. The analyses of PFOS and PFOA are part of the round robin intercomparison program (Professor Dr. med. Hans Drexler, Institute and Outpatient Clinic for Occupational, Social and Environmental Medicine, University of Erlangen-Nuremberg, Germany) with results within the tolerance limits.

Statistical analyses

The Wilcoxon's paired test was used to compare the PFC concentrations between ADHD cases and controls. Conditional logistic regression analysis was used to assess the association between fetal exposure to PFCs and ADHD. The odds ratio was

Table 1. Median concentration (in nanograms/milliliters) of perfluorinated compounds by the maternal and infant demographic characteristics.

Characteristics	Children with ADHD			Control group		
	n (%)	PFOS	PFOA	n (%)	PFOS	PFOA
Group (cases/controls)	203 (49.8%)	6.92	1.80	205 (50.2%)	6.77	1.83
Year of delivery						
1978–1981	2 (1.0)	2.66	0.45	2 (1.0)	8.70	0.85
1982–1985	13 (6.4)	5.69	1.50	10 (4.9)	6.49	1.71
1986–1989	63 (31.0)	6.96	2.0	66 (32.2)	6.71	1.82
1990–1993	86 (42.4)	7.08	1.78	87 (42.4)	6.74	1.82
1994–1997	34 (16.7)	6.65	1.69	35 (17.1)	7.44	1.87
1998–2000	5 (2.5)	7.68	1.64	5 (2.4)	8.11	1.86
Maternal age (years)						
<20	8 (3.9)	6.67	1.57	6 (2.9)	8.65	2.16
20–34	172 (84.7)	6.94	1.81	171 (83.4)	6.74	1.82
≥35	23 (11.3)	6.34	1.64	28 (13.7)	7.05	1.81
Parity						
0 [nulliparous]	97 (47.8)	7.00	2.01	106 (51.7)	7.56	2.13
1	71 (35.0)	6.60	1.55	68 (33.2)	6.22	1.55
≥2	35 (17.2)	6.80	1.68	31 (15.1)	6.17	1.42
Maternal country of origin						
Sweden	168 (83.3)	7.02	1.85	170 (82.9)	7.06	1.89
Other Nordic countries[a]	7 (3.4)	4.28	2.13	7 (3.4)	6.18	1.69
Rest of Europe[b]	8 (3.9)	7.47	1.60	9 (4.4)	5.48	1.48
Sub-Saharan Africa	2 (1.0)	4.23	0.72	2 (1.0)	2.10	0.45
Middle East and North Africa	13 (6.4)	4.42	0.85	12 (5.9)	2.76	0.47
East Asia	1 (0.5)	6.83	1.71	2 (1.0)	9.36	1.43
South America	2 (1.0)	7.58	2.89	2 (1.0)	7.63	13.5
Unknown	1 (0.5)	2.96	0.46	1 (0.5)	2.60	1.70
Maternal body mass index (kg/m²)[c]						
Not available	141 (69.5)	6.85	1.89	142 (69.3)	6.67	1.83
<18.5 (Underweight)	1 (0.5)	10.1	2.64	3 (1.5)	8.75	2.39
18.5–24.9 (Normal)	37 (18.2)	6.83	1.63	42 (20.5)	7.50	1.72
25–29.9 (Overweight)	16 (7.9)	7.27	1.36	14 (6.8)	7.58	2.02
≥30 (Obese)	8 (3.9)	6.06	2.09	4 (2.0)	6.40	2.31
Smoking during pregnancy[d]						
Non-smoker	65 (32.0)	6.54	1.83	85 (41.5)	6.82	1.86
Second-hand smoker	57 (28.1)	7.08	1.71	57 (27.8)	6.91	1.86
Active smoker	81 (39.9)	7.49	1.82	63 (30.7)	6.37	1.72
Infant sex						
Male	180 (88.7)	6.97	1.76	163 (79.5)	6.87	1.84
Female	23 (11.3)	6.32	1.99	42 (20.5)	6.51	1.64
Birth weight (grams)						
<1500	4 (2.0)	5.73	2.31	0		
<2500	9 (4.4)	4.85	1.44	5 (2.4)	6.37	1.84
2500–4000	166 (81.8)	7.12	1.84	152 (74.1)	6.63	1.82
>4000	24 (11.8)	6.41	1.67	48 (23.4)	7.25	1.94
Gestational age (weeks)						
<32	5 (2.5)	4.77	1.44	1 (0.5)	4.71	1.05
<37	6 (3.0)	4.36	1.09	6 (2.9)	4.74	1.97
37–42	176 (86.7)	7.12	1.88	178 (86.8)	6.73	1.82
>42	16 (7.9)	6.54	1.63	20 (9.8)	8.37	1.77

Table 1. Cont.

Characteristics	Children with ADHD			Control group		
	n (%)	PFOS	PFOA	n (%)	PFOS	PFOA
SD scores (for gestational age)						
<-2 (small for gestational age)	9 (4.4)	6.69	2.26	14 (6.8)	7.92	1.94
-2 to -1.1	37 (18.2)	6.60	1.53	29 (14.1)	6.56	1.94
-1.1 to 1	138 (68.0)	7.01	1.79	126 (61.5)	6.67	1.77
1.1 to 2	15 (7.4)	5.48	2.00	32 (15.6)	7.79	2.02
>2 (large for gestational age)	4 (2.0)	7.30	2.40	4 (2.0)	4.05	1.20
Apgar scores						
0–6	5 (2.5)	4.28	1.80	2 (1.0)	10.2	2.44
≥ 7	198 (97.5)	6.94	1.79	203 (99.0)	6.74	1.82

Abbreviations: ADHD, attention deficit hyperactivity disorder; PFOS, perfluorooctane sulfonate; PFOA, perfluorooctanoic acid; Parity, number of previous pregnancies.
[a]Finland, Denmark, and Norway.
[b]Western Europe and former Eastern Europe.
[c]Body mass index was classified according to the standard values of the World Health Organization.
[d]Maternal smoking is based on measured cotinine concentrations in the umbilical cord serum.

calculated for both 1 unit increase (nanogram per milliliter) in the concentrations of PFOS and PFOA and for comparisons between concentrations above and below the 75th percentile for the control group. For PFNA the concentrations above the limit of detection (LOD) were compared to those below LOD (0.2 ng/ml).

The potential confounding variables that were considered in the present study were smoking during pregnancy, parity, and gestational age at birth, since they have been found to be associated with both PFC exposure and ADHD [11,38–46].

Smoking during pregnancy was determined by cotinine levels in umbilical cord serum. Cotinine levels below the LOD (0.2 ng/ml) were related to nonsmoking pregnant women, cotinine levels higher than 15 ng/ml were related to active smokers and levels between 0.2 and 15 ng/ml were related to second-hand smokers [47]. Parity was divided into three groups according to number of previously born children (0 [i.e. nulliparous], 1, or ≥ 2 children). Gestational age was entered in the analyses as class variable divided into three groups; <37, 37–42, and >42 weeks of pregnancy.

The odds ratios were calculated in paired samples (n = 202) using Egret for Windows 2.0 (Cytel Software Corporation). The rest of the analyses were performed in IBM SPSS Statistics version 20 (IBM Corporation 1989, 2011).

The power calculation was based on 206 cases and matched controls. With the current setting, we had an 80% chance of detecting a difference in the levels of 0.20 standard deviations, with α value of 0.05, between cases and controls. For the analysis of the threshold effect, with α value of 0.05 and β value of 0.80, the lowest detectable odds ratio was 1.8.

Results

PFOS and PFOA concentrations were above the LOD in 98% of the samples, whereas for PFNA about 12% were above the LOD. PFOS and PFOA concentrations below the LOD in individual samples (n = 2 for each) were replaced with 0.2 ng/ml.

The demographic characteristics and the umbilical cord PFC concentrations of the study population are presented in Table 1.

Figure 2 shows the distribution of PFOS and PFOA levels in the ADHD cases and the controls. The median concentrations of PFNA above LOD for cases and controls were 0.31 and 0.28 ng/

ml, respectively. Wilcoxon's paired test revealed no differences in cord serum PFC concentrations between children with ADHD diagnosis and controls ($p = 0.72$, 0.44, and 0.48 for PFOS, PFOA, and PFNA respectively).

Conditional logistic regression analyses revealed no significant associations between umbilical cord concentrations of PFCs and ADHD (Table 2). The result did not change after adjusting for smoking during pregnancy, parity, and gestational age at birth.

Discussion

The present study found no statistically significant associations between exposure to PFCs during pregnancy and ADHD diagnosis during childhood, although the measured umbilical cord concentrations of PFOS were among the highest in Europe [42,48,49] and even among the highest in the world [9,50–52]. For PFOA, the levels were higher than those measured in Norway and other non-European countries but lower than those in Denmark and Faraoe Islands [9,42,48,49].

Animal data revealed that neonatal mice that were exposed to high doses of PFOS and PFOA showed behavioral defects which ranged from slight effects at the anxiety level [53] to reduced habituation and hyperactivity in adult mice [17]. It has been suggested that PFOS and PFOA act as developmental neurotoxicants that mediate their effects on normal brain development, with consequences for cognitive and behavioral functions, through different mechanisms. Examples of those mechanisms are alteration in the dopaminergic system [17,18], elevated levels of proteins important for normal neuronal survival, growth and synaptogenesis, such as CaMKII, GAP-43, synaptophysin and tau, in the brain [19], and induction of apoptosis of neuronal cells [54]. Although most of these findings were obtained from experiments on mice or rat derived cell lines that were exposed to extremely high levels of PFOS and PFOA compared to the low levels measured in the present study, other studies found that PFCs were detrimental to neurodevelopment at levels comparable to those observed in humans [17,19].

Our study is primarily comparable to the study by Fei and Olsen [34] because both studies used measures of prenatal rather than postnatal exposure to PFCs. Fei and Olsen [32] found higher levels of PFOS and PFOA compared to those seen among

A

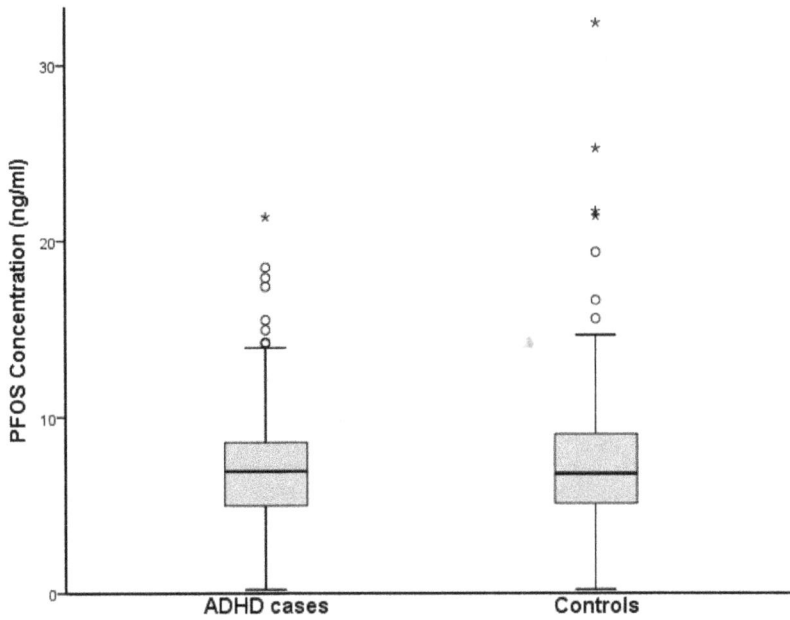

B

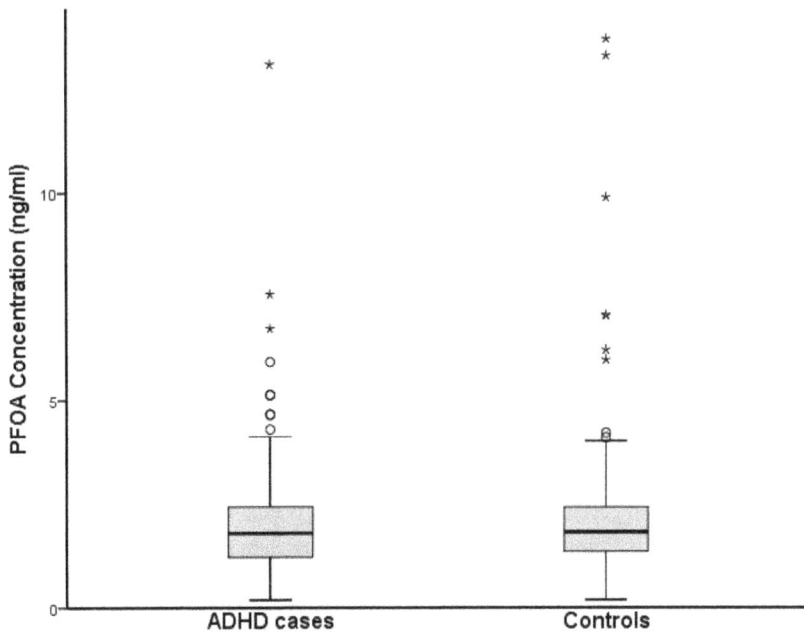

Figure 2. Boxplot of the umbilical cord concentrations of perfluorooctane sulfonate (PFOS) (a) and perfluorooctanoic acid (PFOA) (b) in cases having attention deficit hyperactivity disorder and controls. The extreme values of perfluorooctanoic acid for the ADHD cases, 48 and 36 ng/ml, and for the controls, 66, 49, 31, and 23 ng/ml, are not presented in the boxplot.

pregnant women in other countries including the Nordic countries [11,48]. Consistent with that study, our results provide further indication that fetal exposure to PFCs at the present levels do not play a major role in having ADHD diagnosis at later age. Hoffman et al. [30] found an association between PFC serum concentrations and ADHD in children aged 12 to 15 years. Another study

Table 2. The crude and adjusted odds ratio with 95% confidence interval of attention deficit hyperactivity disorder and exposure to perfluorinated compounds.

	ADHD Diagnosis	
	Crude	**Adjusted[a]**
PFOS[b]	0.98 (0.92–1.03)	0.98 (0.92–1.04)
PFOA[b]	0.98 (0.94–1.02)	0.98 (0.94–1.02)
PFOS[c]		
<75th percentile	1	1
≥75th percentile	0.82 (0.51–1.31)	0.81 (0.50–1.32)
PFOA[c]		
<75th percentile	1	1
≥75th percentile	1.03 (0.65–1.6)	1.07 (0.67–1.7)
PFNA[d]		
<LOD	1	1
≥LOD	1.1 (0.72–1.6)	1.1 (0.75–1.7)

Abbreviations: ADHD, attention deficit hyperactivity disorder; PFOS, perfluorooctane sulfonate; PFOA, perfluorooctanoic acid; PFNA, perfluorononanoic acid; LOD, limit of detection.
[a]Adjusted for maternal active smoking, parity, and gestational age at birth.
[b]Odds ratio is calculated for 1 ng/ml increase in umbilical cord serum concentration.
[c]Odds ratio is calculated for PFOS and PFOA concentrations at or above the 75th percentile (75th percentile for PFOS and PFOA were 9,1 ng/ml and 2,4 ng/ml, respectively).
[d]Odds ratio is calculated for PFNA concentrations at or above the LOD (0.2 ng/ml).

by Stein and Savitz [31] on the relationship between self-reported ADHD and PFC levels in children in the same age range as for those in the study of Hoffman et al. [30] showed an association with PFHxS but not with the other PFCs even though both ADHD prevalence and exposure levels for PFCs were higher in the study by Stein and Savitz. Exposure to PFCs tends to be higher among newborns, toddlers, and children due to high uptake via food consumption, hand-to-mouth transfer of the PFCs from carpets, and through ingestion of dust [55]. If the positive association between PFC exposure and self-reported ADHD found in the study by Hoffman et al. [30] was not due to a chance finding, that might indicate that postnatal exposure to PFCs, rather than prenatal exposure, is associated with ADHD.

The present study has some limitations. Unfortunately, while 419 children with ADHD diagnosis were identified at the Department of Child and Adolescent Psychiatry, there were significant losses to get to the final study sample. The study was restricted to children born in Malmö with available obstetric and demographic information from the SMBR and stored cord blood samples in the biobank. About 50% of the identified children met these two inclusion criteria and were included in the study. The second limitation of the current study is the small number of ADHD cases. Although we would be able to detect an odd ratio of 1.8 and a difference in PFC levels of at least 0.20 standard deviations between cases and controls, it should be emphasized that the statistical power was not high enough to detect some minor associations and we could not accordingly rule out small effects. In addition, we lack information about other exposures that are significant for ADHD, such as exposure to mercury and lead. The PFC levels measured here might also be too low to trigger undesirable effects on the brain development.

During our study period, the clinical diagnostic criteria for ADHD were changed from the definition in DSM-III-R to the definition in DSM-IV, where DSM-IV is regarded as more inclusive. Thus, DSM-IV criteria yield a higher prevalence of

ADHD [56]. Most individuals (93%–97.5%) who fulfill a diagnosis of ADHD according to DSM-III-R also fulfill the diagnostic criteria according to DSM-IV [57-59]. Individuals with ADHD according to DSM-IV that also fulfilled diagnostic criteria according to DSM-III-R were 85% [57] and 60% [59]. Thus the overlap between ADHD diagnoses according to DSM-III-R and DSM-IV is considerable. The more inclusive diagnosis used in the latter part of the study probably includes some less severe cases which might slightly weaken possible statistical associations between exposure to PFCs and having an ADHD diagnosis.

The study has also several important strengths. First, unlike most of the previous studies on the associations between PFC levels and ADHD, our prospective study design is more reliable in the sense that it is based on clinical diagnosis of ADHD made at the Department of Child and Adolescent Psychiatry. Children were diagnosed at the same psychiatric clinic through the whole study period.

Second, the present study is based on analyzed blood samples from the fetal period, which we believe is the most susceptible exposure window, whereas in other studies, which were of a cross-sectional nature, blood samples were collected from school-age children [30,31].

Third, we were able to account for important covariates; smoking during pregnancy, parity, and gestational age at birth that have been found to be associated with both PFC levels in pregnant women or infants [11,38,42,46] and ADHD diagnosis or symptoms [39–41,43–45]. Epidemiological findings have shown that prevalence of ADHD is higher among males [20]. It has also previously been shown that infant sex has no effect on the concentrations of PFCs [11,42,60]. Thus, infant sex was not considered as a potential confounder in the current data set.

In a previous study, we found that fetuses of mothers originating from a country other than Sweden, especially those from Middle East and sub-Saharan Africa, had lower PFC levels in the cord blood than fetuses of native Swedish mothers [11]. Another study found higher odds of having an ADHD diagnosis for native Swedish children compared to children of mothers born outside Sweden [36]. Since the proportion of immigrants is relatively high in Malmö, no matching for the maternal country of birth might result in a false positive relationship between PFCs and ADHD.

Human serum PFC levels showed an increasing pattern from the early 1970s through the late 1990s, followed by leveling out and a decreasing trend right after the phase-out of the production of PFOS and PFOS related compounds in 2002 [8,61–64]. Diagnosis criteria for ADHD have been changed during the study period. We matched for the year of delivery to remove the effect of those differences in PFC levels and diagnosis on the results.

According to our findings, fetal exposure to PFOS, PFOA and PFNA was not associated with ADHD diagnosis in childhood.

Acknowledgments

We would like to thank the laboratory technicians Åsa Amilon and Agneta Kristensen for analyzing the samples.

Author Contributions

Conceived and designed the experiments: KK PG LR SAI PO ARH AO. Performed the experiments: BAGJ CHL. Analyzed the data: AO KK LR ARH. Contributed reagents/materials/analysis tools: BAGJ CHL. Wrote the paper: AO. Wrote the section about the laboratory analysis: BAGJ CHL. Reviewed the manuscript and made comments on it before submission: ARH LR KK PO PG SAI.

References

1. Fromme H, Tittlemier SA, Volkel W, Wilhelm M, Twardella D (2009) Perfluorinated compounds—exposure assessment for the general population in Western countries. Int J Hyg Environ Health 212: 239–270.
2. Kissa E (2001) Fluorinated Surfactants and Repellants. NY, USA: Marcel Dekker Inc.
3. Lemal DM (2004) Perspective on fluorocarbon chemistry. J Org Chem 69: 1–11.
4. Butt CM, Berger U, Bossi R, Tomy GT (2010) Levels and trends of poly- and perfluorinated compounds in the arctic environment. Sci Total Environ 408: 2936–2965.
5. Giesy JP, Kannan K (2001) Global distribution of perfluorooctane sulfonate in wildlife. Environ Sci Technol 35: 1339–1342.
6. Giesy JP, Kannan K (2002) Perfluorochemical surfactants in the environment. Environ Sci Technol 36: 146A–152A.
7. Kannan K, Corsolini S, Falandysz J, Fillmann G, Kumar KS, et al. (2004) Perfluorooctanesulfonate and Related Fluorochemicals in Human Blood from Several Countries. Environmental Science & Technology 38: 4489–4495.
8. Olsen GW, Huang H-Y, Helzlsouer KJ, Hansen KJ, Butenhoff JL, et al. (2005) Historical Comparison of Perfluorooctanesulfonate, Perfluorooctanoate, and Other Fluorochemicals in Human Blood. Environmental Health Perspectives 113: 539–545.
9. Beesoon S, Webster GM, Shoeib M, Harner T, Benskin JP, et al. (2011) Isomer profiles of perfluorochemicals in matched maternal, cord, and house dust samples: manufacturing sources and transplacental transfer. Environ Health Perspect 119: 1659–1664.
10. Inoue K, Okada F, Ito R, Kato S, Sasaki S, et al. (2004) Perfluorooctane Sulfonate (PFOS) and Related Perfluorinated Compounds in Human Maternal and Cord Blood Samples: Assessment of PFOS Exposure in a Susceptible Population during Pregnancy. Environmental Health Perspectives 112: 1204–1207.
11. Ode A, Rylander L, Lindh CH, Kallen K, Jonsson BA, et al. (2013) Determinants of maternal and fetal exposure and temporal trends of perfluorinated compounds. Environ Sci Pollut Res Int 20: 7970–7978.
12. Grandjean P, Landrigan PJ (2006) Developmental neurotoxicity of industrial chemicals. Lancet 368: 2167–2178.
13. Austin ME, Kasturi BS, Barber M, Kannan K, MohanKumar PS, et al. (2003) Neuroendocrine Effects of Perfluorooctane Sulfonate in Rats. Environmental Health Perspectives 111: 1485–1489.
14. Butenhoff JL, Ehresman DJ, Chang SC, Parker GA, Stump DG (2009) Gestational and lactational exposure to potassium perfluorooctanesulfonate (K+ PFOS) in rats: developmental neurotoxicity. Reprod Toxicol 27: 319–330.
15. Cui L, Zhou QF, Liao CY, Fu JJ, Jiang GB (2009) Studies on the toxicological effects of PFOA and PFOS on rats using histological observation and chemical analysis. Arch Environ Contam Toxicol 56: 338–349.
16. Greaves AK, Letcher RJ, Sonne C, Dietz R (2013) Brain region distribution and patterns of bioaccumulative perfluoroalkyl carboxylates and sulfonates in east greenland polar bears (Ursus maritimus). Environ Toxicol Chem 32: 713–722.
17. Johansson N, Fredriksson A, Eriksson P (2008) Neonatal exposure to perfluorooctane sulfonate (PFOS) and perfluorooctanoic acid (PFOA) causes neurobehavioural defects in adult mice. Neurotoxicology 29: 160–169.
18. Slotkin TA, MacKillop EA, Melnick RL, Thayer KA, Seidler FJ (2008) Developmental neurotoxicity of perfluorinated chemicals modeled in vitro. Environ Health Perspect 116: 716–722.
19. Johansson N, Eriksson P, Viberg H (2009) Neonatal exposure to PFOS and PFOA in mice results in changes in proteins which are important for neuronal growth and synaptogenesis in the developing brain. Toxicol Sci 108: 412–418.
20. Faraone SV, Perlis RH, Doyle AE, Smoller JW, Goralnick JJ, et al. (2005) Molecular Genetics of Attention-Deficit/Hyperactivity Disorder. Biological Psychiatry 57: 1313–1323.
21. Gizer IR, Ficks C, Waldman ID (2009) Candidate gene studies of ADHD: a meta-analytic review. Hum Genet 126: 51–90.
22. Faraone SV, Khan SA (2006) Candidate gene studies of attention-deficit/hyperactivity disorder. J Clin Psychiatry 67 Suppl 8: 13–20.
23. Gustafsson P, Thernlund G, Ryding E, Rosen I, Cederblad M (2000) Associations between cerebral blood-flow measured by single photon emission computed tomography (SPECT), electro-encephalogram (EEG), behaviour symptoms, cognition and neurological soft signs in children with attention-deficit hyperactivity disorder (ADHD). Acta Paediatr 89: 830–835.
24. Faraone SV, Biederman J, Mick E (2006) The age-dependent decline of attention deficit hyperactivity disorder: a meta-analysis of follow-up studies. Psychol Med 36: 159–165.
25. Lara C, Fayyad J, de Graaf R, Kessler RC, Aguilar-Gaxiola S, et al. (2009) Childhood predictors of adult attention-deficit/hyperactivity disorder: results from the World Health Organization World Mental Health Survey Initiative. Biol Psychiatry 65: 46–54.
26. Banerjee TD, Middleton F, Faraone SV (2007) Environmental risk factors for attention-deficit hyperactivity disorder. Acta Paediatr 96: 1269–1274.
27. Thapar A, Cooper M, Jefferies R, Stergiakouli E (2012) What causes attention deficit hyperactivity disorder? Arch Dis Child 97: 260–265.
28. Braun JM, Kahn RS, Froehlich T, Auinger P, Lanphear BP (2006) Exposures to environmental toxicants and attention deficit hyperactivity disorder in U.S. children. Environ Health Perspect 114: 1904–1909.
29. Eubig PA, Aguiar A, Schantz SL (2010) Lead and PCBs as risk factors for attention deficit/hyperactivity disorder. Environ Health Perspect 118: 1654–1667.
30. Hoffman K, Webster TF, Weisskopf MG, Weinberg J, Vieira VM (2010) Exposure to polyfluoroalkyl chemicals and attention deficit/hyperactivity disorder in U.S. children 12-15 years of age. Environ Health Perspect 118: 1762–1767.
31. Stein CR, Savitz DA (2011) Serum perfluorinated compound concentration and attention deficit/hyperactivity disorder in children 5–18 years of age. Environ Health Perspect 119: 1466–1471.
32. Gump BB, Wu Q, Dumas AK, Kannan K (2011) Perfluorochemical (PFC) exposure in children: associations with impaired response inhibition. Environ Sci Technol 45: 8151–8159.
33. Fei C, McLaughlin JK, Lipworth L, Olsen J (2008) Prenatal exposure to perfluorooctanoate (PFOA) and perfluorooctanesulfonate (PFOS) and maternally reported developmental milestones in infancy. Environ Health Perspect 116: 1391–1395.
34. Fei C, Olsen J (2011) Prenatal exposure to perfluorinated chemicals and behavioral or coordination problems at age 7 years. Environ Health Perspect 119: 573–578.
35. Pastor PN, Reuben CA (2008) Diagnosed attention deficit hyperactivity disorder and learning disability: United States, 2004–2006. Vital Health Stat 10: 1–14.
36. Gustafsson P, Kallen K (2011) Perinatal, maternal, and fetal characteristics of children diagnosed with attention-deficit-hyperactivity-disorder: results from a population-based study utilizing the Swedish Medical Birth Register. Dev Med Child Neurol 53: 263–268.
37. Lindh CH, Rylander L, Toft G, Axmon A, Rignell-Hydbom A, et al. (2012) Blood serum concentrations of perfluorinated compounds in men from Greenlandic Inuit and European populations. Chemosphere 88: 1269–1275.
38. Apelberg BJ, Goldman LR, Calafat AM, Herbstman JB, Kuklenyik Z, et al. (2007) Determinants of fetal exposure to polyfluoroalkyl compounds in Baltimore, Maryland. Environ Sci Technol 41: 3891–3897.
39. Carballo JJ, Garcia-Nieto R, Alvarez-Garcia R, Caro-Canizares I, Lopez-Castroman J, et al. (2013) Sibship size, birth order, family structure and childhood mental disorders. Soc Psychiatry Psychiatr Epidemiol 48: 1327–1333.
40. Chu SM, Tsai MH, Hwang FM, Hsu JF, Huang HR, et al. (2012) The relationship between attention deficit hyperactivity disorder and premature infants in Taiwanese: a case control study. BMC Psychiatry 12: 85.
41. Desrosiers C, Boucher O, Forget-Dubois N, Dewailly E, Ayotte P, et al. (2013) Associations between prenatal cigarette smoke exposure and externalized behaviors at school age among Inuit children exposed to environmental contaminants. Neurotoxicol Teratol 39C: 84–90.

42. Fei C, McLaughlin JK, Tarone RE, Olsen J (2007) Perfluorinated chemicals and fetal growth: a study within the Danish National Birth Cohort. Environ Health Perspect 115: 1677–1682.

43. Kotimaa AJ, Moilanen I, Taanila A, Ebeling H, Smalley SL, et al. (2003) Maternal smoking and hyperactivity in 8-year-old children. J Am Acad Child Adolesc Psychiatry 42: 826–833.

44. Langley K, Holmans PA, van den Bree MB, Thapar A (2007) Effects of low birth weight, maternal smoking in pregnancy and social class on the phenotypic manifestation of Attention Deficit Hyperactivity Disorder and associated antisocial behaviour: investigation in a clinical sample. BMC Psychiatry 7: 26.

45. Perricone G, Morales MR, Anzalone G (2013) Neurodevelopmental outcomes of moderately preterm birth: precursors of attention deficit hyperactivity disorder at preschool age. Springerplus 2: 221.

46. Washino N, Saijo Y, Sasaki S, Kato S, Ban S, et al. (2009) Correlations between prenatal exposure to perfluorinated chemicals and reduced fetal growth. Environ Health Perspect 117: 660–667.

47. George L, Granath F, Johansson AL, Cnattingius S (2006) Self-reported nicotine exposure and plasma levels of cotinine in early and late pregnancy. Acta Obstet Gynecol Scand 85: 1331–1337.

48. Gutzkow KB, Haug LS, Thomsen C, Sabaredzovic A, Becher G, et al. (2012) Placental transfer of perfluorinated compounds is selective—a Norwegian Mother and Child sub-cohort study. Int J Hyg Environ Health 215: 216–219.

49. Needham LL, Grandjean P, Heinzow B, Jorgensen PJ, Nielsen F, et al. (2011) Partition of environmental chemicals between maternal and fetal blood and tissues. Environ Sci Technol 45: 1121–1126.

50. Kim S, Choi K, Ji K, Seo J, Kho Y, et al. (2011) Trans-placental transfer of thirteen perfluorinated compounds and relations with fetal thyroid hormones. Environ Sci Technol 45: 7465–7472.

51. Kim SK, Lee KT, Kang CS, Tao L, Kannan K, et al. (2011) Distribution of perfluorochemicals between sera and milk from the same mothers and implications for prenatal and postnatal exposures. Environ Pollut 159: 169–174.

52. Liu J, Li J, Liu Y, Chan HM, Zhao Y, et al. (2011) Comparison on gestation and lactation exposure of perfluorinated compounds for newborns. Environ Int 37: 1206–1212.

53. Fuentes S, Vicens P, Colomina MT, Domingo JL (2007) Behavioral effects in adult mice exposed to perfluorooctane sulfonate (PFOS). Toxicology 242: 123–129.

54. Lee HG, Lee YJ, Yang JH (2012) Perfluorooctane sulfonate induces apoptosis of cerebellar granule cells via a ROS-dependent protein kinase C signaling pathway. Neurotoxicology 33: 314–320.

55. Trudel D, Horowitz L, Wormuth M, Scheringer M, Cousins IT, et al. (2008) Estimating consumer exposure to PFOS and PFOA. Risk Anal 28: 251–269.

56. Skounti M, Philalithis A, Galanakis E (2007) Variations in prevalence of attention deficit hyperactivity disorder worldwide. Eur J Pediatr 166: 117–123.

57. Lahey BB, Applegate B, McBurnett K, Biederman J, Greenhill L, et al. (1994) DSM-IV field trials for attention deficit hyperactivity disorder in children and adolescents. Am J Psychiatry 151: 1673–1685.

58. Biederman J, Faraone SV, Weber W, Russell RL, Rater M, et al. (1997) Correspondence between DSM-III-R and DSM-IV attention-deficit/hyperactivity disorder. J Am Acad Child Adolesc Psychiatry 36: 1682–1687.

59. Baumgaertel A, Wolraich ML, Dietrich M (1995) Comparison of diagnostic criteria for attention deficit disorders in a German elementary school sample. J Am Acad Child Adolesc Psychiatry 34: 629–638.

60. Hamm MP, Cherry NM, Chan E, Martin JW, Burstyn I (2010) Maternal exposure to perfluorinated acids and fetal growth. J Expo Sci Environ Epidemiol 20: 589–597.

61. Calafat AM, Wong LY, Kuklenyik Z, Reidy JA, Needham LL (2007) Polyfluoroalkyl chemicals in the U.S. population: data from the National Health and Nutrition Examination Survey (NHANES) 2003–2004 and comparisons with NHANES 1999–2000. Environ Health Perspect 115: 1596–1602.

62. Harada K, Koizumi A, Saito N, Inoue K, Yoshinaga T, et al. (2007) Historical and geographical aspects of the increasing perfluorooctanoate and perfluorooctane sulfonate contamination in human serum in Japan. Chemosphere 66: 293–301.

63. Haug LS, Thomsen C, Becher G (2009) Time trends and the influence of age and gender on serum concentrations of perfluorinated compounds in archived human samples. Environ Sci Technol 43: 2131–2136.

64. Olsen GW, Ellefson ME, Mair DC, Church TR, Goldberg CL, et al. (2011) Analysis of a homologous series of perfluorocarboxylates from American Red Cross adult blood donors, 2000-2001 and 2006. Environ Sci Technol 45: 8022–8029.

10

Temporary Storage or Permanent Removal? The Division of Nitrogen between Biotic Assimilation and Denitrification in Stormwater Biofiltration Systems

Emily G. I. Payne[1]*, **Tim D. Fletcher**[2], **Douglas G. Russell**[3], **Michael R. Grace**[3], **Timothy R. Cavagnaro**[4], **Victor Evrard**[3], **Ana Deletic**[1], **Belinda E. Hatt**[1], **Perran L. M. Cook**[3]

1 Monash Water for Liveability, Department of Civil Engineering, Monash University, Victoria, Australia, **2** Department of Resource Management and Geography, Melbourne School of Land and Environment, The University of Melbourne, Victoria, Australia, **3** Water Studies Centre, School of Chemistry, Monash University, Victoria, Australia, **4** School of Biological Sciences, Monash University, Victoria, Australia

Abstract

The long-term efficacy of stormwater treatment systems requires continuous pollutant removal without substantial re-release. Hence, the division of incoming pollutants between temporary and permanent removal pathways is fundamental. This is pertinent to nitrogen, a critical water body pollutant, which on a broad level may be assimilated by plants or microbes and temporarily stored, or transformed by bacteria to gaseous forms and permanently lost via denitrification. Biofiltration systems have demonstrated effective removal of nitrogen from urban stormwater runoff, but to date studies have been limited to a 'black-box' approach. The lack of understanding on internal nitrogen processes constrains future design and threatens the reliability of long-term system performance. While nitrogen processes have been thoroughly studied in other environments, including wastewater treatment wetlands, biofiltration systems differ fundamentally in design and the composition and hydrology of stormwater inflows, with intermittent inundation and prolonged dry periods. Two mesocosm experiments were conducted to investigate biofilter nitrogen processes using the stable isotope tracer $^{15}NO_3^-$ (nitrate) over the course of one inflow event. The immediate partitioning of $^{15}NO_3^-$ between biotic assimilation and denitrification were investigated for a range of different inflow concentrations and plant species. Assimilation was the primary fate for NO_3^- under typical stormwater concentrations (~1–2 mg N/L), contributing an average 89–99% of $^{15}NO_3^-$ processing in biofilter columns containing the most effective plant species, while only 0–3% was denitrified and 0–8% remained in the pore water. Denitrification played a greater role for columns containing less effective species, processing up to 8% of $^{15}NO_3^-$, and increased further with nitrate loading. This study uniquely applied isotope tracing to biofiltration systems and revealed the dominance of assimilation in stormwater biofilters. The findings raise important questions about nitrogen release upon plant senescence, seasonally and in the long term, which have implications on the management and design of biofiltration systems.

Editor: Marie-Joelle Virolle, University Paris South, France

Funding: This research was supported by funding from the Australian Research Council (LP0990153)(http://www.arc.gov.au/), Melbourne Water (http://www.melbournewater.com.au/Pages/home.aspx), and Western Australia Department of Water (http://www.water.wa.gov.au/). Fletcher is supported by an ARC Future Fellowship (FT100100144). The funders assisted with the study design and discussions of overall project results but had no role in data collection and analysis, decision to publish, or preparation of the manuscript.

Competing Interests: The authors have declared that no competing interests exist.

* E-mail: emily.payne@monash.edu

Introduction

The performance of stormwater biofilters (also known as bioretention systems or raingardens) has traditionally been expressed in terms of simple pollutant removal. Few studies consider the permanency of this removal, yet many processes in such systems may be better described as attenuation - when retention is only temporary and the pollutant is at some point re-released, either in its original or transformed state. The fate of a pollutant between temporary and permanent removal pathways is fundamental to long-term performance. Nitrate is a critical waterway pollutant with possible transformations in both of these categories – biotic assimilation provides temporary immobilization, or denitrification offers permanent removal in gaseous form. While nitrogen transformation and cycling processes have been thoroughly studied in other

characterised across wide natural and engineered environments, they have not been explicitly quantified in the unique conditions of stormwater biofilters. This leaves the long-term efficiency of biofilter nitrogen treatment open to question and constrains the potential for future design improvements.

Biofiltration typically consists of a vegetated layer of sandy loam overlying sand and gravel layers, designed to capture, infiltrate and treat urban stormwater runoff before discharge downstream or into the surrounding environment or collection for harvesting [1,2]. Like wastewater treatment wetlands, biofilters are engineered systems which harness natural biogeochemical processes. However, biofilters differ fundamentally from wetlands as a result of stormwater inflows and infiltration. While biofilters share some common design features with vertical flow wetlands, they are distinguished by being ephemeral, fed by urban intermittent stormwater runoff, which differs substantially from wastewater in composition and inflow hydrology [3,4]. This leads to large, irregular variances in inundation, soil moisture and potentially nutrient, carbon and oxygen availability. As a result, biofilters are typically vegetated with terrestrial and semi-terrestrial plant species. Such differences likely alter the dominant nitrogen processes and drivers between treatment wetlands and stormwater biofilters. Characterising pollutant fate within stormwater biofilters is necessary not only for the optimal design of systems, but also to understand their long-term performance and determine suitable maintenance regimes.

Nitrogen is an essential nutrient in all biomass, but its natural cycling has been substantially altered by anthropogenic inputs and as a result forms a major contaminant of surface and ground waters [5]. Consequently, nitrogen processing has been extensively studied across terrestrial, semi-terrestrial and aquatic environments. This knowledge can be applied to infer possible nitrogen removal pathways in stormwater biofilters. Incoming nitrogen associated with urban stormwater runoff may undergo a range of potential fates, including assimilation, transformation by microbial processes (including nitrification, denitrification, dissimilatory nitrate reduction to ammonium (DNRA)), abiotic processes (including filtration and adsorption), or leaching from the system [6]. Based on research in other environments, the key fates for nitrate, a mobile inorganic form of nitrogen, are expected to be biotic assimilation (uptake by plants, bacteria, fungi or other microbes) and denitrification (conversion into gaseous forms primarily N_2 or N_2O) [7]. Assimilated nitrate is subsequently converted into an array of organic compounds and stored for some period, before return upon cell death or exudation. Decomposition processes act to either lock nitrogen up for longer term storage in recalcitrant components of the soil organic matter or re-release, when it is again available for uptake, transformation or leaching [6]. Hence, in many environments temporary storage from assimilation can occur over days, years, decades and beyond [6,8]. However, in biofiltration systems concentrations of organic matter are initially very minimal in the engineered sandy substrate [9], and it is unknown if a pool will develop to provide significant long-term nitrogen storage. Other key nitrogen processes include nitrification and denitrification, which are both mediated by microbes, but require contrasting redox conditions. Many biofilter designs incorporate an upper drained layer underlain by a saturated layer, maintained using a raised outlet, which may theoretically provide zones for nitrification and denitrification respectively. A supplementary carbon source (e.g. wood chips, straw) is often mixed throughout the saturated zone to provide electrons for denitrification [10]. Despite increased nitrate removal associated with these design features [11,12], to date no study has confirmed that this is due to denitrification. Without definitively relating design features to nitrogen processes, designers may be blind to opportunities for future performance enhancement. While biofilter design has progressed substantially over the past decade [9,13], further improvements may be confined by the current lack of process knowledge. This is particularly the case for nitrogen, where the cyclical nature of assimilation and mineralisation – internal recycling – threatens to eventually overwhelm the demand for incoming nitrogen.

The removal performance quoted by most biofilter studies, which use a simple black-box input-output approach, is a reflection of predominantly short-term processes, ignoring possible long-term changes. While authors acknowledge the potential for re-mobilisation and leaching of previously attenuated nitrogen [14,15], and several have investigated pollutant profiles [2,16] or estimated plant accumulation [15,17], none have explicitly

characterised nitrogen fate to account for re-release. An initial step towards this understanding is determining the immediate fate of incoming nitrogen. While such an assessment focuses on rapid processing in a short timeframe, it has implications on longer term dynamics, given that a system cannot indefinitely accumulate nitrogen [18].

Assimilation and denitrification have been quantified in a range of engineered and natural systems. Denitrification has been reported to dominate processing in wastewater treatment wetlands [7,19], wetlands treating high nitrate groundwater [20] and riparian soils receiving agricultural runoff [21]. In contrast, other studies have noted, sometimes surprisingly, the key role played by plant uptake in environments including streams [22], peat bogs [23], flooded soils planted with wetland species [24] and grassed buffer strips [25]. Factors driving the division may include carbon and nitrogen availability, nitrogen speciation, sediment characteristics, plant species morphology and physiology, plant density and hydrological regime [26,27]. The dominant pathways have not been identified in biofilters, yet these influential factors may differ substantially under the characteristics of stormwater biofilters.

Given the 'black-box' approach of most biofiltration research to date, this experiment was designed to provide an initial investigation into nitrogen processes. The experiment aimed to quantify the immediate nitrate transformation pathways in stormwater biofilters, focusing upon the initial division between assimilation and denitrification, with the specific objectives to investigate the:

(1) Effect of TN influent concentration on nitrate removal pathways in biofilter mesocosms planted with *Carex appressa*

(2) Effect of plant species on nitrate removal pathways within laboratory-scale biofilters

This study uniquely applied isotope tracing techniques to stormwater biofilters in order to quantify processes, an approach which, to the authors' knowledge, has not previously been reported. In addition, few studies in other fields have applied isotope tracer across plant species, despite considerable interaction between species and nitrogen cycling [28]. Quantification across a loading gradient provides a basis to understand how results might vary across systems and inform comparisons. Overall, the study aimed to indicate if nitrogen may accumulate within components of the biofilter over time, which has critical implications for system design, maintenance and life span.

Methods

Experimental overview

This study is made up of two components. The first tested the effect of influent concentration on nitrate removal pathways in mesocosms planted with *Carex appressa*, selected as a high performing species in biofilters [14,29]. In the second, nitrogen pathways were tracked in biofilter columns planted with various species. Both experiments use a laboratory-scale approach to provide insight into biofilter processes under controlled conditions, and include non-vegetated controls.

Key differences between the experiments include their establishment period, depth and configuration of the saturated zone, replicate number, and the mixture of nitrogen species added (Table 1 and Figure 1). The *Carex appressa* influent concentration experiment was conducted under fully saturated conditions (Figure 1(a)), providing a simplified design to specifically investigate the effect of nitrate loading on biofilter saturated zone processes, which have been hypothesized as of primary impor-

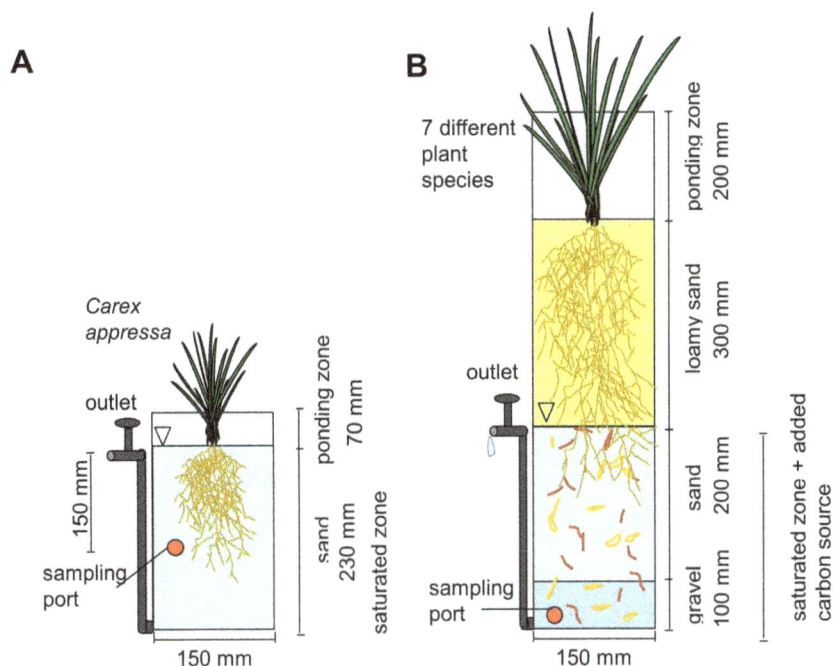

Figure 1. Experiment Configuration. A.) Influent concentration experiment (under fully saturated conditions) and B.) Multiple plant species experiment (with saturated zone overlaid by a non-saturated zone). Note diagrams are not drawn to scale.

tance in biofilter nitrogen removal performance [10,12]. The experiment across plant species was established over a much longer period of plant growth and stormwater application, and incorporated an unsaturated layer above the saturated zone (typical of most vertical-flow stormwater biofilters, Figure 1 (b)).

Experimental setup

(i) Influent concentrations experiment. Twenty mesocosms (4 replicates for each tested nitrogen inflow concentration) were constructed using 150 mm diameter polyvinyl chloride (PVC) pipe and 230 mm of washed landscaping sand. Access holes were drilled and blocked using 12 mm butyl septa to allow sampling. A constant height of water was maintained to saturate the entire sediment. Two week old *Carex appressa* seedlings were planted into 16 of the mesocosms and grown for 8 weeks in a greenhouse alongside 4 non-vegetated controls.

(ii) Multiple plant species experiment. Twenty-four single-plant biofilter columns (3 replicates for each of the 7 tested plant species and non-vegetated control) were grown in an open-air roofed greenhouse for a period of 11 months. The plants were originally sourced as tubestock and established in planter bags (300 mm by 150 mm) using tapwater for 6 months. The columns were constructed using PVC pipe with a clear Perspex ponding zone and filled with loamy sand overlying sand and gravel layers. A raised outlet tap allowed a 300 mm saturated zone with a carbon source of pine chips and sugar cane mulch mixed throughout to form 5% by volume (Figure 1), typical of common stormwater biofilter design guidelines [9]. Seven plant species (Table 1) were selected to cover a range of plant forms and include both high and poor performing species for nitrogen removal in biofilters, determined from previous sampling (not presented here). These columns formed part of a broader 18 month study, which investigated nitrogen cycling across plant species and design configurations.

Dosing and sampling

(i) Influent concentrations experiment. The mesocosms were dosed with varying concentrations of the modified Long-Ashton nutrient solution containing a 1:1 ratio of $Na^{14}NO_3$:-$Na^{15}NO_3$ ($Na^{15}NO_3 > 98\%$ isotopic purity, Cambridge Isotope Laboratories), as described in Cavagnaro et al. [30]. The molar amount of NO_3^- was matched with NH_4^+. Four vegetated mesocosms and four non-vegetated control mesocosms received the equivalent of 1 mg N/L (NO_3^- and NH_4^+) and 0.3 mg P/L. This comprised the 'Low' inflow concentration. The remaining vegetated mesocosms were dosed with 'Medium' = 2 mg N/L, 0.6 mg P/L, 'High' = 10 mg N/L, 2.8 mg P/L and 'Very high' = 20 mg N/L, 5.6 mg P/L concentrations. This range of nutrient concentrations covers the typical range of urban stormwater runoff (1–3 mg N/L) [3] and beyond, towards concentrations more typical of wastewater. A weekly dose of 1.63 L was applied, which flushed the mesocosm by one pore volume based on a sand porosity of 0.4.

Before tracer addition, samples were collected from each inflow solution and the pore water, and analysed for background concentrations of NH_4^+, NO_3^-, N_2 and N_2O. Pore water samples were removed using a 25 mL plastic syringe with an 18-gauge needle. Samples for NH_4^+ and NO_3^- were filtered through a 0.45 μm membrane filter (MicroAnalytix 30PS045AN) into a 12 mL container and frozen for subsequent analysis. Water samples for analysis of N_2 and N_2O concentrations were transferred into a 12.5 mL glass gas-tight vial (Exetainer, Labco). Zinc chloride (250 μL, 50% w/v) was added as a preservative [31]. This sampling regime was repeated 6 hourly for a period of 30 hours. Two pulses of isotope tracer were added, one in July and another in August 2012.

(ii) Multiple species experiment. The columns were dosed with semi-natural stormwater twice weekly, following the method outlined in Bratieres et al. [14] with a target Total Suspended Solids concentration of 150 mg/L and nutrient concentrations

Table 1. Summary of experiment details.

Experiment	1. Influent concentration	2. Multiple plant species
Description	Tested effect of 4 different influent N and P concentrations on NO_3^- partitioning between denitrification, pore water and vegetation, using single plant species *Carex appressa* and non-vegetated controls.	Tested effect of 7 different plant species and non-vegetated control on NO_3^- partitioning between denitrification, pore water and vegetation using constant influent composition of 'typical' stormwater.
	150 mm diameter PVC pipe containing 230 mm washed sand with constant saturation maintained.	150 mm diameter PVC pipe with layers of loamy sand, sand and gravel (Figure 1) across 600 mm. Saturated conditions maintained in lower 300 mm with sugar cane mulch and pine chips mixed throughout.
Stormwater application	Weekly dose of 1.63 L (~one total pore volume)	Twice-weekly dose of 3.7 L (to Vic plant species, see below) and 4.2 L (to WA plant species) in accordance with local rainfall (~one total pore volume)
	1 mg N/L, 0.3 mg P/L (non-vegetated control and 'low' dose)	~2.2 mg N/L, 0.36 mg P/L (all columns)
	2 mg N/L, 0.6 mg P/L ('medium')	
	10 mg N/L, 2.8 mg P/L ('high')	
	20 mg N/L, 5.6 mg P/L ('very high')	
	Modified Long-Ashton nutrient solution [30] – (~50% NO_3^-, 50% NH_4^+ by weight)	Semi-natural urban stormwater with 'typical' components [3,32] – (~45% NO_3^-, 18% NH_4^+, 37% organic N by weight), ~150 mg/L TSS
	Tracer added twice – July and August	Tracer added once - October
Plant species	*Carex appressa* (sedge)	*Carex appressa* (sedge; Vic)
		Palmetto Soft Leaf Buffalo (lawn grass/Vic)
		Dianella revoluta (sedge; Vic)
		Juncus kraussii (reed; WA)
		Allocasurina littoralis (tree; Vic)
		Leptospermum continentale (tree; Vic)
		Hypocalymma angustifolium (tree; WA)
Initial plant growth period	8 weeks	>17 months (includes 11 months in columns with twice-weekly of stormwater application)
Location	Controlled greenhouse	Open-air roofed greenhouse
Number of replicates	4 replicates	3 replicates

based on typical stormwater composition [3,32]. The average nitrogen composition delivered is shown in Table 1 and comprised approximately 2.2 mg N/L and 0.36 mg P/L. The dose volume reflected a biofilter sized to 2.5% of its impervious catchment area, a twice weekly watering frequency and the annual average effective rainfall for Melbourne (Victoria) and Perth (Western Australia). Columns with Victorian (Vic) species were dosed with 3.7 L and Western Australian (WA) species with 4.2 L (Table 1). Buffalo, a common lawn grass, was cut regularly using scissors to simulate mowing. Following 11 months of stormwater dosing, the influent was enriched with $Na^{15}NO_3^-$ in October 2011. Inflow samples were collected before and after tracer addition. Pore water samples were collected approximately 1.5 cm from the base of the saturated zone and processed as described for the previous experiment. An O_2 minisensor (2.5 mm tip diameter) connected to a FireSting O_2 oxygen meter (PyroScience GmbH, Germany) immediately measured dissolved oxygen concentrations in the sample. This sampling procedure was validated in the laboratory by collecting samples from anoxic water (created by 20 minutes of Argon gas bubbling). The anoxic water recorded an average of 1.4% air saturation ($\pm$0.1 standard deviation), but following sample collection using a syringe, plastic tubing and exetainer, the average dissolved oxygen concentration was 5.4% air saturation ($\pm$0.4), indicating the sampling procedure introduced a small amount of O_2. Therefore, samples recording around 7% air

saturation or lower can be considered anoxic. Sampling using a continuous oxygen probe was attempted but the fragile probes were repeatedly damaged when inserted into the biofilters. It should also be noted measurement of pH was not deemed necessary as stormwater influent and effluent from similar laboratory biofilters has been previously measured near neutral [33], CO_2 production acts as a buffer against potentially low pH and denitrifiers operate under a wide range of conditions [34].

Six sets of samples were collected across 5 days from 25th–29th October 2011. Samples were collected initially as each column finished draining, the next day in the morning and afternoon, then daily in the afternoon and following the next stormwater dosing (4 days after tracer addition).

The effluent from each column was sampled on 2nd November 2011 to determine concentrations of total nitrogen (TN), total phosphorus (TP), filterable reactive phosphorus (FRP), total dissolved nitrogen (TDN), NH_4^+ and NO_x, as per the method outlined in Payne et al. [26].

Laboratory analyses - NH_4^+, NO_3^-, N_2 and N_2O

Concentrations of NH_4^+ and NO_x in the pore water were analysed using standard flow injection analysis methods in a NATA (National Association of Testing Authorities) accredited lab. Dissolved N_2O, $^{28}N_2$, $^{29}N_2$ and $^{30}N_2$ were quantified in the water samples after a 4 mL He (99.9%) headspace was placed in the vials and equilibrated by vigorously shaking for 5 minutes. The concentration of N_2O was analysed by injecting a 100 μL sample of the headspace by gas chromatography (Hewlett Packard 5710A Gas Chromatograph). The total amount of N_2O in the vials was calculated using Henry's law [35]. N_2 was analysed on an ANCA GSL2 elemental analyser coupled to a Hydra 20–22 isotope ratio mass-spectrometer (IRMS; Sercon Ltd., UK).

Data analysis

$^{28}N_2$, $^{29}N_2$ and $^{30}N_2$ production and denitrification rates. A linear regression was fitted to the amount of excess $^{29}N_2$ and $^{30}N_2$ over time to calculate a production rate. The rate was adjusted to compensate for the loss rate, determined from a linear regression across the decline in labelled N_2 in the later portion of the time series. The rate of ^{15}N denitrification (D_{15}), ^{14}N denitrification (D_{14}), the proportion of denitrification coupled to nitrification and total denitrification (D_{total}) were calculated from the rates of $^{30}N_2$ and $^{29}N_2$ production, p $^{30}N_2$ and p $^{29}N_2$, respectively and the ratio of $^{14}NO_3^-/^{15}NO_3^-$ [31].

Mass balance. The mass balance was calculated over a period of 12 hours for all treatments, consistent with the period in which denitrification occurred in the Carex appressa mesocosm study (illustrated later by the rapid rise and peak in $^{30}N_2$ and $^{29}N_2$ in Figure 2). The total amount of ^{15}N denitrified was calculated by integrating the rate of denitrification. This estimate of denitrification is expected to be conservative (i.e. an overestimate) given denitrification does not commence until anaerobic conditions establish, but achieved our objective of determining the maximum denitrification occurring within the systems. Both experiments were dosed with approximately one pore volume, resulting in negligible loss of ^{15}N as outflow. The amount of $^{15}NO_3^-$ remaining in the pore water was calculated using the input ratio ^{15}N:$^{14+15}N$ and the pore water NO_3^- concentration at 12 hours. If no sample was collected at this point, a concentration was linearly interpolated across the surrounding sampling time points. The proportion of $^{15}NO_3^-$ assimilated was calculated as the difference between the total amount of $^{15}NO_3^-$ added and the amount denitrified and remaining pore water $^{15}NO_3^-$.

Correlation and Significance Analysis. Pearson's product-moment correlation was used to determine if there was a relationship between the proportions assimilated and denitrified, and the proportions denitrified and remaining in the pore water. As the data were non-normally distributed in some cases, the use of Spearman rank correlation was used to confirm these. We tested for significant differences in assimilation between species, and in denitrification rate between nitrogen loadings using the non-parametric Kruskal-Wallis test. A critical value of $\alpha = 0.05$ was used for hypothesis tests. Analyses were performed using the car Package [36] within the R Software Environment [37]. Michaelis-Menten curves were fitted using the drc Package [38] in R.

Results

Effect of TN influent concentration on nitrate removal pathways in Carex appressa mesocosms

Concentrations of NH_4^+ and NO_3^- in the pore water decreased rapidly within 24 hours in all treatments, except the non-vegetated controls which produced NO_3^-, as a result of significant nitrification. Consistent with the decline in NO_3^- in all vegetated treatments, $^{29}N_2$ and $^{30}N_2$ were produced over the first 12 hours after dosing, followed by a decrease in their concentrations thereafter (Figure 2 demonstrates these patterns for the very high nutrient treatment). Coupled nitrification-denitrification comprised approximately one third of denitrification in the high treatment. The production of N_2O in the pore water represented an insignificant fraction of the nitrogen budget (<1%) and was therefore not considered further.

Rates of denitrification were extremely low in the control and low nutrient dosed columns (<25 μmol m^{-2} h^{-1}), but increased sharply with higher nutrient loading, reaching a maximum rate in the high treatments (600 and 1800 μmol m^{-2} h^{-1}, during July and August respectively) before decreasing in the very high treatment (Figure 3). There were significant differences in denitrification rate between the treatments in both July and August ($p < 0.05$).

There was negligible assimilation or denitrification of nitrate in the non-vegetated control columns; most nitrogen was recovered

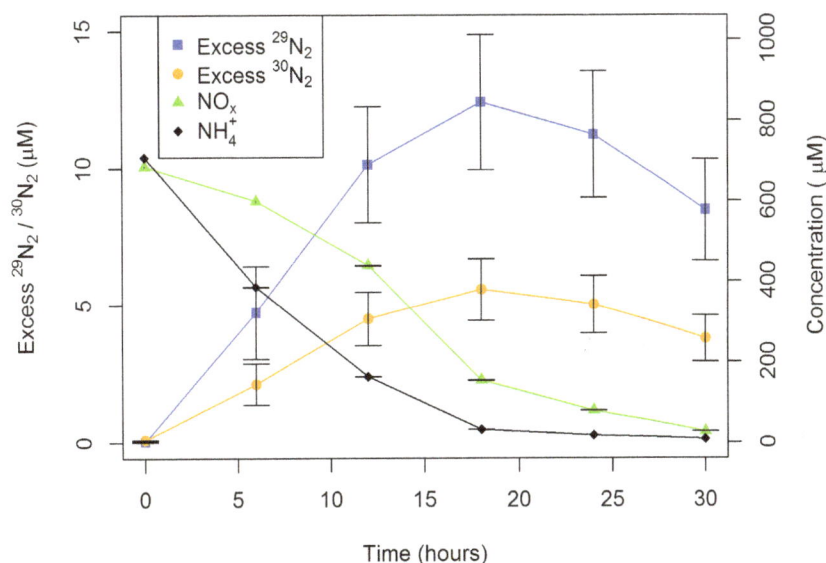

Figure 2. Nitrogen species concentrations. Examples of time series NH_4^+, NO_x, excess $^{29}N_2$ and $^{30}N_2$ concentrations (± standard error (n = 4)) following dosing in the influent concentration experiment under very high nutrient dosing (20 mg N/L) measured in July 2012.

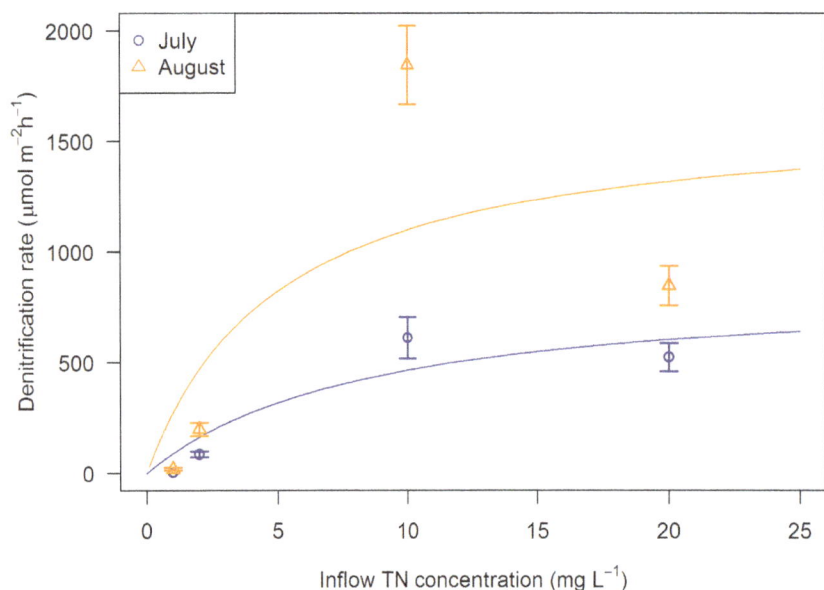

Figure 3. Rates of denitrification ($^{14}N+^{15}N$) against inflow TN concentration. Measured in the influent concentration experiment ($\pm$ standard error (n = 4)) during July and August. Michaelis-Menten curves were fitted to give V_{max} = 861 µmol m^{-2} h^{-1} and K_m = 8.46 mg L^{-1} in July and V_{max} = 1653 µmol m^{-2} h^{-1} and K_m = 5.01 mg L^{-1} in August.

as NO_3^- after 12 hours (Figure 4). In the vegetated columns, assimilation dominated in the low dose mesocosms, but generally decreased with higher loading. The fraction of $^{15}NO_3^-$ denitrified increased with nitrogen loading alongside the proportion remaining as nitrate. However, in the very high treatment denitrification declined and assimilation increased again.

Effect of plant species on nitrate removal pathways within biofilters

The vegetated biofilter columns effectively reduced concentrations of TN and TP in the stormwater from 2.1 mg N/L and 0.31 mg P/L to averages of 0.27 mg N/L and 0.01 mg P/L respectively. NH_4^+ concentration reductions were high irrespective of plant species or the presence of vegetation, reduced from 0.4 mg N/L to <0.05 mg N/L. NO_x removal was also high but more variable; effluent concentrations ranged from 0.001–0.27 mg N/L from an influent concentration of 1.0 mg N/L.

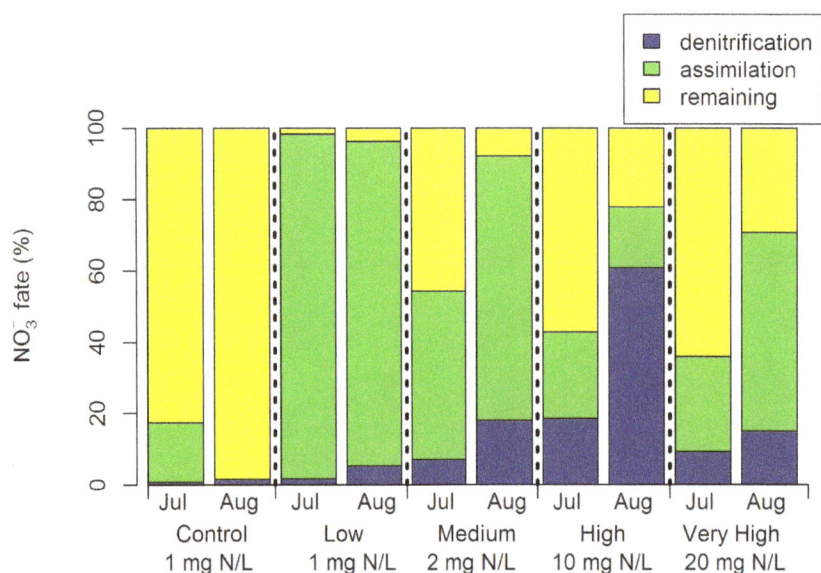

Figure 4. Division of $^{15}NO_3^-$ between denitrification, plant or microbial assimilation and remaining as $^{15}NO_3^-$ within the pore water. Measured 12 hours after dosing during July and August in the control (non-vegetated) and low, medium, high and very high (vegetated) nutrient dosing rates (n = 4).

Removal of organic nitrogen was also effective with dissolved and particulate forms on average reduced from 0.4 mg N/L and 0.3 mg N/L to 0.12 mg N/L and 0.05 mg N/L respectively. The non-vegetated controls were less effective, with outflow concentrations averaging 1.47 mg N/L TN and a net production of NO_x to 1.27 mg N/L.

Similar patterns for NO_3^-, $^{29}N_2$ and $^{30}N_2$ illustrated by Figure 2 for the pore water were also evident for the multiple species column experiment. Concentrations of NO_3^- declined rapidly within 24 hours of tracer addition in all vegetated treatments, but remained elevated with some production in the non-vegetated controls, again indicating nitrification in these systems similarly to the mesocosm experiment. Concentrations of $^{29}N_2$ and $^{30}N_2$ increased rapidly, generally peaking between 25 and 45 hours, but continued to increase in the non-vegetated controls. N_2O production was again minimal (<1% of $^{15}NO_3^-$) and was thus ignored in the mass balance. Concentrations of dissolved oxygen within the saturated zone rapidly declined towards anoxia, with a sharp decline measured across the first 22 hours for all species excluding Buffalo grass and the non-vegetated control which demonstrated a slower decline (Figure 5). Given the introduction of a small amount of oxygen during sample collection (up to ~7% air saturation) and the expected commencement of denitrification <0.5 mg/L (or approximately <5% air saturation) [39], appropriate conditions for denitrification were considered to have occurred for most columns within 22 hours. In addition, the concentrations of dissolved oxygen in vegetated columns were significantly lower than in the non-vegetated columns across the first and second sampling times (collected an average of 5 hours and 22 hours after tracer addition respectively).

Biotic assimilation was the primary fate of $^{15}NO_3^-$ in all vegetated columns; ranging between 58% and 100% of $^{15}NO_3^-$ (Figure 6). While an average of 88% (±7% standard error, n = 3) of $^{15}NO_3^-$ was assimilated, individual species differed significantly (p<0.05). The lowest uptake was associated with columns planted with *Dianella* and *Hypocalymma* which assimilated 58–80% and 69–85% respectively.

Assimilation and denitrification were inversely related (Supplementary information, Figure S1, r = −0.79, p<0.05). Denitrification was only a minor removal mechanism in the vegetated columns, providing a sink for, on average, only 3% (±2%) of $^{15}NO_3^-$ and ranging to a maximum of 5 to 8% across columns planted with *Dianella* and *Hypocalymma* species. These same treatments also had a greater proportion of $^{15}NO_3^-$ remaining in the pore water, demonstrated by a positive relationship between denitrification and nitrate remaining for vegetated columns (Figure S2, r = 0.66, p<0.05). In the non-vegetated controls assimilation was low, accounting for 19 to 26% of $^{15}NO_3^-$ fate. Instead $^{15}NO_3^-$ primarily remained in the pore water, and 3 to 4% was denitrified.

Discussion

Assimilation as a key biofilter pathway

Denitrification only formed a minor removal mechanism at typical stormwater concentrations in the biofilter columns and mesocosms. Biotic assimilation (uptake by plants and microbes) functioned as the major sink for incoming nitrate. The minimal contribution from denitrification is somewhat unexpected, given the focus on designing biofilters to promote denitrification [10,12] and the dominance of denitrification in many treatment wetlands and some aquatic systems [7,40]. Denitrification has been reported to account for 60–95% of removal in wastewater treatment wetlands [19], 89–96% in wetlands treating high nitrate groundwater [20], and up to 61–63% in riparian wetland soils treating agricultural runoff [21]. Denitrification can also be a critical process in semi-aquatic and terrestrial systems, including the soils of urban retention basins and parks [41].

However, the critical role of assimilation has also been noted in many studies, where plant and microbial assimilation make a significant, if not dominant, contribution to the removal of nitrate. This has been observed across riparian zones [42], flooded soils planted with wetland plants [24] and vegetated streams [22]. Assimilation accounted for 75% of nitrate retention in headwater

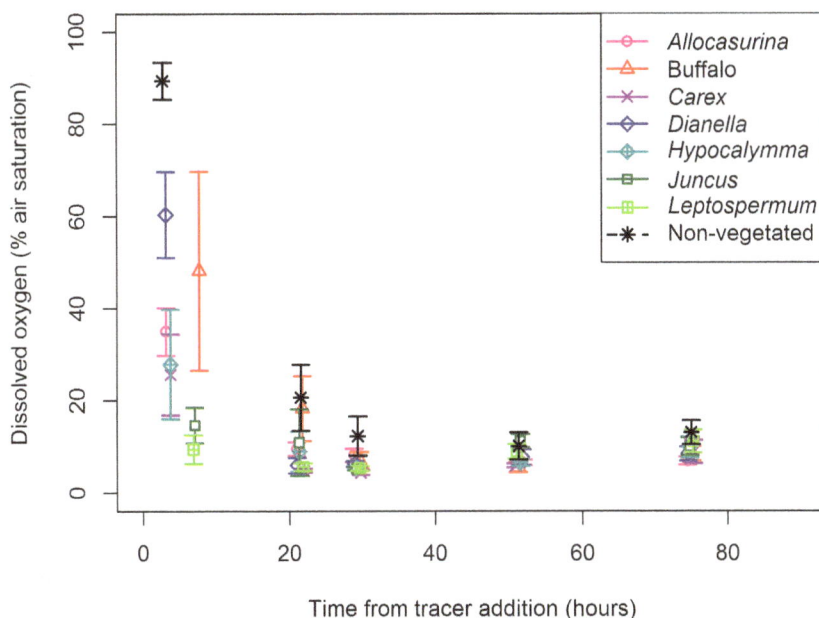

Figure 5. Change in pore water dissolved oxygen. Dissolved oxygen (% air saturated) (± standard error (n = 3) at base of columns across sampling period. Note the sample collection method introduced up to ~7% air saturation.

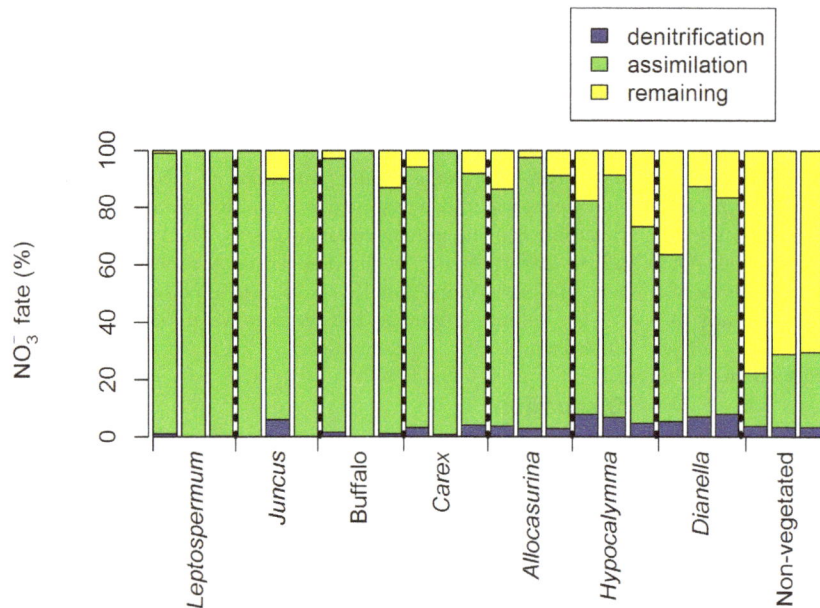

Figure 6. Division of $^{15}NO_3^-$ between denitrification, plant or microbial assimilation or remaining in the pore water. Results for each biofilter column in the multiple species experiment (3 replicates per species, n = 1).

streams [22], almost all nitrate deposited atmospherically on peat bogs [23] and was the primary removal mechanism in grassed buffers treating agricultural irrigation runoff [25].

The variation in nitrate fate between wastewater treatment wetlands and biofilters may result from their fundamental differences. While both are high nutrient, engineered and vegetated systems, biofilters generally experience greater moisture fluctuations, including prolonged drying. Hence, even with an underlying saturated zone, biofilter redox potential is dynamic. For optimal hydraulic and nutrient performance biofilter media is designed to be relatively free-draining with low organic matter or clay content [2]. As a result the system characteristics differ greatly from organic-rich and anoxic wetland sediments. Vertical sub-surface flow wetlands (which operate on similar principles to biofilters) can show particularly low denitrification relative to other treatment wetlands due to oxygenated conditions in the sediment [7], and this may similarly limit denitrification in the saturated zone of biofilters. While anoxic conditions develop (Figure 5), the influent is oxygenated.

It is questionable if assimilation will remain the dominant pathway throughout the biofilter lifespan. Denitrification may increase as organic matter accumulates [41,43] and the uptake capacity of the plant biomass may diminish over time [27,28]. The division between assimilation and denitrification will depend upon the magnitude of nitrogen immobilised in organic material and the availability of carbon, oxygen and nitrate. These dynamics may be sensitive to plant species and nitrate loading, as discussed further in the following sections.

In contrast to nitrate, ammonia reduction from the stormwater was effective regardless of the presence of vegetation or plant species. Removal processes included uptake, nitrification, and there was evidence of some coupled nitrification-denitrification in vegetated treatments.

Adaptability of denitrifying bacteria

The importance of denitrification as a removal mechanism increased as nitrogen concentrations rose towards those of

wastewater (~10 mg N/L). The results suggest that the denitrifying bacteria are largely utilising the nitrate remaining after assimilation and that this portion grows as loading increases (Figure 7). In contrast, plant assimilation diminished in proportion – likely becoming constrained by other growth requirements or uptake rate after a critical point [7,44]. Hence, denitrifying bacteria appear to have the adaptability to increase nitrate processing to a greater extent than plant and microbial assimilation, but remain dependent upon plant-derived carbon. As nitrate loading increases, the key role of plants may shift from assimilation to facilitation of denitrification.

The *Carex apressa* mesocosms represented a simplified experiment designed to investigate the influence of loading rate on the biofilter saturated zone; this experiment lacked the upper unsaturated layer or longer term loading of the multiple species biofilter column experiment, However, the consistency between the two experiments in terms of the proportion denitrified (7–18% mesocosms and 0–8% vegetated biofilter columns) and denitrification rate (88–202 µmol m^{-2} hr^{-1} and <1–635 µmol m^{-2} hr^{-1} respectively), when the same nitrate concentration was applied, suggests the results across the loading gradient can be meaningfully applied to biofiltration systems.

Other studies have also noted increases in denitrification across a loading gradient, but findings differ in the nature of this increase and relative change in assimilation. Denitrification and assimilation may both increase with loading, either linearly [22], or assimilation may increase at a much lower rate than denitrification [43], or denitrification rates may plateau (e.g. at loadings of around 5 mg N/L) [40,44]. In the current study the denitrification rate appeared to increase rapidly, but reach a plateau, before declining at the highest loading rate (20 mg N/L comprising 10 mg NO_3^--N/L). This may be due to a lower investment in root biomass by plants under the nutrient-rich conditions, which may result in less carbon to drive denitrification. Alternatively, the population of denitrifying bacteria may be inhibited by their generally facultative nature or other factors such as plant allelopathy, preventing optimal functioning in response to higher

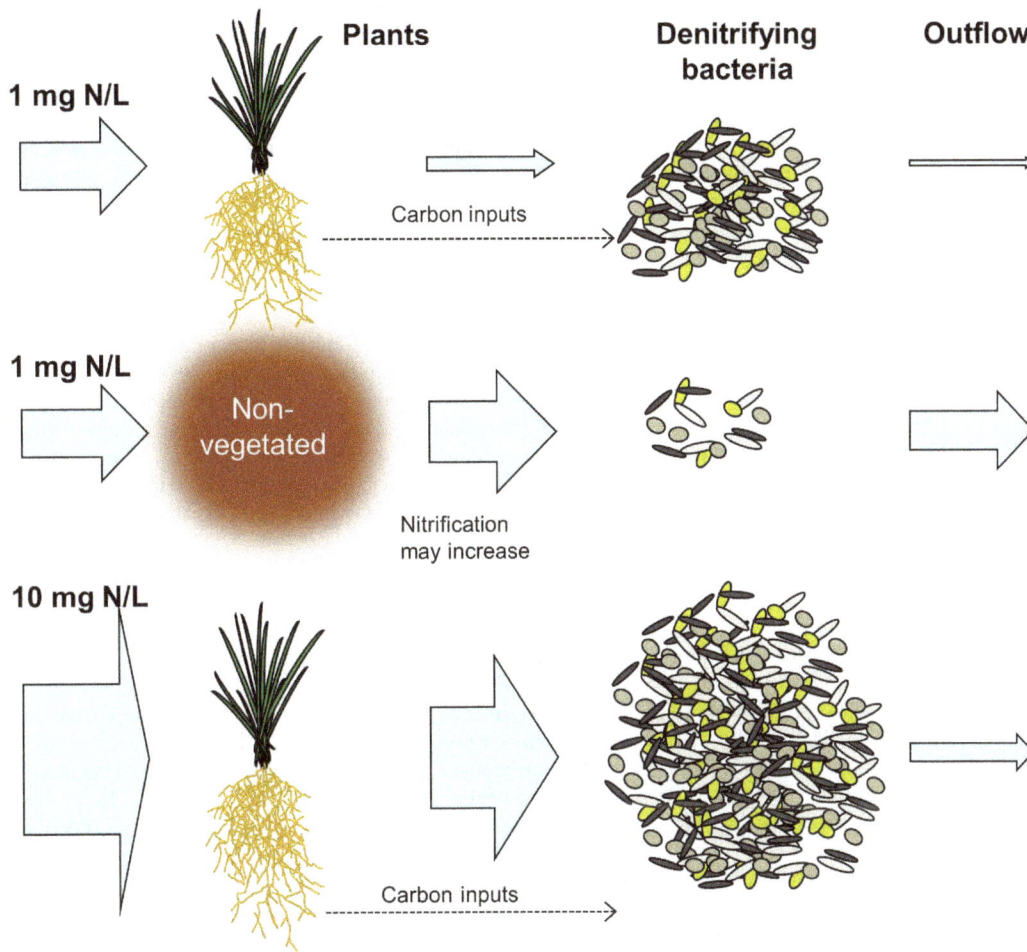

Figure 7. Conceptual diagram illustrating nitrate processing. Removal by assimilation and denitrification at different nitrogen loadings in vegetated and non-vegetated systems. The dependence of denitrifying bacteria on plant-derived carbon is also represented.

nitrate concentrations. However, data from the current study were inconclusive and testing these hypotheses will require further research. Nonetheless, the findings of the current study and others generally suggest that high loading leads to increased processing, although the proportion of nitrate retained by assimilation and denitrification is likely to decline across the loading spectrum as efficiency decreases [20], pathways saturate and mineralisation rates increase [44].

Synergistic relationships

While the discussion has so far focused on the division between biotic assimilation and denitrification, the two processes are not independent. Plants can facilitate denitrification, either directly or coupled to nitrification, by carbon provision (root exudates or sloughed cells) and promoting heterogeneity in redox potential; anoxic (driven by intense heterotrophic decomposition) and oxic (due to root oxygen release by some species in waterlogged conditions) [45,46]. This facilitative role of plants is highlighted by the performance of the non-vegetated control columns. In these treatments net nitrate production was commonly observed, indicating nitrification, which is facilitated by higher nitrate availability in the absence of plant-derived carbon (fuelling heterotrophic respiration) and plant assimilation (the small portion attributed to assimilation (Figure 6) was likely associated with bacteria consuming high C:N ratio organic matter from the

carbon source). Despite the availability of nitrate penetrating the saturated zone and the provision of a carbon source in the non-vegetated treatment, denitrification still failed to dominate processing and was higher in columns planted with *Hypocalymma* or *Dianella*. In the influent concentration experiment, which lacked a carbon source, denitrification was negligible in the non-vegetated controls. This illustrates the importance of root-derived carbon, which acts as a continuous and dynamic source, in driving denitrification, despite the competition from plant and microbial nitrate assimilation [21,24]. This facilitation highlights the contradiction in the close relationship between plants and microbes; it is both synergistic and competitive, but essential and inter-dependent. Root exudation also hastens the onset of anaerobic conditions [46], as observed by the significantly lower concentrations of dissolved oxygen in the vegetated biofilter columns relative to the non-vegetated controls. However, a comparison of species rates of oxygen consumption and denitrification (Figures 5 and 6 respectively) yields no clear relationship (e.g. *Leptospermum*, *Juncus* and *Carex* demonstrate relatively rapid decline in dissolved oxygen but minimal denitrification, while the reverse is apparent for *Dianella*), suggesting available carbon is not exerting primary control on denitrification.

Do differences in kinetics or opportunity dictate the division between assimilation and denitrification? In the multiple species column experiment (and similar to biofilters in the field [9]) plants had the first opportunity to access incoming nitrate in the surface unsaturated layer before it reached the underlying saturated zone with high denitrification potential. In peat bogs this mechanism allowed a 5–10 cm layer of sphagnum moss to assimilate virtually all atmospheric nitrate additions despite favourable conditions for denitrification in the underlying soil [23]. In addition, the stormwater influent oxygenated the saturated zone, delaying the onset of suitable anaerobic conditions for denitrification (Figure 5), which may have advantaged biotic assimilation irrespective of rates. Hence, the kinetic rates of assimilation and denitrification cannot be conclusively compared in the current study. Regardless, biotic assimilation of ammonium appeared more rapid than nitrate (Figure 2), as expected due to general plant preference for ammonium [24].

Plant species variation

The proportion of nitrate assimilated displayed differences among plant species, from an average 99% uptake by *Leptospermum* to 74% for *Hypocalymma* and *Dianella* treatment replicates (Supplementary information, Figure S3). Interestingly, species demonstrating higher nitrate assimilation (83–99% $^{15}NO_3^-$ processing) also tended to be more effective for use in biofilters (i.e. less nitrate remaining in the pore water) (*Leptospermum*, *Carex*, Buffalo lawn grass, *Juncus* and *Allocasurina*), despite minimal denitrification (0–6%) (Figures S4 and S5). Alongside less effective species (*Dianella* and *Hypocalymma*) assimilation was reduced (58–85%), and while denitrification increased (5–8%), it did not compensate entirely, leading to increased nitrate concentrations in the pore water (9–36%). The minimal contribution of denitrification to performance was unexpected, but is compatible with the hypothesis that denitrifying bacteria primarily receive any nitrate left over after assimilation. Many characteristics may contribute to plant species variation including morphological (such as root architecture, biomass and diameter profile) and physiological traits (including root oxygenation, biomass composition, plant strategy, seasonality, assimilation rate, nitrogen preference, mycorrhizal symbiosis and photosynthate partitioning) [28,45,47]. Studies have already identified the importance of the root zone (rhizosphere) [48] or correlated biofilter efficiency for nitrogen with long, deep roots, a high root biomass and rapid growth [29], and further relationships are the subject of a current study.

Organic nitrogen stores – beneficial or false economy?

Plant nitrogen uptake and release follow seasonal cycles and the effect in wetlands has been likened to a 'spiral' [49]. The process acts to attenuate nitrogen within the system [49] and convert inorganic forms to an array of organic compounds. The significant impact of these plant services are recognised across wetlands, aquatic systems and vegetated buffers [7,22,25]. These functions may be particularly beneficial in biofilters, given that inflow events are relatively intermittent and transient. Assimilation will slow nitrogen passage through the biofilter and re-release will occur over a relatively longer timescale, which may facilitate more effective microbial processing. The benefits of temporary nitrogen immobilisation in plants are also well acknowledged in agriculture through the use of cover crops, which are used to minimise nitrate leaching between main crops and increasingly applied to protect groundwater quality [8].

However, the benefits of storage in the biomass may extend well beyond short seasonal or annual cycles if incorporated into recalcitrant compounds. These may be stored over the long term in woody biomass [44] or incorporated into the soil organic matter (SOM), which may exceed the nitrogen storage capacity of the plant biomass by a factor of 10 [50]. Such stores could potentially

endure beyond the biofilter lifespan (generally 15–20 years [9]). In treatment wetlands, the accumulation of organic material, alongside denitrification, can both form significant pathways for nitrogen removal over the long term, even under high loading [43]. However, due to low anaerobic decomposition rates, wetlands are natural sinks for organic material [6] – it is less certain if significant accumulation will occur in the ephemeral environment of biofilters.

On the other hand, the conversion and attenuation function of plants leads to the production of nitrogen forms that require multiple processing steps before permanent removal is possible via denitrification. In particular, loss of dissolved organic nitrogen from the system is a high risk unless efficient mineralisation occurs [25]. In addition, both plants and microbes may over time increasingly source nitrogen from internal cycling processes. Harvesting the plant biomass to remove nitrogen could extend plant nitrogen demand, as has been observed in grassed systems [51]. In biofilters, however, this may be constrained by cost, the likelihood of filter media compaction [2] and concerns over pollution translocation [52].

What does this mean for long term biofilter function?

This experiment was limited by constraints inherent in studies at the laboratory scale, including small biofilter surface area, single-plant columns, regular inflows and the short time frame of the tracer experiment. In light of this, do the findings have any implications on processes in mature field-scale biofilters?

Concentrations of TN in the effluent from vegetated columns in the current study (0.11–0.45 mg N/L) are much lower than previous laboratory and field studies (typically 1 mg N/L at best [1,53]). This difference may in part reflect a design change towards media with a high sand and low nutrient content, which minimises nutrient leaching [9,26]. However, the current results do require validation under field conditions and extension by tracing nitrogen fate over longer periods. Nevertheless, the findings form an initial step in identifying and quantifying biofilter nitrogen processes, and thus represent an important advance on the predominantly "black-box" approach of studies to date.

The experiment quantified denitrification in the first hours following an inflow event. Within the multiple species column experiment the vegetated biofilters functioned effectively and little nitrate remained in the pore water after 24 hours (Figure 6), suggesting rapid initial processing, particularly alongside the most effective plant species. Given the transient nature of biofilter inflows, rapid initial processing may be an inherent characteristic of biofilters, possibly more so than it is for treatment wetlands. Hence, the results of the experiment may have some realistic implications on longer-term nitrogen fate in biofilters.

If denitrification does not form a significant long-term nitrogen sink in biofilters, ongoing performance is heavily dependent upon the capacity and duration of the biomass storage. Studies of ecosystem succession suggest both mature and early-stage systems have limited retentive capacity, and intermediate systems have the greatest potential to capture nitrogen [54]. The storage capacity increases over time as plants and a pool of soil organic matter establish [55,56]. However, growth of these storages will eventually plateau as the system moves towards a steady state [54] and nitrogen saturation [18]. At this point nitrogen returns (e.g. net nitrification) counteract the net biotic uptake, such that system inputs again equal outputs (e.g. leaching) [18]. Sustained high nutrient loading will exacerbate the saturation process and impair long-term functionality [44].

It might be expected that biofilters will similarly display an increase in performance towards this intermediate state, followed by a gradual decline in performance. However, do biofilters follow these same successional patterns and if so, is the point of zero net retention reached within the biofilter lifespan? These succession theories were developed for terrestrial forests where denitrification is negligible. If extrapolated to ephemeral biofiltration systems, the peak in performance may be sustained over a longer period of time if denitrification increases to counteract the decline in net biotic uptake.

Further work

The minor role of denitrification in early biofilter life lends greater urgency to the need to quantify nitrogen processing across the entire biofilter lifespan. If assimilation continues to be a major pathway, the temporal and quantitative dynamics of storage in the biomass or soil organic matter need to be delineated. The return flux of nitrogen from mineralisation should be incorporated into assessments of biofilter lifetime performance. In addition to the techniques employed in this paper, further understanding can also be gained using molecular biology techniques (such as qPCR), which are capable of characterising the bacterial community. As this study has demonstrated, plant species and loading can be critical influences on processes. Hence, the interconnectedness between assimilation, denitrification, plant species and cumulative loading over extended time periods require further research.

Conclusions

This study is the first known to apply a nitrate isotope tracer to the quantification of internal stormwater biofilter processes. Nitrate processing varied significantly with plant species and influent nitrogen concentration. Denitrification was only responsible for processing 0–8% of incoming nitrate in the laboratory-scale stormwater biofilters, suggesting biotic assimilation is the primary sink. Species identified as effective for reducing effluent concentrations (e.g. *Leptospermum, Carex*, Buffalo, *Juncus, Allocasurina*) tended to be associated with higher nitrate assimilation and minimal denitrification. This is surprising, given past efforts in biofilter design to promote denitrification. Instead, the denitrifying bacteria in the underlying saturated zone of biofilters appear to receive only the nitrate remaining after assimilation, such that nitrate plays a more important role in biofilters planted with species shown to be less effective in nitrogen removal. Higher nitrate loads increased the relative contribution of denitrification, implying denitrifying bacteria have greater adaptability to process high concentrations, whereas biotic assimilation becomes overwhelmed. While the results contrast with wastewater treatment wetlands, where microbial processing commonly dominates under higher loading, they are compatible with other studies highlighting the importance of plant assimilation as a nitrogen conversion mechanism and either a temporary or long-term (in soil organic matter or woody biomass) storage. This distinction from wetland functioning may reflect the unique characteristics of biofilters as quasi-terrestrial, engineered, ephemeral and highly-dynamic

systems. The results have significant implications for biofilter design, maintenance and lifespan. With biotic assimilation dominating processing in early biofilter life, the need to characterise long-term organic matter accumulation and decomposition, the influence of plant species and prevalence of denitrification in mature systems becomes far more critical.

Supporting Information

Figure S1   Relationship between percentage of $^{15}NO_3^-$ denitrified and assimilated. Results for each biofilter column in the multiple species experiment (3 replicates per species, n = 1). Linear regression line fitted to the vegetated treatments only.

Figure S2   Relationship between percentage of $^{15}NO_3^-$ remaining in the porewater and denitrified. Results for each biofilter column in the multiple species experiment (3 replicates per species, n = 1). Linear regression line fitted to the vegetated treatments only.

Figure S3   Boxplot comparison of percentage of $^{15}NO_3^-$ assimilated across the vegetated treatments. Results for replicated columns in the multiple species experiment (3 replicates per species, n = 3).

Figure S4   Boxplot comparison of percentage of $^{15}NO_3^-$ denitrified across the vegetated treatments. Results for replicated columns in the multiple species experiment (3 replicates per species, n = 3).

Figure S5   Boxplot comparison of percentage of $^{15}NO_3^-$ remaining in the porewater across the vegetated treatments. Results for replicated columns in the multiple species experiment (3 replicates per species, n = 3).

Acknowledgments

The support of the project sponsors and invaluable assistance from Tracey Pham, Richard Williamson, Frank Winston, Anthony Brosinsky, Jennifer Read, Antonietta Torre, Marion Urrutiaguer, Tania Liaghati, Keralee Browne and Tina Hines are gratefully acknowledged. The authors would also like to thank the anonymous peer reviewers for their helpful comments and suggestions to improve the quality of this paper.

Author Contributions

Conceived and designed the experiments: PLMC TDF MRG AD BEH TRC VE EGIP DGR. Performed the experiments: EGIP DGR VE PLMC. Analyzed the data: DGR EGIP VE PLMC TDF MRG TRC AD BEH. Contributed reagents/materials/analysis tools: PLMC TDF MRG AD BEH TRC VE EGIP DGR. Wrote the paper: EGIP TDF PLMC AD BEH MRG TRC DGR VE.

References

1. Hunt WF, Smith JT, Jadlocki SJ, Hathaway JM, Eubanks PR (2008) Pollutant Removal and Peak Flow Mitigation by a Bioretention Cell in Urban Charlotte, N.C. Journal of Environmental Engineering 134: 403–408.
2. Hatt BE, Fletcher TD, Deletic A (2008) Hydraulic and pollutant removal performance of fine media stormwater filtration systems. Environmental Science and Technology 42: 2535–2541.
3. Taylor GD, Fletcher TD, Wong THF, Breen PF, Duncan HP (2005) Nitrogen composition in urban runoff–implications for stormwater management. Water Research 39: 1982–1989.
4. Gray SR, Becker NSC (2002) Contaminant flows in urban residential water systems. Urban Water 4: 331–346.
5. Vitousek PM, Aber JD, Howarth RW, Likens GE, Matson PA, et al. (1997) Human alteration of the global nitrogen cycle: Sources and consequences. Ecological Applications 7: 737–750.
6. Reddy KR, DeLaune RD (2008) Biogeochemistry of wetlands science and applications. Boca Raton: Taylor & Francis, 2008.
7. Vymazal J (2007) Removal of nutrients in various types of constructed wetlands. Science of the Total Environment 380: 48–65.
8. Powlson DS (1993) Understanding the soil nitrogen cycle. Soil Use and Management 9: 86–93.
9. FAWB (2009) Adoption Guidelines for Stormwater Biofiltration Systems: Facility for Advancing Water Biofiltration, Monash University, June 2009.

10. Kim H, Seagren EA, Davis AP (2003) Engineered bioretention for removal of nitrate from stormwater runoff. Water Environment Research 75: 355.

11. Zinger T, Fletcher TD, Deletic A, Blecken GT, Viklander M (2007) Optimisation of the nitrogen retention capacity of stormwater biofiltration systems. Lyon, France.

12. Zinger Y, Blecken GT, Fletcher TD, Viklander M, Deletic A (2013) Optimising nitrogen removal in existing stormwater biofilters: Benefits and tradeoffs of a retrofitted saturated zone. Ecological Engineering 51: 75–82.

13. Davis AP, Hunt WF, Traver RG, Clar M (2009) Bioretention Technology: Overview of Current Practice and Future Needs. Journal of Environmental Engineering 135: 109–117.

14. Bratières K, Fletcher TD, Deletic A, Zinger Y (2008) Nutrient and sediment removal by stormwater biofilters: A large-scale design optimisation study. Water Research 42: 3930–3940.

15. Davis AP, Shokouhian M, Sharma H, Minami C (2006) Water Quality Improvement through Bioretention Media: Nitrogen and Phosphorus Removal. Water Environment Research 78: 284.

16. Davis AP, Shokouhian M, Sharma H, Minami C (2001) Laboratory study of biological retention for urban stormwater management. Water Environment Research 73: 5.

17. Lucas WC, Greenway M (2011) Hydraulic Response and Nitrogen Retention in Bioretention Mesocosms with Regulated Outlets: Part I - Hydraulic Response. Water Environment Research 83: 692–702.

18. Aber J, McDowell W, Nadelhoffer K, Alison M, Berntson G, et al. (1998) Nitrogen Saturation in Temperate Forest Ecosystems. Bioscience 48: 921–934.

19. Lee C-g, Fletcher TD, Sun G (2009) Nitrogen removal in constructed wetland systems. Engineering in Life Sciences 9: 11–22.

20. Lin Y-F, Jing S-R, Wang T-W, Lee D-Y (2002) Effects of macrophytes and external carbon sources on nitrate removal from groundwater in constructed wetlands. Environmental Pollution 119: 413–420.

21. Matheson FE, Nguyen ML, Cooper AB, Burt TP, Bull DC (2002) Fate of ^{15}N-nitrate in unplanted, planted and harvested riparian wetland soil microcosms. Ecological Engineering 19: 249–264.

22. Cooper AB, Cooke JG (1984) Nitrate loss and transformation in 2 vegetated headwater streams. New Zealand Journal of Marine and Freshwater Research 18: 441–450.

23. Urban NR, Eisenreich SJ, Bayley SE (1988) The Relative Importance of Denitrification and Nitrate Assimilation in Midcontinental Bogs. Limnology and Oceanography 33: 1611–1617.

24. Kirk GJD, Kronzucker HJ (2005) The Potential for Nitrification and Nitrate Uptake in the Rhizosphere of Wetland Plants: A Modelling Study. Annals of Botany 96: 639–646.

25. Bedard-Haughn A, Tate KW, Kessel Cv (2004) Using Nitrogen-15 to Quantify Vegetative Buffer Effectiveness for Sequestering Nitrogen in Runoff. Journal of Environmental Quality 33: 2252–2262.

26. Payne EGI, Pham T, Cook PLM, Fletcher TD, Hatt BE, et al (in press) Biofilter design for effective nitrogen removal from stormwater - influence of plant species, inflow hydrology and use of a saturated zone. Water Science and Technology.

27. Mitsch WJ, Gosselink JG (2000) Wetlands. New York, United States of America: John Wiley & Sons, Inc.

28. Payne EGI, Fletcher TD, Cook PLM, Deletic A, Hatt BE (in press) Processes and drivers of nitrogen removal in stormwater biofiltration. Critical Reviews in Environmental Science and Technology.

29. Read J, Fletcher TD, Wevill T, Deletic A (2010) Plant Traits that Enhance Pollutant Removal from Stormwater in Biofiltration Systems. International Journal of Phytoremediation 12: 34–53.

30. Cavagnaro TR, Smith FA, Lorimer MF, Haskard KA, Ayling SM, et al. (2001) Quantitative Development of Paris-Type Arbuscular Mycorrhizas Formed between Asphodelus fistulosus and Glomus coronatum. New Phytologist 149: 105–113.

31. Nielsen LP (1992) Denitrification in sediment determined from nitrogen isotope pairing. FEMS Microbiology Letters 86: 357–362.

32. Duncan HP (1999) Urban Stormwater Quality: A Statistical Overview. Melbourne, Australia: Cooperative Research Centre for Catchment Hydrology.

33. Blecken G-T, Zinger Y, Deletic A, Fletcher TD, Viklander M (2009) Impact of a submerged zone and a carbon source on heavy metal removal in stormwater biofilters. Ecological Engineering 35: 769–778.

34. Šimek M, Jíšová L, Hopkins DW (2002) What is the so-called optimum pH for denitrification in soil? Soil Biology and Biochemistry 34: 1227–1234.

35. Weiss RF, Price BA (1980) Nitrous oxide solubility in water and seawater. Marine Chemistry 8: 347–359.

36. Fox J, Weisberg S (2011) An {R} Companion to Applied Regression. Thousand Oaks, CA: Sage.

37. R Core Development Team (2012) R: A language and environment for statistical computing. Vienna, Austria: R Foundation for Stastistical Computing

38. Ritz C, Streibig JC (2005) Bioassay Analysis using R. Journal of Statistical Software 12.

39. Seitzinger SP (1988) Denitrification in Freshwater and Coastal Marine Ecosystems: Ecological and Geochemical Significance. Limnology and Ocean-ography 33: 702–724.

40. Kreiling R, Richardson W, Cavanaugh J, Bartsch L (2011) Summer nitrate uptake and denitrification in an upper Mississippi River backwater lake: the role of rooted aquatic vegetation. Biogeochemistry 104: 309–324.

41. Zhu W-X, Dillard N, Grimm N (2004) Urban nitrogen biogeochemistry: status and processes in green retention basins. Biogeochemistry 71: 177–196.

42. Pinay G, Ruffinoni C, Wondzell S, Gazelle F (1998) Change in Groundwater Nitrate Concentration in a Large River Floodplain: Denitrification, Uptake, or Mixing? Journal of the North American Benthological Society 17: 179–189.

43. Craft CB (1997) Dynamics of nitrogen and phosphorus retention during wetland ecosystem succession. Wetlands Ecology and Management 4: 177–187.

44. Bernot M, Dodds W (2005) Nitrogen Retention, Removal, and Saturation in Lotic Ecosystems. Ecosystems 8: 442–453.

45. Reddy KR, Patrick WH Jr, Lindau CW (1989) Nitrification-denitrification at the plant root-sediment interface in wetlands. Limnology & Oceanography 34: 1004–1013.

46. Minett DA, Cook PL, Kessler AJ, Cavagnaro TR (2013) Root effects on the spatial and temporal dynamics of oxygen in sand-based laboratory-scale constructed biofilters. Ecological Engineering 58: 414–422.

47. Tanner CC (1996) Plants for constructed wetland treatment systems — A comparison of the growth and nutrient uptake of eight emergent species. Ecological Engineering 7: 59–83.

48. Lucas W, Greenway M (2008) Nutrient Retention in Vegetated and Nonvegetated Bioretention Mesocosms. Journal of Irrigation and Drainage Engineering 134: 613–623.

49. Kadlec RH, Tanner CC, Hally VM, Gibbs MM (2005) Nitrogen spiraling in subsurface-flow constructed wetlands: Implications for treatment response. Ecological Engineering 25: 365–381.

50. Jenkinson DS (1990) An introduction to the global nitrogen cycle. Soil Use and Management 6: 56–61.

51. Bedard-Haughn A, Tate KW, Kessel Cv (2005) Quantifying the Impact of Regular Cutting on Vegetative Buffer Efficacy for Nitrogen-15 Sequestration. Journal of Environmental Quality 34: 1651–1664.

52. Collins KA, Lawrence TJ, Stander EK, Jontos RJ, Kaushal SS, et al. (2010) Opportunities and challenges for managing nitrogen in urban stormwater: A review and synthesis. Ecological Engineering 36: 1507–1519.

53. Roberts SJ, Fletcher TD, Garnett L (2012) Bioretention saturated zones; do they work at the large-scale? 7th International Conference on Water Sensitive Urban Design 21–23 February 2012. Melbourne, Australia.

54. Vitousek PM, Reiners WA (1975) Ecosystem Succession and Nutrient Retention: A Hypothesis. Bioscience 25: 376–381.

55. Odum EP (1969) The strategy of ecosystem development. Science 164: 262–270.

56. Mitsch WJ, Day JW, Zhang L, Lane RR (2005) Nitrate-nitrogen retention in wetlands in the Mississippi River Basin. Ecological Engineering 24: 267–278.

11

Gene Expression Profiling Analysis of Bisphenol A-Induced Perturbation in Biological Processes in ER-Negative HEK293 Cells

Rong Yin[1⑨], Liang Gu[2⑨], Min Li[2], Cizhong Jiang[2]*, Tongcheng Cao[1]*, Xiaobai Zhang[2]*

1 Department of Chemistry, Tongji University, Shanghai, China, **2** Shanghai Key Laboratory of Signaling and Disease Research, the School of Life Sciences and Technology, Tongji University, Shanghai, China

Abstract

Bisphenol A (BPA) is an environmental endocrine disruptor which has been detected in human bodies. Many studies have implied that BPA exposure is harmful to human health. Previous studies mainly focused on BPA effects on estrogen receptor (ER)-positive cells. Genome-wide impacts of BPA on gene expression in ER-negative cells is unclear. In this study, we performed RNA-seq to characterize BPA-induced cellular and molecular impacts on ER-negative HEK293 cells. The microscopic observation showed that low-dose BPA exposure did not affect cell viability and morphology. Gene expression profiling analysis identified a list of differentially expressed genes in response to BPA exposure in HEK293 cells. These genes were involved in variable important biological processes including ion transport, cysteine metabolic process, apoptosis, DNA damage repair, etc. Notably, BPA up-regulated the expression of ERCC5 encoding a DNA endonuclease for nucleotide-excision repair. Further electrochemical experiment showed that BPA induced significant DNA damage in ER-positive MCF-7 cells but not in ER-negative HEK293 cells. Collectively, our study revealed that ER-negative HEK293 cells employed mechanisms in response to BPA exposure different from ER-positive cells.

Editor: Zhi Xie, Sun Yat-sen University, China

Funding: This study was supported in part by grants from the Ministry of Science and Technology of China (2010CB944901, 2011CB965104, 2012AA020405); and the National Natural Science Foundation of China (91019017, 31271373, 31200952); China Postdoctoral Science Foundation (2013M531207); the Shanghai Postdoctoral Sustentation Fund (13R21416700); Aurora Talent Project of Shanghai (10SG24); the program for Eastern Scholar of Shanghai; the Fundamental Research Funds for the Central Universities (20113048, 20113109); and IRT1168 from the Ministry of Education, China. The funders had no role in study design, data collection and analysis, decision to publish, or preparation of the manuscript.

Competing Interests: The authors have declared that no competing interests exist.

* E-mail: zhangxb@tongji.edu.cn (XZ); czjiang@tongji.edu.cn (CJ); ctc@tongji.edu.cn (TC)

⑨ These authors contributed equally to this work.

Introduction

Bisphenol A (BPA) is an important industrial chemical mainly used as an intermediate in the manufacture of polycarbonate plastics and epoxy resin. BPA has become ubiquitous in the environment due to the extensive use of BPA-containing products including food and beverage packaging, flame retardants, adhesives, building materials, electronic components, and paper coatings. Human bodies are often exposed to BPA that leaches out of containers especially under high temperature and acidic conditions [1,2]. Large-scale surveys have shown that more than 90% of the study population has detectable levels of BPA in urine [3,4,5].

Due to the ubiquity of BPA exposure, more and more attention has been paid to the potential health effects induced by BPA [6,7]. BPA exhibits estrogenic properties, and has been identified as a classical endocrine disrupting chemical that can affect the endocrine system through mimicking or disrupting endogenous estrogens [7,8]. Epidemiologic studies and animal studies showed that BPA exposure contributed to numerous female reproductive disorders, and also suggested that pregnant women, fetuses, infants and children may be most vulnerable to the effects of BPA exposure [9,10,11]. BPA was first declared a toxic substance

excluded from infant formula bottles in Canada in 2010, and then was banned in infant formula bottles in European Union in 2011. Besides the impacts on reproductive system and development, exposure to BPA has been associated with several chronic diseases such as cardiovascular disease, diabetes, liver disease and cancers [5,12,13].

Previous studies on health effects of BPA exposure mainly relied on animal models and epidemiological surveys [14,15,16,17]. While these observations indicate that BPA exposure is potentially harmful to human health, validation of the findings in human remains challenging due to several reasons. For epidemiological studies, there is virtually no unexposed population due to the ubiquity of BPA [2]. The half-life of BPA is short, and the impacts of BPA exposure on human health usually take a long time to emerge. Thus, it is difficult to determine the causal links between BPA exposure and harmful health effects, especially chronic diseases. In vitro experiments were conducted to reveal the direct impacts of BPA exposure on cell viability and gene expression. Due to the estrogen-like properties of BPA, these studies mainly focused on BPA impacts on individual genes of interest in estrogen receptor (ER)-positive cells [18,19,20]. The genome-wide impacts of BPA exposure on gene expression especially in ER-negative cells is yet to be uncovered. To characterize the cellular and molecular effects of BPA on ER-negative cells, we performed RNA-seq to examine perturbation on gene expression exerted by low-dose BPA in HEK293 cells. We did not observe changes in cell morphology and viability. Gene expression profiling analysis identified a list of differentially expressed genes with variable functions. Interestingly, there are on common genes between the differentially expressed genes in ER-negative HEK293 cells and those in ER-positive cells. Particularly, BPA caused DNA damage in MCF-7 cells but not in HEK293 cells. Taken together, BPA affected gene expression in ER-negative HEK293 cells in a manner different from that in ER-positive cells.

Materials and Methods

Cell Culture and BPA Treatment

The human embryonic kidney 293 cells (HEK293) were cultured at $37°C$ in 5% CO_2 as adherent monolayer in Dulbecco modified Eagle medium (DMEM) (Hiclone) supplemented with L-glutamine and 10% fetal bovine serum (FBS) (Hiclone). BPA powder was dissolved in the dimethyl sulfoxide (DMSO) and added to culture medium. The final concentration of BPA and DMSO is 10^{-6} M and 10^{-3} M, respectively. For BPA treatment, cells were treated with 10^{-6} M BPA for 48 h. Meanwhile, cells cultured in BPA-free medium were used as the control.

RNA-seq Experiment

Total RNA was extracted from each sample using Trizol reagent (Invitrogen) according to the manufacturer's instructions. mRNA enrichment, library preparation and sequencing were performed at BGI-Shenzhen (sequencing service provider). 49 bp single-end reads were generated for each sample on Illumina HiSeq2000 platform. RNA-seq data can be accessed through ArrayExpress database (www.ebi.ac.uk/arrayexpress/) under accession number E-MTAB-1959.

Electrochemical Experiment for DNA Damage Detection

After removal of medium, cells were washed in PBS pH 7.4 and then smashed into fractions. DNA damage was detected by using catalytic oxidation with $Ru(bpy)_2dppz^{2+}$. Electrochemical measurements were performed with a CHI660C electrochemical workstation (CH Instruments, Inc., USA). The set-up of the electrochemical system was a conventional three-electrode system consisting of $Ru(bpy)_2dppz^{2+}$ modified glassy carbon electrode (GCE) as working electrode (GCE diameter: 3 mm), a saturated calomel electrode (SCE) as the reference electrode and a platinum wire as the auxiliary electrode. All volumetric flasks, beakers, pipettes, and other glassware were closely washed with chromo-sulfuric acid to remove possible contamination. Differential pulse voltammetry (DPV) was carried out to detect DNA damage from 0.0 to 0.6 V with pulse amplitude of 5 mV.

Bioinformatics Analysis

RNA-seq reads were aligned to the human genome (hg19) using TopHat version 1.3.3 [21]. The reference genome and the corresponding annotation files were downloaded from the UCSC genome browser. Reads alignment was carried out with default parameters except for the following options: "-a 6 -m 2 -i 50–no-novel-juncs". Only the uniquely mapped reads were used for the subsequent analysis. Reads distribution relative to gene structure was statistically analyzed using Ever-seq (http://code.google.com/p/ever-seq/). BEDTools [22] was used to calculate the coverage of reads along transcripts for each sample.

Cuffdiff, a separate program from Cufflinks [23], was used to calculate gene expression levels and test the statistical significance of expression changes between two samples. Reads Per Kilo base per Million mapped reads (RPKM) was used to evaluate gene expression levels. Pearson correlation coefficient was calculated to measure the overall similarity between transcriptome profiles of the BPA-treated sample and the control sample. Expression changes were measured by log2(fold change) with false discovery rate (FDR)-adjusted p-value as an indicator for statistical significance. Genes with |fold change| $> = 1.5$ and FDR-adjusted p-value $< = 0.05$ were defined as the differentially expressed genes as a consequence of BPA exposure.

To investigate the functions and involving pathways of genes affected by BPA exposure, we used Ingenuity Pathway Analysis software (IPA, http://www.ingenuity.com) to perform Gene Ontology (GO) and pathway analysis. To assess the biological relationships among these genes, we built molecular interaction networks among them.

Results and Discussion

Effects of BPA Exposure on Cell Growth and Gene Expression in HEK293 Cells

To assess cellular and molecular effects induced by BPA exposure of environmental relevant concentration, we investigated the toxicity of low-dose (10^{-6} M) BPA exposure on HEK293 cells. After 48 h treatment, we examined the effects of BPA exposure on cell morphology and transcriptome profile. The physiological status and morphology of cells treated with BPA are indistinguishable from those in the control sample (Fig. 1). This is consistent with the observations from previous studies suggesting that BPA exposure do not cause observable effects on viability of ER-negative cells [20,24]. However, BPA exposure decreased cell survival through apoptosis induction in ER-positive MCF-7 cells and ovarian granulosa cells [25,26].

In order to understand the molecular mechanisms by which ER-negative HEK293 cells response to BPA exposure at a dose that is environmentally pertinent. We systematically examined the gene expression changes induced by BPA treatment. High throughput sequencing technology was employed to generate the gene expression profiles for the BPA-treated sample and the control sample, and produce 8,243,449 and 8,547,485 clean reads, respectively (Table 1). More than 93% of these reads from both samples were mapped to the reference genome, and ~85% of the reads were unambiguously mapped to a single location in the genome. These uniquely mapped reads were enriched in the coding regions and untranslational regions in both samples (Fig. S1). This indicated that these reads were from transcribed genes.

We further analyzed the gene coverage by the reads for each sample and found that more than 60% of the annotated genes reached a coverage rate higher than 60%. This suggests that these uniquely mapped reads well represent expressed genes. Interestingly, both read distribution in the genomic regions and gene coverage in the two samples were similar to each other (Fig. S1 & S2). Moreover, the global expression profiles of the two samples showed strong similarity (Fig. 2). The comparison of two gene expression profiles identified 15 differentially expressed genes after BPA treatment, 8 up-regulated and 7 down-regulated, respectively (Table 2). We validated some of the differentially expressed genes using qRT-PCR (Fig. S3). This implied that BPA exposure led to a limited variation in gene expression in ER-negative HEK293 cells. This is likely attributed to the absence of ER in HEK293 cells. In contrast, BPA regulated ER target genes in ER-positive cells and could have more than 300 genes whose expression changed two or more folds [19].

HEK 293 with BPA

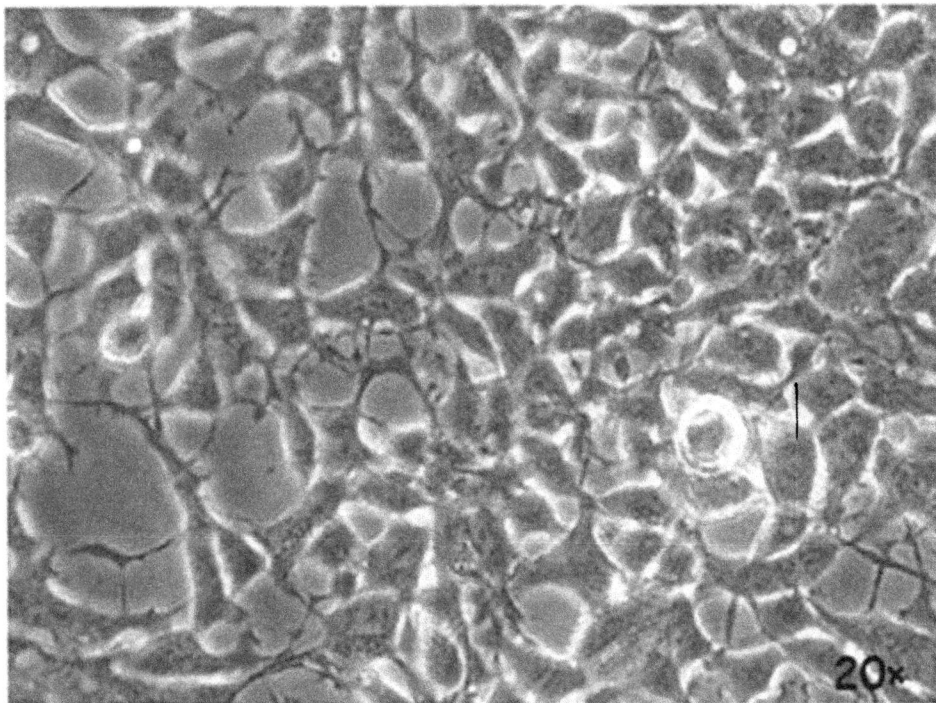

HEK 293 control

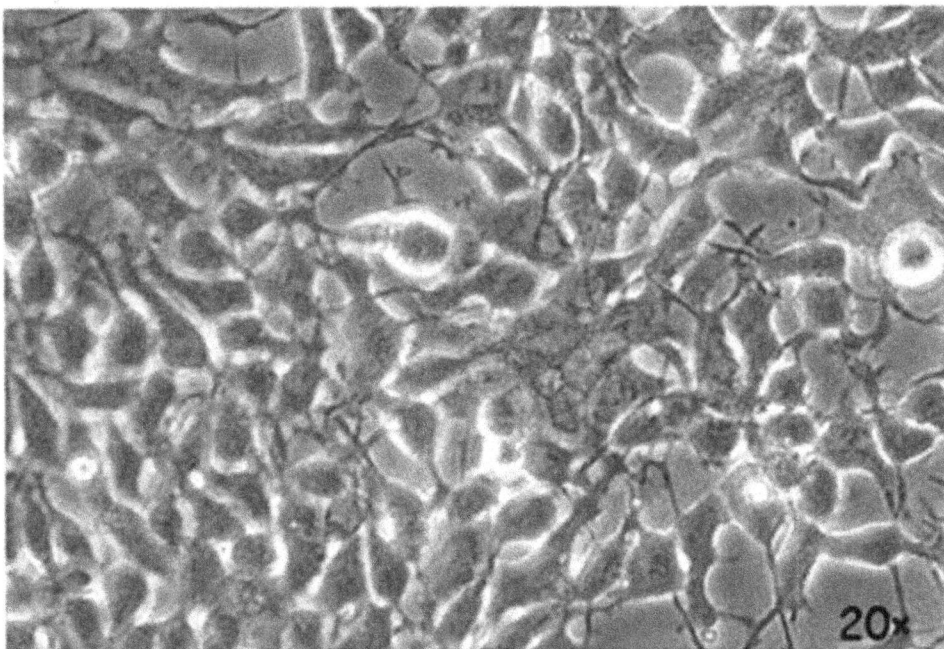

Figure 1. Microscope images showing morphology of HEK293 cells with and without BPA exposure. There are no observable changes in cell morphology of the two samples.

Table 1. RNA-seq read count in HEK293 cells with and without BPA treatment.

	BPA	Control
Total clean reads	8,243,449	8,547,485
Mapped reads (%)	7,684,990 (93.23%)	7,968,367 (93.22%)
Uniquely mapped reads (%)	6,953,073 (84.35%)	7,213,084 (84.39%)

BPA Exposure Leads to Perturbation in Variable Biological Processes in HEK293 Cells

IPA is a web-based functional analysis tool for identification of the most relevant signaling and metabolic pathways, molecular networks, and biological function for a list of genes. Here, we employed IPA to investigate the biological processes in which the differentially expressed genes involve. The results showed that BPA exposure perturbed many metabolic pathways including ion transport (SLCIA4), cysteine metabolic process (CTH, cystathionase), glycogen metabolic process, and toxin metabolic process (AS3MT, arsenic methyltransferase) (Fig. 3). CTH, a cytoplasmic enzyme in the trans-sulfuration pathway, converted cystathione derived from methionine into cysteine. Abnormal expression of CTH may cause cystathioninuria [27]. BPA exposure also influenced biological pathways including aldosterone signaling and glucocorticoid receptor signaling through down regulation of heat shock proteins HSPA8 and HSPA1B [28]. BAX, a pro-apoptotic regulator belonging to the BCL2 protein family, was down-regulated by BPA treatment. This suggests that low-dose BPA may reduce apoptosis of HEK293 cells. Additionally, BPA exposure altered expression of transcription factors TRIM66 and

ZNF460, and regulated the expression of their target genes as a result. Notably, differentially expressed genes of BPA exposure include ERCC5 encoding a DNA endonuclease that involve in nucleotide-excision repair. ERCC5 was up regulated by BPA exposure. It will be interesting to further investigate whether up-regulation of ERCC5 is directly linked to the repair of DNA damage induced by BPA treatment in HEK293 cells. Although BPA mainly functions as an ER agonist and disrupts normal endocrine signaling through regulation of ER target genes, our results suggested that BPA had potential to influence variable physiological processes independent of ER. This may in part explain the facts that BPA exposure was associated with other chronic diseases such as cardiovascular disease, diabetes, liver disease and caners [5,12,13].

BPA Causes DNA Damage in MCF-7 not in HEK293 Cells

It has reported that BPA induced DNA damage in germ cells [16,29]. Since ERCC5 encoding a DNA endonuclease was up regulated in HEK293 cells, it is unclear to what extent the increased expression level of ERCC5 can ameliorate DNA damage by BPA. Therefore, we further used electrochemical

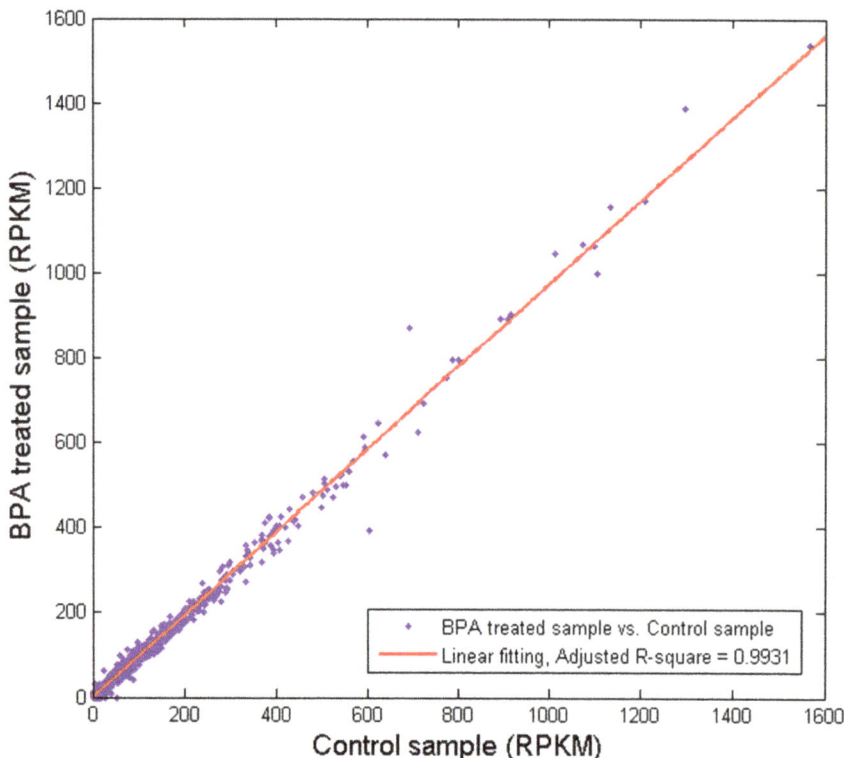

Figure 2. Comparison of gene expression profiles between BPA exposure and control samples. Linear fitting indicates highly similarity between the transcriptome profiles of the BPA-treated sample and the control sample.

Table 2. Significantly differentially expressed genes caused by BPA exposure and their annotated biological process GO terms.

Gene symbol	Description	Biological process	Regulation	Log2(fold change)	q-value
ERCC5	excision repair cross-complementing rodent repair deficiency, complementation group 5	nucleotide excision repair	Up	1.216	0.003
PTCD1	pentatricopeptide repeat domain 1	tRNA 3'-end processing	Up	1.046	0.001
SLC1A4	solute carrier family 1, member 4	glutamate receptor signaling	Up	0.752	0.002
CTH	cystathionase	cysteine metabolic process	Up	0.631	0.005
TRIM66	tripartite motif containing 66	transcription regulation	Up	0.629	0.036
BSN	bassoon presynaptic cytomatrix protein	cytoskeleton organization	Up	0.624	0.016
H6PD	hexose-6-phosphate dehydrogenase	pentose phosphate pathway	Up	0.623	0.005
PPP1R3E	protein phosphatase 1, regulatory subunit 3E	glycogen metabolic process	Up	0.619	0.039
HSPA8	heat shock 70 kDa protein 8	glucocorticoid receptor signaling	Down	−0.624	3.988e-7
BAX	BCL2-associated X protein	apoptosis signaling	Down	−0.635	0.037
PNO1	partner of NOB1 homolog	-	Down	−0.742	0.002
NOL9	nucleolar protein 9	phosphorylation; rRNA processing	Down	−0.753	4.125e-8
HSPA1B	heat shock 70 kDa protein 1B	glucocorticoid receptor signaling	Down	−1.132	0
ZNF460	zinc finger protein 460	regulation of transcription	Down	−1.277	0.029
AS3MT	arsenic methyltransferase	toxin metabolic process	Down	−2.019	0.021

method to detect DNA damage in ER-negative HEK293 cells and ER-positive MCF-7 cells both with BPA treatment. Adenine and guanine bases in DNA undergo electrochemical oxidations when DNA is chemically damaged. Damaged DNA reacts more rapidly than intact ds-DNA with $Ru(bpy)_2dppz^{2+}$ [30]. Thus, DPV method is able to detect damaged DNA on GCE by using catalytic oxidation with $Ru(bpy)_2dppz^{2+}$. Under pH 7.4 condition, DPVs of DNA-damage were recorded as shown in Figure 4. An

obvious oxidation peak was observed at 0.110 V (vs. SCE) in BPA-treated MCF-7 cells but not in HEK293 cells. The peak clearly indicated DNA damage in MCF-7 cells. The absence of the peak, i.e., no detectable DNA damage, is likely due to the up regulation of ERCC5 (Table 2) in HEK293 cells by BPA exposure that promoted DNA repair. Consistently, ERCC5 was not up regulated in the previous studies on ER-positive cells [18,19]. Interestingly, a previous study reported that BPA caused DNA

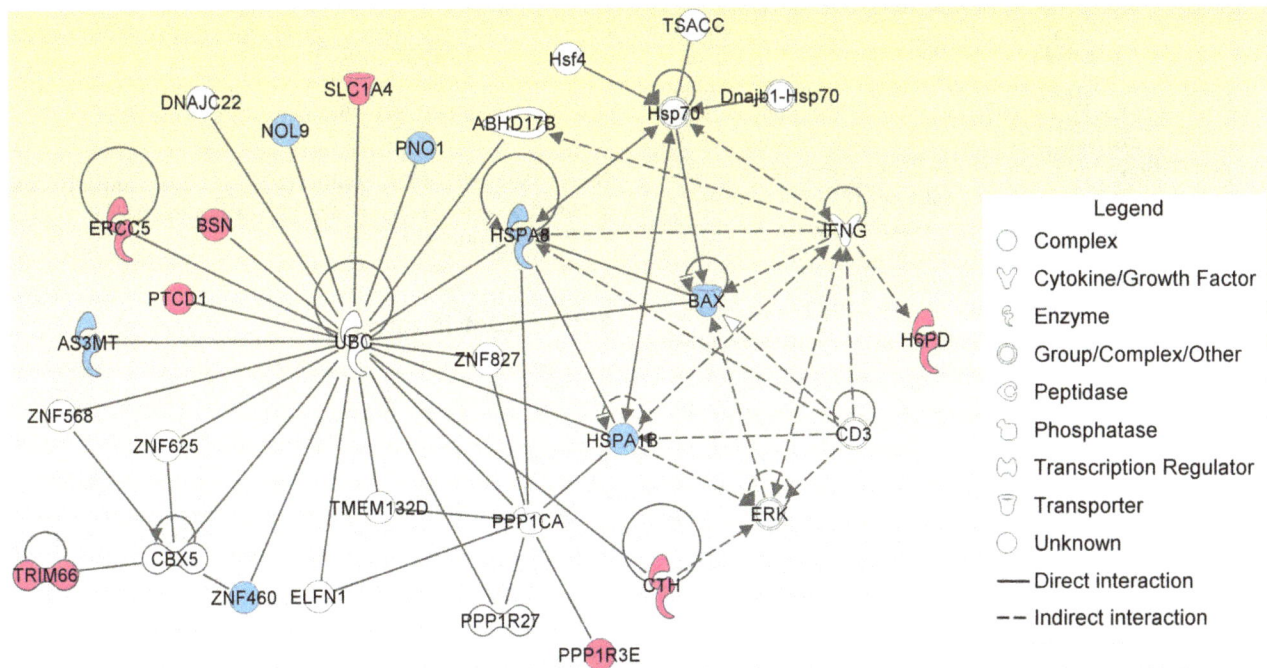

Figure 3. Interaction network of differentially expressed genes caused by BPA exposure. Genes are represented as nodes with various shapes for different types. Red and blue nodes indicate up-regulated and down-regulated genes induced by BPA, respectively. Connective edges represent various types of biological relationships among these genes.

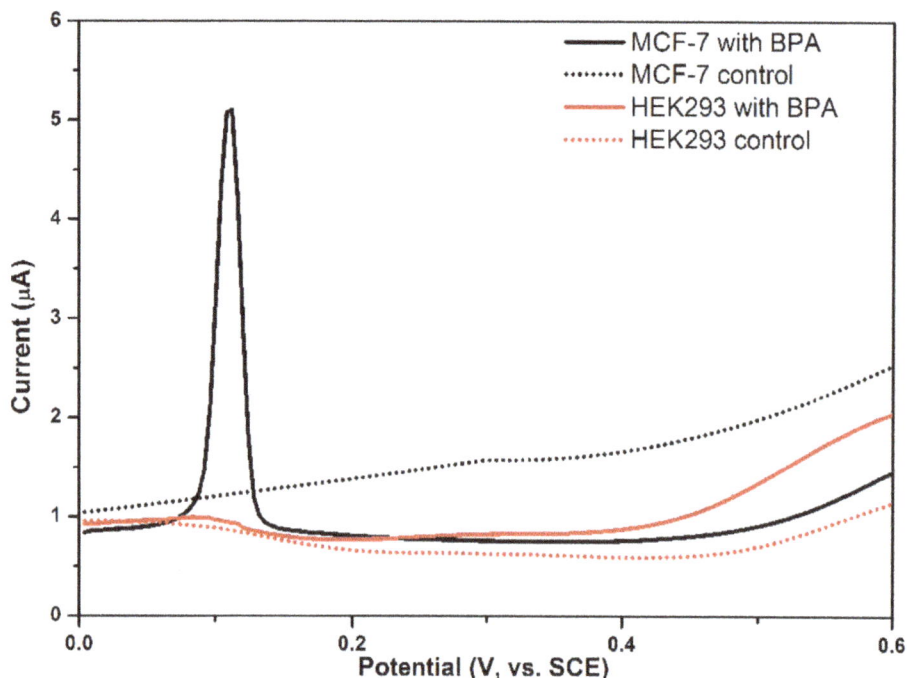

Figure 4. BPA induced significant DNA damage in ER-positive MCF-7 cells but not in ER-negative HEK293 cells. DPVs show voltammetric response to BPA-induced DNA damage in the range of 0.0 to 0.6 V (vs. SCE). The peak indicates DNA damage in MCF-7 cells. Solid lines indicate DPV signals in HEK293 cells (red) and MCF-7 cells (black) treated with 10^{-6} M BPA for 48 h, while dotted lines represent DPV signals in the corresponding control sample.

damage through estrogenic activity [31]. Thus, we speculated that BPA maybe induced DNA damage in a cell type-specific manner.

BPA functions as an ER agonist and regulates ER target genes. Roles of BPA in disrupting normal endocrine signaling have been studied in ER-positive MCF-7 cells. Two recent gene expression profiling studies in ER-positive cells with BPA treatment identified a list of genes differentially expressed in response to BPA exposure [18,19]. Interestingly, comparison analysis found no common genes between the list of genes and those in our study. A study using endometrial cancer cell line ECC-1 and breast cancer cell line T-47D showed that BPA induced binding of estrogen receptor 1 to the target sites dependent of cell type [18]. BPA also altered expression of thyroid specific transcription factors in thyroid follicular cells [24]. Collectively, this suggested that impacts of BPA on regulation of gene expression was likely cell type-specific. It also implied that BPA could disrupt variable physiological processes in ER-negative cells, though it mainly disrupted normal endocrine signaling as an ER agonist.

Supporting Information

Figure S1   Distribution of uniquely mapped reads in genomic features.

Figure S2   Distribution of gene coverage in BPA-treated and normal HEK293 cells. Gene coverage is the proportion of a gene region covered by RNA-seq reads and binned with an interval of 10%. Each gene belongs to one of the bin.

Figure S3   qRT-PCR validation of differentially expressed genes. There are three replicates for each gene. The p-value was from t-test.

Author Contributions

Conceived and designed the experiments: CJ TC. Performed the experiments: LG RY. Analyzed the data: XZ ML. Wrote the paper: XZ CJ.

References

1. Flint S, Markle T, Thompson S, Wallace E (2012) Bisphenol A exposure, effects, and policy: a wildlife perspective. J Environ Manage 104: 19–34.
2. Geens T, Aerts D, Berthot C, Bourguignon JP, Goeyens L, et al. (2012) A review of dietary and non-dietary exposure to bisphenol-A. Food Chem Toxicol 50: 3725–3740.
3. Calafat AM, Kuklenyik Z, Reidy JA, Caudill SP, Ekong J, et al. (2005) Urinary concentrations of bisphenol A and 4-nonylphenol in a human reference population. Environ Health Perspect 113: 391–395.
4. Calafat AM, Ye X, Wong LY, Reidy JA, Needham LL (2008) Exposure of the U.S. population to bisphenol A and 4-tertiary-octylphenol: 2003–2004. Environ Health Perspect 116: 39–44.
5. Huang YQ, Wong CK, Zheng JS, Bouwman H, Barra R, et al. (2012) Bisphenol A (BPA) in China: a review of sources, environmental levels, and potential human health impacts. Environ Int 42: 91–99.
6. Tsai WT (2006) Human health risk on environmental exposure to Bisphenol-A: a review. J Environ Sci Health C Environ Carcinog Ecotoxicol Rev 24: 225–255.
7. Rogers JA, Metz L, Yong VW (2013) Review: Endocrine disrupting chemicals and immune responses: a focus on bisphenol-A and its potential mechanisms. Mol Immunol 53: 421–430.
8. Erler C, Novak J (2010) Bisphenol a exposure: human risk and health policy. J Pediatr Nurs 25: 400–407.
9. Foster WG (2008) Fetal and early postnatal environmental contaminant exposures and reproductive health effects in the female. Fertil Steril 89: e53–54.
10. Golub MS, Wu KL, Kaufman FL, Li LH, Moran-Messen F, et al. (2010) Bisphenol A: developmental toxicity from early prenatal exposure. Birth Defects Res B Dev Reprod Toxicol 89: 441–466.

11. Crain DA, Janssen SJ, Edwards TM, Heindel J, Ho SM, et al. (2008) Female reproductive disorders: the roles of endocrine-disrupting compounds and developmental timing. Fertil Steril 90: 911–940.

12. Nadal A (2013) Obesity: Fat from plastics? Linking bisphenol A exposure and obesity. Nat Rev Endocrinol 9: 9–10.

13. Krieter DH, Canaud B, Lemke HD, Rodriguez A, Morgenroth A, et al. (2013) Bisphenol A in chronic kidney disease. Artif Organs 37: 283–290.

14. Wang T, Lu J, Xu M, Xu Y, Li M, et al. (2013) Urinary bisphenol a concentration and thyroid function in Chinese adults. Epidemiology 24: 295–302.

15. Bhandari R, Xiao J, Shankar A (2013) Urinary bisphenol A and obesity in U.S. children. Am J Epidemiol 177: 1263–1270.

16. Li DK, Miao M, Zhou Z, Wu C, Shi H, et al. (2013) Urine bisphenol-A level in relation to obesity and overweight in school-age children. PLoS One 8: e65399.

17. Weber Lozada K, Keri RA (2011) Bisphenol A increases mammary cancer risk in two distinct mouse models of breast cancer. Biol Reprod 85: 490–497.

18. Gertz J, Reddy TE, Varley KE, Garabedian MJ, Myers RM (2012) Genistein and bisphenol A exposure cause estrogen receptor 1 to bind thousands of sites in a cell type-specific manner. Genome Res 22: 2153–2162.

19. Singleton DW, Feng Y, Yang J, Puga A, Lee AV, et al. (2006) Gene expression profiling reveals novel regulation by bisphenol-A in estrogen receptor-alpha-positive human cells. Environ Res 100: 86–92.

20. Chepelev NL, Enikanolaiye MI, Chepelev LL, Almohaisen A, Chen Q, et al. (2013) Bisphenol A activates the Nrf1/2-antioxidant response element pathway in HEK 293 cells. Chem Res Toxicol 26: 498–506.

21. Trapnell C, Pachter L, Salzberg SL (2009) TopHat: discovering splice junctions with RNA-Seq. Bioinformatics 25: 1105–1111.

22. Quinlan AR, Hall IM (2010) BEDTools: a flexible suite of utilities for comparing genomic features. Bioinformatics 26: 841–842.

23. Trapnell C, Williams BA, Pertea G, Mortazavi A, Kwan G, et al. (2010) Transcript assembly and quantification by RNA-Seq reveals unannotated transcripts and isoform switching during cell differentiation. Nat Biotechnol 28: 511–515.

24. Gentilcore D, Porreca I, Rizzo F, Ganbaatar E, Carchia E, et al. (2013) Bisphenol A interferes with thyroid specific gene expression. Toxicology 304: 21–31.

25. Diel P, Olff S, Schmidt S, Michna H (2002) Effects of the environmental estrogens bisphenol A, o,p'-DDT, p-tert-octylphenol and coumestrol on apoptosis induction, cell proliferation and the expression of estrogen sensitive molecular parameters in the human breast cancer cell line MCF-7. J Steroid Biochem Mol Biol 80: 61–70.

26. Xu J, Osuga Y, Yano T, Morita Y, Tang X, et al. (2002) Bisphenol A induces apoptosis and G2-to-M arrest of ovarian granulosa cells. Biochem Biophys Res Commun 292: 456–462.

27. Wang J, Huff AM, Spence JD, Hegele RA (2004) Single nucleotide polymorphism in CTH associated with variation in plasma homocysteine concentration. Clin Genet 65: 483–486.

28. Rajapandi T, Greene LE, Eisenberg E (2000) The molecular chaperones Hsp90 and Hsc70 are both necessary and sufficient to activate hormone binding by glucocorticoid receptor. J Biol Chem 275: 22597–22604.

29. Allard P, Colaiacovo MP (2010) Bisphenol A impairs the double-strand break repair machinery in the germline and causes chromosome abnormalities. Proc Natl Acad Sci U S A 107: 20405–20410.

30. Zhou L, Yang J, Estavillo C, Stuart JD, Schenkman JB, et al. (2003) Toxicity screening by electrochemical detection of DNA damage by metabolites generated in situ in ultrathin DNA-enzyme films. J Am Chem Soc 125: 1431–1436.

31. Iso T, Watanabe T, Iwamoto T, Shimamoto A, Furuichi Y (2006) DNA damage caused by bisphenol A and estradiol through estrogenic activity. Biol Pharm Bull 29: 206–210.

12

Does Mortality Risk of Cigarette Smoking Depend on Serum Concentrations of Persistent Organic Pollutants? Prospective Investigation of the Vasculature in Uppsala Seniors (PIVUS) Study

Duk-Hee Lee[1,2], Lars Lind[3], David R. Jacobs Jr.[4,5], Samira Salihovic[6], Bert van Bavel[6], P. Monica Lind[7]*

1 Department of Preventive Medicine, School of Medicine, Kyungpook National University, Daegu, Korea, **2** BK21 Plus KNU Biomedical Convergence Program, Department of Biomedical Science, Kyungpook National University, Daegu, Korea, **3** Department of Medical Sciences, Cardiovascular Epidemiology, Uppsala University Hospital, Uppsala, Sweden, **4** Division of Epidemiology and Community Health, School of Public Health, University of Minnesota, Minneapolis, Minnesota, United States of America, **5** Department of Nutrition, University of Oslo, Oslo, Norway, **6** MTM Research Center, School of Science and Technology, Örebro University, Örebro, Sweden, **7** Department of Medical Sciences, Occupational and Environmental Medicine, Uppsala University, Uppsala, Sweden

Abstract

Cigarette smoking is an important cause of preventable death globally, but associations between smoking and mortality vary substantially across country and calendar time. Although methodological biases have been discussed, it is biologically plausible that persistent organic pollutants (POPs) like polychlorinated biphenyls (PCBs) and organochlorine (OC) pesticides can affect this association. This study was performed to evaluate if associations of cigarette smoking with mortality were modified by serum concentrations of PCBs and OC pesticides. We evaluated cigarette smoking in 111 total deaths among 986 men and women aged 70 years in the Prospective Investigation of the Vasculature in Uppsala Seniors (PIVUS) with mean follow-up for 7.7 years. The association between cigarette smoking and total mortality depended on serum concentration of PCBs and OC pesticides (P value for interaction = 0.02). Among participants in the highest tertile of the serum POPs summary score, former and current smokers had 3.7 (95% CI, 1.5–9.3) and 6.4 (95% CI, 2.3–17.7) times higher mortality hazard, respectively, than never smokers. In contrast, the association between cigarette smoking and total mortality among participants in the lowest tertile of the serum POPs summary score was much weaker and statistically non-significant. The strong smoking-mortality association observed among elderly people with high POPs was mainly driven by low risk of mortality among never smokers with high POPs. As smoking is increasing in many low-income and middle-income countries and POPs contamination is a continuing problem in these areas, the interactions between these two important health-related issues should be considered in future research.

Editor: Keitaro Matsuo, Kyushu University Faculty of Medical Science, Japan

Funding: This study was supported by the Swedish Research Council (VR) and the Swedish Research Council for Environment, Agricultural Sciences and Spatial Planning (FORMAS), Korea Health technology R&D Project, Ministry of Health & Welfare, Republic of Korea (A111716), and BK21 PLUS KNU Biomedical Convergence Program for Creative Talent. The funders had no role in study design, data collection and analysis, decision to publish, or preparation of the manuscript.

Competing Interests: The authors have declared that no competing interests exist.

* E-mail: monica.lind@medsci.uu.se

Introduction

Cigarette smoking, as a major risk factor for cancer, cardiovascular diseases, and pulmonary diseases, is an important cause of preventable death globally [1]. However, the magnitude of associations between cigarette smoking and mortality varies across countries and calendar time [2,3,4,5]. To date, this variation has been attributed mainly to methodological biases such as misclassification of smoking status, cohort effects, or random variation [2,3,4,5].

In our previous study [6], we formulated a novel hypothesis that persistent organic pollutants (POPs) modify the risk of cigarette smoking on death. POPs are lipophilic chemicals that accumulate in adipose tissue and are associated with the risk of various chronic diseases [7,8,9]. The biological plausibility for our hypothesis was that experimental studies in mice reported that pretreatment with some POPs increased toxicity of important chemicals contained in cigarette smoke, like benzopyrene, dimethylnitrosamine, and N-nitrosodiethlyamine [10,11].

Supporting our prior hypothesis, we observed different associations between cigarette smoking and total mortality depending on serum concentrations of POPs among the elderly in the U.S. [6]. In that study, one surprising finding was that cigarette smoking did not increase the risk of mortality in the lowest category of POPs. Despite the possibility that smoking-related diseases risks are lower among the elderly than in younger persons due to selective survival [12], finding any subgroup with no association between cigarette smoking and mortality was unexpected.

Our earlier study is the only one published on this topic and the number of current smokers was too small [6]. Thus, we studied

Does Mortality Risk of Cigarette Smoking Depend on Serum Concentrations of Persistent Organic...

107

here whether the finding was replicable in another dataset. We evaluated if there are similar interactions of serum concentrations of POPs with cigarette smoking on the risk of mortality among men and women aged 70 years, living in the community of Uppsala, Sweden (Prospective Investigation of the Vasculature in Uppsala Seniors (PIVUS) study). Because our previous study found that, among various POPs, polychlorinated biphenyls (PCBs) or organochlorine (OC) pesticides showed clear interactions with cigarette smoking and mortality [6], we used POPs burden calculated based on serum concentrations of both OC pesticides and PCBs in this study.

Materials and Methods

Study subjects at baseline were 1,016 men and women, residents of Uppsala, Sweden and age 70 years at time of examination between April 2001 and June 2004. Among participants at baseline, 986 subjects had valid measurement of PCBs and OC pesticides at baseline. The study was approved by the Ethics Committee of the University of Uppsala. The participants gave written informed consent.

All subjects were investigated in the morning after an overnight fast, with no medication or smoking allowed after midnight. The participants were asked about their health behaviors, medical history, and regular medication. Body mass index (BMI) was derived from measured height and weight (kg/m²). Serum cholesterol and triglyceride concentrations were determined in an enzymatic assay (Abbott, Abbott Park, IL, USA).

POPs were measured in stored plasma samples collected at baseline. Analyses of POPs were performed using a Micromass Autospec Ultima (Waters, Mildford, MA, USA) high resolution chromatography coupled to high resolution mass spectrometry (HRGC/HRMS) system based on the method by Sandau et al [13] with some modifications. All details of POPs analyses were provided elsewhere [14]. Among 16 PCB congeners and 5 OC pesticides, 2 OC pesticides (trans-chlordane and cis-chlordane) with detection rate <10% were not included in the final analyses. An established summation formula based on serum cholesterol and serum triglyceride concentrations was used to calculate the total amount of lipid in each plasma sample [15]. Thereafter the wet-weight concentrations of the POPs were divided by this lipid estimate to obtain lipid-normalized concentrations. As models based on wet-weight concentrations adjusted for serum cholesterol and triglyceride as covariates showed similar results, we presented the results based on lipid-normalized concentrations for the consistency of analytic strategy with the previous study [6].

The fact of death was ascertained through linkage to the Swedish Register of Death Causes at the National Board of Health and Welfare. Causes of death data were not available. Follow-up time for each person was calculated as the difference between the first examination date and the last known date alive or censored. Persons who survived the entire follow-up period were censored on Jan 1, 2012. Median follow-up time was 7.7 years (range 0.3–9.8 years) and we documented 111 deaths.

For the analysis, smoking status was expressed as never, former and current. First, we calculated the summary measure of 16 PCB congeners and 5 OC pesticides by summing the rank orders of the individual POPs for subjects with detectable values of each POP, assigning rank 0 to not detectable values. Three summary measures were made based on both PCBs and OC pesticides, PCBs only, and OC pesticides only.

Next, we checked if the associations between cigarette smoking and mortality differed by tertiles of the summary measures of POPs in predicting total mortality using Cox proportional hazard models. P values for interaction were calculated based on three categories of each of POPs and cigarette smoking. Adjusting covariates were gender, physical activity (none, moderate, and vigorous), BMI (kg/m²), and alcohol consumption (g/day). We further considered medication for diabetes or hypertension and history of myocardial infarction or stroke as possible confounders.

In addition to stratified analyses, adjusted HRs with the common reference group of never smokers within the 1st tertile of POPs were presented. Also, the same analyses were applied to the U.S. elderly within our previous study [6] because only results stratified analyses by POPs levels were previously reported in that study. Methodologic details have been presented [6]. All statistical analyses were performed with PC-SAS version 9.1.

Results

Baseline characteristics and history of chronic diseases according to cigarette smoking status were shown in Table 1. Compared to never or current smokers, former smokers tended to be men, more obese, alcohol drinkers, and more under diabetes medication. Current smokers were less obese and more physically inactive than never or former smokers. There were no statistical significant differences across smoking categories for prevalence of hypertension medication or history of myocardial infarction or stroke.

Table 2 indicates mean concentrations of individual POPs depending on cigarette smoking status. Among 16 PCBs, PCB074, PCB105, and PCB118 were statistically significant or marginally significantly lower in current vs. never smokers. These three PCBs have weak dioxin activity, however, other PCBs with dioxin activity like PCB126, PCB156, and PCB169 did not show any trend. On the contrary, among 3 OC pesticides, p,p'-DDE had higher levels among current than never smokers.

When POPs were not considered in analyses, adjusted hazard ratios (HRs) for all-cause mortality were 1.3 for former smokers and 2.1 (95% CI: 1.2–3.7) for current smokers, compared with never smokers. However, the associations were substantially different depending on summary measures of POPs (Table 3). Compared with 70 year old people within the 1st or 2nd tertiles of summary measures of PCBs and OC pesticides, those within the 3rd tertile showed strong associations between cigarette smoking and total mortality. Adjusted HRs were 3.7 (1.5–9.3) for former smokers and 6.4 (2.3–17.7) for current smokers. Further adjustment for medication for diabetes or hypertension and history of myocardial infarction or stroke did not change the result. Summary measures calculated from PCBs or OC pesticides separately showed similar associations (Table S1).

Figure 1 shows adjusted HRs when never smokers within the 1st tertile of PCBs and OC pesticides were used as the common reference group. Never smokers within the 3rd tertile of PCBs and OC pesticides showed a statistically significantly lower risk of mortality with adjusted HR of 0.3 (0.1–0.9) while current smokers with the same levels of POPs concentrations had a statistically significantly higher risk of mortality with adjusted HR of 2.3 (1.0–5.0). When we applied the same analyses to the U.S. elderly who were included in our previous study, patterns were similar to those in the current study, although no individual HR reached statistical significance (Figure S1). In that study, the summary measures for PCBs and for OC pesticides were separate because PCB vs OC pesticide measurements were performed in different participants.

To better understand possible confounding, Table 4 compared baseline characteristics of 4 subgroups (never smokers & low POPs, current smokers & high POPs, never smokers & low POPs, and current smokers & high POPs), in particular focusing on never smokers with high POPs who had the lowest mortality. Compared

Table 1. Baseline characteristics according to the status of cigarette smoking among 986 elderly aged 70, Prospective Investigation of the Vasculature in Uppsala Seniors (PIVUS) study.

Characteristics	Status of cigarette smoking			p value
	Never smokers (N = 471)	Former smokers (N = 410)	Current smokers (N = 105)	
Men (%)	44.4	57.6	47.6	0.01
BMI (kg/m², %)				<0.01
<25	34.4	26.6	51.4	
25-<30	44.8	48.1	34.3	
≥30	20.8	25.4	14.3	
Exercise (%)				0.01
No	8.5	11.7	18.1	
Mild	64.1	63.4	66.7	
Moderate or vigorous	27.4	24.9	15.2	
Alcohol consumption (g/day, %)				<0.01
0	17.2	11.5	14.3	
1–14	75.2	74.8	80.0	
≥15	7.6	13.7	5.7	
Diabetes medication (%)	4.7	8.8	4.8	0.03
Hypertension medication (%)	29.5	33.4	27.6	0.34
History of myocardial infarction (%)	5.7	8.3	10.7	0.13
History of stroke (%)	3.2	3.9	4.8	0.69

to other subgroups, 70 year old participants within the never smokers & high POPs group were the most physically active and had the lowest prevalence of history of myocardial infarction.

Discussion

Generally, the present study replicated our report in a U.S. elderly population [6]. The association between cigarette smoking and total mortality depended on serum concentration of summary scores reflecting the background mixture of PCBs and/or OC pesticides. When POPs were not considered in the analyses, the risk of mortality among current-smokers was about two times higher than never smokers. However, among elderly with relatively high POPs, former or current smokers had about 4 to 7 times higher mortality than never smokers while the association between cigarette smoking and total mortality much weaker and statistically non-significant among elderly with relatively low serum concentrations of POPs. In addition to the main results, a subsidiary finding of lower concentrations of dioxin-like PCBs among smokers than never smokers was also similar to finding in the previous study [6].

These findings suggested that different concentrations of POPs among populations may partly explain variability in smoking-related total mortality association across previous epidemiological studies [2,3,4,5]. More importantly, the similar weak association between cigarette smoking and mortality among the elderly people in the lowest category of POPs from these two studies could indicate that the presence of certain levels POPs is a necessary factor for cigarette smoking to increase the risk of death in the elderly.

At first, we had expected the interaction between these two factors to be driven by the high risk of death among current smokers with high POPs. However, these two studies showed that the strong association between cigarette smoking and mortality among the elderly with high POPs had a strong component of low mortality in never smokers as well as of high mortality in former or current smokers.

The very low mortality among the 150 never smokers in PIVUS with high POPs is a provocative finding. A similar pattern of reduced mortality was apparent in the National Health and Nutrition Examination Survey (NHANES) (Figure S1), suggesting that this finding is not simply bias or chance. An ecologic finding in the cohort study of male British doctors may be viewed as concordant. In that study, the excess mortality associated with smoking was greater during the second half of follow-up (1971~1991) than the first half (1951~1971) [2], largely because the mortality rate among non-smokers had decreased substantially over time while the mortality rate among smokers had remained about constant [2]. Historically, PCBs and OC pesticides were widely used after World War II and periods with the highest body burden of POPs were 1960s and 1970s [16]. Therefore, any effect due to high POPs may be more strongly reflected in the latter part of cohort study.

It is difficult to explain very low mortality among never smokers with high POPs. One possible explanation is survival bias. However, for survival bias to explain this finding, never smokers with high POPs must have had a higher death rate before reaching age 70 than other subgroups like active smokers with high POP, leaving healthier survivors to participate in the PIVUS and NHANES studies. This would be an odd pattern. Also, compared to other subgroups, the elderly never smokers with high POPs were the most physically active and had the lowest prevalence of history of myocardial infarction, both of which might predispose to reduced mortality rate. Therefore, this type of survival bias would seem to be unlikely even though we cannot totally exclude it.

Table 2. Adjusted* serum concentrations (geometric means, ng/g lipid) of individual polychlorinated biphenyls (PCBs) or organochlorine (OC) pesticides according to the status of cigarette smoking, Prospective Investigation of the Vasculature in Uppsala Seniors (PIVUS) study.

Analytes	Status of cigarette smoking			p for trend
	Never smokers (N = 471)	Former smokers (N = 410)	Current smokers (N = 105)	
Polychlorinated biphenyls (PCBs)				
PCB074[†]	14.1	13.6	12.9	0.09
PCB099	13.4	13.7	14.3	0.38
PCB105[†]	5.0	5.0	4.1	0.02
PCB118[†]	31.0	30.7	25.0	<0.01
PCB126[†]	5.9	5.9	6.5	0.44
PCB138	123.7	124.5	135.4	0.15
PCB153	213.7	216.5	228.3	0.19
PCB156[†]	23.7	23.1	24.7	0.78
PCB157	4.4	4.3	4.5	0.70
PCB169[†]	25.8	25.9	26.7	0.52
PCB170	75.5	75.3	80.1	0.30
PCB180	177.6	177.3	187.1	0.37
PCB189	3.2	3.2	2.9	0.34
PCB194	16.0	15.2	17.2	0.88
PCB206	4.2	3.9	4.1	0.23
PCB209	4.0	3.8	4.0	0.36
Organochlorines pesticides (OCPs)				
p,p′-DDE	270.1	296.4	322.4	0.04
Trans-nonachlor	20.6	21.9	21.8	0.14
Hexachlorobenzene	40.1	39.1	39.5	0.46

*Adjusted for gender, BMI, exercise, and alcohol consumption.
[†]POPs with dioxin activity.

Table 3. Adjusted hazard ratios (HRs)* and 95% confidence intervals (CIs) for all-cause mortality rate by summary measures[†] of polychlorinated biphenyls (PCBs) or organochlorine (OC) pesticides, Prospective Investigation of the Vasculature in Uppsala Seniors (PIVUS) study.

		Status of cigarette smoking			p for trend	p for interaction
		Never smokers	Former smokers	Current smokers		
All subjects						
	Cases/No	40/471	50/410	21/105		
	Adjusted HR(95%CI)	Referent	1.3(0.9–2.0)	2.1(1.2–3.7)	<0.01	
Summary measure of 16 PCBs and 3 OC pesticides						
1st tertile	Cases/No	16/160	19/138	5/30		
	Adjusted HR(95%CI)	Referent	1.2 (0.6–2.4)	1.4 (0.5–4.0)	0.46	0.02
2nd tertile	Cases/No	18/161	10/134	5/34		
	Adjusted HR(95%CI)	Referent	0.7 (0.3–1.6)	1.2 (0.5–3.4)	0.97	
3rd tertile	Cases/No	6/150	21/138	11/41		
	Adjusted HR(95%CI)	Referent	3.7 (1.5–9.3)	6.4 (2.3–17.7)	<0.01	

*Hazard Ratios (HRs) adjusted for gender, BMI, exercise, and alcohol consumption.
[†]Values of compounds belonging to in each summary measure were individually ranked; the rank orders of the individual POPs were summed to calculate summary measures and the summaries were divided into tertiles.

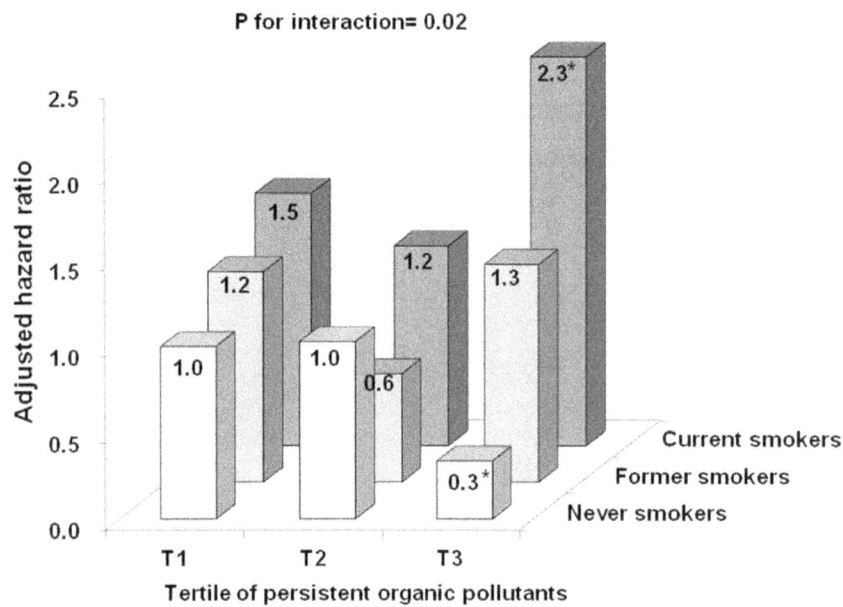

Figure 1. Interactions of cigarette smoking with summary measure of polychlorinated biphenyls (PCBs) and organochlorines (OC) pesticides on mortality. Hazard ratios (HRs) were estimated using a common reference group (never smokers & T1), adjusted for gender, BMI, exercise, and alcohol consumption. T1, first tertile; T2, second tertile; T3, third tertile, Prospective Investigation of the Vasculature in Uppsala Seniors (PIVUS) study.

In addition, low mortality in relation to high POPs is biologically plausible under some conditions, a case in point relating to associations between POPs and telomeres. Telomere length has been proposed as a marker of mitotic cell age and as a general index of human aging [17]. In our previous cross-sectional study among apparently healthy Koreans [18], telomere length was increasing across serum concentrations of POPs within the lower range of POPs. In that study, the interpretation was that low dose POPs may act as a tumor promoter in carcinogenesis based on experimental findings on arsenic. For example, low concentrations of arsenic relevant to current human exposure elongated telomeres in vitro and increased *myc* and *ras* oncogenes while high concentrations of arsenic decreased telomere length [19]. As *myc* oncogene can activate telomerase [20], the authors interpreted these experimental findings on arsenic as a role of tumor promoter of low dose arsenic in human.

However, an opposite interpretation of how POPs relate to telomere length is also possible. It is well-known that shorter telomere length is associated with higher risk of early death [21]. There is a higher mortality rate, especially from heart disease and infectious disease, among elderly people who have shorter telomeres in blood DNA [22]. Therefore, in PIVUS and NHANES persons with relatively high serum POPs concentrations within background exposure levels may have had a longer survival than persons with lower serum POPs concentrations. As cigarette smoking is reported to decrease telomere length [23], longer survival with higher POPs could be observed only in never smokers, as we observed in these two studies.

Although molecular mechanisms for low dose POPs to increase telomere length have not been studied, one speculation is that certain levels of POPs might excite production of cytoprotective and restorative proteins including growth factors, phase II and antioxidant enzymes, and protein chaperones, known as hormetic effects [24]. Increased telomerase activity or slow-down of age-dependent telomere shortening has been regarded as one marker of activation of cytoprotective and restorative proteins [25]. Some

previous human and experimental studies support possible beneficial effects of certain levels of POPs. For example, decreased risk of soft tissue sarcoma was reported with increased concentrations of dioxins or PCBs with dioxin activity in the general population[26]. Also, in some animal studies of low doses of TCDD or DDT, there was a tendency towards fewer tumors or altered hepatic foci than in controls, indicating an anti-carcinogenic process [27,28]. Furthermore, decreased lipid peroxidation level in rats treated with low dose DDT was reported, compared to control rats [29].

Studies of the levels of POPs in the global environment show that emission sources of a number of POPs in the last 20 years have shifted from industrialized countries of the Northern Hemisphere to less developing countries in tropical and subtropical regions [30]. This is due to a late production ban on OC pesticides: some OC pesticides are still being used in agriculture and for the control of diseases, such as malaria [31]. Also, there is another active exposure source of PCBs through e-waste recycling in these countries [32]. Although the burden of cigarette smoking use is currently greatest in high-income countries, rates of smoking are increasing in many low-income and middle-income countries. Therefore, the interaction between these two important health-related issues, POPs and cigarette smoking, should be further studied from a variety of viewpoints including molecular mechanisms.

The strengths of our study include the homogenous age and community-based sampling of study subjects. There were some limitations to our study. Due to a limited sample size, we could not consider more detailed information on cigarette smoking like total pack-years or duration of smoking cessation in analyses. In addition, analyses focusing on cause-specific mortality were not possible as cause of death information was not available. Finally, even though the consistency of findings across two studies lends credence to our findings and seems to reduce the likelihood that our finding is explainable by chance or bias, our findings still require replication by other cohort studies.

Table 4. Comparison of baseline characteristics among 4 subgroups, Prospective Investigation of the Vasculature in Uppsala Seniors (PIVUS) study.

	Low POPs		High POPs		
	Never smokers (N = 160)	**Current smokers (N = 30)**	**Never smokers (N = 150)**	**Current smokers (N = 41)**	**p value**
Dead (%)	10.0	16.7	4.0	26.8	<0.01
Men (%)	31.9	40.0	56.0	53.7	<0.01
BMI (kg/m^2, %)					<0.01
<25	26.3	46.7	36.0	48.8	
25-<30	40.6	36.7	51.3	46.3	
≥30	33.1	16.7	12.7	4.9	
Moderate or vigorous exercise (%)					<0.01
No	10.0	10.0	10.7	26.8	
Mild	63.8	80.0	58.7	61.0	
Moderate or vigorous	26.3	10.0	30.7	12.0	
Alcohol consumption (g/day, %)					0.03
0	20.6	20.0	14.0	7.3	
1-14	76.3	76.7	74.0	85.4	
≥15	3.1	3.3	12.0	7.3	
Diabetes medication (%)	3.8	0%	7.3	9.8	0.17
Hypertension medication (%)	31.3	30.0	31.3	22.0	0.68
History of myocardial infarction (%)	6.9	13.8	4.0	15.0	0.05
History of stroke (%)	1.9	6.7	5.3	4.9	0.35

This study has confirmed a strong association between cigarette smoking and mortality among elderly people with relatively high POPs, but a much weaker and not statistically significant smoking-mortality association among those with relatively low POPs. However, the observation that the lowest mortality was seen among never smokers with high POPs is provocative and requires further studies on the role of POPs in longevity. In addition more prospective studies in human, in-vitro and in-vivo experimental studies would help to elucidate potential molecular mechanisms.

Supporting Information

Figure S1 Interactions of cigarette smoking with summary measure of polychlorinated biphenyls (PCB) or organochlorine (OC) pesticides predicting mortality in the U.S. general population (reanalyses of published results using the same National Health and Nutrition Examination Survey (NHANES) datasets (Lee, 2013). Number of study subjects and deaths were 610 and 142 for

PCB analyses and 702 and 157 for OC pesticide analyses. Hazard ratios (HRs) were estimated using a common reference group (never smokers & T1), adjusted for gender, BMI, exercise, and alcohol consumption. T1, first tertile; T2, second tertile; T3, third tertile. Although none of the HRs was statistically significantly different from the reference group, as previously reported, interaction p-values were 0.008 for PCBs and 0.024 for OC pesticides.

Author Contributions

Conceived and designed the experiments: DHL LL PML. Performed the experiments: DHL LL PML. Analyzed the data: DHL. Contributed reagents/materials/analysis tools: SS BB. Wrote the paper: DHL LL DRJ SS BB PML.

References

1. Murray CJ, Lopez AD (1997) Mortality by cause for eight regions of the world: Global Burden of Disease Study. Lancet 349: 1269–1276.
2. Doll R, Peto R, Wheatley K, Gray R, Sutherland I (1994) Mortality in relation to smoking: 40 years' observations on male British doctors. Bmj 309: 901–911.
3. Hunt D, Blakely T, Woodward A, Wilson N (2005) The smoking-mortality association varies over time and by ethnicity in New Zealand. Int J Epidemiol 34: 1020–1028.
4. Jacobs DR Jr, Adachi H, Mulder I, Kromhout D, Menotti A, et al. (1999) Cigarette smoking and mortality risk: twenty-five-year follow-up of the Seven Countries Study. Arch Intern Med 159: 733–740.
5. van de Mheen PJ, Gunning-Schepers LJ (1996) Differences between studies in reported relative risks associated with smoking: an overview. Public Health Rep 111: 420–426; discussion 427.
6. Lee YM, Bae SG, Lee SH, Jacobs DR, Lee DH (2013) Associations between cigarette smoking and total mortality differ depending on serum concentrations

of persistent organic pollutants among the elderly. J Korean Med Sci 28: 1122–1128.
7. Carpenter DO (2011) Health effects of persistent organic pollutants: the challenge for the Pacific Basin and for the world. Rev Environ Health 26: 61–69.
8. Ha MH, Lee DH, Jacobs DR (2007) Association between serum concentrations of persistent organic pollutants and self-reported cardiovascular disease prevalence: results from the National Health and Nutrition Examination Survey, 1999–2002. Environ Health Perspect 115: 1204–1209.
9. Lee DH, Lee IK, Song K, Steffes M, Toscano W, et al. (2006) A strong dose-response relation between serum concentrations of persistent organic pollutants and diabetes: results from the National Health and Examination Survey 1999–2002. Diabetes Care 29: 1638–1644.
10. Diwan BA, Ward JM, Kurata Y, Rice JM (1994) Dissimilar frequency of hepatoblastomas and hepatic cystadenomas and adenocarcinomas arising in hepatocellular neoplasms of D2B6F1 mice initiated with N-nitrosodiethylamine

and subsequently given Aroclor-1254, dichlorodiphenyltrichloroethane, or phenobarbital. Toxicol Pathol 22: 430–439.

11. Hutton JJ, Meier J, Hackney C (1979) Comparison of the in vitro mutagenicity and metabolism of dimethylnitrosamine and benzo[a]pyrene in tissues from inbred mice treated with phenobarbital, 3-methylcholanthrene or polychlorinated biphenyls. Mutat Res 66: 75–94.

12. Psaty BM, Koepsell TD, Manolio TA, Longstreth WT Jr, Wagner EH, et al. (1990) Risk ratios and risk differences in estimating the effect of risk factors for cardiovascular disease in the elderly. J Clin Epidemiol 43: 961–970.

13. Sandau CD, Sjodin A, Davis MD, Barr JR, Maggio VL, et al. (2003) Comprehensive solid-phase extraction method for persistent organic pollutants. Validation and application to the analysis of persistent chlorinated pesticides. Anal Chem 75: 71–77.

14. Salihovic S, Lampa E, Lindstrom G, Lind L, Lind PM, et al. (2012) Circulating levels of persistent organic pollutants (POPs) among elderly men and women from Sweden: results from the Prospective Investigation of the Vasculature in Uppsala Seniors (PIVUS). Environ Int 44: 59–67.

15. Phillips DL, Pirkle JL, Burse VW, Bernert JT Jr, Henderson LO, et al. (1989) Chlorinated hydrocarbon levels in human serum: effects of fasting and feeding. Arch Environ Contam Toxicol 18: 495–500.

16. Solomon GM, Weiss PM (2002) Chemical contaminants in breast milk: time trends and regional variability. Environ Health Perspect 110: A339–347.

17. Blackburn EH (2005) Telomeres and telomerase: their mechanisms of action and the effects of altering their functions. FEBS Lett 579: 859–862.

18. Shin JY, Choi YY, Jeon HS, Hwang JH, Kim SA, et al. (2010) Low-dose persistent organic pollutants increased telomere length in peripheral leukocytes of healthy Koreans. Mutagenesis 25: 511–516.

19. Zhang TC, Schmitt MT, Mumford JL (2003) Effects of arsenic on telomerase and telomeres in relation to cell proliferation and apoptosis in human keratinocytes and leukemia cells in vitro. Carcinogenesis 24: 1811–1817.

20. Wu KJ, Grandori C, Amacker M, Simon-Vermot N, Polack A, et al. (1999) Direct activation of TERT transcription by c-MYC. Nat Genet 21: 220–224.

21. Bakaysa SL, Mucci LA, Slagboom PE, Boomsma DI, McClearn GE, et al. (2007) Telomere length predicts survival independent of genetic influences. Aging Cell 6: 769–774.

22. Cawthon RM, Smith KR, O'Brien E, Sivatchenko A, Kerber RA (2003) Association between telomere length in blood and mortality in people aged 60 years or older. Lancet 361: 393–395.

23. Valdes AM, Andrew T, Gardner JP, Kimura M, Oelsner E, et al. (2005) Obesity, cigarette smoking, and telomere length in women. Lancet 366: 662–664.

24. Mattson MP (2008) Hormesis defined. Ageing Res Rev 7: 1–7.

25. Yokoo S, Furumoto K, Hiyama E, Miwa N (2004) Slow-down of age-dependent telomere shortening is executed in human skin keratinocytes by hormesis-like-effects of trace hydrogen peroxide or by anti-oxidative effects of pro-vitamin C in common concurrently with reduction of intracellular oxidative stress. J Cell Biochem 93: 588–597.

26. Tuomisto J, Pekkanen J, Kiviranta H, Tukiainen E, Vartiainen T, et al. (2005) Dioxin cancer risk—example of hormesis? Dose Response 3: 332–341.

27. Sukata T, Uwagawa S, Ozaki K, Ogawa M, Nishikawa T, et al. (2002) Detailed low-dose study of 1,1-bis(p-chlorophenyl)-2,2,2- trichloroethane carcinogenesis suggests the possibility of a hormetic effect. Int J Cancer 99: 112–118.

28. Viluksela M, Bager Y, Tuomisto JT, Scheu G, Unkila M, et al. (2000) Liver tumor-promoting activity of 2,3,7,8-tetrachlorodibenzo-p-dioxin (TCDD) in TCDD-sensitive and TCDD-resistant rat strains. Cancer Res 60: 6911–6920.

29. Shutoh Y, Takeda M, Ohtsuka R, Haishima A, Yamaguchi S, et al. (2009) Low dose effects of dichlorodiphenyltrichloroethane (DDT) on gene transcription and DNA methylation in the hypothalamus of young male rats: implication of hormesis-like effects. J Toxicol Sci 34: 469–482.

30. Tanabe S, Minh TB (2010) Dioxins and organohalogen contaminants in the Asia-Pacific region. Ecotoxicology 19: 463–478.

31. Wong MH, Leung AO, Chan JK, Choi MP (2005) A review on the usage of POP pesticides in China, with emphasis on DDT loadings in human milk. Chemosphere 60: 740–752.

32. Someya M, Ohtake M, Kunisue T, Subramanian A, Takahashi S, et al. (2010) Persistent organic pollutants in breast milk of mothers residing around an open dumping site in Kolkata, India: specific dioxin-like PCB levels and fish as a potential source. Environ Int 36: 27–35.

13

Short-Term Effects of the Particulate Pollutants Contained in Saharan Dust on the Visits of Children to the Emergency Department due to Asthmatic Conditions in Guadeloupe (French Archipelago of the Caribbean)

Gilbert Cadelis[1]*, Rachel Tourres[1], Jack Molinie[2]

1 Department of Pulmonary Medicine, Universitary Hospital of Pointe-a-Pitre, Pointe-a-Pitre, Guadeloupe, French West Indies, **2** Laboratory of Research in Geoscience and Energy, University of Antilles and Guyane, Pointe-a-Pitre, Guadeloupe, French West Indies

Abstract

Background: The prevalence of asthma in children is a significant phenomenon in the Caribbean. Among the etiologic factors aggravating asthma in children, environmental pollution is one of the main causes. In Guadeloupe, pollution is primarily transported by Saharan dust including inhalable particles.

Methods: This study assesses, over one year (2011), the short-term effects of pollutants referred to as PM_{10} (PM_{10}: particulate matter <10 μm) and $PM_{2.5-10}$ ($PM_{2.5-10}$: particulate matter >2.5 μm and <10 μm) contained in Saharan dust, on the visits of children aged between 5 and 15 years for asthma in the health emergency department of the main medical facility of the archipelago of Guadeloupe. A time-stratified case-crossover model was applied and the data were analysed by a conditional logistic regression for all of the children but also for sub-groups corresponding to different age classes and genders.

Results: The visits for asthma concerned 836 children including 514 boys and 322 girls. The Saharan dust has affected 15% of the days of the study (337 days) and involved an increase in the average daily concentrations of PM_{10} (49.7 μg/m^3 vs. 19.2 μg/m^3) and $PM_{2.5-10}$ (36.2 μg/m^3 vs. 10.3 μg/m^3) compared to days without dust. The excess risk percentages (IR%) for visits related to asthma in children aged between 5 and 15 years on days with dust compared to days without dust were, for $PM_{10,}$ ((IR %: 9.1% (CI95%, 7.1%–11.1%) versus 1.1%(CI95%, −5.9%–4.6%)) and for $PM_{2.5-10}$ (IR%: 4.5%(CI95%, 2.5%–6.5%) versus 1.6% (CI95%, −1.1%–3.4%). There was no statistical difference in the IR% for periods with Saharan dust among different age group of children and between boys and girls for PM_{10} and $PM_{2.5-10}$.

Conclusion: The PM_{10} and $PM_{2.5-10}$ pollutants contained in the Saharan dust increased the risk of visiting the health emergency department for children with asthma in Guadeloupe during the study period.

Editor: Aimin Chen, University of Cincinnati, United States of America

Funding: The authors have no support or funding to report.

Competing Interests: The authors have declared that no competing interests exist.

* E-mail: gilbert.cadelis@chu-guadeloupe.fr

Introduction

The prevalence of asthma, especially in children, increased in the world over the last decade [1]. In the Caribbean, this disease represents a preoccupying public health problem [2]. In Guadeloupe (16° North latitude and 61° West longitude), the prevalence of asthma in children is higher than in France (14% versus 9%) [3]. Among the many etiologic factors causing asthma in children, pollution, especially particulate pollution, plays an important part [4].

The archipelago of Guadeloupe is periodically exposed to Saharan dust generating peak exposures to fine particles, which can last several days. This particulate pollution contributes to exceeding the particle thresholds set in relation to health protection. The desert sand dust is generated above the Sahara desert. The production of dust is at its highest level between April and June. It is assessed at around 500 to 1000 tonnes per year [5]. The dust particles, captured by the winds at the ground surface, are driven to tropospheric altitudes. These particles are transported as suspended matter, at an altitude between 1500 and 6000 m of the African desert, towards the West, over the Atlantic Ocean, and they reach the United States due to the influence of the maritime trade winds. They therefore pass via the Caribbean, between April and October, and settle through wet or dry processes [6]. The granulometric measurements performed on the particles, at more than 100 km from the dust source, showed that the median diameter rapidly decreases below 10 μm [7]. During a dust episode, the concentration in particles can reach 2000 μg/m^3 as a maximum hourly average [7]. The majority of the dust particles measure less than 10 μ/m of aerodynamic diameter and are thus inhalable, as the smallest particles can easily penetrate inside the respiratory tracts and therefore reach the bronchi and the small airways. The Saharan particles are of mineral origin and result from the progressive abrasion of rocks. They are essentially made of quartz, silicon oxide, clay, and carbonates. They contain iron and are also covered with organic matter (bacteria and viable spores, grains of pollens) [8]. The experimental studies performed on rats revealed the toxic and inflammatory potential of desert dust for airways [9]. Epidemiologic studies have shown that this desert particulate pollution increased morbidity and mortality as well as aggravated the condition of patients suffering from chronic respiratory diseases [10], [11]. Few studies exist on the connections between Saharan dust and asthma. For example, in the Caribbean, only two studies have focused on the effects of Saharan dust on asthma in children and have provided contradictory results [12], [13].

This study concerns the city of Pointe-a-Pitre and its suburbs located on the archipelago of Guadeloupe, which, each year, is exposed to Saharan dust during several months.

We have studied, over one year, the effects of particulate pollutants (PM_{10}, $PM_{2.5-10}$) contained inside Saharan dust on the aggravation of asthma in children, by considering as a criterion, the number of visits of asthmatic children to the paediatric emergency department of the main medical facility of the archipelago. Our main assumption was based on a possible association between the intrusion of dust from the Sahara on the territory and the visits of asthmatic children to the emergency department.

Materials and Methods

A Ethics Statement

This study has been approved by the Institutional Review Board of the French Learned Society for Respiratory Medicine (Société de Pneumologie de Langue Française; CEPRO: 2013/018). Due to the fact that the data file has been anonymised, the name of the participants was not necessary for the analysis, and therefore we did not collect the participants' names. The evaluation committee for observational research (CEPRO) of the Institutional Review Board of the French Learned Society for Respiratory Medicine (Respiratory Society of French Language) estimated that this study was purely observational and consent written and informed consent of participants was not necessary because the research involves no intervention or contact with the patient.

B/Study Area

The study area concerned the suburbs of the city of Pointe-a-Pitre, including all of the Grande Terre area, a region of the archipelago of Guadeloupe (16° North latitude and 61° West longitude), which is a French department located in the Caribbean. This area has a regular relief and is a large limestone plateau. It has a surface of 588 km^2 with 197,603 inhabitants (Nation Institute of Statistics and Economic Surveys (INSEE); 2011) and a density estimated at 336 inhabitants/km^2. The suburbs are crossed by roads but do not include any heavy industries.

C/Population Under Study

The sanitary data were collected by the university hospital centre of Pointe-a-Pitre, the main medical structure of the department, which receives more than 65,000 visits in the emergency department each year, including 20,000 children. This centre is equipped with IT tools (IT extraction software) which are used for collecting medical data (visits to the emergency

department). The daily visits due to the aggravation of asthma were codified according to the international classification of diseases (CIM 10th edition) (J45–J46). The information provided by data extraction corresponded to the number of visits per day for asthma conditions to the paediatric emergency department, as well as administrative data concerning the age and gender of the children admitted in the health emergency structure. The study was carried out from January 1, 2011 to December 31, 2011 and concerned asthmatic children aged between 5 and 15 years old (included).

D/Exposure Data

The days on which the dust intruded on the territory (index days) could be detected thanks to American meteorological satellite data available in real time on the following website: "Aerosol looper" [14]. The exposure data for the pollutants were provided by the regional agency approved by the public authorities for the quality of air, established at Pointe-a-Pitre. The agency has 4 measuring stations (3 fixed urban and peri-urban stations and 1 mobile stations), which regularly measure the following pollutants: PM_{10} (particles suspended in the air, with a median diameter lower than 10 micrometres, $PM_{2.5}$ (particles suspended in the air, with a median diameter lower than 2.5 micrometres, sulphur dioxide (SO_2), nitrogen dioxide (NO_2), the nitric oxide (NO) and ozone (O_3).

The $P_{2.5-10}$ particles were calculated by subtracting the values of the PM_{10} and $PM_{2.5}$ particles. For each pollutant, the daily average was determined by calculating the arithmetic average of the time values measured between 0.00 and 24.00. For ozone, the maximum value of the rolling averages over 8 hours was chosen.

The daily climatic parameters were provided by the regional meteorological station located at the airport of the city of Pointe-a-Pitre. The periods of maximum pollen emissions were determined according to the pollinic calendar of the department. The pollen calendar for the Pointe-à Pitre region (Guadeloupe) is a provisional calendar. The pollen characterization and count was carried out in 2004 by the Palynology lab of the Higher National School of Aerobiology (ENSA) in Montpellier, France. The responsible taxa corresponded to the Poaceae family (cereal) for 43% and to the Mimosaceae family for 16% of the cases. The pollination, taking into account the climate, is perennial in Guadeloupe. The responsible taxa represent the majority of emissions. The maximum emissions correspond to a period of around one week over a month. In the absence of a daily count of pollen on our territory, we only considered periods involving maximum emission of taxa.

The data on influenza epidemics were obtained by consulting the regional agency for monitoring influenza on the territory. The Regional Flu Monitoring Agency uses the number of flu consultations collected by the network of sentinel physicians practicing in the territory of Guadeloupe and from all the emergency cases at the region's hospitals. An epidemic is declared when the epidemic threshold set by the agency on the percentage increase in the number of flu consultations is reached (2 to 3% per week).

E/Statistical Analysis

Descriptive statistics were employed to describe all of the variables of this study for periods with and without the intrusion of Saharan dust.

The averages of the daily concentrations in pollutants, the daily average of climatic variables and the average number of visits for asthma per day to the emergency department were compared by means of the Student's t-test or Mann-Whitney's test for days with and without intrusions of Saharan dust.

Pearson's correlations were calculated for the PM_{10}, $PM_{2.5-10}$ and $PM_{2.5}$.

The chronological series, for particulate and gaseous pollutants, were produced graphically in order to observe their temporal distribution during the study period.

The frequency of visits due to asthmatic conditions per month during the study period was illustrated graphically.

The association between the daily concentrations in PM_{10} and $PM_{2.5-10}$ and the daily visits due to asthmatic conditions in children were analysed by means of a time-stratified case-crossover study [15].

In this type of approach, each case has its own control: the exposure of a subject during the sanitary event (case period) is compared to the exposure of the same subject during one or several different moments (control period) where, a priori, the subject did not present the medical condition [16].

In agreement with the methods of this approach, the control periods were selected so as to correspond to the same day of the week and to the same month as the case day in order to minimise possible bias concerning trends and seasonality of the time series [16].

The effects of an exposure to the PM_{10} and $PM_{2.5-10}$ were examined the same day (lag0) and up to 2 days before the exposure (lag2) but also by averaging the pollutant concentrations corresponding to two days before, one day before until the day of the event (lag0–2), (lag0–1).

We carried out a multivariate conditional logistic regression to estimate the odds ratio (ORs) by adjusting, on the climatic variables (temperature, humidity), the days of influenza epidemics (binary variable), and the days of maximum taxa emissions (binary variable) on bank holidays and during holidays (binary variable).

A binary variable was created for the days with and without Saharan dust. An interaction term between the average concentrations in PM_{10} and $PM_{2.5-10}$ and the presence of Saharan dust was introduced into the modelling.

The analysis of sub-groups was based on the age categories (5 to 8 years old, 9 to 11 years old and 12 to 15 years old) and on the gender of the children (male or female).

The results were presented in the form of an excess risk percentage (IR %) with a confidence interval at 95% (CI 95%) for visits due to asthmatic conditions to the emergency department for an increase of 10 µg/m^3 of PM_{10} and $PM_{2.5-10}$ pollutants. These results were produced during two periods, with and without Saharan dust. The calculation of the IR % for an increase of 10 µg/m^3 for PM_{10} and $PM_{2.5-10}$ has been carried out using the following formula: $(\exp^{(\beta*10)} - 1) \times 100\%$, where ß is the model estimate. We also tested a bi-pollutant model with $PM_{2.5}$ and $PM_{2.5-10}$.

The statistical processing and analysis of data were carried out from an anonymised file using version 2.1.3.0 of R software. The significance threshold was set at 5% for all of the statistical tests.

Results

1/Population Under Study

The study period included 337 days of observation, including 52 days (15% of the days of observation) involving the presence of Saharan dust (index days) and 285 days without Saharan dust.

During the study period, 836 visits to the emergency department took place in relation to asthmatic conditions in children aged between 5 and 15 years old. This figure included 58% (n = 489) of children aged between 5 and 8 years old, 27% (n = 222) of children aged between 9 and 11 years old and 15% (n = 125) of children aged between 12 and 15 years old. There were more boys (n = 514) than girls (n = 322). The sex ratio was of 1.6 in favour of the boys.

The ratio between boys and girls was comparable during the periods with and without Saharan dust.

Figure 1 shows the frequency of visits due to asthmatic conditions per month during the study period. The number of visits to the emergency department was higher from May to September.

Table 1 indicates the average and the number of visits per day to the emergency department due to asthmatic conditions for all children (5 to 15 years old) and for each age section being studied (5 to 8 years old, 9 to 11 years old, 12 to 15 years old) during periods with and without Saharan dust. The number of visits to the emergency department due to asthmatic conditions amounted to 220 visits for all of the children for the 52 days during which the presence of Saharan dust was detected and to 616 visits for the 285 days without Saharan dust.

For all the children (5 to 15 years old), the average number of visits per day was higher during periods with Saharan dust compared to periods without Saharan dust (4.2±1.9 visits/day versus 2.1±1.8 visits/day; p = 0.02).

2/Data Concerning the Levels of Pollutants under Study and the Other Parameters of the Study: Climatic Variables, Periods of Influenza, Periods of Maximum Pollen Emissions

Figure 2 illustrates the temporal distribution of the average daily concentrations in pollutants during the months of the study for the following pollutants: PM_{10}, $PM_{2.5}$, NO_2, NO, SO_2, O_3.

Table 2 describes the average daily concentrations in pollutants and average measurements of climatic parameters for periods with Saharan dust (n = 52 days) and periods without Saharan dust (n = 285 days).

The average daily concentrations in particulate pollutants were higher during days with Saharan dust compared to days without Saharan dust: PM_{10} (49.7±13.4 µg/m^3 versus 19.2±5.6 µg/m^3; Student's t tests, p = 0.001), $PM_{2.5-10}$ (36.2±14.1 µg/m^3 vs.10.3±5.3 µg/m^3; p = 0.001). $PM_{2.5}$ (14.4±10.5 µg/m^3 vs.8.8±2.4 µg/m^3; p = 0.001).

The $PM_{2.5}$/PM_{10} ratio was of 0.2 on average on days with Saharan dust and of 0.4 on days without Saharan dust.

For periods involving Saharan dust, the percentage of days with an average daily concentration in PM_{10} exceeding 50 µg/m^3 was of 38% (n = 20 days).

The average daily concentrations in NO_2, SO_2 and O_3 pollutants were lower during days with Saharan dust compared to days without Saharan dust (p = 0.001, p = 0.001, p = 0.02, respectively).

$PM_{2.5}$ and $PM_{2.5-10}$ were moderately correlated (Pearson' S correlation coefficient = 0.24) while $PM_{2.5}$ and PM_{10} and $PM_{2.5-10}$ and PM_{10} were highly correlated (r = 0.64 and r = 0.96, respectively).

The temperature was higher during the days with Saharan sand (p = 0.001), while the humidity was significantly comparable during the two study periods (p = 0.11).

The periods involving maximum pollen emissions represented 48% (n = 163 days) of all of days of observation (n = 337 days) and corresponded to 50% (n = 26 days) of the days with Saharan dust and 47% (n = 134 days) of the days without Saharan dust, without any significant difference in proportions between the two periods

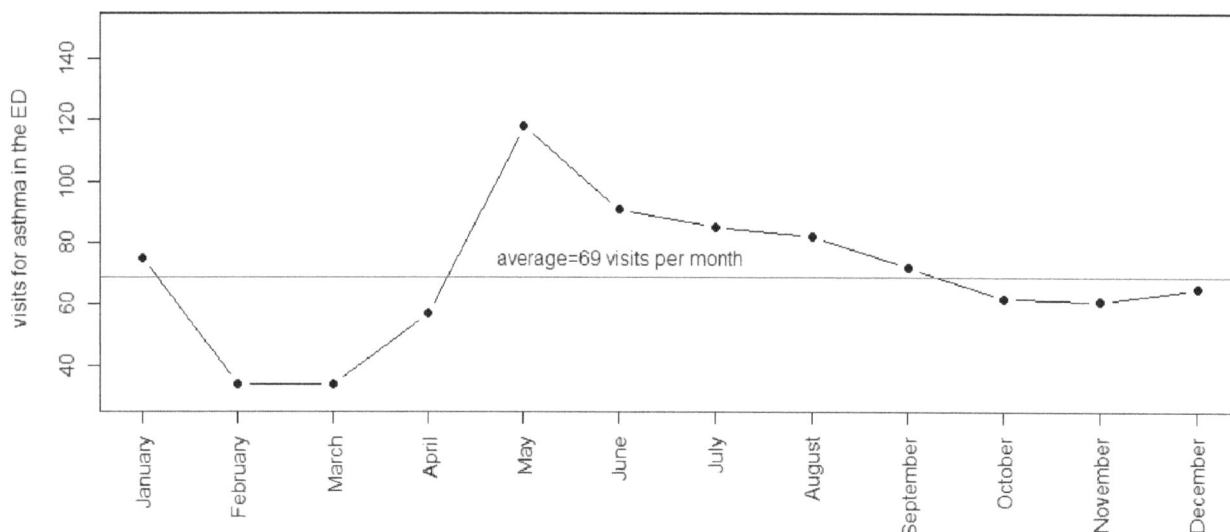

Figure 1. Frequency of visits for asthmatic conditions per month in the pediatric emergency department (ED) during the study period (n = 337 days).

(p = 0.55). The relevant taxa were from the Poaceae family for 43% and from the Mimosaceae family for 16%.

Periods involving influenza represented 20% (n = 69 days) of days of observation. They concerned 7% (n = 4 days) of days with Saharan dust and 22% (n = 65 days) of days without Saharan dust with a significant difference in proportions between the two periods (p = 0.01).

3/Relation between Saharan Dust Episodes and the Number of Visits to the Paediatric Emergency Department Due to Asthmatic Conditions

Figures 3 and 4 graphically represent the IR% of daily visits with a CI 95% for asthmatic conditions in sub-groups of children aged between 5 and 8 years old, 9 and 11 years old, 12 and 15 years old and all of the children aged between 5 and 15 years old for an increase of 10 $\mu g/m^3$ of pollutants PM_{10} and $PM_{2.5-10}$ at lag 0 and at lag (0–1) during periods with and without intrusions of Saharan dust.

A statistically significant association is determined at lag0 and lag (0–1) for PM_{10} and $PM_{2.5-10}$ and the visits due to asthmatic conditions during periods involving intrusions of Saharan dust, adjusted in relation to climatic parameters, periods of influenza and maximum pollen emissions, bank holidays and school holidays.

No statistically significant association was determined with pollutants PM_{10} and $PM_{2.5-10}$ at lag0 and lag (0–1) for the visits of children suffering from asthma during the period without Saharan dust.

The IR% with a CI95% for an increase of 10 $\mu g/m^3$ in pollutants PM_{10} and $PM_{2.5-10}$ for periods with and without Saharan dust presented the following values (table S1 and S2).

For PM_{10} at lag 0:

For all children aged between 5 and 15 years old: the IR% was, during periods with and without Saharan dust, of 9.1% (CI95%, 7.1%–11.1%) versus 1.1% (CI95%, −5.9%–4.6%), respectively. The interaction between the index days and the association between the pollutant and the sanitary variable was significant (p value = 0.0012).

For the sub-groups of children during the periods with and without Saharan dust respectively: The IR% was, for children aged between 5 and 8 years old, of 9.5% (CI95%, 6.8%–12.2%) versus 0.1% (CI95%, −1.4%–4.6%), for children aged between 9 and 11 years old, of 8.4% (CI95%, 5.2%–11.7%) versus 1.9% (CI95%, −5.5%–9.7%), and for children aged between 12 and 15 years old of 8.0% (CI95%, 6.4%–9.6%) versus 1.1% (CI95%, −5.9%–4.6%).

For PM_{10} at lag (0–1):

For all children aged between 5 and 15 years old:

The IR % was, during periods with and without Saharan dust, of 5.1% (CI95%, 1.8%–7.7%) versus 2.4% (CI95%, −0.3%–5%), respectively.

For the sub-groups of children during the periods with and without Saharan dust respectively: The IR% for children aged between 5 and 8 years old was of 5.7% (CI95%, 4.4%–7.1%) versus 2.0% (CI95%, − −0.7%–7.2%), for children aged between 9 and 11 years old, of 4.0% (CI95%, 2.2%–5.8%) versus 2.2% (CI95%, −6.8%–5.3%), and for children aged between 12 and 15 years old, of 7.5% (CI95%, 5.0%–10.3%) versus 1.7% (CI95%, −0.4%–3.7%).

For $PM_{2.5-10}$ at lag 0:

For all children aged between 5 and 15 years old: the IR% was, during the periods with and without Saharan dust, of 4.5% (CI95%, 3.3%–5.7%) versus 1.6% (CI95%, −6.5%–10.4%), respectively. The interaction between the index days and the association between the pollutant and the sanitary variable was significant (p value = 0.002).

For sub-groups of children during periods: with and without Saharan dust, the IR% was, for children aged between 5 and 8 years old, of 6.2% (CI95%, 4.4%–8.1%) versus 2.9% (CI95%, − 2.5%–8.7%), for children aged between 9 and 11 years old, of 5.7% (CI95%, 3.3%–8.2%) versus 4.3% (CI95%, −3.7%–12.8%), for children aged between 12 and 15 years old, of 4.8% (CI95%, 3.8%–5.9%) versus 4.3% (CI95%, −3.7%–12.8%), respectively.

For $PM_{2.5-10}$ at lag (0–1):

For all children aged between 5 and 15 years old: the IR% was, during periods with and without Saharan dust, of 4.7% (CI95%, 2.5% −6.5%) versus 1.8% (CI95%, −1.1%–3.4%), respectively.

Table 1. Daily visits for asthma in emergency department (ED) during Saharan dust-affected days and Saharan dust-free days in a study period.

Daily visits for asthma in ED during Saharan dust-affected days (52 days)					
	Mean	**(SD)**	**Min**	**Median**	**Max**
For all children 5–15 years **n = 220 Male = 132** **Female = 88**	4.2	(1.9)	0.0	3.0	10.0
For children 5–8 years **n = 97 Male = 58** **Female = 39**	1.8	(1.4)	0.0	1.0	7.0
For children 9–11 years **n = 82 Male = 49** **Female = 33**	1.5	(0.9)	0.0	0.0	5.0
For children 12–15 years **n = 41 Male = 25** **Female = 16**	0.7	(0.9)	0.0	0.0	4.0
Daily visits for asthma in ED during Saharan dust-free days (285 days)					
	Mean	**(SD)**	**Min**	**Median**	**Max**
For all children 5–15 years **n = 616 Male = 362** **Female = 254**	2.1	(1.8)	0.0	2.0	8.0
For children 5–8 years **n = 392 Male = 231** **Female = 161**	1.3	(1.5)	0.0	1.0	6.0
For children 9–11 years **n = 140 Male = 87** **Female = 53**	0.4	(0.6)	0.0	0.0	2.0
For children 12–15 years **n = 84 Male = 46** **Female = 38**	0.2	(0.5)	0.0	0.0	2.0

For sub-groups of children during periods with and without Saharan dust: the IR% was, for children aged between 5 and 8 years old, of 5.9% (CI95%, 5.0%–7.2%) versus 1.7% (CI95%, –0.1%–4.4%), for children aged between 9 and 11 years old, of 4.9% (CI95%, 3.0%–6.9%) versus 1.8% (CI95%, −0.2%–3.8%), for children aged between 12 and 15 years old, of 4.4% (CI95%, 2.8%–7.0%) versus 1.4% (CI95%, −0.7%–3.6%), respectively.

Figure 5 graphically represents the IR% of daily visits with a CI 95% due to asthmatic conditions per gender and for all children (5 to 15 years old) and for an increase of 10 µg/m^3 in pollutants PM_{10} and $PM_{2.5-10}$ at lag 0 for periods with Saharan dust:

For PM_{10}: the IR% was of 7.2% (CI95%, 3.1%–11.4%) for boys and of 4.6% (CI95%, 1.6%–7.6%) for girls.

For $PM_{2.5-10}$: the IR% was of 7.8% (CI95%, 4.2%–11.5%) for boys and of 5.1% (CI95%, 2.5%–7.8%) for girls.

There was no statistical difference in the IR% for periods with Saharan dust among different age group of children (Table S3, S4) and between boys and girls (Figure 5) for PM_{10} and $PM_{2.5-10}$ at lag0 and at lag (0–1).

The other delays (lag1, lag2, lag0, 2) tested for pollutants PM_{10} and $PM_{2.5-10}$ during the study period for all children and for the sub-groups were not contributory. Results are reported in the table S5.

No significant effect was revealed for the $PM_{2.5}$ during periods with and without Saharan dust at different delays. (Table S5).

A bi-pollutant model was tested with pollutants $PM_{2.5}$ and $PM_{2.5-10}$. Excess risks for $PM_{2.5-10}$ were not significantly modified for all children at lag 0 and at lag (0–1).

Discussion

This study highlighted a statistically significant association between the PM_{10} and $PM_{2.5-10}$ pollutants contained in the Saharan dust and the visits made to the emergency department due to asthmatic conditions in children aged between 5 and 15 years old during the period involving Saharan dust intrusions.

This association resulted in an excess risk, on the actual day of exposure to the pollutants, of 9.1% (CI95%, 7.1%–11.1%) for the PM_{10} and of 4.5% (CI95%, 2.5%–6.5%) for the $PM_{2.5-10}$ on days involving Saharan dust intrusions with a significant interaction with the pollutants.

The level of average daily concentrations in PM_{10} in this study was comparable to that of the studies performed on pollution due to Saharan dust in Madrid (Spain) [11] or in Rome (Italy) [17]. The level of average concentrations in $PM_{2.5-10}$ on days with Saharan dust was relatively higher in our study compared to the study carried out in Madrid (average = 36.2 µg/m^3 versus 24.2 µg/m^3) [11], because the levels of concentration in $PM_{2.5}$ had lower values in our study (14.4 µg/m^3 versus 24.4 µg/m3). The former studies carried out in the Caribbean did not provide measurements of the concentrations in pollutants contained in the Saharan dust on days of pollution [12], [13]. The study performed by Gyan et al. only measured optical visibility whereas the study carried out by Prospero et al. was based on the measurement of the total rate of dust during Saharan dust intrusions.

As regards the level of gas pollution, for example for pollutant NO_2 in our study, the level of average daily concentrations was 10

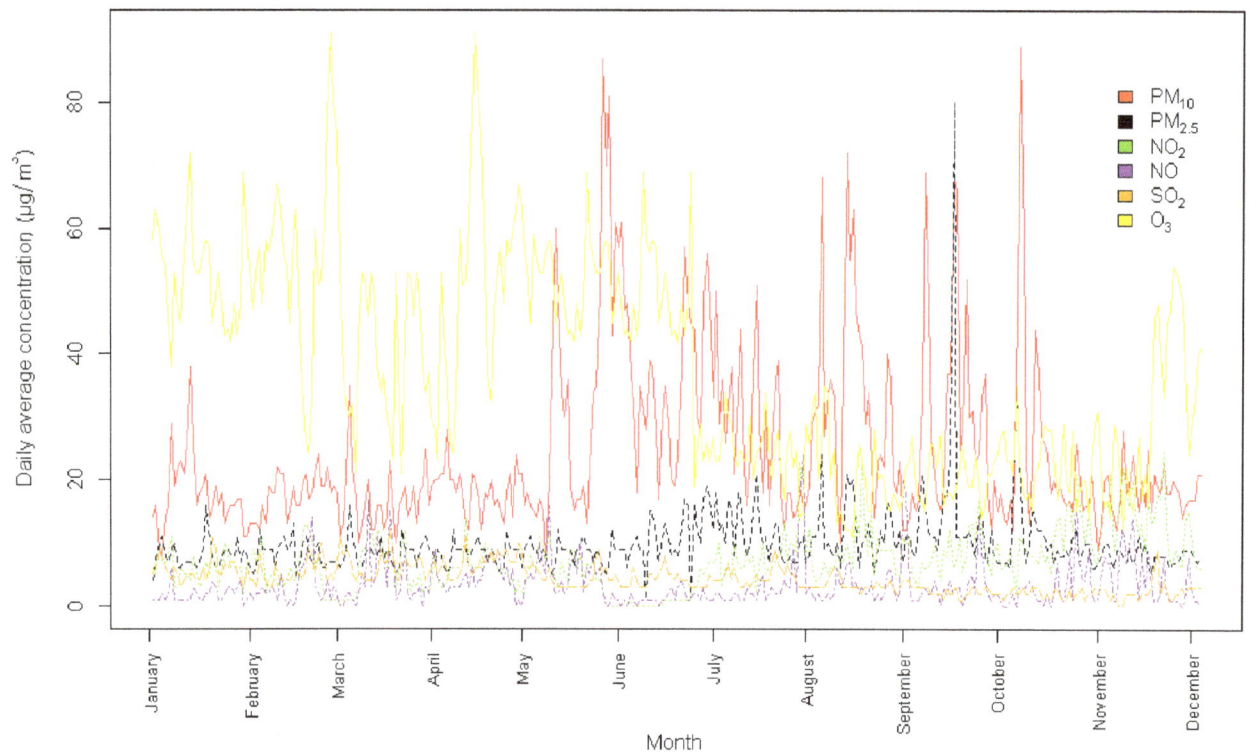

Figure 2. Temporal distribution of pollutants (PM$_{10}$, PM$_{2.5}$, NO$_2$, NO, SO$_2$, O$_3$) during the study period (n = 337 days) in Guadeloupe.
Abbreviations: PM$_{10}$ particles with an aerodynamic diameter of 10 μm or less, PM$_{2.5}$ particles with an aerodynamic diameter of 2.5 μm or less, NO$_2$ nitrogen dioxide, NO nitrogen oxide, SO$_2$ sulphur dioxide, O$_3$ ozone.

times lower than the level recorded in Madrid (5.3 μg/m^3 versus 60.5 μg/m^3) [11]. Throughout the study period, gas pollution (NO$_2$, SO$_2$) and chemical pollution (O$_3$) remained at low levels of daily concentrations and the thresholds set by the WHO were never exceeded. This can be explained by the absence of heavy industries on the archipelago and by the all-year-round presence of trade winds, which dispersed the gaseous and chemical pollutants.

The chemical composition of the Saharan dust includes mineral elements: primarily quartz (60%), oxides (SiO$_2$, FeO$_2$) and carbonates (CaCO$_3$), but also iron, titanium and vanadium [7]. In certain cases, the dust intrusion episodes can involve peaks in sulphate resulting from the chemical reaction between the carbonates contained in the dust and the gaseous pollutants present locally (NO$_2$, SO$_2$) [7]. Toxicological studies reported that the quartz, crystalline silica, aluminium and oxides contained in desert dust could cause an inflammation of the bronchi and lungs in rats due to a hyperproduction of cytokines [9], [18].

Studies have shown that the Saharan dust introduced, into the atmosphere, viable micro-organisms and other microbiological matter (pollen, lipopolysaccharide (LPS), viable mushrooms, mould, viruses, and bacteria) [8]. Several authors have reported an excess risk of mortality due to respiratory causes during Saharan dust intrusions [11], [17]. It has been known for a long time that pollutants, and especially particulate pollutants, aggravate asthmatic conditions in children [19]. It was reported that asthmatic children were more sensitive to air pollutants than non-asthmatic children [20].

In the Caribbean, children's asthma is significantly predominant and the reasons are still unknown [21]. In Guadeloupe, the ISAAC study reported a prevalence of asthma in children of

14.1%, which was higher for boys than for girls (15% versus 13.1%) [3].

In our study, the population being analysed was important (n = 836 children) with a prevalence of boys (60% overall), which complied with the epidemiologic studies on asthma in children performed in our area. These results are also in agreement with the epidemiologic studies on asthma in children, which indicated a major incidence of asthma among boys [22].

The determination of an association between the visits to the emergency department due to asthmatic conditions and Saharan dust intrusions was carried out in this study based on a time-stratified case-crossover methodology. This approach offers several advantages:

It can be used for controlling the individual confounding factors as each case is its own control [16], which is interesting in relation to asthma where individual susceptibilities are to be considered and controlled. In addition, it also makes it possible to take into account individual characteristics such as age or gender to explore the effects of pollutants on the different sub-groups [23]. Moreover, it can be used to control trends and seasonality [24].

This study has highlighted an increase in risks with pollutants PM$_{10}$ and PM$_{2.5-10}$ during periods involving Saharan dust intrusions. The effects were more marked with the PM$_{10}$ than with the PM$_{2.5-10}$ with a significant interaction between these particles and the index days. These effects were maximum and significant on the actual day of pollution. However, we did not reveal any effect with the PM$_{2.5}$ on the visits for asthmatic conditions regardless of the period being analysed; this could be explained by the low level of concentrations in PM$_{2.5}$ in this study, adjusted to a low relative anthropic pollution observed on the archipelago [25].

Table 2. Descriptive statistics of particulate matter, gaseous pollutants and meteorological variables during a period with Saharan dust-affected days and Saharan dust-free days.

Environmental variable	Mean	(SD)	Min.	p25	Median	p 75	Max
Period with Saharan dust-affected days (n = 52 days)							
Particulate matter							
PM$_{10}$ (µg/m^3)	49.7	(13.4)	36.0	38.0	47.7	57.0	89.0
PM$_{2.5}$ (µg/m^3)	14.4	(10.5)	1.0	9.0	13.0	17.0	70.0
PM$_{2.5-10}$ (µg/m^3)	36.2	(14.1)	12.0	27.0	32.0	43.0	81.0
Other pollutants							
NO$_2$ (µg/m^3)	5.3	(1.6)	0.0	1.2	6.0	8.0	23.0
SO$_2$ (µg/m^3)	3.7	(1.4)	2.0	3.0	3.0	4.0	9.0
O$_3$ (µg/m^3)	33.4	(10.5)	11.5	19.6	26.3	48.0	72.0
Weather							
Temperature (C°)	28.9	(0.9)	26.9	28.0	29.0	30.0	31.0
Relative humidity (%)	76.8	(4.5)	69.0	73.0	76.8	79.0	92.0
Period with Saharan dust-free days (n = 285 days)							
Particulate matter							
PM$_{10}$ (µg/m^3)	19.2	(5.6)	8.0	16.0	18.0	21.5	34.0
PM$_{2.5}$ (µg/m^3)	8.8	(2.4)	1.0	7.0	9.0	10.0	21.0
PM$_{2.5-10}$ (µg/m^3)	10.3	(5.3)	0.0	7.0	10.0	13.0	29.0
Other pollutants							
NO$_2$ (µg/m^3)	7.6	(4.5)	0.0	5.0	7.0	10.0	25.0
SO$_2$ (µg/m^3)	4.3	(2.2)	2.0	3.0	4.0	6.0	12.0
O$_3$ (µg/m^3)	38.5	(10.2)	12.5	23.2	37.1	53.0	91.0
Weather							
Temperature (C°)	26.7	(1.6)	22.0	26.0	27.0	28.0	30.0
Relative humidity (%)	77.5	(5.5)	57.0	74.0	77.5	82.0	93.0

p25:25th percentile.
p75:75th percentile.
SD: standard deviation.

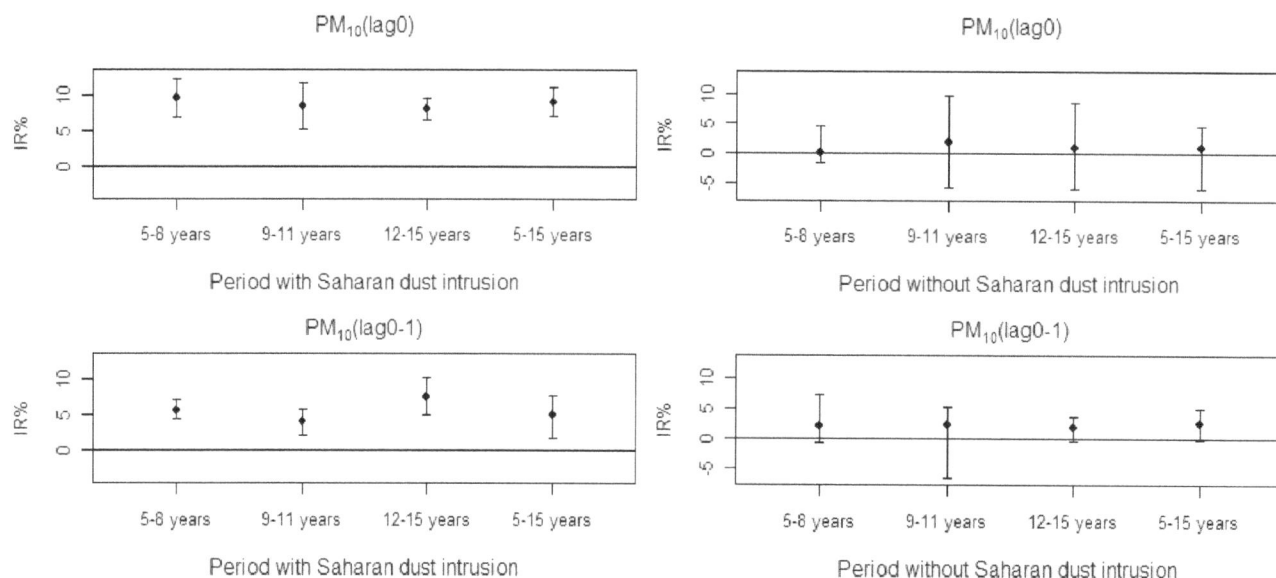

Figure 3. Percentage increase (IR %) of asthma-related visits to the ED for an increase of 10 μg/m³ of PM₁₀ on the day of the visit (lag 0) or the previous 0 to 1 days (lag (0–1)) in each subgroup of children (aged 5 to 8 years old, 9 to 11 years old, 12 to 15 years old and 5 to 15 years old) during periods with and without Saharan dust intrusions. Error bars represent 95% confidence intervals.

Samoli et al. objectified, in Athens, an increase in admissions for asthmatic conditions of 2.54% with the PM_{10} during periods involving Saharan dust intrusions although the interaction between the presence of desert dust and the concentrations in PM_{10} was not significant. This author also observed a maximum effect of the PM_{10} on the actual day of exposure [26].

A study performed in Toyama, in Japan, observed an association between desert dust from Mongolia and China and the hospitalisation of children due to aggravated asthmatic conditions, by using a quantitative measurement of mineral dust in air [27]. The risk of hospitalisation for asthmatic conditions was high for boys and the youngest children, but in this study, only 6

days involved desert dust intrusions [27]. In our study, the effects of the PM_{10} were marked, on days involving Saharan dust intrusions, on young children (5 to 8 years old) while the $PM_{2.5-10}$ had a pronounced effect on teenagers (12 to 15 years old). Both pollutants had an effect on boys and girls. Several explanations can be proposed: young children have a greater tendency to breathe through their mouth than their nose and we know that the breathing process (nasal, oral) is an important factor in the deposit and concentration of pollutants at the level of the bronchial tree [28]. In addition, their respiratory frequency is high and they have a greater pulmonary surface per weight unit than an adult [29]. Moreover, they spend more time outside and still have an

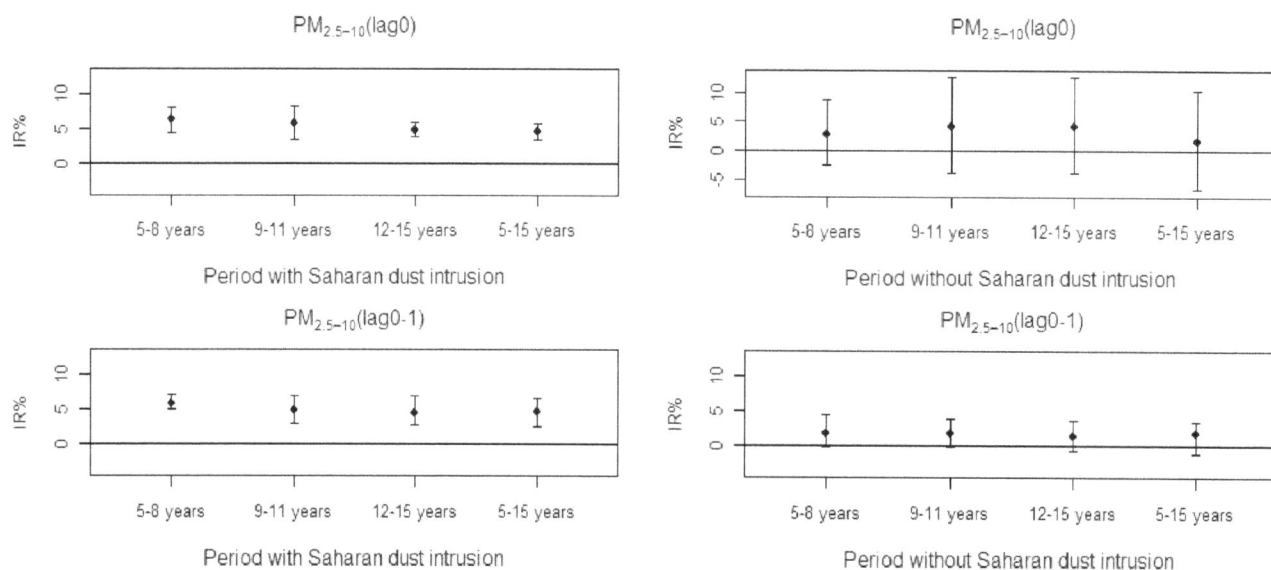

Figure 4. Percentage increase (IR%) of asthma-related visits to the ED for an increase of 10 μg/m³ of PM₂.₅₋₁₀ on the day of the visit (lag 0) or the previous 0 to 1 days (lag (0–1)) in each subgroup of children (aged 5 to 8 years old, 9 to 11 years old, 12 to 15 years old and 5 to 15 years old) during periods with and without Saharan dust intrusions. Error bars represent 95% confidence intervals.

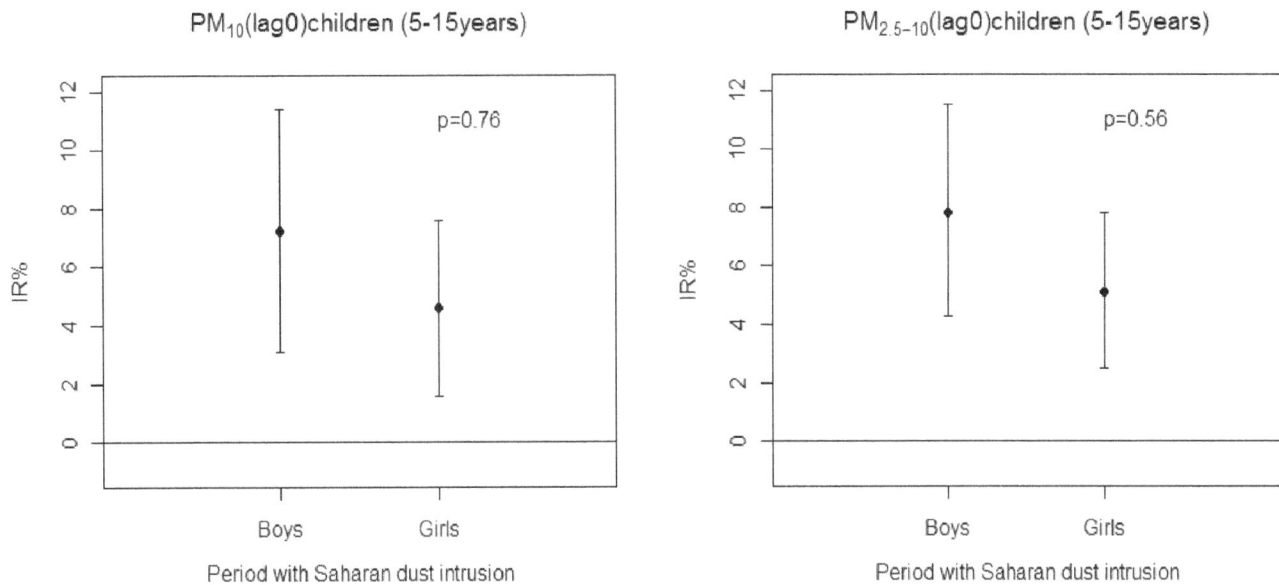

Figure 5. Percentage increase (IR %) of asthma-related visits to the ED for an increase of 10 μg/m³ of PM$_{10}$ and PM$_{2.5-10}$ on the day of the visit (lag 0) in each subgroup of children (boys and girls) during a period with Saharan dust intrusions. Error bars represent 95% confidence intervals.

immature immune system. In addition, boys, especially the youngest, are hyperactive [30]. Our results are consistent with other studies, which reported that the effects of the pollutants were significant among boys and young children [22].

The difference between the effects of the PM$_{10}$ and PM$_{2.5-10}$ contained in dust on different age sub-groups is difficult to explain. A possible assumption concerns the location and quantity of dust deposit through the respiratory system. Venkataraman et al. showed that, on the basis of the mass of the particles, their deposit in the lungs and in the bronchi, was much higher with the PM$_{10}$ than with the PM$_{2.5-10}$ [31]. The inflammatory potential of particles PM$_{2.5-10}$ is as experimentally high as for the PM$_{10}$ [32]. Lin et al. observed an increase in hospitalisations for asthmatic conditions in the presence of the PM$_{2.5-10}$ [33].

In the Caribbean, two studies with the contradictory results were undertaken with Saharan dust:

The first study was carried out on the island of Trinidad, it used optical visibility to measure the density of dust and revealed, by using a Poisson regression model, a significant association between the decrease in visibility and the admissions of children due to asthmatic conditions adjusted to climatic variables [12]. The second study carried out in Barbados did not show any association between Saharan dust and the number of admissions for asthmatic conditions in the main hospital of the island [13].

The disparity of the observations in the Antilles seems to suggest a strong dependency on the surrounding environment. The polluting industries, automobile traffic, and natural presence of pollen or aerosols in suspension in the atmosphere create local particularities specific to each island of the Caribbean. The intrusion of Saharan dust can therefore have different impacts, which are more or less significant, on the health of the population being exposed [34].

In the United States, a study determined an association between the PM$_{10}$ from desert dust and hospital admissions due to respiratory causes in Washington [35]. Another study performed in Anchorage (USA) highlighted an increase in hospital visits due

to asthma, bronchitis and high respiratory infections with PM$_{2.5-10}$ particles from desert dust [36].

In Asia, two studies carried out in Korea [37], [38], showed the significant effects of desert dust on asthmatic children, while another study, carried out in Taiwan [39], did not find any connection. In Australia, an impact on severe asthma conditions was highlighted during dust periods [40].

There still are a limited number of studies, which have been carried out on desert dust. A majority of studies support the possibility that pollutants contained in desert dust have an effect on asthmatic children. However, these studies are difficult to compare because the exposure measurements are not sufficiently described and the confounding factors are not controlled sufficiently [41].

The strong point of our study is that it provides a complete description of the pollutants during periods with and without Saharan dust intrusions and it analyzes the effects of Saharan dust among groups of children of different age and gender in addition to pollutants. As far as we know, this study showed, for the first time, that particles PM$_{2.5-10}$ contained in Saharan dust had an impact on the visits of children to the emergency department due to aggravated asthmatic conditions during the study period.

However, our study presents a certain number of limitations:

Firstly, this type of study is based on the assumption that all of the population under study was exposed to the same amount of pollutant, which cannot be verified. Moreover, it did not take into account the concentrations of pollutants actually inhaled by each child.

Furthermore, this study was carried out on a single site, which could lead to a selection bias.

In addition, the visits to the emergency department were counted by data-processing extraction according to the international classification of diseases (CIM 10th edition) J45–J46 and we cannot affirm the absence of errors in the codes.

Concerning pollen emissions, we used the pollinic calendar of the region, which is definitely less accurate than pollinic counting, which has not been implemented on the archipelago.

A bi-pollutant model was tested with the $PM_{2.5}$ and $PM_{2.5-10}$ but not between the PM_{10} and $PM_{2.5}$ or $PM_{2.5-10}$ due to the strong correlation existing between these pollutants. In addition, due to the fact that the gaseous and chemical pollutants did not show any connection with the sanitary variable, they were not tested by a bi-pollutant model.

Lastly, we did not obtain any data on the chemical or biological nature of the particles contained in the dust to confirm their mineral or anthropic origin.

Conclusion

This study showed that on days involving Saharan dust intrusions, during the study period, the PM_{10} and $PM_{2.5-10}$ particulate pollutants contained in the dust were responsible for an excess risk for visits to the emergency department due to aggravated asthmatic conditions in children aged between 5 and 15 years old. These results deserve to be confirmed by other studies on this topic and could have interesting repercussions, especially concerning the implementation of preventive or therapeutic strategies aiming to improve the treatment of asthmatic children during days involving Saharan dust intrusions.

Supporting Information

Table S1   The excess risk percentages (IR %) with 95% confidence intervals (CI) of visits to the pediatric emergency department due to asthmatic conditions (stratified by age of the children) for an increase of 10 µg/m^3 of pollutants (PM_{10}, $PM_{2.5-10}$) on the day of visit (lag 0) during periods with and without Saharan dust.

Table S2   The excess risk percentages (IR %) with 95% confidence intervals (CI) of visits to the pediatric emergency department due to asthmatic conditions (stratified by age of the children) for an increase of 10 µg/m^3 of pollutants (PM_{10}, $PM_{2.5-10}$) on the previous 0 to 1 days of visit (lag 0–1) during periods with and without Saharan dust.

Table S3   Comparison of IR% (excess risk percentages) between groups of children 5–8 years and 9–11 years during periods of Saharan dust for PM_{10} and $PM_{2.5-10}$ at lag 0 and lag (0–1).

Table S4   Comparison of IR% (excess risk percentages) between groups of children 5–8 years and 12–15 years during periods of Saharan dust for PM_{10} and $PM_{2.5-10}$ at lag 0 and lag (0–1).

Table S5   The excess risk percentages (IR %) with 95% confidence intervals (CI) of visits to the pediatric emergency department due to asthmatic conditions (for all the children (5–15 years) for an increase of 10 µg/m^3 of pollutants (PM_{10} and $PM_{2.5-10}$ (lag1, lag2, lag (0, 2), $PM_{2.5}$ (lag0, lag1, lag (0, 1) lag2, lag0, 2)) during periods with and without Saharan dust.

Author Contributions

Conceived and designed the experiments: GC RT JM. Performed the experiments: GC RT JM. Analyzed the data: GC RT JM. Contributed reagents/materials/analysis tools: GC RT JM. Wrote the paper: GC.

References

1. D'amato G, Baena-cagnani CE, Cecchi L, Annesi-Maesano I, Nunes C et al. (2013) Climate change, air pollution and extreme events leading to increasing prevalence of allergic respiratory diseases. Multidiscip Respir Med 1: 8–12.
2. Monteil MA, Joseph G, Changkrit C, Wheeler G, Antoine RM. (2005) Comparaison of prevalence and severity of asthma among adolescents in the Caribbean islands of Trinidad and Tobago: results of a nationwide cross-sectional survey. BMC Public Health 14: 5–96.
3. Mounouchy MA, Cordeau L, Raherison C. (2009) Prevalence of asthma and related symptoms among adolescents in Guadeloupe: phase I of the ISAAC survey 2003. Rev Mal Respir 9: 944–51.
4. Mortimer KM, Neas LM, Dockery DW, Redline S, Tager IB. (2002) The effect of air pollution on inner-city children with asthma. Eur Respir J 4: 699–705.
5. Taylor DA. (2002) Dust in the wind. Environ Health Perspect 110: 80–7.
6. Middleton NJ, Goudie AS. (2001) Saharan dust: sources and trajectories. Trans inst Br Geogr 26: 165–181.
7. Coz E, Gomez-Moreno JG, Pujadas M, Casucio GS, Lersch TL et al. (2009) Individual particle characteristics of North African dust under different long-range transport scenarios. Atmospheric Environment 43: 1850 1863.
8. Griffin DW.(2007) Atmospheric movement of microorganisms in clouds of desert dust and implications for human health. Clin Microbiol Rev 20: 459–77.
9. Ichinose T, Yoshida S, Sadakane K, Takano H, Yanagisawa R et al. (2008) Effects of Asian sand dust, Arizona sand dust, amorphous silica and aluminum oxide on allergic inflammation in the murine lung. Inhal Toxicol 20: 685–694.
10. Middleton N, Yiallouros P, Kleanthous S, Kolokotroni O, Schwartz J, et al. (2008) 10-year time series analysis of respiratory and cardiovascular morbidity in Nicosia, Cyprus. Environ Health 1: 7–39.
11. Tobias A, Pérez L, Diaz J, Linares C, Pey J et al. (2011) Short-term effects of particulate matter during Saharan dust outbreaks: A case – crossover in Madrid (Spain). Sci Total Environ 412: 386–389.
12. Gyan K, Henry W, Lacaille S, Laloo A, Lamsee-Ebanks C et al. (2005) African dust clouds are associated with increased pediatric asthma accident and emergency admissions on the Caribbean's island of Trinidad. Int J Biometeorol 49: 371–6.
13. Prospero JM, Blades E, Naidu R, Mathison G, Thanie Haresh et al. (2008) Relation between African dust carried in the Atlantic trade winds and surges in pediatric asthma attendances in the Caribbean. Int J Biometeorol 52: 823–822.
14. Aerosollooper. Available:http://www.nrlmry.navy.mil/aerosol_web/loop_html/globaer_centam_loop.html#.Accessed September 2013.
15. Maclure M, Mittleman MA. (2000) Should we use a case-crossover design? Annu Rev Public Health 21: 193–221.
16. Janes H, Sheppard L, Lumley T. (2005) Case-crossover analyzes of air pollution exposure data: referent selection strategies and their implications for bias. Epidemiology 16: 717–26.
17. Alessandrini ER, Stafoggia M, Faustini A, Gobi GP, Forastiere F. (2013) Saharan dust and the association between particulate matter and daily hospitalizations in Rome, Italy. Occup Environ Med 70: 432–434.
18. Mancino D, Vuotto ML, Minucci M. (1984) Effects of a crystalline silica on antibody production to T-dependent and T- independent antigens in Balb/c mice. Int Arch Allergy Appl Immuno 73: 10–13.
19. Tecer LH, Alagha O, Karaca F, Tuncel G, Eldes N. (2008) Particulate matter ($PM_{2.5}$, $PM_{10-2.5}$ and PM_{10}) and children's hospital admission for asthma and respiratory disease: a bidirectional case-crossover study. J Toxicol Environ Health 71: 512–520.
20. Vedal S, Petkau J, White R, Blair J. (1998) Acute effects of ambient inhalable particles in asthmatic and non asthmatic children. Am J Respir Crit Care Med 157: 1034–43.
21. Howitt ME.(2000) Asthma management in the Caribbean- an update. Post grad Doctor Caribb 16: 86–104.
22. Dougherty RH, Fahy JV. (2009) Acute exacerbations of asthma: epidemiology, biology, and the exacerbation-prone-phenotype. Clin Exp Allergy 39: 193 202.
23. Jaakkola JJ. (2003)' Case-crossover design in air pollution epidemiology. Eur Respi J Suppl 40: 81–85.
24. Bateson T F, Schwartz J. (1999) Control for seasonal variation and time trend in case-cross studies of acute effects of environmental exposures. Epidemiology 10: 539–544.
25. Cadelis G, Tourres R, Molinie J, Petit RH. (2013) Exacerbations of asthma in Guadeloupe (French West Indies) and volcanic eruption in Montserrat (70 km from Guadeloupe). Rev Mal Respir 30: 203–14.
26. Samoli E, Nastos PT, Paliastos AG, Katsouyanni K, Priftis KN. (2011) Acute effects of air pollution on pediatric exacerbation: Evidence of association and effect modification. Environ Res 3: 418–24.
27. Kanatani KT, Ito I, Al-Delaimy WK, Adachi Y, Mathews WC et al. (2010) Desert dust exposure is associated with increased risk of asthma hospitalization in children. Am J Resp Crit Care med 182: 1475–1481.
28. Bennett WD, Zeman KL, Jarabek AM. (2003) Nasal contribution to breathing with exercice: effect of race and gender. J Appl Physiol 95: 497–503.

29. Arcus-Arth A, Blaisdell RJ. (2007) Statistical distribution of daily breathing rates for narrow age groups of infants and children. Risk Anal 27: 97–110.

30. Bateson TF, Schwartz J. (2007) Children's response to air pollutants. J Toxicol Environ Health A 71: 3,238–243.

31. Venkataraman C, Kao AS. (1999) Comparaison of particle lung doses from the fine and coarse fractions of urban PM_{10} aerosols. Inhal Toxicol 11: 151–169.

32. BruneKreef B, Forsberg B. (2005) Epidemiological evidence of effects of coarse airborne particles on health. Eur respire J 26: 309–318.

33. Lin M, Chen Y, Burnett RT, Villeneuve PJ, Krewski D. (2002) The influence of ambient coarse particulate matter on asthma hospitalization in children : case-crossover and times-series analyses. Environ Health Perspect 110: 575–581.

34. Monteil MA. (2008) Saharan dust clouds and human health in the English-speaking Caribbean: what we know and don't know. 30: 439–43.

35. Schwartz J. (1996) Air pollution and hospital admissions for respiratory disease. Epidemiology 7: 20–28.

36. Gordian ME, Ozkaynak H, Xue J, Moriss SS, Spengler JD. (1996) Particulate air pollution and respiratory disease in Anchorage, Alaska. Environ Health Perspect 104: 290–297.

37. Park JW, Lim YH, Hyung SY, An CH, Lee SP et al. (2005) Effects of ambient particulate matter on peak expiratory flow rates and respiratory symptoms of asthmatics during Asian dust periods in Korea. Respirology 10: 470–476.

38. Hong YC, Pan XC, Kim SY, Park K,Park EJ, et al. (2010) Asian dust storm and pulmonary function of school children in Seoul. Sci Total Environ 408: 754–759.

39. Yang CY, Tsai SS, Chang CC, Ho SC. (2005) Effects of Asian dust storm events on daily admission for asthma in Tapei, Taiwan. Inhal Toxicol 17: 817–821.

40. Rutherford S, Clark E, Mc Tainsh G, Simson R, Mitchell C. (1999) Characteristics of rural dust events shown to impact on asthma severity in Brisbane, Australia. Int J Biometeorol 42: 217–225.

41. Hashizume M, Ueda K, Nishwaki Y, MichiKawa T, Onozuka D. (2010) Health effects of Asia dusts events : a review of the littérature. Nippon Eiseigaku Zasshi 65: 413–21.

14

Traffic-Related Air Pollution and the Onset of Myocardial Infarction: Disclosing Benzene as a Trigger? A Small-Area Case-Crossover Study

Denis Bard[1]*, Wahida Kihal[1], Charles Schillinger[2], Christophe Fermanian[1], Claire Ségala[3], Sophie Glorion[1], Dominique Arveiler[4], Christiane Weber[5]

1 Department of Epidemiology and Biostatistics, École des Hautes Études en Santé Publique, Rennes and Sorbonne Paris Cité, Paris, France, 2 Association pour la Surveillance de la Qualité de l'Air en Alsace-ASPA, Schiltigheim, France, 3 SEPIA-Santé, Baud, France, 4 Department of Epidemiology and Public Health (EA3430), University of Strasbourg, Strasbourg, France, 5 Laboratoire Image, Ville, Environnement (LIVE UMR7362 CNRS), Faculté de géographie et d'aménagement, University of Strasbourg, Strasbourg, France

Abstract

Background and Objectives: Exposure to traffic is an established risk factor for the triggering of myocardial infarction (MI). Particulate matter, mainly emitted by diesel vehicles, appears to be the most important stressor. However, the possible influence of benzene from gasoline-fueled cars has not been explored so far.

Methods and Results: We conducted a case-crossover study from 2,134 MI cases recorded by the local Coronary Heart Disease Registry (2000–2007) in the Strasbourg Metropolitan Area (France). Available individual data were age, gender, previous history of ischemic heart disease and address of residence at the time of the event. Nitrogen dioxide, particles of median aerodynamic diameter <10 μm (PM_{10}), ozone, carbon monoxide and benzene air concentrations were modeled on an hourly basis at the census block level over the study period using the deterministic ADMS-Urban air dispersion model. Model input data were emissions inventories, background pollution measurements, and meteorological data. We have found a positive, statistically significant association between concentrations of benzene and the onset of MI: per cent increase in risk for a 1 μg/m^3 increase in benzene concentration in the previous 0, 0–1 and 1 day was 10.4 (95% confidence interval 3–18.2), 10.7 (2.7–19.2) and 7.2 (0.3–14.5), respectively. The associations between the other pollutants and outcome were much lower and in accordance with the literature.

Conclusion: We have observed that benzene in ambient air is strongly associated with the triggering of MI. This novel finding needs confirmation. If so, this would mean that not only diesel vehicles, the main particulate matter emitters, but also gasoline-fueled cars –main benzene emitters–, should be taken into account for public health action.

Editor: Stephania Ann Cormier, University of Tennessee Health Science Center, United States of America

Funding: This work was supported by French Agency for Food, Environmental and Occupational Health & Safety (ANSES); Institute for Public Health Research (IRESP); Fondation Coeur et Artères; French Environment and Energy Management Agency (ADEME); and SITA Corporation. The funders had no role in study design, data collection and analysis, decision to publish, or preparation of the manuscript.

Competing Interests: The authors have the following interests. Wahida Kihal's Ph.D. was financially supported by both ADEME and the SITA Corporation. Charles Schillinger is employed by Association pour la Surveillance de la Qualité de l'Air en Alsace-ASPA, and Claire Ségala by SEPIA-Santé. There are no patents, products in development or marketed products to declare.

* Email: denis.bard@ehesp.fr

Introduction

The effects of traffic-related air pollution on cardiorespiratory mortality have been consistently established since the late 1980s [1]. Further studies specifically investigated the association between exposure to traffic and the onset of myocardial infarction (MI), one of the most frequent causes of death. Since the publication of the seminal paper by Peters *et al.* (2001) [2], which shows an association between exposure to traffic and the onset of myocardial infarction (MI), many studies have been published on the issue. The latest review and meta-analysis to date considered the association between short-term exposure to traffic-related air pollutants and subsequent MI risk [3]. The authors retained 34 studies, considering the effects of various air pollutants, either alone or in association, *e.g.* particles with aerodynamic diameter < 10 μm (PM_{10}), particles with diameter <2.5 μm ($PM_{2.5}$), black carbon/black smoke, ozone (O_3), carbon monoxide (CO), nitrogen oxides and sulphur dioxide. Using a random-effect model to estimate the meta-relative risk and 95% confidence interval, a significant, positive association appeared between all analyzed pollutants, with the exception of ozone, and MI risk, although published studies are quite inconsistent regarding both the direction of the association and statistical significance for all pollutants. Thus, the present state of knowledge strongly supports the role of exposure to traffic-related air pollution in triggering the

Table 1. Myocardial infarction events (ICD-9: 410) collected by the Bas-Rhin Coronary Heart Disease Register, Strasbourg Metropolitan Area, France, 2000–2007.

Age group	Females (n = 492) n (%)	Males (n = 1,642) n (%)	Total (N = 2,134) n (%)
35–54	136 (27.7)	637 (38.8)	773 (36.2)
55–74	356 (72.3)	1,005 (61.2)	1,361 (63.8)

onset of MI. In addition, PM_{10} is the air pollutant most consistently associated with myocardial infarction onset [4,5].

None of the above reviews mentions benzene, although gasoline- fueled engines emit this pollutant [6,7]. The literature addressing the acute cardiovascular effects of benzene in occupational settings is scarce, *e.g.* [8,9] but lends some support to an association between exposure to benzene and arrhythmias. However, in the Kotseva and Popov (1998) paper, benzene concentrations seem to have been very high (up to 65 mg/m^3). The authors provide few details on the study population. In addition, probable co-exposures to various stressors are poorly discussed. To our knowledge, a single group investigated the cardiovascular effects of traffic-related air pollution in a mortality study in Taiwan, addressing exposure to benzene [10]. The authors found a significant association between benzene concentration and cardiovascular mortality (ICD-9-CM 410–411, 414, 430–437, same-day association: lag 0). However, this study considered only fatal cases of cardiovascular diseases. Furthermore, exposure was loosely defined on an ecological basis (air concentrations were measured at a single monitoring station for a study population defined as those living within a 10 km radius). Thus, the potential for error in exposure assessment was high.

The aim of the present study is to investigate the possible association between traffic-related benzene emissions as well as 'classical' traffic-related pollutants (NO_2, PM_{10}, O_3, CO, SO_2) and the onset of myocardial infarction. Study design was time-stratified case-crossover, using a very small area as the statistical unit.

Methods

Setting

The Strasbourg Metropolitan Area (SMA), an urban area of 28 municipalities (316 km^2), is located in the Bas-Rhin district in northeastern France, with a population of about 450,000 inhabitants. It is subdivided into 190 census blocks of average population 2,000 (range 2–4,885) and a median area of 0.45 km^2 (range 0.05–19.60). These blocks are the smallest administrative geographic unit in France for which socioeconomic and demographic information from the national census is available. They are devised as to be homogeneous for population size, socioeconomic characteristics and land use. Sixteen blocks of population size <250 were excluded of the dataset (0.8% of the SMA population), for the sake of compliance with French confidentiality regulations.

Cases

Cases were all MI events (ICD-9: 410) either fatal or non-fatal, occurring in the age group 35–74 years between January 1, 2000 and December 31, 2007, ascertained by the local Bas-Rhin Coronary Heart Disease, a collaborating center of the WHO MONICA Project [11]. MIs were documented events, which have definitively been diagnosed as such whether clinically or at necropsy. Individual data available were age, gender, previous history of ischemic heart disease, and address of residence at the time of the event. Cases were geocoded to their census block of residence using ArcGIS version 9.1 (ESRI, Redlands, CA).

Assessment of exposure to air pollution

Nitrogen dioxide (NO_2), particles of median aerodynamic diameter <10 µm (PM_{10}), ozone (O_3), carbon monoxide (CO), sulphur dioxide (SO_2) and benzene air concentrations were modeled on an hourly basis at the census block level over the whole study period using the deterministic ADMS-Urban air dispersion model (Atmospheric Dispersion Modeling System) [12]. Model input data comprised of emissions inventories, background pollution measurements and meteorological data. More details can be found elsewhere [13]. Model performance assessment took place on two occasions. First, we compared predictions to measures at monitoring stations on a yearly basis for NO_2, O_3, and SO_2. However, there was no routine station measurement for CO. Mean differences between measured and modeled values were −2% (range −10% to 9%) for NO_2, −4% (−10% to 10%) for O_3, −1% (−2% to 8%) for PM_{10}. Second, we used passive samplers to measure benzene, PM_{10}, and NO_2 concentrations at the census block level (61 measurements points for NO_2 and benzene, four occasions throughout the year, seven points for PM_{10} on eight occasions). Measurement points were selected as to cover the SMA and compared to the predicted value for the census block, both on a yearly and hourly basis. For both circumstances and whatever the pollutant, the mean error was small: −1% (range −39% to 42%) for NO_2, 0% (−26% to 11%) for PM_{10}, −10% (−33% to 30%) for benzene. One exception was SO_2, of which concentrations were low (maximum 11 µg/m^3), and modeling performance was poor. Accordingly, we did not consider SO_2 further in the analysis.

Proven or likely confounders

Daily meteorological variables (temperature, atmospheric pressure, and relative humidity) were obtained from the French meteorological service (Météo France); weekly influenza-like case counts came from the Sentinelles network [14] of the French National Institute of Health and Medical Research.

Statistical analysis

Associations between MI events and air pollution were assessed with a case-crossover model [15]. Control days were defined according to a monthly time-stratified design [16]. For a MI occurring on a given weekday (*e.g.*, a Monday), control days were the same days of the week throughout the rest of the month (thus, three or four days; here, the other Mondays of the month). Associations between MI events and ambient air pollution concentrations modeled by census block were estimated, adjusting for holidays, meteorological variables (daily maximum temperature, maximum

Table 2. Air pollutants daily concentrations ($\mu g/m^3$) and meteorological parameters in the Strasbourg (France) Metropolitan Area, 2000–2007.

Pollutant	Mean	SD	Minimum	Q1	Median	Q3	Maximum
NO_2	33.4	13.45	2.6	23.3	32.4	42.5	120.2
O_3	63.3	36.85	1.1	35.45	59.41	85.0	228.3
PM_{10}	21.1	9.94	1.65	14.1	19.3	26.1	107.5
CO	596.7	83.5	501.1	540.4	573.5	626.8	1800.5
Benzene	1.8	1.1	0.1	0.9	1.5	2.4	19.6
Minimum temperature (°C)	7.1	-	−15.3	2.0	7.5	12.5	21.8
Maximum atmospheric pressure (hPa)	1007.9	-	970.3	999.4	1005.9	1016.2	1043.1

atmospheric pressure, and mean relative humidity), and influenza epidemics. We tested the influence of the various lags reported in the literature between air pollution indicators, treated as continuous variables, and MI events: average of the day of the event: (lag 0), average of the day of the event and the 1st previous day (lag 0–1), and average of the previous day (lag 1).

The daily air pollution indicator considered for NO_2, PM_{10}, CO and benzene was the 24-hour average concentration, and for ozone, it was the maximum daily value of the 8-hour moving average. The analysis for ozone considered MI events occurring between April 1 and September 30 of each year, because of the very low concentrations of this pollutant in winter. Associations were assessed for cases of all ages and then for cases aged 35–54 and 55–74 years, respectively (categorizing was only for these two age groups, for ensuring confidentiality of the health data geocoded by census blocks).

We employed conditional logistic regression for analyses. Results are expressed as the percent increase in MI risk per 10 $\mu g/m^3$ increase in pollutant concentrations for NO_2, PM_{10} and O_3, per 100 $\mu g/m^3$ increase for CO, and per 1 $\mu g/m^3$ increase for benzene. All statistical analyses were performed with SAS 9.1 software (SAS Institute, Cary, NC).

Ethics statement

The French data protection authority (CNIL) approved the study.

Results

Over the 8 years of the study period, the Bas-Rhin Coronary Heart Disease Register ascertained 2,141 MI events. Seven (0.3%) events could not be geocoded and were excluded from the analyses. Thus, 2,134 cases were analyzed (Table 1), among whom 21.9% of females and 24.0% of males had a previous history of ischemic disease (ICD-9: 410–414). The small number of subjects with a previous history of ischemic disease precluded a specific analysis of this group, because of a very limited statistical power.

Air pollution data appear in Table 2. The mean and range of air pollutants concentrations in the SMA for the study period were comparable to those observed in many metropolitan areas of Europe [17] or in the US [18]. All these pollutants were well correlated, as expected: Pearson's r for daily mean concentrations (Table 3) were in the range reported in the literature, the lowest being −0.16 (O_3 and PM_{10}) and the highest being 0.73 (NO_2 and PM_{10}), all statistically significant, in agreement with results reported in the literature [19,20].

We have found a positive, statistically significant association between incremental concentrations of benzene and the onset of MI for our study population (base model), for all lags tested (0, 0–1 and 1 day), slightly more marked for the first two lags studied (Table 4). The associations between the other individual pollutants and outcome were essentially inconclusive (although a negative, statistically significant association appeared for ozone, lag 1).

When examining the risk (excess odds ratio –eOR–, that is, per cent increase in risk for a 1 $\mu g/m^3$ increase in benzene concentration) associated to incremental exposure to benzene in specific population segments, we observed that men aged 35–54 years were particularly at risk (lag 0, eOR = 15.3; 95% confidence interval [1.0–31.7]), as was the case for older women (age group 55–74 years) for all lags, all statistically significant but more marked for lag 0–1 (eOR = 29.6 [10.3–52.2]) (Table 5).

As regards to the effects of other ambient air pollutants, we have found a higher risk for younger men for NO_2 (lag 0, eOR = 9.3 [0–19.4]). In older women (55–74 years), NO_2 was strongly associated

Table 3. Ambient air pollution daily mean value correlation coefficients, Strasbourg (France) Metropolitan Area, 2000–2007.

Pollutants	Pearson's r				
	Benzene	NO$_2$	O$_3$	PM$_{10}$	CO
Benzene	1.00				
NO$_2$	0.64	1.00			
O$_3$	−0.51	−0.34	1.00		
PM$_{10}$	0.63	0.73	−0.16	1.00	
CO	0.60	0.72	−0.34	0.54	1.00

with MI risk for lag 0–1 (eOR = 15.0 [0.9–31.2]) and lag 1 (eOR = 15.4 [3–29.3]), as were PM$_{10}$ (lag 0–1, eOR = 16.8 [2–33.7]; lag 1, eOR = 17.8 [4.2–33.1]). Models considering the May-September period for ozone were very similar to those covering the April-September period. Incorporating mean relative humidity and holidays as covariates led to results very similar to the above. In addition, influenza epidemics did not influence the results.

Discussion

Our finding of an association between ambient benzene and the triggering of MI has never been reported before. A possible explanation is that benzene is inconsistently measured in standard air monitoring systems, and so far essentially associated to long-term effects, such as cancer.

Among the strengths of our work is the accurate air pollution modeling at a very fine scale (census block) over the study period, diminishing as much as possible the potential for exposure misclassification. Another robust feature is case collection from a specialized registry, using internationally validated diagnosis ascertainment procedures. In addition, using a proven, robust case-crossover design, we observed associations (Table 4) between classically studied traffic-related pollutants (PM$_{10}$, NO$_2$, CO and ozone) and outcomes that were within the range reported in the literature [3], although non-significant for the study population as a whole, due to our limited sample size. Nonetheless, it appears a striking effect of benzene. However, this observation could be partially confounded by ultrafine particulate matter [21], unmeasured in this study.

Ambient air pollutants, in particular those produced by traffic, are systematically found being highly correlated. This is the case

for PM$_{10}$ and PM$_{2.5}$, the latter (unmeasured in our study) being a sizeable constituent of the PM$_{10}$ fraction [22]. In addition, unmeasured pollutants may confound associations [23]. Available methods aiming at disentangling the separate effects of individual pollutants do not provide so far a gold standard [24] and results remain difficult to interpret [25]. As most other authors, we assessed excess risks for each pollutant under study.

We observed the expected [2,26] baseline risk differences between genders, with a 3.3/1 male/female ratio in our study population. In subgroups analysis, younger males appear more at risk (benzene and NO$_2$, lag 0) than older ones, perhaps because the conditions or cumulative risk factors that contribute to a MI in this age group make them especially sensitive, to effects of benzene in particular. Older females appear also at higher risk with benzene, and with NO$_2$ and PM$_{10}$ (Table 5). Such results have already been reported in the literature for the latter two pollutants [23,27] although not convincingly explained so far.

No data were available at the individual level on tobacco smoking and lifestyle, but these factors contribute concurrently to long term susceptibility, not to the very short-term circumstances triggering a MI.

Altogether, the overall consistency between our results and those published in the literature, associating exposure to usual traffic-related air pollutants and the triggering of MI, lends support to our finding of an association between this outcome and exposure to benzene.

As for limitations of this study, we acknowledge that there remains some room for exposure misclassification, since exposure was assimilated to the levels of air pollutants in the subjects' census block of residence. We have no data on the mobility of our study population. However, the lags showing a positive, statistically

Table 4. Exposure to air pollution and the onset of a myocardial infarction (MI) in the Strasbourg (France) Metropolitan Area, 2000–2007, base model[a].

Pollutant	Lag 0		Lag 0–1		Lag 1	
	eOR (95%CI)	p value	eOR (95%CI)	p value	eOR (95%CI)	p value
Benzene	10.4 (3.0, 18.2)	0.005	10.7 (2.7, 19.·2)	0.008	7.2 (0.3, 14.5)	0.04
PM$_{10}$	2.6 (−2.7, 8.2)		3.5 (−2.3, 9.7)		3.1 (−2.0, 8.5)	
NO$_2$	4.7 (−0.2, 9.9)	0.06	5.4 (−0.1, 11.2)	0.05	3.6 (−1.0, 8.5)	
CO	3.2 (−6.1, 13.3)		4.4 (−6.6, 16.7)		3.0 (−6.2, 13.1)	
O$_3$	−1.3 (−3.8, 1.3)		−2.7 (−5.5, 0.2)	0.07	**−3.1 (−5.7, −0.5)**	**0.02**

[a]Associations observed for different lag times; excess odds ratios (eOR) are expressed as per cent (95% confidence interval) increase for i) a 1 µg/m^3 increase in benzene concentrations; ii) a 10 µg/m^3 in NO$_2$, O$_3$ and PM$_{10}$ concentrations and iii) a 100 µg/m3 increase in CO concentrations. Adjusted for the previous day maximum atmospheric pressure, same day minimum temperature and influenza epidemics.

Table 5. Exposure to air pollution and the onset of a myocardial infarction in the Strasbourg (France) Metropolitan Area, 2000–2007, by subgroups[a].

Gender (age group)	Pollutant	Lag 0 eOR (95% CI)	Lag 0–1 eOR (95% CI)	Lag 1 eOR (95% CI)
Males (35–54)	Benzene	15.3 (1.0, 31.7)*	11.1 (−3.5, 27.9)	3.8 (−8.2, 17.4)
	PM_{10}	1.1 (−8.4, 11.6)	1.3 (−8.9, 12.6)	1.0 (−8.0, 10.9)
	NO_2	9.3 (0, 19.4)*	7.0 (−2.9, 17.8)	2.1 (−6.0, 10.9)
	CO	10.3 (−7.6, 31.7)	4.4 (−15.0, 28.1)	−3.0 (−18.1, 14.9)
	O_3	−0.4 (−4.9, 4·3)	−2.1 (−7.0, 3.2)	−3.0 (−7.5, 1.7)
Males (55–74)	Benzene	6.6 (−3.7, 18.0)	5.3 (−5.9, 17.8)	2.2 (−7.7, 13.1)
	PM_{10}	3.8 (−4.1, 12.2)	3.7 (−4.9, 13.0)	2.2 (−5.3, 10.2)
	NO_2	4.4 (−2.7, 12.0)	4.4 (−3.5, 13.1)	2.4 (−4.3, 9.7)
	CO	1.2 (−11.8, 16.0)	0.6 (−14.8, 18.7)	−0.3 (−13.3, 14.5)
	O_3	−0.5 (−4.2, 3.4)	−1.4 (−5.5, 3.0)	−1.7 (−5.5, 2.2)
Females (35–54)	Benzene	−13.9 (−36.8, 17.2)	−3.9 (−30.0, 31.8)	6.2 (−18.6, 38.7)
	PM_{10}	−15.8 (−32.8, 5.6)	−17.9 (−35.7, 4.7)	−13.7 (−30.1, 6.5)
	NO_2	−15.1 (−29.5, 2.3)	−12.9 (−29.0, 6.8)	−5.9 (−21.4, 12.6)
	CO	−30.5 (−54.2, 5.3)	−14.3 (−46.1, 36.3)	11.9 (−23.9, 64.5)
	O_3	−5.7 (−15.7, 5.5)	−8.5 (−19.4, 3.9)	−8.5 (−18.6, 2.9)
Females (55–74)	Benzene	21.5 (4.6, 41.0)*	29.6 (10.3, 52.2)**	27.1 (9.8, 47.1)**
	PM_{10}	9.5 (−3.1, 23.9)	16.8 (2.0, 33.7)*	17.8 (4.2, 33.1)*
	NO_2	7.2 (−4.6, 20.5)	15.0 (0.9, 31.2)*	15.4 (3.0, 29.3)*
	CO	9.1 (−11.6, 34.6)	22.1 (−5.5, 57.8)	23.3 (−1.7, 54.6)
	O_3	−3.8 (−10.1, 2.9)	−6.0 (−13.0, 1.5)	−5.9 (−12.2, 0.9)

[a]Associations observed for different lag times; excess odds ratios (eOR) are expressed as per cent (95% confidence interval) increase for i) a 1 µg/m³ increase in benzene concentrations; ii) a 10 µg/m³ in NO_2, O_3 and PM_{10} concentrations and iii) a 100 µg/m³ increase in CO concentrations. Adjusted for the previous day maximum atmospheric pressure, same day minimum temperature and influenza epidemics.
*$p < 0.05$.
**$p < 0.001$.

significant association between benzene exposure and MI onset span from the same day to the previous day. Thus, exposure misclassification as regards time spent out of area of residence is limited, since people usually spend the major part of their time at home. We feel highly unlikely that such relative misclassification could account for the sizeable associations we have observed.

Mechanisms of action

The literature extensively addresses the underlying mechanisms of action of ambient air pollutants involved in the triggering of MI. Overall, it appears that changes in the synthesis or reactivity of nitric oxide that may be caused by environmental oxidants [28] or an increased endogenous production of reactive oxygen species are candidate mechanisms [29]. A recent review (although targeted to benzene-induced mutation mechanisms) indicates that oxidative stress is one mechanism-of-action of benzene as well [30], but the real contribution of such mechanism to the association we have observed remains to be assessed. In addition, short-term associations with ambient benzene have also been shown for asthma exacerbation [18], although providing no clues for a mechanism of action. The development of epigenetics may shed some light on intimate mechanisms [31].

Public health impact

In the above cited meta-analysis [3], the authors estimated the population attributable fractions (PAF) for those pollutants.

Assuming a 100% prevalence of exposure, the PAFs were of 4.5% for an incremental exposure of 1 mg/m³ carbon monoxide, and ranging between 0.6% and 2.5% for a 10 µg/m³ incremental exposure to the other pollutants. In an earlier analysis of the relative importance of triggers of MI [32], calculated from 14 studies an overall PAF for air pollution of 7.4%. That is, such PAF is of a magnitude similar to that of other well-documented triggers such as physical exertion, alcohol, and coffee. If the whole community were exposed, such a relatively limited PAF would have considerable public health impact. Provided our findings were replicated, this would be the case for benzene but with a much higher effect. However, we felt that calculating a PAF for benzene would be irrelevant in the absence of convergent studies.

Conclusion

We have observed a benzene-associated risk for the triggering of myocardial infarction, using a robust characterization of cases and of exposure. This association has not been documented previously. In addition, the strength of the association was greater for benzene as compared to traffic-related pollutants usually investigated, such as particulate matter. Of course, these results may be the product of an unmeasured confounder (at least partially for ultrafine particles, which are strongly correlated to PM_{10}). If our findings were confirmed by others, this would mean that not only diesel vehicles, the main particulate matter emitters [21,33] but also gasoline-fueled cars –main benzene emitters-, should be taken into account for public health action.

Acknowledgments

We warmly thank Dr William Sherlaw and Ms Kristina Parkins for their editorial assistance.

References

1. Peters A, Pope CA 3rd (2002) Cardiopulmonary Mortality And Air Pollution. Lancet 360: 1184–1185.
2. Peters A, Dockery DW, Muller JE, Mittleman MA (2001) Increased Particulate Air Pollution And The Triggering Of Myocardial Infarction. Circulation 103: 2810–2815.
3. Mustafic H, Jabre P, Caussin C, Murad MH, Escolano S, et al. (2012) Main Air Pollutants And Myocardial Infarction: A Systematic Review And Meta-Analysis. Jama 307: 713–721.
4. Brook RD, Rajagopalan S, Pope CA 3rd, Brook JR, Bhatnagar A, et al. (2010) Particulate Matter Air Pollution And Cardiovascular Disease: An Update To The Scientific Statement From The American Heart Association. Circulation 121: 2331–2378.
5. Baccarelli A, Benjamin EJ (2011) Triggers Of Mi For The Individual And In The Community. Lancet 377: 694–696.
6. Keenan JJ, Gaffney SH, Galbraith DA, Beatty P, Paustenbach DJ (2010) Gasoline: A Complex Chemical Mixture, Or A Dangerous Vehicle For Benzene Exposure? Chem Biol Interact 184: 293–295.
7. Faroon O, Wilbur S (2008) Special Issue On The Toxicology And Epidemiology Of Benzene. Toxicol Ind Health 24: 261–262.
8. Kotseva K, Popov T (1998) Study Of The Cardiovascular Effects Of Occupational Exposure To Organic Solvents. Int Arch Occup Environ Health 71 Suppl: S87–91.
9. Kurppa K, Hietanen E, Klockars M, Partinen M, Rantanen J, et al. (1984) Chemical Exposures At Work And Cardiovascular Morbidity. Atherosclerosis, Ischemic Heart Disease, Hypertension, Cardiomyopathy And Arrhythmias. Scand J Work Environ Health 10: 381–388.
10. Tsai DH, Wang JL, Chuang Kj, Chan CC (2010) Traffic-Related Air Pollution And Cardiovascular Mortality In Central Taiwan. Sci Total Environ 408: 1818–1823.
11. Tunstall-Pedoe H, Kuulasmaa K, Amouyel P, Arveiler D, Rajakangas AM, et al. (1994) Myocardial Infarction And Coronary Deaths In The World Health Organization Monica Project. Registration Procedures, Event Rates, And Case-Fatality Rates In 38 Populations From 21 Countries In Four Continents. Circulation 90: 583–612.
12. Carruthers D, Edmunds H, Lester A, McHugh C, Singles R (1998) Use And Validation Of Adms-Urban In Contrasting Urban And Industrial Locations. Int J Environ Pollut 14: 2000.
13. Havard S, Deguen S, Zmirou-Navier D, Schillinger C, Bard D (2009) Traffic-Related Air Pollution And Socioeconomic Status: A Spatial Autocorrelation Study To Assess Environmental Equity On A Small-Area Scale. Epidemiology 20: 223–230.
14. Flahault A, Blanchon T, Dorléans Y, Toubiana L, Vibert JF, et al. (2006) Virtual Surveillance Of Communicable Diseases: A 20-Year Experience In France. Stat Methods Med Res 15: 413–421.
15. Maclure M, Mittleman MA (2000) Should We Use A Case-Crossover Design? Annu Rev Public Health 21: 193–221.
16. Janes H, Sheppard L, Lumley T (2005) Case-Crossover Analyses Of Air Pollution Exposure Data: Referent Selection Strategies And Their Implications For Bias. Epidemiology 16: 717–726.
17. European Environment Agency (2012) Air Quality In Europe – 2012 Report. Copenhagen: European Environment Agency. 108 P.
18. Delfino RJ, Gong H Jr, Linn WS, Pellizzari ED, Hu Y (2003) Asthma Symptoms In Hispanic Children And Daily Ambient Exposures To Toxic And Criteria Air Pollutants. Environ Health Perspect 111: 647–656.
19. Fusco D, Forastiere F, Michelozzi P, Spadea T, Ostro B, et al. (2001) Air Pollution And Hospital Admissions For Respiratory Conditions In Rome, Italy. Eur Respir J 17: 1143–1150.
20. Schwartz J (1999) Air Pollution And Hospital Admissions For Heart Disease In Eight U.S. Counties. Epidemiology 10: 17–22.
21. Huang C, Lou D, Hu Z, Feng Q, Chen Y, et al. (2013) A Pems Study Of The Emissions Of Gaseous Pollutants And Ultrafine Particles From Gasoline- And Diesel-Fueled Vehicles. Atmospheric Environment 77: 703–710.
22. Marczazzan GM, Vaccaro S, Valli G, Vecchi R (2001) Characterisation Of Pm_{10} And $Pm_{2.5}$ Particulate Matter In The Ambient Air Of Milan (Italy). Atmospheric Environment 35: 4639–4650.
23. Bhaskaran K, Hajat S, Haines A, Herrett E, Wilkinson P, et al. (2009) Effects Of Air Pollution On The Incidence Of Myocardial Infarction. Heart 95: 1746–1759.
24. Billionnet C, Sherrill D, Annesi-Maesano I (2012) Estimating The Health Effects Of Exposure To Multi-Pollutant Mixture. Ann Epidemiol 22: 126–141.
25. Sacks JD, Ito K, Wilson We, Neas LM (2012) Impact Of Covariate Models On The Assessment Of The Air Pollution-Mortality Association In A Single- And Multipollutant Context. Am J Epidemiol 176: 622–634.
26. Bhaskaran K, Armstrong B, Hajat S, Haines A, Wilkinson P, et al. (2012) Heat And Risk Of Myocardial Infarction: Hourly Level Case-Crossover Analysis Of Minap Database. Bmj 345: E8050.
27. Peters A, Von Klot S, Heier M, Trentinaglia I, Hormann A, et al. (2004) Exposure To Traffic And The Onset Of Myocardial Infarction. N Engl J Med 351: 1721–1730.
28. Murugesan K, Baumann S, Wissenbach DK, Kliemt S, Kalkhof S, et al. (2013) Subtoxic And Toxic Concentrations Of Benzene And Toluene Induce Nrf2-Mediated Antioxidative Stress Response And Affect The Central Carbon Metabolism In Lung Epithelial Cells A549. Proteomics 13: 3211–3221.
29. Bhatnagar A (2006) Environmental Cardiology: Studying Mechanistic Links Between Pollution And Heart Disease. Circ Res 99: 692–705.
30. Mc Hale CM, Zhang L, Smith MT (2012) Current Understanding Of The Mechanism Of Benzene-Induced Leukemia In Humans: Implications For Risk Assessment. Carcinogenesis 33: 240–252.
31. Bollati V, Baccarelli A, Hou L, Bonzini M, Fustinoni S, et al. (2007) Changes In Dna Methylation Patterns In Subjects Exposed To Low-Dose Benzene. Cancer Research 67: 876–880.
32. Nawrot TS, Perez L, Kunzli N, Munters E, Nemery B (2011) Public Health Importance Of Triggers Of Myocardial Infarction: A Comparative Risk Assessment. Lancet 377: 732–740.
33. Pant P, Harrison RM (2013) Estimation Of The Contribution Of Road Traffic Emissions To Particulate Matter Concentrations From Field Measurements: A Review. Atmospheric Environment 77: 78–97.

Author Contributions

Conceived and designed the experiments: DB WK C. Ségala DA CW. Performed the experiments: WK. Analyzed the data: WK C. Schillinger CF SG C. Ségala. Contributed reagents/materials/analysis tools: C. Schillinger. Wrote the paper: DB WK CW.

15

Exposure to Organochlorine Pollutants and Type 2 Diabetes

Mengling Tang[1], Kun Chen[2], Fangxing Yang[1], Weiping Liu[1]*

1 MOE Key Laboratory of Environmental Remediation and Ecosystem Health, College of Environmental and Resource Sciences, Zhejiang University, Hangzhou, China,
2 Department of Epidemiology & Health Statistics, School of Public Health, Zhejiang University, Hangzhou, China

Abstract

Objective: Though exposure to organochlorine pollutants (OCPs) is considered a risk factor for type 2 diabetes (T2DM), epidemiological evidence for the association remains controversial. A systematic review and meta-analysis was applied to quantitatively evaluate the association between exposure to OCPs and incidence of T2DM and pool the inconsistent evidence.

Design and Methods: Publications in English were searched in MEDLINE and WEB OF SCIENCE databases and related reference lists up to August 2013. Quantitative estimates and information regarding study characteristics were extracted from 23 original studies. Quality assessments of external validity, bias, exposure measurement and confounding were performed, and subgroup analyses were conducted to examine the heterogeneity sources.

Results: We retrieved 23 eligible articles to conduct this meta-analysis. OR (odds ratio) or RR (risk ratio) estimates in each subgroup were discussed, and the strong associations were observed in PCB-153 (OR, 1.52; 95% CI, 1.19–1.94), PCBs (OR, 2.14; 95% CI, 1.53–2.99), and *p,p'*-DDE (OR, 1.33; 95% CI, 1.15–1.54) based on a random-effects model.

Conclusions: This meta-analysis provides quantitative evidence supporting the conclusion that exposure to organochlorine pollutants is associated with an increased risk of incidence of T2DM.

Editor: Noel Christopher Barengo, University of Tolima, Colombia

Funding: This study was supported by the National Natural Science Foundations of China (21177112, 21320102007). The funders had no role in study design, data collection and analysis, decision to publish, or preparation of the manuscript.

Competing Interests: The authors have declared that no competing interests exist.

* Email: wliu@zju.edu.cn

Introduction

Organochlorine pollutants (OCPs), represented by DDT (Dichlorodiphenyltrichloroethane) and PCBs (Polychlorinated biphenyls), are environmental contaminates of global concern because of their potential for bio-accumulate and bio-magnify in ecosystems and hazardous effects on human health. Though DDT and PCBs were forbidden in most countries in the 1970s [1] and 1980s [2], and the concentrations of these chemicals in the environment, organisms and human tissues were decreasing over the past 30 years, they can still be detected due to their characteristics of persistency, semi-volatility, lipid solubility, bioaccumulation and biomagnification [1]. In Ghana, DDE (Dichlorodiphenyldichloroethylene) was detected at the highest levels among DDT isomers at 44.8 and 7.1 ng/g in breast milk and serum, respectively [3]. In China, PCBs were detected at 0.9 ng/g in lipid in the placentas of women who had pregnancies affected by neural tube defects and at levels of 0.87 ng/g in lipid controls [4].

Type 2 diabetes mellitus (T2DM), formerly called adult-onset diabetes, is a noninsulin-dependent diabetes that accounts for 90–95% of all diabetes cases [5]. As a result of a metabolic disorder of glucose, T2DM has become a major global epidemic in recent years, and its prevalence will likely double over the next 20 years

[6]. World Health Organization (WHO) projects that diabetes will be the 7th leading cause of death in 2030. The prevalence of T2DM may be affected by the interaction of conventional risk factors and a combination of genetic susceptibility [7], metabolic syndromes such as obesity [8] and hypertension [9], age, race, and poor diet. In addition, the accumulation of environmental pollutants in the human body has been suggested to have a significant contribution to the disease [10].

Within different populations, the positive associations were observed in the epidemiological studies about T2DM risk exposure to OCPs [11,12]. The associations may be attributed to certain mechanisms of the active ingredients of OCPs, such as γ-aminobutyric acid, which affect the neurotransmitter or ion channel systems involved in regulating pancreatic function and then influence glucose homeostasis [13]. Toxic effects through direct binding and activation of the aryl hydrocarbon receptor (AhR) pathway [14] and mediation through AhR-independent oxidative stress and mitochondrial dysfunction [15] have also been reported as biological mechanisms. Furthermore, toxic effects on estrogen receptor, peroxisome proliferator-activated receptor γ (PPARγ), and progesterone receptor were considered other mechanisms. However, the pathogenesis of exposure to OCPs is currently obscure.

To our knowledge, there have been many epidemiological studies regarding the association between exposure to OCPs and the prevalence of T2DM. However, the results showed contradictory. In order to fully evaluate and characterize the association and fill the vacancy of epidemiological evidence in the comprehensive summary, we performed a sub-group meta-analysis of the results of T2DM risk from exposure to OCPs. We systematically analyzed all studies on T2DM risk from exposure to OCPs up to August 2013.

Methods

Study Identification

We reported the meta-analysis according to the Preferred Reporting Items for Systematic Reviews and Meta-analyses (PRISMA) [16] (Checklist S1). Publications about epidemiological evidence of T2DM risk from exposure to OCPs were identified by a search on MEDLINE (National Library of Medicine, Bethesda, MD) and WEB OF SCIENCE databases. A preliminary total of 116 related studies published up to August 2013 were selected using various combinations of the following keywords: "diabetes", "DDE", "DDT", "PCB", and "organochlorine" with no restriction of publication type and date. The reference lists of the relevant publications identified were checked for additional studies and the recent articles in relevant journals were also scanned to identify other potential studies. The whole search was limited to studies published in English in the open literature in peer-reviewed journals.

Criteria for Inclusion

The systematic review and identification of eligible studies was performed. The titles and abstracts were screened to determine their relevance to the diabetes effects of humans when exposed to OCPs. The full text of potentially relevant studies was then examined and the eligibility criteria were applied to select the included studies.

A publication was considered eligible for review if it fulfilled the following six inclusion criteria. (1) It must be an original epidemiologic study using a case-control, cross-sectional, or prospective study design and other types of reviews, meta-analysis, case-reports, comments, letters, editorials, abstracts were excluded. (2) Papers should be written in the English language. (3) OCP exposure levels had to be measured in actual tissue samples (serum or serum lipid), not by environmental data or other indirect ways. (4) It unequivocally reported measures of association, including odds ratios (OR) and relative risk (RR) and confidence intervals (CIs) for diabetes risk and also considered papers that did not report these measures directly, but were able to extrapolate the relevant values. (5) Studies should use biomarkers of OCPs within our selected ones including PCB-153, PCBs, and p,p'-DDE, while others OCPs, for instance, using PCB-126 as the biomarker, were not included. (6) In addition, T2DM was confirmed by self-report or hospital diagnosis, and diseases related to T2DM as insulin resistence were excluded. Finally, 23 epidemiological studies were extracted for further systematic analysis.

Data Extraction

The authors examined the articles and independently extracted and tabulated the information. A standard data abstraction form was created to record the following information for each suitable article: first author name, year of publication, geographic region of the studies, epidemiologic design, subject selection, exposure pathways, type of OCP, biologic specimens, number of cases and controls, and a risk index calculated with the categories of the

exposure and referent, corresponding 95% CI for T2DM. The risk indexes, adjusted for different confounding such as sex, BMI, cigarette smoking, and the ones stratified by age and sex were all extracted. The results of this abstraction were compared between the authors and consensus was obtained before the meta-analysis.

Stratification of the data were performed focusing on several variables that could influence the results, including exposure levels (background or high concentration exposure), study design (case-control, perspective, or cross-sectional study), population selection (general population or women), and biologic specimen (serum or serum lipid).

Quality Assessment

In order to assess evidence, all included studies underwent an independent quality assessment modified from the versions of the 1998 Downs and Black [17] and Wigle et al. checklists [18]. We discussed the individual items on the checklist to clarify their interpretation before conducting the quality assessment. The same to the version made by the latter group, we also added exposure measurement as the internal validity assessment to the checklist of quality assessment. However, some items that were either related only to reporting or were not applicable were removed from the checklist. No attempt was made to blind the reviewers of the authorship or publication status of the original studies. The evaluated factors including the representativeness of the selected participants, bias, and confounding were given a mark to assess the article quality. Finally, a total of 13 items and 16 scores were listed (Table S1). The results with higher scores were considered to be of superior quality. Differences in quality assessment were resolved by consensus.

Data Analysis

The heterogeneity across individual studies was quantified by the Q-test and I^2-test: when the result of the Q-test showed evidence of the heterogeneity (p<0.1), we used the random-effect analysis; otherwise, the I^2-test, which interpreted I^2 values of 25%, 50%, 75% as low, moderate, and high degrees of heterogeneity respectively, was used to assess heterogeneity. This is because the Q-test has low statistical power with few studies [19] and the fixed-effect analysis was conducted when $I^2 < 25\%$. p<0.10 or $I^2 > 25\%$ was considered significant heterogeneity which questions the validity of pooled estimates. The I^2 describes the percentage of total variation across studies due to heterogeneity rather than chance [20]. When heterogeneity exists, subgroup analyses were conducted to investigate potential sources of heterogeneity.

The risk estimates of OR or RR were combined for the evaluation of the dose-response relation between OCP exposure and T2DM prevalence. We assumed similarity between the OR and RR. When combined these binary variables, we aimed to choose the ones calculated between the highest exposed group and the references and the ones with the most adjusted variables. We attempted to combine adjusted OR or RR from primary studies, but if not possible, we pooled raw outcome data to yield unadjusted OR. In addition, we combined the risk estimates which were calculated by OCPs concentrations tested in serum lipid, otherwise we choose the values tested directly in serum. We considered all the OR stratified by ages and by BMI in each study. To conduct meta-analyses, we defined the least group as 4 articles with risk estimates [21], which corresponded to a minimum of 100 cases of T2DM.

We performed meta-analyses using Review Manager (RevMan) version 5.0 (Nordic Cochrane Centre, Cochrane Collaboration, Copenhagen, Denmark) to evaluate the overall risk of T2DM caused by exposure to OCPs. For the risk estimates presented as a binary variable, such as OR and RR, the inverse of variance for fixed-effects models using the Mantel-Haenszel [22] method which assumes that results across studies differ only by sampling error. The DerSimonian and Laird method [23] for random-effects models were used to combine the overall binaries and their corresponding 95% CI. The results of meta-analysis including all subgroup analysis were illustrated by forest plots.

Publication bias due to study size was investigated by visual inspection of funnel plots which showed the natural logarithm of the estimate of RR (lnRR) versus the inverse of standard error (1/ SE). Funnel plot asymmetry can be illustrated by factors as the non-publication of small studies with negative results, differences in study quality and study heterogeneity.

To determine whether some of the decisions we made had a major impact on the results of the review, sensitivity analyses were conducted by (1) removing studies with the highest and lowest percentage weight in all included studies, (2) deleting studies with highest and lowest quality scores, (3) excluding studies reporting the lowest or highest estimator of binary variables.

Results

Study Characteristics

The searching process and selection studies was performed in Figure 1 and the characteristics of the 23 included studies [12,24,25,26,27,28,29,30,31,32,33,34,35,36,37,38,39,40,41,42,43-,44,45] are summarized in Table 1. Among the 32 related epidemiological studies, 2 were excluded because of their examination of other diseases related to T2DM [46,47]; 4 were removed because of the absence of dichotomous variables, OR or RR, and the CI [48,49,50,51]; 2 were excluded because the biomarkers were p,p'-DDT, PCB-126 [52] and PCB-170 [10], and not the ones (p,p'-DDE, PCBs, PCB-153) we selected for this study; and 1 was excluded because of the combination of both type 1 and type 2 diabetes [53].

From the 23 remaining studies, 1 were case-control studies [54], 18 were cross-sectional studies [24,25,26,27,28,29,30,31,32,33, 35,36,37,38,39,40,41,42,44], and only 4 study was a prospective study [12,34,43,45]. 10 studies were conducted in the United States [12,25,26,32,34,38,40,41,45], 4 in Sweden [30,36,39,42], and the rest were conducted in Japan [27], the Faroe Islands [28], Korea [31], Slovakia [33], Canada [35], Belgium [43], Taiwan, China [37], Spain [24,44], and Finland [29].

13 studies used PCB-153 [24,26,27,29,30,34,35,36,38,39, 41,42,45], 11 studies used PCBs [12,24,25,26,28,33,35,37, 38,44,45] and 18 studies used p,p'-DDE as a biomarker [24,28,29,30,31,32,33,35,36,38,39,40,41,42,44,45]. Among these studies, different models were established to assess the OR or RR, and the related CI was adjusted by confounders such as age, body mass index (BMI), sex, or other factors. 2 studies estimated body burden levels of OCPs in both wet weight values (serum sample) and lipid-standardized values (serum lipid sample) [35,38]. In addition, 2 studies estimated the OR or RR of both men and women [25,37], and 1 study discussed the groups under and over the age of 55 years and also the total group separately [25]. The risk estimates from Wu et al. were obtained from two independent study [45].

In most studies, the study population was selected by background exposure to OCPs [24,27,29,30,31,32,34,36,38, 40,41,43,44,45]. However, aquatic product exposure [12,28,35, 39,42], heavy pollution area exposure [25,26,33], and specific diet exposure, such as rice-bran oil exposure [37], were also estimated in some studies. As a susceptible population, women were selected [36,39,45] in 3 studies. 6 studies collected serum as biologic

specimens [12,24,25,30,36,37], while 12 chose serum lipids [26,28,29,32,33,34,39,41,42,43,44,45]. 5 studies detected the pollutants in both serum and serum lipid specimens [27,31,35,38,40].

Quality Assessment

The quality factor scores for the 23 studies are listed (Table S2). From the results of the quality assessment, all the included epidemiological studies accorded with most of the quality criteria we listed, but the items of participation rate, blind laboratory testing, data dredging, specific exposure measurement, and adequate adjustment for confounding were different among the original studies. The total quality scores were in the range from 9 to 14 with a possible maximum score of 16, reflecting the existing of study design limitations. More recent studies tended to have higher quality scores. Among these studies, only 4 studies had the external validity that participation rate for cases and controls reaches 70% [34,36,40,43]. Only 3 studies reported having made an attempt to blind those measuring the main outcomes of the OCPs exposure [29,31,37]. Most of the included studies got the scores of other 11 items.

Main Analysis

23 studies, contributing a total of 73 OR or RR estimators met the inclusion criteria and were taken into consideration. When combining the main data of all studies, the exposure to all 3 biomarkers showed positive associations with the prevalence of T2DM. The combined OR estimate of PCB-153 was 1.52 (95% CI, 1.19–1.94), for PCBs was 2.14 (95% CI, 1.53–2.99), and for p,p'-DDE was 1.33 (95% CI, 1.15–1.54) based on a random-effects model. Forest plots of the 3 organochlorine biomarkers, which show the weight of each study and the combined OR estimates, are provided in Figure 2. Considering the high evidence of heterogeneity for PCB-153 ($I^2 = 64\%$), PCBs ($I^2 = 59\%$) and p,p'-DDE ($I^2 = 56\%$), subgroup meta-analyses were conducted for the OR combining and further analyses of sources of heterogeneity. The results for the meta-analyses of 3 organochlorine biomarkers (PCB-153, PCBs, and p,p'-DDE) and their subgroups were analyzed and are summarized in Table 2.

Subgroup and Sensitivity Analyses

From the characteristics summary of the epidemiological studies, exposure levels (background or high concentration exposure), study design (case-control or cross-sectional study), population selection (general population or women), and biologic specimen (serum or serum lipid) were chosen as the stratifications for subgroup analyses to find the sources of heterogeneity.

For PCB-153, the exposure subgroup analyses may not be a heterogeneity source from the increased results of I^2 test. When the studies were stratified by the study design, heterogeneity and inconsistency among the epidemiological studies were eliminated in the perspective subgroup ($I^2 = 0\%$). When the studies were divided by the sex of the population, the consistency was observed among the women subgroup ($I^2 = 0\%$) and the heterogeneity between studies remained high for general population group ($I^2 = 71\%$). Finally, when the studies were stratified by biologic specimen, the high inconsistencies in both the serum ($I^2 = 74\%$) and serum lipid subgroup ($I^2 = 67\%$) also existed.

For PCBs, the background exposure subgroup ($I^2 = 30\%$) the subgroup of serum specimen exposure ($I^2 = 8\%$) resulted in a statistically decreasing heterogeneities from the total studies ($I^2 = 64\%$). The subgroup analysis of the case-control subgroup ($I^2 = 55\%$, n = 2 only), general population ($I^2 = 59\%$) and women

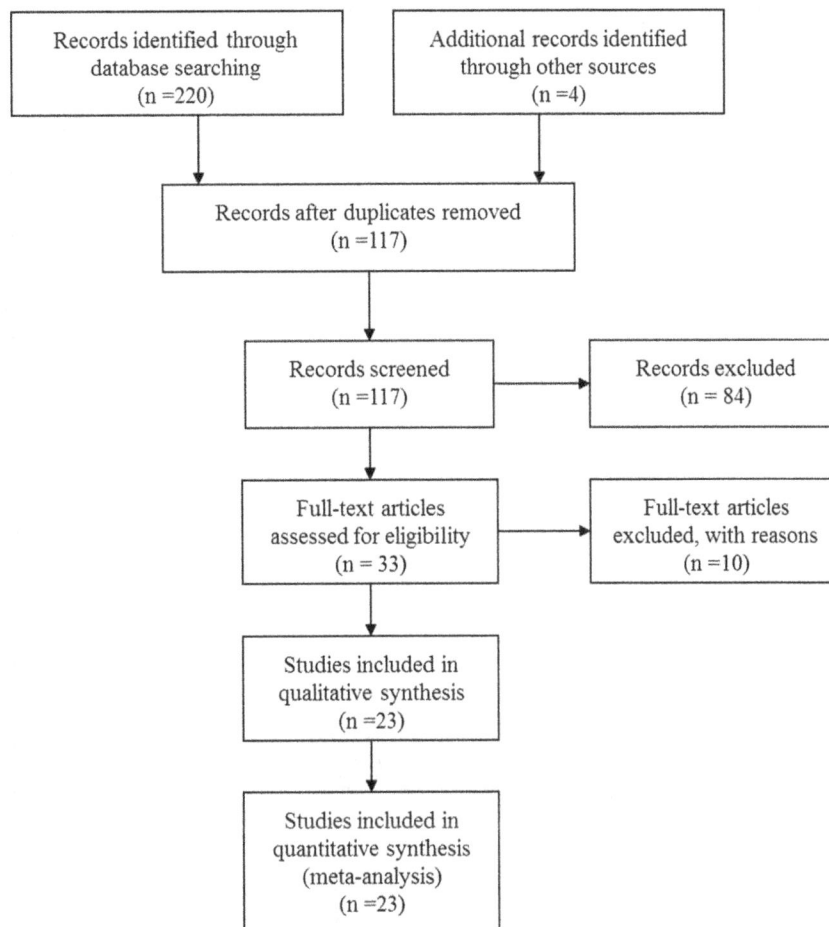

Figure 1. The study search and selection process.

population ($I^2 = 74\%$) gave the result that they still had relatively high heterogeneity.

The subgroup of background exposure from p,p'-DDE showed decreased heterogeneity ($I^2 = 25\%$) compared with total studies ($I^2 = 56\%$), and a risk factor was found from the combined OR (OR, 1.49; 95% CI, 1.18–1.88). Decreased heterogeneity was also found in the perspective group ($I^2 = 0\%$) and women subgroup ($I^2 = 0\%$) with a combined OR of 1.16 (95% CI, 0.87–1.55) and 1.27 (95% CI, 1.10–1.46). The heterogeneities in subgroups classified by biologic specimen were still in relatively high levels.

In general, findings from each sensitivity analysis did not substantially alter the results of the overall pooled estimate OR using the random effects model in direction and magnitude. Exclusion of the studies with the highest and lowest percentage weight, the highest and lowest quality scores, and the lowest or highest estimator of OR performed consistently with the pooled estimator OR for all indicators, including PCB-153, PCBs, and p,p'-DDE (data not shown).

Publication Bias

A funnel plot of standard error (SE) versus ln(OR) for the meta-analyses of the relationships between OCPs and T2DM, in which the number of studies was more than 10, are presented in Figure 3. Visual inspection of the funnel plot suggests that risk estimates stemmed mostly from large, precise studies, which are distributed in the superior part of the figure; however, possible

publication bias was found from the evidence of asymmetry of some subgroup meta-analyses. Additionally, exclusion of studies published with non-English and other factors such as differences in study quality or heterogeneity, sample size, and study design may be other reasons for the asymmetries of the funnel plots.

Discussion

While a large number of studies about the association between OCPs exposure and the prevalence of T2DM were published, only few cohort studies are available. Among these studies, inconsistency was found among most of these epidemiological studies in different populations and different sources of exposure. However, our combined estimates of meta-analyses demonstrated a modest but statistically significant increase in the odds of T2DM with exposure to OCPs. For instance, a 52% increase of T2DM resulted from an exposure to PCB-153. In addition, all subgroup analyses stratified by exposure levels, design of the studies, study subjects and biologic specimen resulted in positive correlations. The consistency in the magnitude of increased risk indicates that this is unlikely to be a chance finding and these increased risks support the suggestion that exposure to OCPs may be a potential causal factor for prevalence of T2DM.

To determine the sources of heterogeneity of the studies and obtain the pooled estimates of PCBs, PCB-153, and p,p'-DDE in subgroups, subgroup analysis was conducted by stratified exposure levels, study designs, study subjects and biologic specimens.

Table 1. Study characteristics.

Reference	Country	Study design	age	Exposure pathway	biomarker	Biologic specimen	Cases/controls	OR or RR (95%CI)
Arrebola2013	Spain	Cross-sectional	16-	Background	p,p'-DDE	Serum lipid	34/352	1.69(0.54–5.22)[a]
Wu2012	USA	Prospective	30–55	Nurses	PCB-153	Serum lipid	10/214	2.19(0.72–6.68)[a]
			30–55	Nurses	PCB-153	Serum lipid	6/135	0.76(0.22–2.60)[a]
			30–55	Nurses	PCBs	Serum lipid	9/215	1.30(0.43–3.93)[a]
			30–55	Nurses	PCBs	Serum lipid	7/134	0.79(0.23–2.71)[a]
			30–55	Nurses	p,p'-DDE	Serum lipid	9/215	1.32(0.41–4.27)[a]
			30–55	Nurses	p,p'-DDE	Serum lipid	10/131	1.79(0.54–5.86)[a]
Persky2012	USA	Cross-sectional	35-	Electrical utilities company men	PCB-153	Serum lipid	7/56	3.1(1.2–7.8)[a]
			35-	Electrical utilities company men	PCBs	Serum lipid	7/56	3.0(1.3–7.2)[a]
Gasull2012	Spain	Cross-sectional	18–74	Background	PCB-153	Serum	77/192	1.6(1.2–2.4)[a]
			18–74	Background	PCBs	Serum	77/193	1.7(1.1–2.6)[a]
			18–74	Background	p,p'-DDE	Serum	73/207	1.1(0.7–1.7)[a]
Silverstone2012	USA	Cross-sectional	18-	PCBs plant area	PCBs	Serum	78/157	2.78(1.00–7.73)
			18–55	PCBs plant area	PCBs	Serum	27/111	4.78(1.11–20.6)
			55-	PCBs plant area	PCBs	Serum	51/47	4.19(0.26–68.12)
Lee 2011	Sweden	Prospective	70	background	PCB-153	Serum	12/277	1.7(0.5–6.2)
			70	background	p,p'-DDE	Serum	16/271	2.1(0.7–6.3)
Airaksinen2011	Finland	Cross-sectional	70	background	PCB-153	Serum lipid	69/398	1.64(0.92–2.93)[b]
			70	background	PCB-153	Serum lipid	10/121	1.03(0.25–4.18)[b]
			70	background	PCB-153	Serum lipid	23/184	1.97(0.75–5.23)[b]
			70	background	PCB-153	Serum lipid	36/93	2.3(0.87–6.11)[b]
			70	background	p,p'-DDE	Serum lipid	62/398	1.75(0.96–3.19)[b]
			70	background	p,p'-DDE	Serum lipid	7/120	0.88(0.18–4.35)[b]
			70	background	p,p'-DDE	Serum lipid	18/175	1.91(0.69–5.27)[b]
			70	background	p,p'-DDE	Serum lipid	37/103	1.82(0.71–4.65)[b]
Tanaka 2011	Japan	Cross-sectional	40–64	background	PCB-153	Serum	32/85	0.95(0.90–1.00)[a]
			40–64	background	PCB-153	Serum lipid	32/85	0.73(0.51–1.07)
Grandjean2011	Faroe Islands	Cross-sectional	70–74	Aquatic product	PCBs	Serum lipid	168/544	1.11(0.91–1.35)[a]
			70–74	Aquatic product	p,p'-DDE	Serum lipid	168/544	1.01(0.87–1.16)[a]
Son2010	Korea	Cross-sectional	40-	Background	p,p'-DDE	Serum	27/26	26.6(2.0–349.1)
			40-	Background	p,p'-DDE	Serum lipid	28/26	12.7(1.9–83.7)
Everett2010	USA	Cross-sectional	20-	Background	p,p'-DDE	Serum lipid	334/2715	1.08(0.58–2.03)[a]
Ukropec2010	Slovakia	Cross-sectional	21–75	Heavy pollution	PCBs	Serum lipid	120/699	1.86(1.09–3.17)

Table 1. Cont.

Reference	Country	Study design	age	Exposure pathway	biomarker	Biologic specimen	Cases/controls	OR or RR (95%CI)
			21-75	Heavy pollution	p,p'-DDE	Serum lipid	125/694	1.94(1.11-3.78)
Lee2010	USA	Perspective	18-30	Background	PCB-153	Serum lipid	39/45	0.8(0.2-2.6)[a]
			18-30	Background	p,p'-DDE	Serum lipid	47/45	0.7(0.2-1.9)[a]
Philibert2009	Canada	Cross-sectional	15-86	Aquatic product	PCB-153	Serum	25/101	4.91(1.27-19.01)
			15-86	Aquatic product	PCB-153	Serum lipid	25/101	6.46(2.07-36.63)
			15-86	Aquatic product	PCBs	Serum	25/101	4.91(1.27-19.01)
			15-86	Aquatic product	PCBs	Serum lipid	25/101	5.51(1.26-24.07)
			15-86	Aquatic product	p,p'-DDE	Serum	25/101	6.11(1.37-27.3)
			15-86	Aquatic product	p,p'-DDE	Serum lipid	25/101	3.56(0.97-13.08)
Rignell-Hydbom2009	Sweden	Prospective	50-59	Background women	PCB-153	Serum	371/371	0.99(0.71-1.4)[c]
			50-59	Background women	PCB-153	Serum	208/208	0.91(0.59-1.4)[c]
			50-59	Background women	PCB-153	Serum	163/163	1.1(0.66-1.9)[c]
			50-59	Background women	PCB-153	Serum	107/107	1.4(0.72-2.6)[c]
			50-59	Background women	PCB-153	Serum	74/74	1.4(0.67-3.1)[c]
			50-59	Background women	PCB-153	Serum	39/39	1.6(0.61-4.0)[c]
			50-59	Background women	p,p'-DDE	Serum	371/371	1.1(0.76-1.5)[c]
			50-59	Background women	p,p'-DDE	Serum	208/208	0.9(0.57-1.4)[c]
			50-59	Background women	p,p'-DDE	Serum	163/163	1.3(0.78-2.2)[c]
			50-59	Background women	p,p'-DDE	Serum	107/107	1.5(0.8-2.8)[c]
			50-59	Background women	p,p'-DDE	Serum	74/74	2.5(0.97-6.4)[c]
			50-59	Background women	p,p'-DDE	Serum	39/39	5.5(1.2-25)[c]
Turyk2009	USA	Prospective	25-76	Aquatic product	p,p'-DDE	Serum	24/285	7.1(1.6-31.9)
			25-76	Aquatic product	PCBs	Serum	21/293	1.8(0.6-5.0)
Wang2008	Taiwan, China	Cross-sectional	30-	Rice-bran oil men	PCBs	Serum	155/152	1.7(0.7-4.6)[a]
			30-	Rice-bran oil women	PCBs	Serum	233/218	5.5(2.3-13.4)[a]
Codru2007	USA	Cross-sectional	30-	Background	PCBs	Serum		2.8 (0.7-10.8)[a]
			30-	Background	PCBs	Serum lipid		2.6 (0.8-8.1)[a]
			30-	Background	PCB-153	Serum		3.0 (0.7-12.8)[a]
			30-	Background	PCB-153	Serum lipid		1.4 (0.4-4.8)[a]
			30-	Background	p,p'-DDE	Serum		2.6 (0.8-8.8)[a]
			30-	Background	p,p'-DDE	Serum lipid		2.4 (0.7-8.3)[a]
Rignell-Hydbom2007	Sweden	Cross-sectional	29-59	Aquatic product women	PCB-153	Serum lipid	15/528	1.4(0.8-2.5)[a]
			29-59	Aquatic product women	p,p'-DDE	Serum lipid	15/528	1.3(1.1-1.5)[a]

Table 1. Cont.

Reference	Country	Study design	age	Exposure pathway	biomarker	Biologic specimen	Cases/controls	OR or RR (95%CI)
Cox2007	USA	Cross-sectional	20–74	Background	p,p'-DDE	Serum	45/768	2.63(1.2×5.8)[a]
			20–74	Background	p,p'-DDE	Serum lipid	35/560	1.5(0.8–2.9)[a]
Lee2006	USA	Cross-sectional	12–	Background	PCB-153	Serum lipid	52/598	6.8(3.0–15.5)
			12–	Background	p,p'-DDE	Serum lipid	69/704	4.3(1.8–10.2)
Rylander2005	Sweden	Cross-sectional	49–84	Aquatic product	PCB-153	Serum lipid	22/358	1.16(1.03–1.32)
			49–84	Aquatic product	p,p'-DDE	Serum lipid	22/358	1.05(1.01–1.09)
Fierens2003	Belgium	Case-control	50.3–59.4	Background	PCBs	Serum lipid	9/248	7.6(1.58–36.3)

[a]Different models adjusted by confounding, such as sex, age, BMI, total cholesterol and triglycerides, and various compounds.
[b]stratified by BMI.
[c]stratified by the years diagnosed after the baseline investigation.

Specific exposure, such as seafood consumption and living in a high exposure area, did not show an increased risk of T2DM compared with background exposure studies. This result may be attributed to the limited studies of specific exposure and the discrepant results of exposure concentration in serum levels such as 70 to 70000 ng/g and 44.33 to 8863 ng/g PCBs found in a serum lipid in two aquatic food consumption studies [28,35], and 148 to 101413 ng/g PCBs found in a study on serum lipid in a heavily polluted area [33]. Additionally, the heterogeneity analysis indicated that background exposure was a group with smaller heterogeneity and specific exposure with higher heterogeneity for the significant differences of the exposure. With regard to study design and subjects, from the consistent I^2 test results of the subgroup analyses and the limited studies of the perspective study subgroup and women's subgroup, it may not be an obvious heterogeneity source of the meta-analyses for all organochlorine biomarkers in this study. The biologic specimen was considered a heterogeneity source for the decreasing I^2 test results from the serum subgroup.

Some studies preferred express OCP concentrations per weight of lipid rather than on a whole weight for the lipophilic character of the pollutants. Total lipid was defined using different formulas as total lipids $(mg/dL) = 2.27 \times$ total cholesterol (mg/dL)+triglycerides (mg/dL)+62.3 [31,38] and total lipids $= 1.13+1.13 \times$(cholesterol+ triglycerides) [39]. Various definitions of total lipids may be one of the reasons that serum lipid is an obviously heterogeneity source. The gravimetric analysis of cholesterol and triglycerides with different detected method is another labile factor [55]. While, OCPs is lipophilic and likely to concentrate in serum lipid, so many studies used the concentrations based on serum lipid to present the residual levels in humans.

Furthermore, the different exposure contrast used in each study population may also be a larger source of heterogeneity. For instance, some studies [24,33] evaluated the increase in the odds comparing the 80th and the 20th percentiles of OCP concentrations, some studies used the quartiles for the comparison [25,34], while Airaksinen et al. calculated OR on the basis of percentile intervals <10th, 10th to <50th, 50th to <90th, and ≥90th. The diverse categories may cause the great heterogeneity among studies [29].

The original studies may be subject to limitations mainly related to the quality of potential sources of bias, exposure assessment, confounding, and the validity of the enrolled data. The bias in internal validity is most likely attributed to the misclassification of diabetes outcomes that only rely on self-reporting the prevalence of T2DM, the use of oral antidiabetic drugs or insulin, or that patients were on a specific diet [25,30,31,34,37,38,42,43] but lack of an accurate fasting glucose. Publication bias was also among the potential limitations from the evidence of asymmetry of the funnel plot. Some of the studies included in the meta-analysis were based on the same data, meaning that data for some subjects were included twice. For instance, some subjects from NHANES 1999–2004 was used both in the study of Everett et al. [32] and Lee et al. [41] and in both two studies, p,p'-DDE concentrations in serum were used as the indicator of the OCPs exposure. This may be a reason of the publication bias. However, this was not enough to negate the overall conclusion of an increased risk based on limited evidence for the deficit in small negative studies with effect sizes smaller than those from larger studies and in non-English published original studies.

It has been argued that exposure measurement is a typical effect factor for the quality of environmental epidemiology studies [56]. Though most studies gave a sufficient exposure gradient in the T2DM risk assessment (a dose-dependent manner across quantiles

A

Study or Subgroup	Weight	Odds Ratio IV, Random, 95% CI
Airaksinen 2011	2.5%	1.03 [0.25, 4.23]
Airaksinen 2011	7.6%	1.63 [0.91, 2.94]
Airaksinen 2011	4.3%	2.29 [0.86, 6.11]
Airaksinen 2011	4.3%	1.97 [0.74, 5.26]
Codru 2007	3.1%	1.40 [0.41, 4.83]
Gasull 2012	10.6%	1.60 [1.12, 2.28]
Lee 2006	5.4%	6.82 [2.99, 15.54]
Lee 2010	2.9%	0.80 [0.22, 2.87]
Lee 2011	3.7%	2.10 [0.70, 6.28]
Persky2012	4.5%	3.10 [1.21, 7.93]
Philibert 2009	2.4%	6.49 [1.55, 27.13]
Rignell-Hydbom 2007	7.8%	1.40 [0.80, 2.48]
Rignell-Hydbom 2009	10.8%	0.99 [0.71, 1.38]
Rylander 2005	13.1%	1.16 [1.03, 1.31]
Tanaka 2011	10.3%	0.73 [0.51, 1.06]
Wu2012	3.1%	0.76 [0.22, 2.62]
Wu2012	3.6%	2.18 [0.71, 6.67]
Total (95% CI)	100.0%	1.52 [1.19, 1.94]

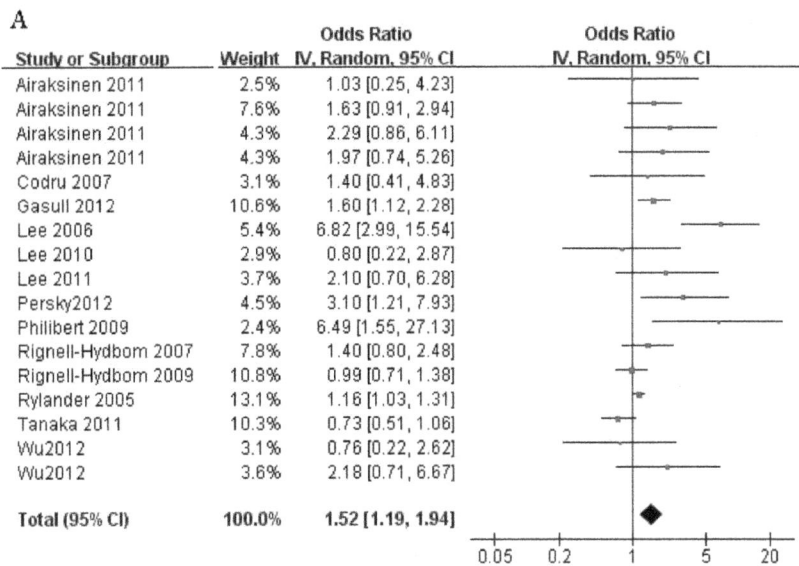

B

Study or Subgroup	Weight	IV, Random, 95% CI
Codru 2007	5.4%	2.61 [0.82, 8.30]
Fierens 2003	3.5%	7.61 [1.59, 36.52]
Gasull 2012	12.1%	1.70 [1.10, 2.61]
Grandjean 2011	14.4%	1.11 [0.91, 1.34]
Persky2012	7.5%	3.00 [1.27, 7.12]
Philibert 2009	3.8%	5.53 [1.27, 24.05]
Silverstone 2012	6.3%	2.77 [1.00, 7.68]
Silverstone 2012	3.8%	4.76 [1.09, 20.70]
Silverstone 2012	1.3%	4.18 [0.26, 67.57]
Turyk 2009	6.0%	1.80 [0.63, 5.20]
Ukropec 2010	11.0%	1.86 [1.10, 3.16]
Wang 2008	6.9%	1.70 [0.66, 4.35]
Wang 2008	7.4%	5.47 [2.27, 13.22]
Wu2012	4.9%	0.79 [0.23, 2.70]
Wu2012	5.7%	1.30 [0.43, 3.89]
Total (95% CI)	100.0%	2.14 [1.53, 2.99]

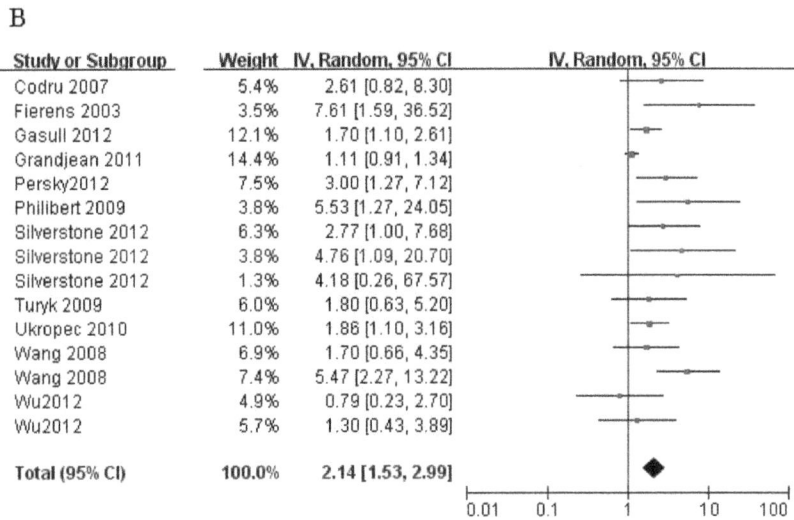

C

Study or Subgroup	Weight	IV, Random, 95% CI
Airaksinen 2011	2.1%	1.82 [0.71, 4.67]
Airaksinen 2011	0.8%	0.88 [0.18, 4.30]
Airaksinen 2011	4.4%	1.75 [0.95, 3.21]
Airaksinen 2011	1.8%	1.92 [0.69, 5.31]
Arrebola2013	1.5%	1.68 [0.54, 5.24]
Codru 2007	1.3%	2.41 [0.70, 8.29]
Cox 2007	4.0%	1.51 [0.79, 2.88]
Everett 2010	4.2%	1.08 [0.58, 2.03]
Gasull 2012	6.7%	1.11 [0.70, 1.73]
Grandjean 2011	15.6%	1.01 [0.88, 1.16]
Lee 2006	2.5%	4.31 [1.82, 10.20]
Lee 2010	1.6%	0.70 [0.23, 2.13]
Lee 2011	1.6%	2.10 [0.70, 6.28]
Philibert 2009	1.2%	3.56 [0.98, 12.98]
Rignell-Hydbom 2007	14.9%	1.30 [1.11, 1.52]
Rignell-Hydbom 2009	9.3%	1.11 [0.79, 1.54]
Rylander 2005	17.8%	1.05 [1.01, 1.09]
Son 2010	0.6%	12.68 [1.89, 84.87]
Turyk 2009	0.9%	7.10 [1.60, 31.49]
Ukropec 2010	4.4%	1.93 [1.05, 3.55]
Wu2012	1.4%	1.79 [0.54, 5.90]
Wu2012	1.4%	1.32 [0.41, 4.29]
Total (95% CI)	100.0%	1.33 [1.15, 1.54]

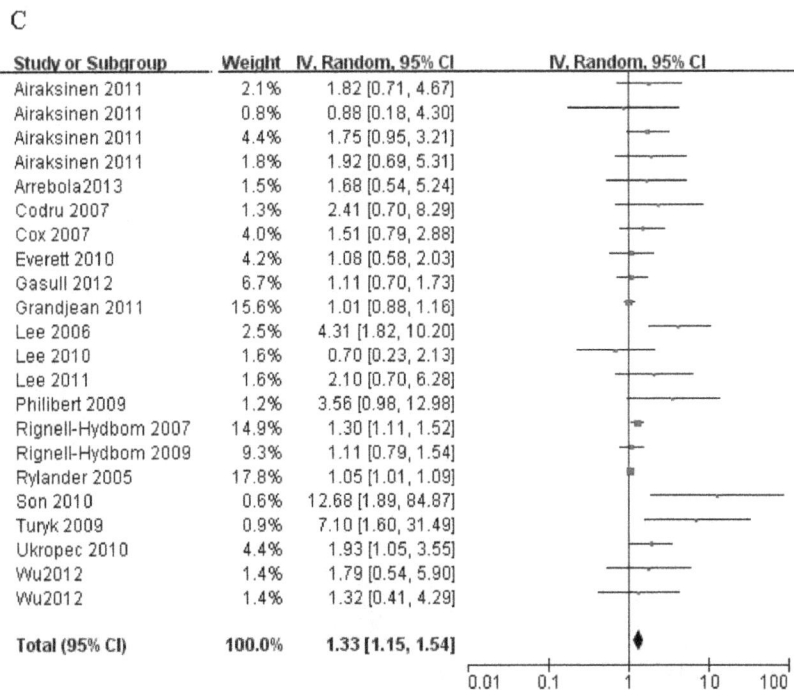

Figure 2. Subgroup analysis forest plots of the studies on T2DM risk from exposure to all three biomarkers. (A) Result of exposure to PCB-153. (B) Result of exposure to PCBs. (C) Result of exposure to p,p'-DDE.

of the exposure levels), multiple quantile categories, such as tertile [31,44,45], quartile [24], quintile [25], percentile intervals [29] were used for the OR estimate via setting the lowest quantile or percentile intervals as the reference. Another limitation is the unspecific exposure measurements. In the present meta-analysis, we selected representative PCB-153, PCBs, and p,p'-DDE as the biomarkers. For PCBs, 209 congeners existed in the environment, and the selection of the representatives PCBs varied among studies. For instance, 15 PCB congeners were selected in a study on high level exposure [33], while another study conducted in a

Table 2. Subgroup analysis of the included epidemiological studies.

Subgroup	NO. of studies	Weight	Summary OR (95%CI)	Q test (p)	I^2
PCB-153					
Background exposure	10	76.6%	1.57(1.13–2.19)	0.0004	65%
Specific exposure	3	23.4%	-	0.05	66%
Total	13	100%	1.52(1.19–1.94)	0.0002	64%
Perspective study	4	23.5%	1.05(0.78–1.40)	0.60	0%
Cross-sectional study	9	76.5%	1.69(1.24–2.31)	<0.0001	72%
Total	13	100%	1.50(1.18–1.92)	0.0002	63%
General population	10	74.7%	1.69(1.23–2.33)	<0.0001	71%
Women	3	25.3%	1.11(0.85–1.46)	0.41	0%
Total	13	100%	1.52(1.19–1.94)	0.0002	64%
Serum	6	38.8%	1.36(0.95–1.95)	0.002	74%
Serum lipid	10	61.2%	1.62(1.18–2.24)	0.0002	67%
Total	13	100%	1.44(1.18–1.76)	<0.00001	74%
PCBs					
Background exposure	4	31.6%	1.74(1.22–2.48)[a]	0.22	30%
Specific exposure	7	68.4%	2.39(1.52–3.77)	0.001	67%
Total	11	100%	2.14(1.53–2.99)	0.002	59%
Case-control study	2	10.9%	2.83(1.18–6.81)[a]	0.14	55%
Cross-sectional study	8	89.1%	2.28(1.55–3.34)	0.001	66%
Total	10	100%	2.36(1.64–3.41)	0.0008	64%
General population	9	80.4%	2.18(1.50–3.16)	0.006	59%
Women	2	19.6%	1.88(0.56–6.26)	0.02	74%
Total	11	100%	2.19(1.54–3.13)	0.001	62%
Serum	6	48.3%	2.31(1.71–3.11)[a]	0.36	8%
Serum lipid	7	51.7%	1.91(1.20–3.04)	0.009	62%
Total	11	100%	2.24(1.62–3.10)	0.001	58%
p,p'-DDE					
Background exposure	12	45.2%	1.39(1.16–1.67)[a]	0.18	25%
Specific exposure	6	54.8%	1.20(1.01–1.43)	0.001	75%
Total	18	100%	1.33(1.15–1.54)	0.0007	56%
Perspective study	4	14.7%	1.16(0.87–1.55)[a]	0.64	0%
Cross-sectional study	13	85.3%	1.33(1.13–1.56)	0.0006	62%
Total	17	100%	1.29(1.12–1.48)	0.003	52%
General population	15	72.9%	1.41(1.17–1.71)	0.0008	59%
Women	3	27.1%	1.27(1.10–1.46)[a]	0.79	0%
Total	18	100%	1.33(1.15–1.54)	0.0007	56%
Serum	8	24.4%	2.22(1.32–3.73)	0.005	66%
Serum lipid	14	75.6%	1.34(1.14–1.57)	0.001	58%
Total	18	100%	1.45(1.24–1.70)	<0.0001	62%

[a]Based on fixed model, others based on random model.

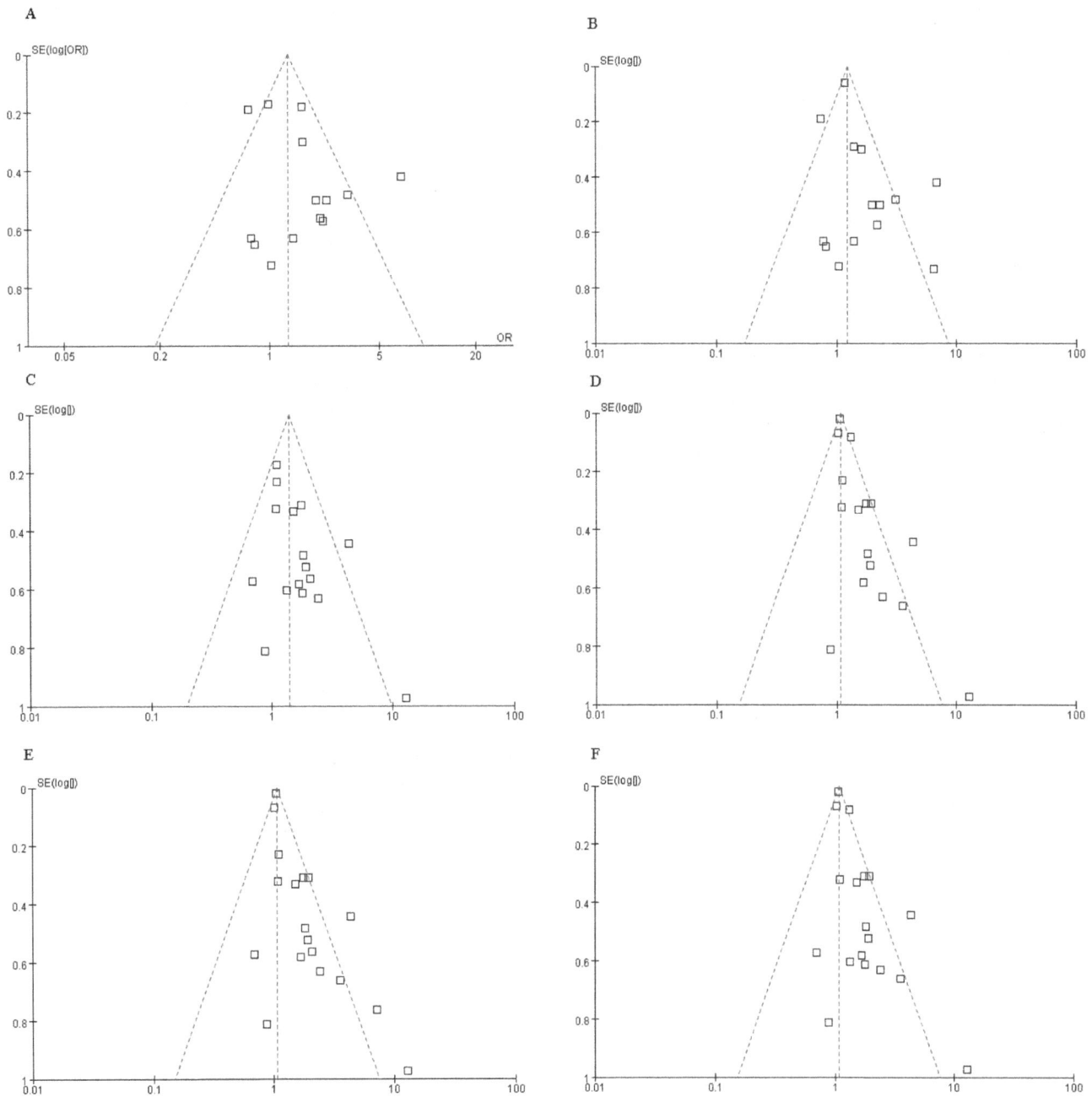

Figure 3. A funnel plot of SE versus ln(OR) for the meta-analyses. (*A*) Funnel plot for the meta-analysis on T2DM from background exposure to PCB-153. (*B*) Funnel plot for the subgroup analysis of serum lipid on T2DM from exposure to PCB-153. (*C*) Funnel plot for the subgroup analysis on T2DM from background exposure to *p,p'*-DDE. (*D*) Funnel plot for the subgroup analysis of the cross-sectional study on T2DM from exposure to *p,p'*-DDE. (*E*) Funnel plot for the subgroup analysis of the general population on T2DM from exposure to *p,p'*-DDE. (*F*) Funnel plot for the subgroup analysis of serum lipid on T2DM from exposure to *p,p'*-DDE.

fish consumption group selected only 8 PCB congeners as representatives of PCBs [35].

Confounding is also a potentially limited factor. Relevant confounders were selected from different adjusted models including basic demographics such as age, sex, and BMI and other major covariates, such as serum lipids, serum triglycerides, total cholesterol, fish consumption, smoking and alcohol. Models that were adjusted differently, including crude estimates, models with basic demographic variables, and models with all major confounders, led to discrepancies in the estimated OR. In the

present meta-analyses, we selected the models with the most confounders which may give more accurate effects values to create pooled OR estimates. For instance, in a marine food consumption study, the model with basic demographic and all of the major confounders, was set and adjusted to evaluate the OR estimates [28].

Other than bias, exposure measurement, and confounding, the effect of the validity of the enrolled data is important. When outcome of interest was rare, such as the prevalence of cancer or birth defect, one can generally ignore the distinctions among the various measures of relative risk [57]. According to statistical data from the Ministry of Health of The People's Republic of China, the prevalence of T2DM were about 5 times the prevalence of cancer. In this study, we combined the binary variables of OR and RR. Considering the high prevalence of T2DM, the data processing may impact the consequence of the meta-analysis.

From the systematic screening of the relevant studies about the prevalence of T2DM exposure to OCPs, we found several other systematic reviews [58,59,60]. However, this study was the first meta-analysis to our knowledge to evaluate the pooled effect values. A previous systematic review assessed the risk for developing T2DM from exposure to organochlorine pesticides [58] but only analyzed the results at the qualitative and quantitative levels. Although the positive relationships were the same as our results, many limitations and uncertainties were proposed in that review. Exposure to OCPs cannot be concluded as being the only contributor to the prevalence of T2DM, and many factors other than exposure to OCPs may be causative for T2DM, such as obesity, race, gender, age, genetic susceptibility, dietary habits and lifestyle. From another review that discussed the impacts of OCPs on metabolic health [60], no associations between OCPs exposures and stages of glucose intolerance or markers of insulin resistance were observed [61]. This specific result is not in our meta-analysis because of the misclassification of T2DM outcomes, which may cause bias in our study. Additionally, a cross-sectional study conducted in Swedish [51] found that none of the PCB congeners selected were significantly associated with diabetes in age, BMI, weight change and region adjusted analyses. However, because of the lack of OR or RR estimates, we removed this study in our meta-analysis. Another cross-sectional study of 380 Swedish fishermen and their wives found significantly increased risk from exposure to PCB-153 congener in men but not in women [42]. For the subjects were either women or all overall population, it was not included. In the review of the relationship of PCBs with T2DM and hypertension, the author attributed these two results to hypothesis generating [59].

Overall, the findings from the present meta-analysis provide quantitative evidence consistent with the hypothesis that exposure to OCPs is a contributing risk factor for the prevalence of T2DM. From the heterogeneity analysis, the specific exposure and biologic specimen of serum lipid may be the heterogeneity sources for the large disparities of the concentration of this class of environmental pollutants. Based on our sensitivity analysis, sources of bias, exposure assessment, and confounding are unlikely to significantly affect the results. In regards to the possible observed publication bias, more studies with small samples and adverse results should be included in future research. Apart from the conventional etiologies that include genetic susceptibility, metabolic disorder and obesity, the finding of the meta-analysis indicates that environmental factors, especially exposure to OCPs, may also be a risk factor of T2DM.

Supporting Information

Table S1 Modified Downs and Black checklist for the quality assessment of epidemiological studies.

Table S2 Quality assessment of the included epidemiological studies.

Checklist S1 PRISMA (Preferred Reporting Items for Systematic Reviews and Meta-Analyses) 2009 Checklist.

Acknowledgments

We would like to thank the authors of the original studies included in this meta-analysis.

Author Contributions

Conceived and designed the experiments: MT KC WL. Performed the experiments: MT FY. Analyzed the data: MT KC WL. Wrote the paper: MT WL.

References

1. Govarts E, Nieuwenhuijsen M, Schoeters G, Ballester F, Bloemen K, et al. (2012) Birth Weight and Prenatal Exposure to Polychlorinated Biphenyls (PCBs) and Dichlorodiphenyldichloroethylene (DDE): A Meta-analysis within 12 European Birth Cohorts. Environ Health Perspect 120: 162–170.

2. Lopez-Cervantes M, Torres-Sanchez L, Tobias A, Lopez-Carrillo L (2004) Dichlorodiphenyldichloroethane burden and breast cancer risk: A meta-analysis of the epidemiologic evidence. Environ Health Perspect 112: 207–214.

3. Ntow WJ, Tagoe LM, Drechsel P, Kelderman P, Gijzen HJ, et al. (2008) Accumulation of persistent organochlorine contaminants in milk and serum of farmers from Ghana. Environ Res 106: 17–26.

4. Ren AG, Qiu XII, Jin L, Ma J, Li ZW, et al. (2011) Association of selected persistent organic pollutants in the placenta with the risk of neural tube defects. Proc Natl Acad Sci U S A 108: 12770–12775.

5. Navas-Acien A, Silbergeld EK, Streeter RA, Clark JM, Burke TA, et al. (2006) Arsenic exposure and type 2 diabetes: A systematic review of the experimental and epidemiologic evidence. Environ Health Perspect 114: 641–648.

6. Chiasson JL, Josse RG, Gomis R, Hanefeld M, Karasik A, et al. (2002) Acarbose for prevention of type 2 diabetes mellitus: the STOP-NIDDM randomised trial. Lancet 359: 2072–2077.

7. Sladek R, Rocheleau G, Rung J, Dina C, Shen L, et al. (2007) A genome-wide association study identifies novel risk loci for type 2 diabetes. Nature 445: 881–885.

8. Kahn SE, Hull RL, Utzschneider KM (2006) Mechanisms linking obesity to insulin resistance and type 2 diabetes. Nature 444: 840–846.

9. Gress TW, Nieto FJ, Shahar E, Wofford MR, Brancati FL (2000) Hypertension and antihypertensive therapy as risk factors for type 2 diabetes mellitus. Atherosclerosis risk in communities study. N Engl J Med 342: 905–912.

10. Patel CJ, Bhattacharya J, Butte AJ (2010) An environment-wide association study (EWAS) on type 2 diabetes mellitus. PLoS One 5: e10746.

11. Everett CJ, Frithsen IL, Diaz VA, Koopman RJ, Simpson WM, et al. (2007) Association of a polychlorinated dibenzo-p-dioxin, a polychlorinated biphenyl, and DDT with diabetes in the 1999–2002 National Health and Nutrition Examination Survey. Environ Res 103: 413–418.

12. Turyk M, Anderson H, Knobeloch L, Imm P, Persky V (2009) Organochlorine exposure and incidence of diabetes in a cohort of Great Lakes sport fish consumers. Environ Health Perspect 117: 1076–1082.

13. Thayer KA, Heindel JJ, Bucher JR, Gallo MA (2012) Role of environmental chemicals in diabetes and obesity: a national toxicology program workshop review. Environ Health Perspect 120: 779–789.

14. Remillard RB, Bunce NJ (2002) Linking dioxins to diabetes: epidemiology and biologic plausibility. Environ Health Perspect 110: 853–858.

15. Wallace DC (2005) A mitochondrial paradigm of metabolic and degenerative diseases, aging, and cancer: a dawn for evolutionary medicine. Annu Rev Genet 39: 359–407.

16. Liberati A, Altman DG, Tetzlaff J, Mulrow C, Gøtzsche PC, et al. (2009) The PRISMA statement for reporting systematic reviews and meta-analyses of studies that evaluate health care interventions: explanation and elaboration. PLoS Med 6(7): e1000100.

17. Downs SH, Black N (1998) The feasibility of creating a checklist for the assessment of the methodological quality both of randomised and non-randomised studies of health care interventions. J Epidemiol Community Health 52: 377–384.

18. Wigle DT, Turner MC, Krewski D (2009) A systematic review and meta-analysis of childhood leukemia and parental occupational pesticide exposure. Environ Health Perspect 117: 1505–1513.

19. Hardy RJ, Thompson SG (1998) Detecting and describing heterogeneity in meta-analysis. Stat Med 17: 841–856.

20. Higgins J, Thompson SG, Deeks JJ, Altman DG (2003) Measuring inconsistency in meta-analyses. Br Med J 327: 557–560.

21. Vrijheid M, Martinez D, Manzanares S, Dadvand P, Schembari A, et al. (2011) Ambient air pollution and risk of congenital anomalies: a systematic review and meta-analysis. Environ Health Perspect 119: 598–606.

22. Mantel N, Haenszel W (1959) Statistical aspects of the analysis of data from retrospective studies of disease. J Natl Cancer Inst 22: 719–748.
23. Dersimonian R, Laird N (1986) Meta-analysis in clinical-trials. Controlled Clin Trials 7: 177–188.
24. Gasull M, Pumarega J, Tellez-Plaza M, Castell C, Tresserras R, et al. (2012) Blood concentrations of persistent organic pollutants and prediabetes and diabetes in the general population of catalonia. Environ Sci Technol 46: 7799–7810.
25. Silverstone AE, Rosenbaum PF, Weinstock RS, Bartell SM, Foushee HR, et al. (2012) Polychlorinated biphenyl (PCB) exposure and diabetes: results from the Anniston Community Health Survey. Environ Health Perspect 120: 727–732.
26. Persky V, Piorkowski J, Turyk M, Freels S, Chatterton RJ, et al. (2012) Polychlorinated biphenyl exposure, diabetes and endogenous hormones: a cross-sectional study in men previously employed at a capacitor manufacturing plant. Environ Health 11: 57.
27. Tanaka T, Morita A, Kato M, Hirai T, Mizoue T, et al. (2011) Congener-specific polychlorinated biphenyls and the prevalence of diabetes in the Saku Control Obesity Program. Endocr J 58:589–596.
28. Grandjean P, Henriksen JE, Choi AL, Petersen MS, Dalgard C, et al. (2011) Marine food pollutants as a risk factor for hypoinsulinemia and type 2 diabetes. Epidemiology 22: 410–417.
29. Airaksinen R, Rantakokko P, Eriksson JG, Blomstedt P, Kajantie E, et al. (2011) Association between type 2 diabetes and exposure to persistent organic pollutants. Diabetes Care 34: 1972–1979.
30. Lee DH, Lind PM, Jacobs DJ, Salihovic S, van Bavel B, et al. (2011) Polychlorinated biphenyls and organochlorine pesticides in plasma predict development of type 2 diabetes in the elderly: the prospective investigation of the vasculature in Uppsala Seniors (PIVUS) study. Diabetes Care 34: 1778–1784.
31. Son HK, Kim SA, Kang JH, Chang YS, Park SK, et al. (2010) Strong associations between low-dose organochlorine pesticides and type 2 diabetes in Korea. Environ Int 36: 410–414.
32. Everett CJ, Matheson EM (2010) Biomarkers of pesticide exposure and diabetes in the 1999–2004 national health and nutrition examination survey. Environ Int 36: 398–401.
33. Ukropec J, Radikova Z, Huckova M, Koska J, Kocan A, et al. (2010) High prevalence of prediabetes and diabetes in a population exposed to high levels of an organochlorine cocktail. Diabetologia 53: 899–906.
34. Lee DH, Steffes MW, Sjodin A, Jones RS, Needham LL, et al. (2010) Low dose of some persistent organic pollutants predicts type 2 diabetes: a nested case-control study. Environ Health Perspect 118: 1235–1242.
35. Philibert A, Schwartz H, Mergler D (2009) An exploratory study of diabetes in a First Nation community with respect to serum concentrations of p,p'-DDE and PCBs and fish consumption. Int J Environ Res Public Health 6: 3179–3189.
36. Rignell-Hydbom A, Lidfeldt J, Kiviranta H, Rantakokko P, Samsioe G, et al. (2009) Exposure to p,p'-DDE: a risk factor for type 2 diabetes. PLoS One 4: e7503.
37. Wang SL, Tsai PC, Yang CY, Guo YL (2008) Increased risk of diabetes and polychlorinated biphenyls and dioxins - A 24-year follow-up study of the Yucheng cohort. Diabetes Care 31: 1574–1579.
38. Codru N, Schymura MJ, Negoita S, Rej R, Carpenter DO (2007) Diabetes in relation to serum levels of polychlorinated biphenyls and chlorinated pesticides in adult Native Americans. Environ Health Perspect 115: 1442–1447.
39. Rignell-Hydbom A, Rylander L, Hagmar L (2007) Exposure to persistent organochlorine pollutants and type 2 diabetes mellitus. Hum Exp Toxicol 26: 447–452.
40. Cox S, Niskar AS, Narayan KM, Marcus M (2007) Prevalence of self-reported diabetes and exposure to organochlorine pesticides among Mexican Americans: Hispanic health and nutrition examination survey, 1982–1984. Environ Health Perspect 115: 1747–1752.
41. Lee DH, Lee IK, Song K, Steffes M, Toscano W, et al. (2006) A strong dose-response relation between serum concentrations of persistent organic pollutants and diabetes - Results from the National Health and Examination Survey 1999–2002. Diabetes Care 29: 1638–1644.

42. Rylander L, Rignell-Hydbom A, Hagmar L (2005) A cross-sectional study of the association between persistent organochlorine pollutants and diabetes. Environ Health 4: 28.
43. Fierens S, Mairesse H, Heilier JF, De Burbure C, Focant JF, et al. (2003) Dioxin/polychlorinated biphenyl body burden, diabetes and endometriosis: findings in a population-based study in Belgium. Biomarkers 8: 529–534.
44. Arrebola JP, Pumarega J, Gasull M, Fernandez MF, Martin-Olmedo P, et al. (2013) Adipose tissue concentrations of persistent organic pollutants and prevalence of type 2 diabetes in adults from Southern Spain. Environ Res 122: 31–37.
45. Wu HY, Bertrand KA, Choi AL, Hu FB, Laden F, et al. (2013) Persistent organic pollutants and type 2 diabetes: a prospective analysis in the nurses' health study and meta-analysis. Environ Health Perspect 121: 153–161.
46. Lee DH, Jacobs DJ, Steffes M (2008) Association of organochlorine pesticides with peripheral neuropathy in patients with diabetes or impaired fasting glucose. Diabetes 57: 3108–3111.
47. Lee DH, Lee IK, Jin SH, Steffes M, Jacobs DJ (2007) Association between serum concentrations of persistent organic pollutants and insulin resistance among nondiabetic adults: results from the National Health and Nutrition Examination Survey 1999–2002. Diabetes Care 30: 622–628.
48. Turyk M, Anderson HA, Knobeloch L, Imm P, Persky VW (2009) Prevalence of diabetes and body burdens of polychlorinated biphenyls, polybrominated diphenyl ethers, and p,p'-diphenyldichloroethene in Great Lakes sport fish consumers. Chemosphere 75: 674–679.
49. Montgomery MP, Kamel F, Saldana TM, Alavanja MC, Sandler DP (2008) Incident diabetes and pesticide exposure among licensed pesticide applicators: Agricultural Health Study, 1993–2003. Am J Epidemiol 167: 1235–1246.
50. Vasiliu O, Cameron L, Gardiner J, Deguire P, Karmaus W (2006) Polybrominated biphenyls, polychlorinated biphenyls, body weight, and incidence of adult-onset diabetes mellitus. Epidemiology 17: 352–359.
51. Glynn AW, Granath F, Aune M, Atuma S, Darnerud PO, et al. (2003) Organochlorines in Swedish women: determinants of serum concentrations. Environ Health Perspect 111: 349–355.
52. Everett CJ, Frithsen IL, Diaz VA, Koopman RJ, Simpson WJ, et al. (2007) Association of a polychlorinated dibenzo-p-dioxin, a polychlorinated biphenyl, and DDT with diabetes in the 1999–2002 National Health and Nutrition Examination Survey. Environ Res 103: 413–418.
53. Longnecker MP, Klebanoff MA, Brock JW, Zhou H (2001) Polychlorinated biphenyl serum levels in pregnant subjects with diabetes. Diabetes Care 24: 1099–1101.
54. Fierens S, Mairesse H, Heilier JF, De Burbure C, Focant JF, et al. (2003) Dioxin/polychlorinated biphenyl body burden, diabetes and endometriosis: findings in a population-based study in Belgium. Biomarkers 8: 529–534.
55. Bernert JT, Turner WE, Patterson DG, Needham LL (2007) Calculation of serum "total lipid" concentrations for the adjustment of persistent organo-halogen toxicant measurements in human samples. Chemosphere 68: 824–831.
56. Turner MC, Wigle DT, Krewski D (2010) Residential pesticides and childhood leukemia: a systematic review and meta-analysis. Environ Health Perspect 118: 33–41.
57. Greenland S (1987) Quantitative methods in the review of epidemiologic literature. Epidemiol Rev 9: 1–30.
58. Henley P, Hill J, Moretti ME, Jahedmotlagh Z, Schoeman K, et al. (2012) Relationships between exposure to polyhalogenated aromatic hydrocarbons and organochlorine pesticides and the risk for developing type 2 diabetes: a systematic review and a meta-analysis of exposures to 2,3,7,8-tetrachlorodiben-zo-p-dioxin (TCDD). Toxicol Environ Chem 94: 814–845.
59. Everett CJ, Frithsen I, Player M (2011) Relationship of polychlorinated biphenyls with type 2 diabetes and hypertension. J Environ Monit 13: 241–251.
60. Langer P (2010) The impacts of organochlorines and other persistent pollutants on thyroid and metabolic health. Fron Neuroendocrinol 31: 497–518.
61. Jorgensen ME, Borch-Johnsen K, Bjerregaard P (2008) A cross-sectional study of the association between persistent organic pollutants and glucose intolerance among Greenland Inuit. Diabetologia 51: 1416–1422.

16

Zebrafish Transgenic Line *huORFZ* Is an Effective Living Bioindicator for Detecting Environmental Toxicants

Hung-Chieh Lee[1]◙, Po-Nien Lu[1,2]◙, Hui-Lan Huang[1,2]◙, Chien Chu[3,4], Hong-Ping Li[3], Huai-Jen Tsai[1]*

1 Institute of Molecular and Cellular Biology, National Taiwan University, Taipei, Taiwan, **2** Liver Disease Prevention & Treatment Research Foundation, Taipei, Taiwan, **3** Taiwan Agricultural Chemicals and Toxic Substances Research Institute Council of Agriculture, Executive Yuan, Taichung, Taiwan, **4** Department of Soil and Environmental Sciences, National Chung Hsing University, Taichung, Taiwan

Abstract

Reliable animal models are invaluable for monitoring the extent of pollution in the aquatic environment. In this study, we demonstrated the potential of *huORFZ*, a novel transgenic zebrafish line that harbors a human upstream open reading frame of the *chop* gene fused with GFP reporter, as an animal model for monitoring environmental pollutants and stress-related cellular processes. When *huORFZ* embryos were kept under normal condition, no leaked GFP signal could be detected. When treated with hazardous chemicals, including heavy metals and endocrine-disrupting chemicals near their sublethal concentrations (LC50), *huORFZ* embryos exhibited different tissue-specific GFP expression patterns. For further analysis, copper (Cu^{2+}), cadmium (Cd^{2+}) and Chlorpyrifos were applied. Cu^{2+} triggered GFP responses in skin and muscle, whereas Cd^{2+} treatment triggered GFP responses in skin, olfactory epithelium and pronephric ducts. Moreover, fluorescence intensity, as exhibited by *huORFZ* embryos, was dose-dependent. After surviving treated embryos were returned to normal condition, survival rates, as well as TUNEL signals, returned to pretreatment levels with no significant morphological defects observed. Such results indicated the reversibility of treatment conditions used in this study, as long as embryos survived such conditions. Notably, GFP signals decreased along with recovery, suggesting that GFP signaling of *huORFZ* embryos likely reflected the overall physiological condition of the individual. To examine the performance of the *huORFZ* line under real-world conditions, we placed *huORFZ* embryos in different river water samples. We found that the *huORFZ* embryos correctly detected the presence of various kinds of pollutants. Based on these findings, we concluded that such uORF^{chop}-based system can be integrated into a first-line water alarm system monitoring the discharge of hazardous pollutants.

Editor: Christoph Winkler, National University of Singapore, Singapore

Funding: This work was supported by the National Science Council, Taiwan (http://web1.nsc.gov.tw/mp.aspx?mp = 7) with grant number: NSC -01-2321-B-002-018-. This study was also partially supported by the Liver Disease Prevention & Treatment Research Foundation, Taipei, Taiwan. The funders had no role in study design, data collection and analysis, decision to publish, or preparation of the manuscript.

Competing Interests: The authors have declared that no competing interests exist.

* E-mail: hjtsai@ntu.edu.tw

◙ These authors contributed equally to this work.

Introduction

Among the threats endangering aquatic environments, the discharge of industrial and domestic wastewater has the most significant impacts on freshwater ecosystems, as well as agricultural production and human health [1–4]. Thus, constant monitoring is essential to ensure timely response whenever damaging waste discharge events happen.

Chemical analysis is typically used to detect traces of known toxins in aquatic environments. However, this technique relies heavily on pre-established standards governing which chemicals and which concentrations are considered dangerous; therefore, it cannot be used to detect the existence of unexpected hazardous chemicals. In contrast, a biomonitoring system may reflect the subtle cellular and physiological changes occurring in living organisms when challenged by a variety of environmental pollutants.

Fish have been considered an ideal biomonitoring organism based on their biodiversity, population and health status. Using fish as an organism for *in vivo* toxicity assays has previously been proposed [5]. Normal conditions, such as growth, survival rates and egg hatchability, can be used as monitoring parameters. Quantifying the activity of enzymatic defenses in fish is also a common approach to assess water quality. However, the interpretation of the data obtained from these methods is limited by the fact that multiple physiological, genetic, and metabolic factors may simultaneously affect these multifunctional enzymes [6–12]. For example, mixed-function oxygenase (MFO), or mono-oxygenase, are important components of many metabolic systems and have been validated in a large number of field studies worldwide. However, the enzyme activities of MFO components, which contain cytochrome P450, cytochrome b5 and NADPH-cytochrome C reductase, must be measured individually to obtain the biomonitoring index. Moreover, tissue samples must be handled with great care to guard against denaturation and/or proteolysis. To overcome these limitations, transgenic fish lines have been developed by using native gene promoters, including the *cyp1a1* promoter, which is induced by polycyclic aromatic hydrocarbons [13,14], or the *heat-shock* promoter, which is induced by heat and other stressors [15]. However, since these promoters only respond to specific forms of stress, their advantages over traditional chemical analysis are not particularly significant. Apart from this consideration, it is also true that a given stress with little harm to the animal may still induce the expression of a reporter gene controlled by the *heat-shock* promoter [15]. Under these circumstances, the reporter activity would have little relationship to the actual physiological stresses. Hence, for an animal model to be a practical biomonitor, it must 1) respond to a wide range of pollutants with accuracy and sensitivity and 2) dynamically trace physiological stresses.

In recognition of these objectives, we took advantage of a zebrafish transgenic line *huORFZ*, which harbors a GFP transgene regulated by the upstream open reading frame (uORF) fragment of human *DNA-damage-inducible transcript 3* (*ddit3*, previously named *chop*) cDNA [16]. Recent studies have shown that uORF-based translational regulation plays significant, if not primary, roles in the production of CHOP protein [17-19], while the *chop* gene is one of the most commonly used biomarkers for endoplasmic reticulum (ER) stress [20,21]. We found that embryos derived from the *huORFZ* line only display fluorescent signals upon encountering stresses, with no detectable leakage under normal condition. Thus, *huORFZ* embryos can give a faithful account of cellular stresses. Using *in vivo* imaging, we further demonstrated that this line could be used to detect various environmental contaminants, including heavy metals and endocrine-disrupting chemicals (EDCs). Depending upon the treatment time, the limits of detection (LODs) for several common pollutants examined in this study were equal to, or below, World Health Organization (WHO) drinking water standard [22]. Importantly, different stresses were found to cause different GFP expression patterns in a dose-dependent manner. Moreover, after surviving treated embryos were returned to normal condition, survival rates, as well as TUNEL signals, returned to pretreatment levels with no significant morphological defects observed. Such results indicated the reversibility of treatment conditions used in this study, as long as embryos survived such conditions. Notably, GFP signals decreased along with recovery, suggesting that GFP signaling of *huORFZ* embryos likely reflected the overall physiological condition of the individual. Therefore, since time-consuming and complex analysis in various physiological conditions may not be necessary, the use of the *huORFZ* embryos holds considerable promise as a novel fluorescent biomonitoring method.

Materials and Methods

Ethics Statement

The animal protocol, which was strictly followed in this study, was reviewed and approved by the IACUC, National Taiwan University, Taiwan, with approval number NTU-102-EL-19.

Animal husbandry

All wild-type zebrafish (*Danio rerio*) were AB/TU strains, and transgenic lines of zebrafish *Tg(-2.9krt18:RFP)* [23] and *huORFZ* [16] were crossed into the AB/TU background. All fish were maintained at a temperature of 28.5°C with a photoperiod of 14 hr light:10 hr dark. All fish were bred according to guidelines outlined in *The Zebrafish Book* [24]. Embryos were raised in embryo medium (140 mM NaCl, 5.4 mM KCl, 0.25 mM Na$_2$HPO$_4$, 0.44 mM KH$_2$PO$_4$, 1.3 mM CaCl$_2$, 1.0 mM MgSO$_4$, and 4.2 mM NaHCO$_3$ at pH 7.2) until 24 hr post-fertilization (hpf), followed by incubation in embryo medium containing 0.003% 1-phenyl-2-thiourea (Sigma) to prevent pigment formation.

Chemicals

Lithium chloride (Sigma), copper sulfate pentahydrate (Sigma), nickel (II) sulfate hexahydrate (Sigma), zinc sulfate (Merck),

aluminum chloride hexahydrate (Sigma), cobalt chloride hexahydrate (Sigma), lead chloride (Sigma) and acrylamide (Sigma) were used for the toxicity tests. Cadmium chloride (Merck), arsenic trioxide (Merck), atrazine (EQ Laboratories), chlorpyrifos (EQ Laboratories), carbofuran (Chem Service), dimethoate (Chem Service), glyphosate (EQ Laboratories), and methoxychlor (Fluka) were provided by Taiwan Agricultural Chemicals and the Toxic Substances Research Institute, Council of Agriculture, Taiwan. Heavy metals, acrylamide, dimethoate and glyphosate were dissolved in sterile distilled water to make stock solutions, whereas other chemicals were either dissolved or diluted with dimethyl sulfoxide (DMSO) to the desired concentration as stock solutions.

Stress treatment

All embryos were raised in 10-cm Petri dishes and incubated at 28.5°C. All of the toxicity tests began at 72 hpf. Treatments were performed in 3-cm Petri dishes containing 20 embryos. The embryos were first washed once with 3 mL of distilled water. After washing, the liquid was removed, leaving the least amount possible in the dish. Then 3mL of working solution were added to each dish. Control embryos were treated with the equivalent amount of sterile distilled water or DMSO-containing water. The embryos were then returned to the incubator set to standard zebrafish embryo culture condition (28°C). Unless otherwise stated, each experiment consisted of 60 healthy embryos divided into three groups (n = 20 embryos per group) for each treatment condition. Each experimental design was repeated three times independently, and the results were pooled to calculate the percentage. Lethal concentrations for 10%, 50% and 90% mortality were calculated from a linear regression of log probit transformations of the dose response data [25]. The final chemical concentrations of heavy metals and EDCs used in this study are listed in Table S1. In the survival rate experiment, five independent repetitions of 20 embryos for each treatment condition were performed. Data presented as mean±SD in the manuscript were the averages of these five independent experiments, which were carried out using different clutches of embryos on different dates. For the recovery experiment, one dish of *huORFZ* embryos remained under stress throughout the experiment in each repetition for each kind of stress, while the other dish was washed twice with sterile distilled water and returned to normal condition at the end of each specific treatment.

Chlorpyrifos was chosen as a representative of chronic toxicity for testing in *huORFZ* larvae. Embryos at 72 hpf were treated with chlorpyrifos at 86 nM, according to the values for drinking water quality, as specified in the WHO guidelines, and observed for three days. Solution was refreshed daily. For the semi-quantitative experiment, two independent repetitions of 20 embryos for each treatment condition were performed. Images of three to four randomly selected live embryos were taken from each treatment condition and each repetition.

Western blotting

Zebrafish embryos were lysed by 1 x whole cell extract buffer (15 mM Tris-HCl pH 7.5, 250 mM sucrose, 2 mM EDTA, and 0.2 mM PMSF), and proteins were separated by 10% SDS-PAGE, followed by transfer to PolyScreen PVDF Hybridization Transfer Membrane (PerkinElmer). Subsequently, the membranes were probed with antiserum, including 1:1000 diluted anti-GFP (Chemicon Millipore), 1:500 diluted anti-GADD153 (mouse CHOP homologue) (Abcam ab11419) and 1:5000 diluted anti-α-tubulin (Sigma). Membranes were washed with TBST solution (0.2 M Tris, 1.37 M NaCl and 0.1% Tween-20, pH 7.6) and probed with 1:5000 diluted horseradish peroxidase-conjugated goat anti-mouse antibody (Santa Cruz Biotechnology). The bound antibody was detected by Western Lightning ECL Pro (PerkinElmer) and then exposed to X-ray film (Fujifilm).

Immunohistochemistry

huORFZ larvae at 4-days post-fertilization (dpf) were fixed with 4% paraformaldehyde (PFA) overnight at 4°C and cryoprotected by 30% sucrose before sectioning at 20 μm horizontally on a cryostat. Slides were washed with phosphate-buffered saline (PBS) for 30 min. They were subsequently incubated in blocking solution (PBS, 2% Bovine serum albumin, 0.2% Triton X-100) for 30 min. The following primary antibodies were used: anti-GFP (Abcam; 1:200) and mouse glutamine synthetase (GS) (clone GS-6; 1:500). The immunohistochemical signals were detected with FITC-conjugated secondary antibody (Santa Cruz Biotechnology) for GFP and Cy3-conjugated secondary antibody (Millipore) for GS. Nuclei were counterstained with 4′,6-diamidino-2-phenylindole (DAPI) (Sigma).

TUNEL staining

huORFZ larvae were fixed and cryoprotected following the same protocol described above. Whole mount TUNEL was carried out on Cu^{2+}-treated embryos. Cd^{2+}-treated embryos were coronally sliced, and TUNEL staining was performed following the manufacturer's protocol (In Situ Cell Death Detection Kit TMR red, Roche). Slides were washed three times with PBST for 5 min each, and the images were examined by confocal microscopy.

Microscopy and imaging

Fluorescent images were captured by a fluorescent stereomicroscope (MZ FLIII, Leica) coupled with Nikon D3 digital camera or a confocal spectral microscope (LSM 780, Zeiss). Unless otherwise indicated, representative images used in this study represent more than 75% of the embryos in an experimental group.

Semi-quantification analysis based on fluorescent images

The analysis was modified from the procedure described by Noche (2011) [26]. Briefly, green fluorescent images were taken under a Leica MZ FLIII fluorescent stereomicroscope with objective set to 4X. All images were lateral view with ISO 3200 and 4 seconds of exposure. Images were saved as 24-bit RGB TIFF files.

The images were then opened under ImageJ 1.47 [27]. "Split channel" function was then used to extract the green channel, while the blue and red channels were discarded. The outline of the one embryo to be analyzed was loosely selected using the "Polygon selections" tool. Only the area more anterior to the most posterior end of the yolk stalk was selected. This arbitrary selection rule was applied to address the size of embryos in relation to the width of view under 4× magnitude. The "Rolling Ball Background" filter (algorithm based on Sternberg, 1983 [28]) with the rolling ball radius set to 35 was then used to subtract the background. After background correction, under the Brightness/Contrast window, the "Minimum" value was arbitrarily set to 10 to further eliminate remaining background value. A previously selected area was then recalled using the "Restore selection" function, and the "Measure" function was used to obtain the "Integrated density" value for further analysis.

River water samples

River samples collected from Station 1, 2 and 3, designated as Sample 1, 2 and 3, respectively, were purchased from CENPRO Technology Co., Ltd. These samples were collected by Water Quality Monitoring Stations of the Environmental Protection Administration of Taiwan. Sample 4 was collected by H.L. Huang. Details on the sampling station coordinate and data are listed in Table S2. All samples were refrigerated for one to three days before biological toxicity assessments were conducted using *huORFZ* embryos and chemical analyses. River water samples for chemical analysis were additionally treated with concentrated nitric acid to digest the heavy metals.

Quantification of the concentrations of heavy metals

Heavy metal measurements were carried out using Inductively Coupled Plasma-Mass Spectroscopy (ICP-MS, Agilent 7500 Series). Arsenic was determined by inductively coupled plasma atomic emission spectroscopy (ICP-AES, JY 138 UL Trace, France) equipped with a hydride generator. A hot plate was used for wet digestion of solution. All glassware and plastic were cleaned by Trace Clean (Milestone). Sampling cans and bottles were rinsed with deionized water and soaked in 3% HNO_3 for 24 hr. After acid bath, the bottles for storage of precipitation samples were rinsed twice, filled with 1% HNO_3, and plugged. Other containers and instruments were rinsed twice with deionized water, dried, plugged and packed in two clean plastic bags with zip locks. The rings and filter supports from the filter packs were soaked in 1% HNO_3 for 12 hr and rinsed properly with deionized water. Autosampler tubes and cups were also rinsed with deionized water, soaked in 1% HNO_3 at least 12 hr and rinsed twice with deionized water before use. Nitric acid (Ultrapure Reagent, 69%) was supplied by J.T. Baker. The metal standard was supplied by Merck ICP Multi Element Standard Solution XVI. For medium solution detection, approximately 5 mL of sample were digested with 10 mL of HNO_3 on a hot plate. The temperature of the hot plate was maintained at 190°C for 1.5 hr. After cooling, the sample was diluted to 25 mL with distilled water. Metal contents of the final solution were determined by ICP-MS, as well as ICP-AES. The water samples were filtrated on membrane filters of 0.45 mm pore size, followed by determination of Pb^{2+}, Zn^{2+}, Cd^{2+}, Cu^{2+}, Ni^{2+} and As^{3+} concentrations by ICP-MS. The ions were separated by mass-to-charge ratio (m/z) and measured by a channel electron multiplier (Pb^{2+}: 208 m/z, Zn^{2+}: 66 m/z, Cd^{2+}: 111 m/z, Cu^{2+}: 63 m/z, Ni^{2+}: 60 m/z, As^{3+}: 75 m/z). The instrumental operating conditions for ICP-MS were as follows: Radio frequency power: 1500 W; RF matching: 1.75 V; Sample depth: 9 mm; Sample skimmer cones: Ni; Peristaltic pump: 0.10 rps; Argon plasma flow rate: 15 L/min; Auxiliary: 0.32 L/min; Nebulizer: 0.87 L/min; Spray chamber temperature: 2°C; Integration time: 0.1 s. The instrumental operating conditions for ICP-AES were as follows: Plasma power: 1100 W; Argon flow rate: 12 L/min; Nebulizer flow: 0.4 L/min; Nebulizer P: 2.5–3.5 bar; Wavelength: As 193.696 nm. An ICP multi-element standard solution (100 mg/L), containing the analyzed elements, was used in the preparation of calibration stock solutions. Working standard calibration solutions were prepared daily by dilution of the stock solutions, containing 50, 20, 10, 5, and 1 μg/L heavy metal solutions. The correlation coefficient r^2 obtained for all cases was 0.9995. The LODs were calculated as the concentrations of an element that gave the standard deviation of a series of ten consecutive measurements of blank solutions.

Statistical analysis

Unless otherwise indicated, each experiment was repeated three times or more. All bar graphs are presented as mean values with error bar indicating ± SD. Student's *t*-test was used for statistical analyses. Significance was determined at $P<0.05$ (*), $P<0.01$ (**) and $P<0.001$ (***).

Results

The GFP expression patterns in *huORFZ* embryos are tissue-specific responses to various stresses

To evaluate whether *huORFZ* zebrafish embryos could be used to detect environmental pollutants *in vivo*, the heterozygous *huORFZ* embryos at 72 hpf were treated with common pollutants found in freshwater bodies, including heavy metals and EDCs. To determine the effective pollutant monitoring range, we first determined the lethal concentrations as 10%, 50% and 90% of each pollutant for 72 hpf embryos in a 24 hr treatment (Table S1). We then used a concentration equal to, or lower than, the LC_{50} of each pollutant for subsequent experiments. Control embryos in distilled water exhibited no GFP signals during the experimental period (Figure 1A, control). Intriguingly, among treated embryos, the GFP expression patterns varied according to the type of pollutants. Representative GFP expression patterns under each treatment condition are shown in Figures 1A and S1. Detailed distribution of GFP-responsive cells/tissues of each experimental group is shown in Figure 1B-K and summarized in Table 1. With heavy metal treatments, GFP signals were detected in the brain (Figure 1B) and muscle (Figure 1C) of the embryos treated with either 2.7 mg/L (0.1 mM) of Al^{3+} (from $AlCl_3$) or 58.93 mg/L (1 mM) of Co^{2+} (from $CoCl_2$). All of the GFP-positive cells in the brain reacted positively to the antibody against GS (Figure 1B'; red labeling), indicating a glial cell identity. When the embryos were incubated in 50 mg/L (0.67 mM) of As^{3+} (from As_2O_3), GFP signals were strongly expressed in the lateral line system. With 0.56 mg/L (0.005 mM) of Cd^{2+} ($CdCl_2$) treatment, GFP signals were strongly expressed in the olfactory epithelium (Figure 1D), pronephric ducts (Figure 1E), skin and lateral line neuromasts (Figure 1F). When 58.69 mg/L (1 mM) of Ni^{2+} (from $NiSO_4$) was used, GFP signals were detected in the olfactory epithelium, brain and muscle of the treated embryos. GFP was presented in a scattered form on the skin of embryos incubated with 0.1 mg/L (0.0015 mM) of Cu^{2+} (from $CuSO_4$). The stress-induced GFP-expressing skin cells on the trunk colocalized with the expression of *keratin18* (*krt18*), as indicated by the expression of RFP in the embryos obtained from crossing *huORFZ* to *Tg(-2.9krt18:RFP)* (Figure 1G; red labeling), indicating the keratinocyte lineage of the GFP-expressing cells. GFP was also highly expressed in the intestine of embryos incubated with 242.9 mg/L (35 mM) of Li^+ (from LiCl) (Figure 1H), and GFP responses were significant in the brain, spinal cord (Figure 1I) and kidney (Figure 1J) of embryos treated with 0.76 mg/L (0.0036 mM) of Pb^{2+} (from $PbCl_2$). Finally, in embryos treated with 22.24 mg/L (0.34 mM) of Zn^{2+} (from $ZnSO_4$), GFP was highly expressed in the brain, but only weakly detected in the spinal cord and yolk sac skin. The *huORFZ* embryos also responded to EDCs. For example, GFP was highly expressed in the brain of embryos treated with 26.98 mg/L (0.38 mM) of acrylamide, and while it was also strongly apparent in the brain and muscle of embryos treated with 32.35 mg/L (0.15 mM) of atrazine, GFP was only weakly apparent in the heart of these embryos (Figure 1K). GFP was also highly expressed in the brain and muscle of embryos treated with either 1.34 mg/L (0.0021 mM) carbofuran or 1375 mg/L (6 mM) dimethoate. Finally, GFP was strongly detected in the brain and some muscle fibers of embryos treated with 1.65 mg/L (0.0047 mM) chlorpyrifos. However, when treated with either 18.6 mg/L (0.11 mM) of glyphosate or 4.87 mg/L (0.0141 mM) of methoxychlor, GFP signals were only detected in the brain.

The intensity of GFP signal in *huORFZ* embryos is correlated to the strength of stresses

When *huORFZ* embryos were exposed to higher concentrations of toxic reagents, stronger fluorescence signals could be detected. The GFP signals also became detectable in more tissues (Figure 2A). For example, *huORFZ* embryos that were treated with 0.06 mg/L (0.5 mM) of Cd^{2+} for 24 hr displayed strong fluorescence signals in the olfactory epithelium, but only weak signals in the skin cells. However, when treated with 0.56 mg/L (5 mM) of Cd^{2+} for 24 hr, the *huORFZ* embryos displayed strong fluorescence signals in both olfactory epithelium and pronephric ducts. When the Cd^{2+} concentration was increased to 1.12 mg/L (10 mM), more skin cells began to exhibit GFP response. Also, when treatment time was extended to 48 hr, the GFP expression levels were correspondingly increased. Similarly, when the concentration of Cu^{2+} was increased from 0.06 mg/L (1 mM) to 0.19 mg/L (3 mM), more epithelial cells became GFP-positive. When the treatment time was extended to 48 hr, GFP signals became detectable in tissues other than epithelium. Specifically, the brain started to express GFP in the group treated with 0.13 mg/L (2 mM) of Cu^{2+} for 48 hr, while the muscle cells started to express GFP in the group treated with 0.19 mg/L (3 mM) of Cu^{2+} for 48 hr. A similar dose dependency was observed in chlorpyrifos-treated embryos. Again, as the concentration of chlorpyrifos was increased from 1.47 mg/L (4.2 mM) to 2.04 mg/L (5.83 mM), more tissues became GFP-positive. To better demonstrate the dose- and time-dependent effects of GFP signal strength in *huORFZ*, we conducted semi-quantitative analysis using fluorescent imaging. Three to four images from each treatment condition and each treatment time were taken and analyzed. The results confirmed our descriptional observation in that increased GFP signal follows increased dosage and treatment time (Figure 2C).

Furthermore, we found that the emergence of GFP signals over time (Figure 2B, C) correlated with the endogenous expression level of Ddit3, which is a known indicator of ER stress (Figure 2D). Such results suggest that the GFP response in *huORFZ* embryos may be seen as an indicator of ER stress. Thus, the induced GFP expression pattern in *huORFZ* embryos could be used to identify the type and dosage of the pollutant presented. In addition, based on the parameters suggested above, *huORFZ* embryos could potentially be employed to monitor the presence of poorly studied or previously unknown contaminants that induce ER stress.

The GFP signals in *huORFZ* embryos indicate cells responding to acute and chronic toxic stress

To confirm the correlation between the fluorescence performance of *huORFZ* embryos and their physiological responses to acute and chronic toxic challenge, embryos at 72 hpf were first incubated in solutions containing either Cd^{2+} (Figure 3A-I) or Cu^{2+} (Figure 3J-R) for at least 24 hr. Afterwards, a portion of the GFP-positive embryos were kept in the toxic reagents, while others were transferred to clean distilled water for recovery. The mortality of such recovering embryos was significantly lower than that of embryos left under continuous toxic challenge (Figure 3A, 3J). Additionally, the GFP signals decreased gradually after the GFP-expressing embryos were removed to clean distilled water, and GFP fluorescence could hardly be detected at 72 hr post-treatment of either Cd^{2+} (Figure 3B-E) or Cu^{2+} (Figure 3K-N). Apoptosis was next examined using the TUNEL assay following Cd^{2+} (Figure 3F-I) and Cu^{2+} treatments (Figure 3O-R). In the treatment group exposed over a period of 24 hr with an increased concentration of Cd^{2+} (0.56 mg/L, 5 μM), apoptosis was primarily

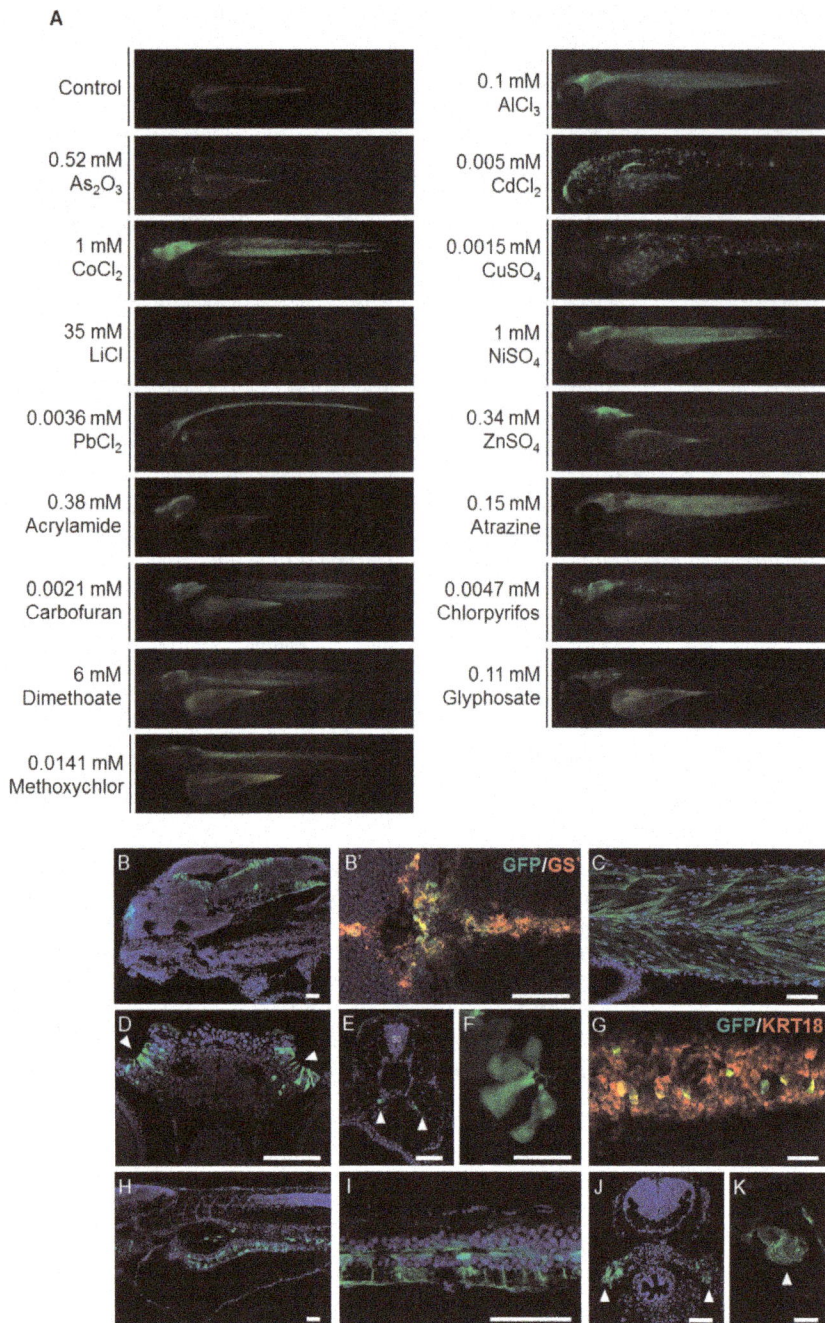

Figure 1. The GFP signals shown in *huORFZ* embryos are distinctly responsive to various stresses. (A). Various heavy metal-containing chemicals, including $AlCl_3$, As_2O_3, $CdCl_2$, $CoCl_2$, $CuSO_4$, LiCl, $NiSO_4$, $PbCl_2$ and $ZnSO_4$, or different EDCs, including acrylamide, atrazine, carbofuran, chlorpyrifos, dimethoate, glyphosate and methoxychlor, were used individually to treat *huORFZ* embryos at 72 hpf. The concentrations of each heavy metal ion or EDC used were indicated on each panel. GFP expression patterns were observed at 96 hpf. The percentages of the representative patterns among the treatment groups are labeled on the images. (B-K). Selected GFP expression patterns were imaged in detail under confocal microscopy. For $AlCl_3$-treated *huORFZ* embryos, GFP signals were observed in the brain (B) and muscle (C). Through immunostaining of antibody against GS, it was observed that only the glial cell lineages in the brain of embryos expressed GFP (B'; red labeling). For $CdCl_2$-treated *huORFZ* embryos, GFP signals were detected in the olfactory epithelium (D; arrowhead), pronephric ducts (E; arrowhead), skin and lateral line system (F). For $CuSO_4$-treatment, embryos obtained from crossing *huORFZ* to *Tg(-2.9krt18:RFP)* were used. When treated with $CuSO_4$, GFP-expressing skin cells were always also *krt18*:RFP-expressing, indicating a keratinocyte lineage (G). For $PbCl_2$-treated *huORFZ* embryos, the GFP-positive tissues included the spinal cord (I) and kidney (J; arrowhead). Interestingly, when either LiCl or Atrazine was used to treat the *huORFZ* embryos, GFP signals were observed in the intestine (H) and heart (K; arrowhead), respectively. (A, B, C, F – I, K) are lateral views with anterior to the left. (B') is dorsal view with anterior to the left. (D, E, J) are transverse view with posterior to the top. SC: spinal cord. BR: brain. The scale bar in F is 20 μm; all other scale bars are 50 μm.

restricted to the olfactory epithelium, as determined by transverse section through the forebrain (Figure 3H). In contrast, TUNEL signals were undetectable in the olfactory organ of larvae from the control groups (Figure 3F, 3G) and in larvae recovering 72 hr after Cd^{2+} exposure (Figure 3I). Similarly, in the treatment group exposed over a period of 24 hr with a concentration of 0.1 mg/L

Table 1. GFP-expressing tissues in huORFZ embryos treated with different heavy metals and endocrine-disrupting chemicals (EDCs).

Treatments	Concentration (mg/L)	Olfactory epithelium	Brain	Spinal cord	Heart	Muscle	Skin	Lateral line system	Kidney	Pronephric duct	Intestine
Heavy Metals											
Al(III)	2.7	-	+*	+	-	+	-	-	-	-	-
As(III)	50	-	-	-	-	-	-	+	-	-	-
Cd(II)	0.56	+*	-	-	-	-	+	+	-	+	-
Co(II)	58.93	-	+	+	-	+	-	-	-	-	-
Cu(II)	0.1	-	-	-	-	-	+*	+	-	-	-
Li(I)	242.9	-	-	-	+	-	-	-	-	-	+*
Ni(II)	58.69	+*	+	+	-	+	-	-	-	-	-
Pb(II)	0.76	+	+*	+	-	-	+	+	+	-	-
Zn(II)	22.24	-	+*	+	-	-	+	-	-	-	-
EDCs											
Acrylamide	26.98	-	+*	-	-	-	-	-	-	-	-
Atrazine	32.35	-	+*	-	-	+	-	-	-	-	-
Carbofuran	1.34	-	+*	-	-	+	-	-	-	-	-
Chlorpyrifos	1.65	+	+*	-	-	+	-	-	-	-	-
Dimethoate	1375	-	+	-	-	+	-	-	-	-	-
Glyphosate	18.6	-	+*	-	-	-	-	-	-	-	-
Methoxychlor	4.87	-	+*	+	-	-	-	-	-	-	-

+:GFP signals are present; -: GFP signals are absent; *: Tissues that responded with highest sensitivity.

(1.5 μM) of Cu^{2+}, TUNEL signals increased in skin cells when compared with control group (Figure 3Q vs. 3O). However, larvae recovering 72 hr after Cu^{2+} exposure showed few TUNEL signals (Figure 3R), and no significant difference was observed when compared with untreated larvae (Figure 3P), suggesting the abatement of stress-induced apoptosis as long as larvae were not continuously exposed to the metal contaminant and allowed to recover. Therefore, we concluded that the GFP expression of huORFZ embryos is likely a distress signal of cells responding to an external contaminant.

To examine the utility of huORFZ embryos as an organismal tool for monitoring chronic environmental stresses, embryos at 72 hpf were treated with 0.03 mg/L (86 nM) of chlorpyrifos for at least 72 hr. Then, a portion of the GFP-positive embryos were kept in the toxic reagent, while others were transferred to clean water to recover. The mortality of such recovering embryos was significantly lower than that of the group challenged by chronic toxic stress (Figure 3S). Additionally, the GFP signals decreased gradually after the GFP-expressing embryos were removed to clean distilled water, becoming nearly undetectable at 72 hr post-treatment (Figure 3T, 3U). These findings indicate that the GFP signals exhibited by huORFZ embryos most likely represent the presence or absence of stress-causing metals before morphological defects of metal-treated embryos could be observed. It should also be noted that the concentration used for chlorpyrifos, 0.03 mg/L, or 30 ppb, is in accordance with WHO drinking water guidelines. Therefore, huORFZ embryos could potentially be employed as a novel organismal tool to monitor chronic environmental stresses.

Field monitoring using the transgenic line huORFZ embryos

To assess the real-world performance of huORFZ, we first determined the LODs for several common heavy metals and EDCs. For aluminum, copper, lead and acrylamide, we found that one-day treatments at, or below, the WHO drinking water guidelines were sufficient for the huORFZ to exhibit detectable tissue-specific GFP patterns. However, for cadmium and chlorpyrifos, four and three days, respectively, were required (Figure 4).

As a proof of concept, we tested the response of huORFZ embryos against water samples collected from different local river basins. For comparison, the zinc, copper, cadmium, arsenic, nickel and/or lead contamination in these samples was also analyzed using conventional methods (Figures 5 and S3). Results showed the absence of GFP fluorescence in embryos incubated with water from Sample 1, indicating that the water quality in this location contains no toxicant whose levels reach beyond the WHO standard. This result was confirmed by conventional analysis. However, distinct GFP expression patterns were observed in embryos incubated separately with water from Samples 2, 3 or 4. Specifically, huORFZ embryos exhibited a scattered skin GFP expression pattern when treated with Sample 2, and they displayed GFP signals in brain after exposure to Sample 3. In the case of Sample 2 and Sample 3, the source of GFP expression in the huORFZ embryos was conclusively attributed to the respective pollutants tested. The primary source for Sample 4 was highly polluted and led to 100% mortality within 24 hr and thus required dilution. Accordingly, we diluted the sample from 4% to 5%, 10%, 20%, 40% and 80% and then used these diluted samples to treat the huORFZ embryos for 24 hr (data not shown). The sample diluted to 20% was then selected for use in the

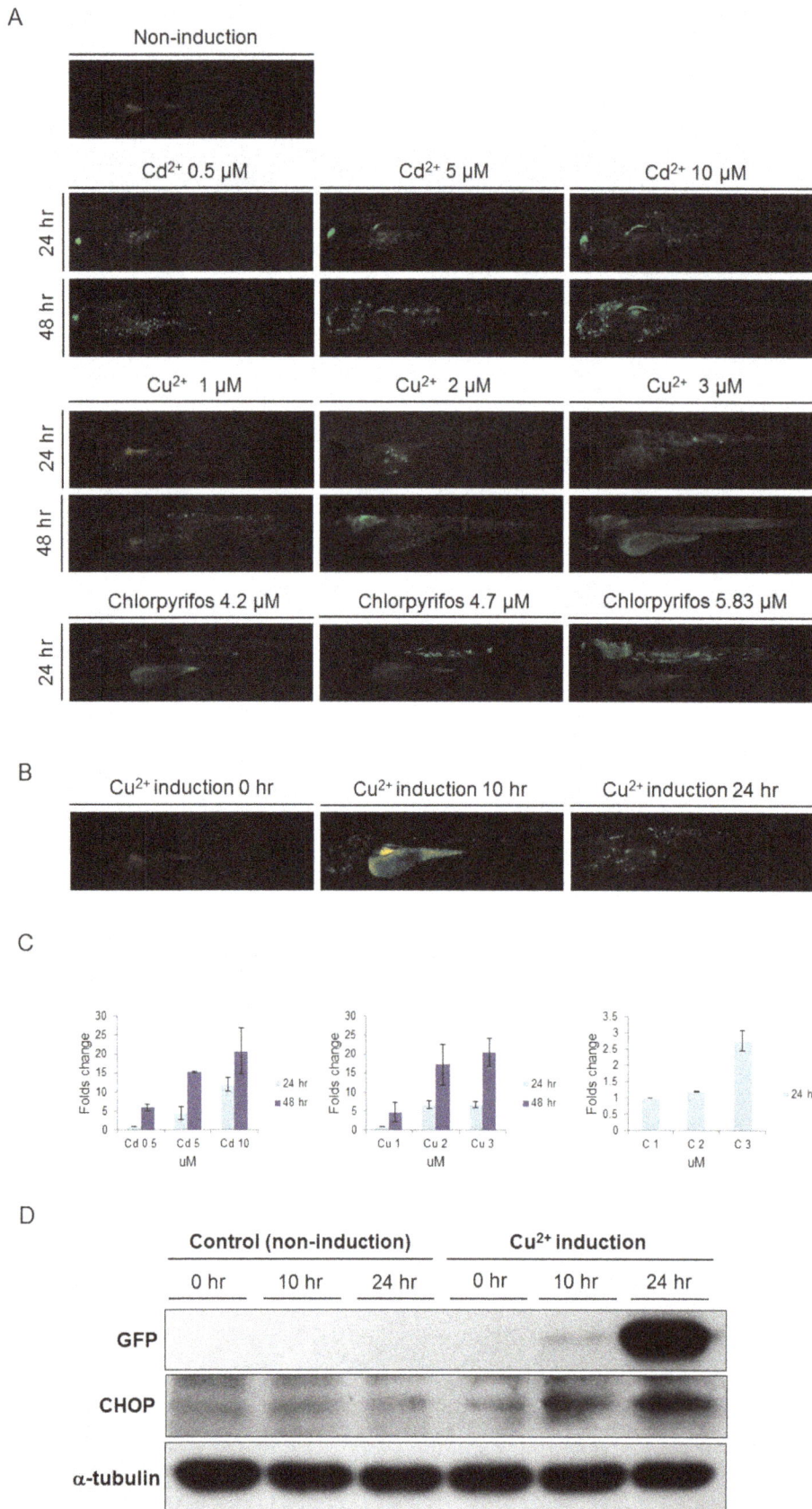

Figure 2. The intensity of GFP signal was positively correlated with the strength of stress and the expression of endogenous Ddit3.
(A). At 72 hpf, *huORFZ* embryos were exposed to different concentrations of Cadmium (Cd^{2+}), Copper (Cu^{2+}), and chlorpyrifos, as indicated, and fluorescence signals were observed at 96 and 120 hpf. Mock group was treated with water containing DMSO which was added to the concentration

representing the DMSO in the chlorpyrifos treatment group. As the chemical concentrations and incubation times increased, the GFP signals also increased. All images are lateral views with anterior to the left. (B). GFP expression patterns of *huORFZ* embryos after treatment with 0.1 mg/L (1.5 μM) of Cu^{2+} from 72 hpf for 10 hr and 24 hr, as indicated. All images are representative with the percentage among treatment groups labeled. All images are lateral views with anterior to the left. (C). The semi-quantification analysis based on fluorescent images. Both increased toxicant concentration and prolonged treatment time resulted in increased GFP signal intensity in *huORFZ* embryos. Note that the readings of each chemical treatment were normalized to the lowest dosage group of the same chemical. Thus, the signal values from different chemical treatments were not comparable. (D). The expression of endogenous Ddit3 in *huORFZ* embryos positively correlates to the signal strength of stress-induced GFP. Total cell lysates were prepared and analyzed by Western blot with specific antibodies to exogenous GFP and to the ER stress protein Ddit3. α-tubulin served as a loading control.

following study. The GFP response in the CNS of *huORFZ* embryos demonstrated a synergistic reaction to the toxicity of Zn^{2+} (0.068 mg/L) and Pb^{2+} (0.013 mg/L) in Sample 4. These results support the hypothesis that GFP expression in *huORFZ* embryos gives faithful reflections of multiple aquatic pollutants. To confirm that the expression pattern we observed in Sample 4 was indeed caused by the metal pollutants we detected, instead of unknown and untested chemicals in the water sample, we treated the *huORFZ* embryos with water containing individual ions according to their concentrations found in Sample 4. Specifically, we treated the embryos separately with As^{3+} (0.069 mg/L), Ni^{2+} (0.0003 mg/L), Pb^{2+} (0.013 mg/L) or Zn^{2+} (0.068 mg/L), as well as a mixture of all four metals. We observed that Pb^{2+} alone, Zn^{2+} alone and the mixture of four metals were all capable of inducing a GFP expression pattern similar to that of Sample 4 in *huORFZ* embryos (Figure S2). These results suggested that *huORFZ* embryos are responsive to a complex sample containing mixed pollutants.

Discussion

The advantages and disadvantages of traditional methods

Traditionally, chemical analysis is accurate in terms of quantification of trace amount chemicals. However, it is often costly, labor-intensive and time-consuming. Such characteristics make this method ideal for producing detailed results for a limited amount of samples, but unrealistic for first-line monitoring that covers vast water bodies in large areas. Enzyme immunoassay is a relatively practical tool for on-site environmental screening [29]. However, it can only detect the targets for which detection kits are designed. The use of bioindicators, on the other hand, offers numerous advantages over the traditional methods. Most significantly, the tolerance of bioindicators for toxicants provides a biologically meaningful picture of pollution levels (Holt and Miller, 2011). To date, a growing number of identified and characterized DNA motifs that can respond to environmental stresses have emerged, including estrogen response elements (EREs) [16,18,30] and heat-shock protein promoters [31–33]. However, limited detection targets and leaked signal of the reporter gene are still problematic for these systems.

The advantages of the *huORFZ* system

In this study, we demonstrated that the zebrafish transgenic line *huORFZ*, with the reporter gene regulated by a human *chop* uORF cassette, exhibits no detectable leakage under normal condition. The *huORFZ* system is a rapid, sensitive, and simple bioindicator able to provide a fluorescent, and, therefore, visible, signal of environmental toxicants. Since the GFP signals exhibited by *huORFZ* do not directly respond to the presence of any hazardous chemical, but rather reflect the cellular or physiological condition, *huORFZ* can be used as a first-line alarm system to detect the presence of stress-inducing chemicals, even when the chemical is not included in the standard water quality guidelines. Also, *huORFZ* embryos can be used to assess the effects of pollutants on

living organisms exposed to chronic stress significantly below lethal dosages. More importantly, the expression of GFP is reversible once the exposed embryos are returned to normal physiological conditions, further supporting *huORFZ* embryos as bioindicators of stress states at the cellular level. To date, a growing number of structurally and functionally diverse groups of chemicals are being generated for industrial use. Some of these chemicals are suspected of having EDC activity, while others remain understudied. Under these circumstances, the ability of *huORFZ* to detect a broad spectrum of pollutants is an unprecedented characteristic for a transgenic animal model. Importantly, we demonstrated that one day's treatment is sufficient to identify the LODs of *huORFZ* embryos for many common pollutants, even at values lower than those specified in the drinking water guidelines recommended by WHO. For other pollutants, including Cd^{2+} and chlorpyrifos, three to four days of treatment were required for the *huORFZ* to detect WHO guideline concentration.

The sensitivity, specificity, reproducibility and confounding factors of using huORFZ as a biomonitor

Using *huORFZ* embryos GFP signals for monitoring chemicals in water, we have to concern its sensitivity and specificity. Regarding sensitivity, in this study we have demonstrated that *huORFZ* can easily detect pollutants with near LC_{10} concentration. The LOD of *huORFZ* can even be pushed to reach WHO guideline values for various heavy metals and EDCs, even though signal strength was significantly weaker than optimal, and in several cases, the treatment time needed to be extended. Regarding specificity, the mechanism of GFP signaling in *huORFZ* is not a specific response to target toxicants. Rather, it most likely reflects the level of physiological stress in the embryo's cells or tissues. Notably, when applied as a biomonitoring system, the detection range of *huORFZ* spans multiple designated pollutants. Also, upon challenge by different pollutants, the correspondingly different GFP patterns can indicate what groups of pollutants are likely present in the water, thus improving the efficiency of subsequent chemical analysis. In this sense, we anticipate that *huORFZ* is suitable of being used as another tool supporting chemical analysis. Since many pollutants can cause similar GFP pattern in *huORFZ* embryos, the effects of mixed pollutants on *huORFZ* are not fully predictable.

Moreover, in the development of biomarker/bioindicator for pollution monitoring, we have to concern the reproducibility and the confounding factors interfering the transgenic fish's GFP responses to chemicals. Regarding reproducibility, as demonstrated in Figure S1, individual variation does exist among *huORFZ* embryos. However, as demonstrated by the percentage indicated in Figures 1A, 2A and 2B, in most the cases, more than 70% of the embryos were responsive and exhibited similar GFP patterns. We did observe that sometimes an abnormally large percentage of the embryos died under treatments with intermediate concentrations. This may have been caused by suboptimal parental health, and the results of those repetitions were discarded. Otherwise, we did not observe any significant different between repetitions. In

A

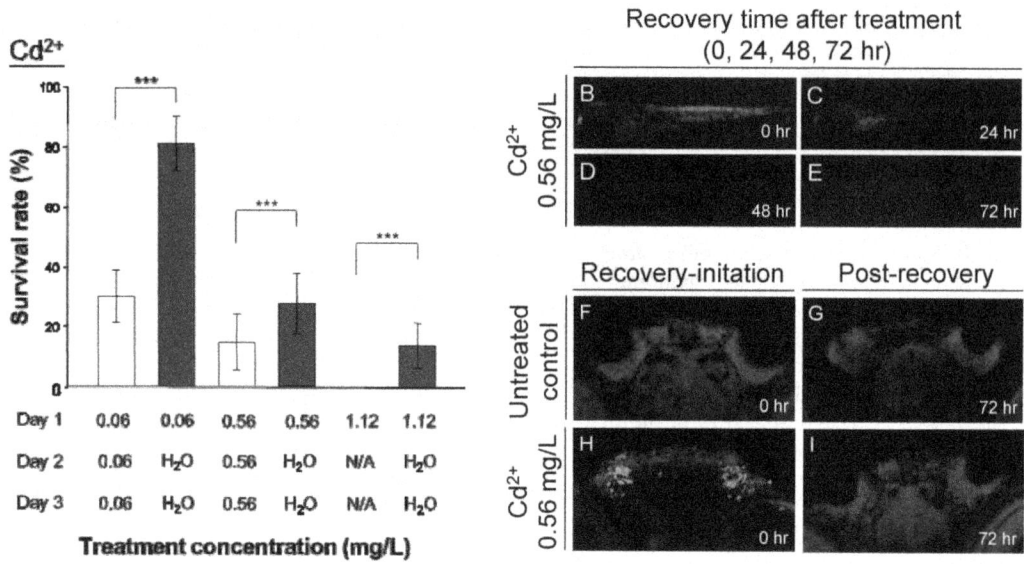

J

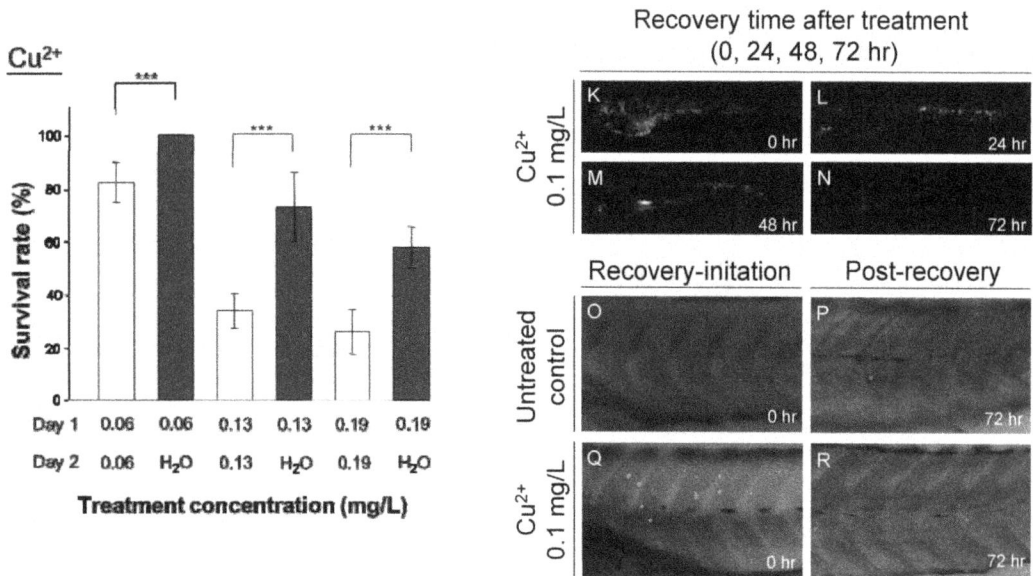

S

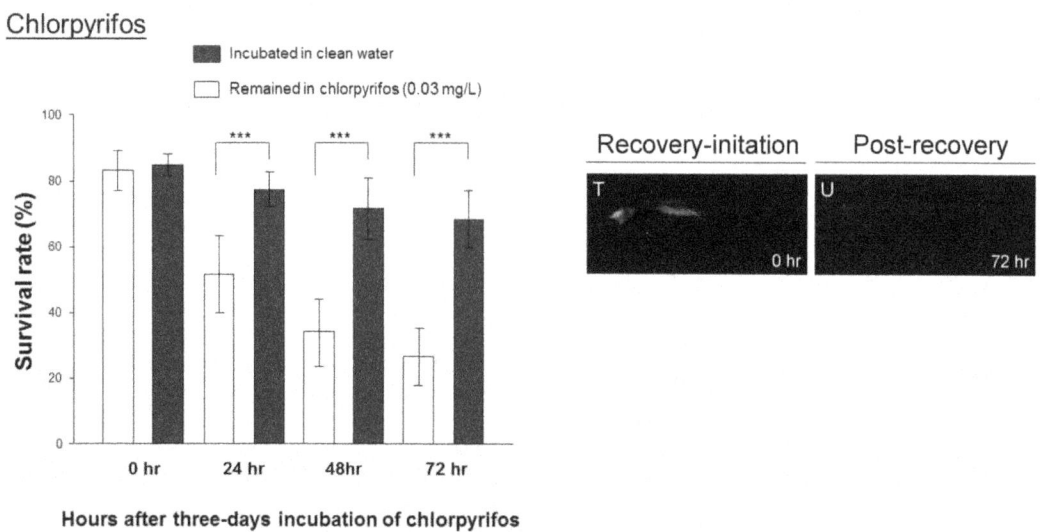

Figure 3. GFP expression in *huORFZ* as signals of cell under acute and chronic toxic stresses. At 72 hpf, *huORFZ* embryos were treated with Cadmium (Cd^{2+}; A-I) or Copper (Cu^{2+}; J-R). (A). The survival rate of *huORFZ* embryos incubated in Cd^{2+} with different concentrations and treatment time. Open bars indicate *huORFZ* embryos that remained in the Cd^{2+} solution throughout the experiment; black bars indicate the embryos removed for recovery in clean water after 24 hr of Cd^{2+} treatment. N/A indicates that prolonged treatment was not conducted as a result of 100% lethality. (B-E). GFP signals immediately or 24, 48 and 72 hr after 24 hr of Cd^{2+} treatment. (F-I). TUNEL assay at the olfactory epithelium immediately or 72 hr after 24 hr of Cd^{2+} treatment. Red color represents TUNEL assay; Green color represents GFP signal; and Blue color represents DAPI staining. (J). The survival rates of *huORFZ* embryos incubated in Cu^{2+} with different concentrations and treatment times. The experimental strategy was the same as A-I. (K-N). GFP signals immediately or 24, 48 and 72 hr after 24 hr of Cu^{2+} treatment. (O-R). TUNEL assay at the skin immediately or 72 hr after 24 hr of Cu^{2+} treatment. (S-U). For the chronic toxicity test, 72 hpf *huORFZ* embryos were treated with chlorpyrifos with a concentration (86 nM) below WHO guidelines. (S). The survival rates of the embryos treated with chlorpyrifos for zero to three days. (T, U). GFP signals immediately or 72 hr after three days of chlorpyrifos treatment. (F-I). are transverse section with dorsal to the top. All other images are lateral views with anterior to the left. For each treatment (each bar in A, J and S), n = 100 embryos evenly distributed in five repetitions.

addition, as demonstrated in Figure S2, the GFP patterns were the same in *huORFZ* exposed to Sample 4 river water as those responding to an artificial mixture containing heavy metals identical to those in the river water sample. This result also indicates that the GFP pattern of *huORFZ* is reproducible. Regarding confounding factors, the practical use of *huORZF* as a biomonitoring system could encounter a situation where certain combination of the pollutants may alter its GFP expression, leading to biased results. Also, since GFP expression likely reflects stresses at the cellular level, it is possible that certain harmful pollutants will not be detected since they do not cause direct cellular stress. To better address these concerns, we will focus our future study on understanding the mechanism and regulation of $uORF^{chop}$ in order to elucidate the precise physiological meaning of *huORFZ* response, as indicated by the GFP pattern.

Physiological significance of the stress-specific expression pattern found in *huORFZ*

While the *chop* gene is generally regarded as an indicator of ER stress [17–18], the mechanism and upstream regulation of the translation inhibitory activity of $uORF^{chop}$ remains unclear. The fact that the GFP signals of *huORFZ* can be observed in tissues without known endogenous *ddit3* expression (skin, muscle and pronephric duct) suggests that the mechanism upstream of $uORF^{chop}$ may play roles beyond simply regulating *chop*. Thus, the detailed physiological meaning of the GFP signals of *huORFZ* remains to be further investigated.

Chemicals	WHO drinking water guideline value (mg/L)	LOD (mg/L)	Typical GFP expression pattern and exposure time*
Aluminum	0.9[a]	0.9	1day
Cadmium	0.003	0.0025	 4day
Copper	2	0.05	1day
Lead	0.01	0.0074	1day
Acrylamide	0.0005	0.0005	 1day
Chlorpyrifos	0.03	0.03	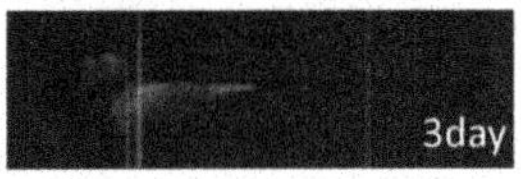 3day

[a]: Guildline level has not been set by WHO. Health based limitation is used here instead.

*: indicates the time elapsing before GFP signals become visible in stress-treated
huORFZ embryos at 72 hpf.

Figure 4. The limit of detection (LOD) of *huORFZ* embryos can reach WHO guideline values for various heavy metals and endocrine-disrupting chemicals (EDCs). LOD was defined as the lowest tested concentration that led to detectable GFP signals in more than 80% of the treated embryos after one to four days of treatment. For each kind of treatment, a representative image of a *huORFZ* embryo treated with the chemical at the LOD concentration for the period of time indicated was presented. All images are lateral views with anterior to the left.

Field monitoring using the transgenic line huORFZ embryos: the expression of GFP in embryos treated with river samples from different sources.

	Control	Sample 1	Sample 2	Sample 3	Sample 4
Tissues with GFP responses	100%	100%	100%	87 %	95 %
Locations Metal constitute	Laboratory prepared	Zengwun River	Erren River	Agongdian Creek	Huang Gang Creek
Cd (mg/L)	–	–	<0.001	–	0.0002
Cu (mg/L)	–	0.0004	**0.104**	0.0033	0.0007
Zn (mg/L)	–	–	0.052	**26.1**	**0.068**
Ni (mg/L)	–	–	–	0.0054	0.0003
As (mg/L)	–	0.0053	0.0124	0.0031	0.069
Pb (mg/L)	–	–	–	–	**0.013**

%: indicates the number of GFP-expression embryos among the total embryos we examined (n=60) in percentage.
Control: embryos were incubated with deionized distilled water from lab.
Sample 1 was taken from the Zengwun River Bridge Station (23°9'18.82"N, 120°20'20.92"E) on January 7, 2013.
Sample 2 was taken from the Geetan Bridge Station (22°59'28.07"N, 120°15'1.72"E) on September 7, 2010. Sample 3 was taken from the Agongdian Bridge Station (22°47'7.10"N, 120°17'49.57"E) on January 2, 2013. Sample 4 was taken from the Huang Gang Creek Bridge Station (25°8'2.53"N, 121°30'7.62"E) on January 15, 2013.
– : undetectable.

Figure 5. Embryos derived from transgenic line *huORFZ* provide true signals of a contaminated aquatic environment. The GFP fluorescent signal intensities induced in *huORFZ* embryos showed responses relative to different river samples collected from local waterways. In Sample 1, no GFP signal was observed in *huORFZ* embryos, consistent with WHO water safety standards. In Sample 2, the GFP response in skin tissue of *huORFZ* embryos indicated potential copper pollution. In Sample 3, GFP signals in the brain of *huORFZ* embryos corresponded to embryonic toxicity consistent with high Zn^{2+} levels. Finally, in Sample 4, the GFP response shown in CNS of *huORFZ* embryos was attributed to the presence of multiple pollutants, such as Zn^{2+} and Pb^{2+}. All images are lateral views with anterior to the left.

Limitations of the *huORFZ* system and future study

Currently, the utility of *huORFZ* embryos does have certain limitations. First, the reporter gene used in *huORFZ* is fluorescent-based. While the cost of fluorescent microscopy instrumentation is reasonable by the standard of a modern biology laboratory, field stations cannot be expected to bear such costs, thus potentially voiding this advantage of the *huORFZ* system. However, it is expected that different reporter genes or proteins may be used to eliminate the requirement of fluorescent microscopes. Secondly, when concentrations of the chemical treatment were reduced to near WHO drinking water guidelines, a GFP signal could still be detected, albeit at a significantly lower intensity than results otherwise obtained from treating the embryos with higher chemical concentrations. Such results are not completely unexpected since we suspect that the GFP expression of *huORFZ* reflects cellular or physiological stress, while the WHO drinking water guideline levels are generally considered relatively "safe" and are unlikely to induce significant cellular stress responses. Thirdly, the GFP patterns expressed in the arsenic-exposed *huORFZ* embryos are not always consistent. Specifically, when embryos were incubated with arsenic for 24 hr, they exhibited strong GFP signals at the lateral line system, whereas the expression patterns changed randomly when they were treated for 48 hr (data not shown). The reason for this irregularity should be the subject of further study.

Conclusion

In this study, we demonstrated the sensitivity and versatility of the *huORFZ* system as a bioindicator for various kinds of stress-inducing pollutants in water. We also demonstrated that the *huORFZ* system performed well under real-world conditions. When fully developed, we anticipate that this *uORF^{chop}*-based system can be integrated into a first-line water security system monitoring fresh water bodies and seawater against the discharge of hazardous pollutants.

Supporting Information

Figure S1 Images of larger field of view are used to demonstrate the general patterns and individual variability of *huORFZ* embryos treated with different heavy. (A) huORFZ embryos were treated with embryo media, ddH2O, or DMSO for 24 and 48 hr starting at 72 hpf. (B-D) The effects of different treatment times (24 and 48 hr) with different concentrations of cadmium (0.5, 1, and 5 µM), copper (1, 2 and 3 µM) and chlorpyrifos (4.2, 4.7, and 5.83 µM) on huORFZ embryos. For chlorpyrifos treated group, at 48 hr treatments are 100% lethal. All images were taken under the Leica MZ FLIII microscope with 2x objective. All images were taken under the same exposure time, iso value and other camera settings.

Figure S2 The four major metal pollutants found in the river water sample 4 are sufficient to induce *huORFZ* embryos to express the GFP signal similar to what was caused by river water sample 4. *huORFZ* embryos were treated with water containing (A, A') Zinc, (B, B') Nickel, (C, C') Arsenic, or (D, D') Lead ion individually, or (E, E') the water containing all four pollutants. The left panel (A, B, C, D, E) demonstrates group images taken under 2x objective while the right panel (A', B', C', D', E') contains the images of one representative embryo of each group, taken under 4x objective. Right panel images are lateral views with anterior to the left. All scale bars are 1 mm.

Figure S3 Higher resolution images of the images presented in Figure 5. All images are exactly the same as in Figure 5, only in larger format.

Table S1 Lethal concentrations for 10%, 50% and 90% mortality of 72-hpf huORFZ zebrafish embryos treated with heavy metals and endocrine-disrupting chemicals (EDCs) for 24-hr.

Table S2 River water sampling record.

Acknowledgments

We are grateful to Ms. Yi-Chun Chuang and Ms. Ya-Chan Yang, College of Life Science, NTU, for helping with the confocal laser scanning microscopy.

Author Contributions

Conceived and designed the experiments: HCL PNL HLH HJT. Performed the experiments: HCL PNL HLH CC HPL. Analyzed the data: HCL PNL HLH. Contributed reagents/materials/analysis tools: CC HPL. Wrote the paper: HCL PNL HLH HJT.

References

1. Berg M, Tran HC, Nguyen TC, Pham HV, Schertenleib R, et al. (2001) Arsenic contamination of groundwater and drinking water in Vietnam: A human health threat. Environmental Science & Technology 35: 2621–2626.
2. Sorg O, Zennegg M, Schmid P, Fedosyuk R, Valikhnovskyi R, et al. (2009) 2,3,7,8-tetrachlorodibenzo-p-dioxin (TCDD) poisoning in Victor Yushchenko: identification and measurement of TCDD metabolites. Lancet 374: 1179–1185.
3. Klecka G, Persoon C, Currie R (2010) Chemicals of Emerging Concern in the Great Lakes Basin: An Analysis of Environmental Exposures. Reviews of Environmental Contamination and Toxicology, Vol 207 207: 1–93.
4. Schwarzenbach RP, Egli T, Hofstetter TB, von Gunten U, Wehrli B (2010) Global Water Pollution and Human Health. Annual Review of Environment and Resources, Vol 35 35: 109–136.
5. Cardwell RD, Foreman DG, Payne TR, Wilbur DJ (1976) Acute toxicity of selenium dioxide to freshwater fishes. Arch Environ Contam Toxicol 4: 129–144.
6. Haasch ML, Wejksnora PJ, Stegeman JJ, Lech JJ (1989) Cloned Rainbow-Trout Liver P1450 Complementary-DNA as a Potential Environmental Monitor. Toxicology and Applied Pharmacology 98: 362–368.
7. Payne JF (1976) Field Evaluation of Benzopyrene Hydroxylase Induction as a Monitor for Marine Petroleum Pollution. Science 191: 945–946.
8. Payne JF, Fancey LL, Rahimtula AD, Porter EL (1987) Review and Perspective on the Use of Mixed-Function Oxygenase Enzymes in Biological Monitoring. Comparative Biochemistry and Physiology C-Pharmacology Toxicology & Endocrinology 86: 233–245.
9. Gill TS, Tewari H, Pande J (1990) Use of the Fish Enzyme-System in Monitoring Water-Quality - Effects of Mercury on Tissue Enzymes. Comparative Biochemistry and Physiology C-Pharmacology Toxicology & Endocrinology 97: 287–292.
10. Goksoyr A, Larsen HE, Husoy AM (1991) Application of a cytochrome P-450 IA1-ELISA in environmental monitoring and toxicological testing of fish. Comp Biochem Physiol C 100: 157–160.
11. Goksoyr A, Andersson T, Buhler DR, Stegeman JJ, Williams DE, et al. (1991) Immunochemical Cross-Reactivity of Beta-Naphthoflavone-Inducible Cytochrome P450 (P450ia) in Liver-Microsomes from Different Fish Species and Rat. Fish Physiology and Biochemistry 9: 1–13.
12. Rodriguezariza A, Dorado G, Navas JI, Pueyo C, Lopezbarea J (1994) Promutagen Activation by Fish Liver as a Biomarker of Littoral Pollution. Environmental and Molecular Mutagenesis 24: 116–123.
13. Hung KWV, Suen MFK, Chen YF, Cai HB, Mo ZX, et al. (2012) Detection of water toxicity using cytochrome P450 transgenic zebrafish as live biosensor: For polychlorinated biphenyls toxicity. Biosensors & Bioelectronics 31: 548–553.
14. Kim KH, Park HJ, Kim JH, Kim S, Williams DR, et al. (2013) Cyp1a reporter zebrafish reveals target tissues for dioxin. Aquat Toxicol 134–135C: 57–65.
15. Halloran MC, Sato-Maeda M, Warren JT, Su F, Lele Z, et al. (2000) Laser-induced gene expression in specific cells of transgenic zebrafish. Development 127: 1953–1960.
16. Lee HC, Chen YJ, Liu YW, Lin KY, Chen SW, et al. (2011) Transgenic zebrafish model to study translational control mediated by upstream open reading frame of human chop gene. Nucleic Acids Res 39: e139.
17. Jousse C, Bruhat A, Carraro V, Urano F, Ferrara M, et al. (2001) Inhibition of CHOP translation by a peptide encoded by an open reading frame localized in the chop 5'UTR. Nucleic Acids Res 29: 4341–4351.
18. Chen H, Hu J, Yang J, Wang Y, Xu H, et al. (2010) Generation of a fluorescent transgenic zebrafish for detection of environmental estrogens. Aquat Toxicol 96: 53–61.
19. Palam LR, Baird TD, Wek RC (2011) Phosphorylation of eIF2 facilitates ribosomal bypass of an inhibitory upstream ORF to enhance CHOP translation. J Biol Chem 286: 10939–10949.
20. Wang XZ, Ron D (1996) Stress-induced phosphorylation and activation of the transcription factor CHOP (GADD153) by p38 MAP Kinase. Science 272: 1347–1349.
21. Dalton LE, Clarke HJ, Knight J, Lawson MH, Wason J, et al. (2013) The endoplasmic reticulum stress marker CHOP predicts survival in malignant mesothelioma. Br J Cancer 108: 1340–1347.
22. World Health Organization. (2011) Guidelines for drinking-water quality. Geneva: World Health Organization. xxiii, 541 p. p.
23. Wang YH, Chen YH, Wu TN, Lin YJ, Tsai HJ (2006) A keratin 18 transgenic zebrafish Tg (k18 (2.9): RFP) treated with inorganic arsenite reveals visible overproliferation of epithelial cells. Toxicology letters 163: 191–197.
24. Westerfield M (2000) The zebrafish book: a guide for the laboratory use of zebrafish (Danio rerio): Institute of Neuroscience. University of Oregon.
25. Sakuma M (1998) Probit analysis of preference data. Applied entomology and zoology 33: 339–348.
26. Noche RR, Lu PN, Goldstein-Kral L, Glasgow E, Liang JO (2011) Circadian rhythms in the pineal organ persist in zebrafish larvae that lack ventral brain. BMC Neurosci 12: 7.
27. Rasband WS (1997–2012) ImageJ. Bethesda, Maryland, USA: U. S. National Institutes of Health.
28. Sternberg SR (1983) Biomedical Image-Processing. Computer 16: 22–34.
29. Morozova V, Levashova A, Eremin S (2005) Determination of pesticides by enzyme immunoassay. Journal of Analytical Chemistry 60: 202–217.
30. Lee O, Takesono A, Tada M, Tyler CR, Kudoh T (2012) Biosensor Zebrafish Provide New Insights into Potential Health Effects of Environmental Estrogens. Environmental health perspectives 120: 990–996.
31. Blechinger SR, Warren Jr JT, Kuwada JY, Krone PH (2002) Developmental toxicity of cadmium in living embryos of a stable transgenic zebrafish line. Environmental health perspectives 110: 1041.
32. Mukhopadhyay I, Nazir A, Saxena D, Chowdhuri DK (2003) Heat shock response: hsp70 in environmental monitoring. Journal of biochemical and molecular toxicology 17: 249–254.
33. Wu YL, Pan X, Mudumana SP, Wang H, Kee PW, et al. (2008) Development of a heat shock inducible gfp transgenic zebrafish line by using the zebrafish hsp27 promoter. Gene 408: 85–94.

17

Concentrations, Source and Risk Assessment of Polycyclic Aromatic Hydrocarbons in Soils from Midway Atoll, North Pacific Ocean

Yuyi Yang[1], Lee Ann Woodward[2], Qing X. Li[3]*, Jun Wang[1,3]*

1 Key Laboratory of Aquatic Botany and Watershed Ecology, Wuhan Botanical Garden, Chinese Academy of Sciences, Wuhan, Hubei Province, China, **2** U.S. Fish and Wildlife Service, Pacific Reefs NWRC, Honolulu, Hawaii, United States of America, **3** Department of Molecular Biosciences and Bioengineering, University of Hawaii at Manoa, Honolulu, Hawaii, United States of America

Abstract

This study was designed to determine concentrations of polycyclic aromatic hydrocarbons (PAHs) in soil samples collected from Midway Atoll and evaluate their potential risks to human health. The total concentrations of 16 PAHs ranged from 3.55 to 3200 $\mu g\ kg^{-1}$ with a mean concentration of 198 $\mu g\ kg^{-1}$. Higher molecular weight PAHs (4–6 ring PAHs) dominated the PAH profiles, accounting for 83.3% of total PAH mass. PAH diagnostic ratio analysis indicated that primary sources of PAHs in Midway Atoll could be combustion. The benzo[a]pyrene equivalent concentration (BaP_{eq}) in most of the study area (86.5%) was less than 40 $\mu g\ kg^{-1}$ BaP_{eq} and total incremental lifetime cancer risks of PAHs ranged from 1.00×10^{-10} to 9.20×10^{-6} with a median value of 1.24×10^{-7}, indicating a minor carcinogenic risk of PAHs in Midway Atoll.

Editor: Jonathan H. Freedman, NIEHS/NIH, United States of America

Funding: This work was supported in part by The U.S. Fish and Wildlife Service (QXL), the National Institute on Minority Health and Health Disparities grant G12 MD007601 (QXL), and Open Funding Project of the Key Laboratory of Aquatic Botany and Watershed Ecology (JW). The funders had no role in study design, data collection and analysis, decision to publish, or preparation of the manuscript.

Competing Interests: The authors have declared that no competing interests exist.

* E-mail: qingl@hawaii.edu (QXL); wangjun@wbgcas.cn (JW)

Introduction

Polycyclic aromatic hydrocarbons (PAHs) are an important group of environmental pollutants. They are introduced into the environment from both natural (e.g., oil seeps, forest fires and volcanic activity) and anthropogenic sources (e.g., petrochemical industrial effluents, coal tar processing wastes, combustion processes) [1–3]. PAHs may accumulate in the organisms due to their low solubility and high octanol-water partition coefficient and undergo long-range transport [4–6]. Furthermore, PAHs present potential carcinogenic risks to residents [7]. Thus, 16 PAHs are selected as the priority pollutants due to their frequency and/or risk by the U.S. Environmental Protection Agency [8].

Soil is the primary steady reservoir and sinks for PAHs in the terrestrial environment, because PAHs are readily absorbed by organic matter in soil and difficult to degrade [9]. Furthermore, the accumulation of PAHs in soil may lead to contamination of food chains, which could cause a potential risk to human health [10,11]. Therefore, concentrations of PAHs in soil have been widely investigated in urban, rural, industrial and agricultural areas of mainland [1,12,13]. However, less data on concentrations of PAHs in soils of atolls and islands have been reported. Such data are required for understanding the potential risk to biota inhabiting the island and global distribution of PAHs.

Midway Atoll is located in the North Pacific Ocean, approximately 1100 miles northwest of Oahu, Hawaii. The atoll is comprised of two main islands, Sand and Eastern, and one smaller islet, enclosed within a reef approximately 8 km long. It is the home to a variety of seabirds, Hawaiian green sea turtles, Hawaiian monk seals, and spinner dolphins. It played a historical role in World War II and was altered very heavily by the military during the war and afterwards. This study intends to be a comprehensive study on PAHs in soils of Midway Atoll. The main objectives were: (1) to determine concentrations and compositions of PAHs; (2) to elucidate potential sources by PAHs diagnostic ratio analysis; and (3) to evaluate the possible carcinogenic risk of PAHs in the soil of Midway Atoll.

Materials and Methods

Study Area and Sample Collection

Midway Atoll is located at the northwest end of the Hawaiian Islands archipelago, at 28.208 °N latitude and -177.379°W longitude (Fig. 1). Midway Atoll had the land area of 5 km^2 with northwest monsoon in winter. Soil has been augmented on Sand Island using naturally occurring guano from seabirds, as well as a shipment of 9,000 tons of soil in the early 1900's from Oahu and Guam. The latter soil augmentation was done to facilitate growing vegetables on the island, and to extend the runway. The main textural class in Midway Atoll is sandy soil. One hundred and eleven samples of surface layers of soil (0–15 cm) from Midway Atoll by grid sampling strategies were collected in June 2009. Samples were hand dug with a trowel that was cleaned off between samples. A minimum of 50 g of soil per sample was collected for analysis. All samples were lyophilized, ground to pass through a sieve of 2 mm openings, and stored in an amber glass container at

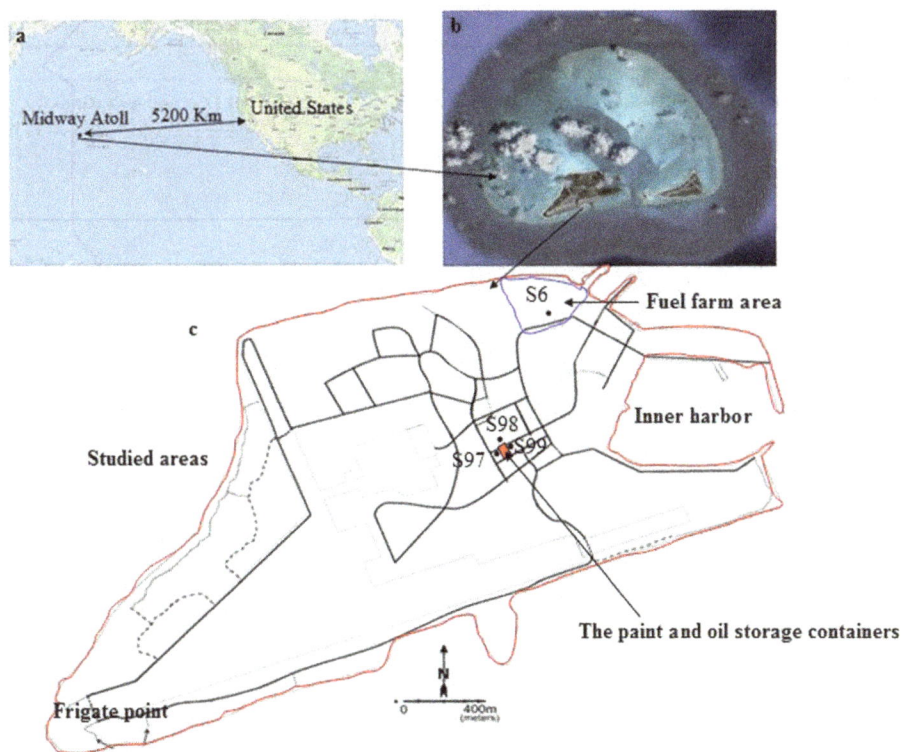

Figure 1. Locations of study area on Midway Atoll, the Pacific Ocean (a: The location of Midway Atoll in the Pacific Ocean; b: The Sand Island and Eastern Island of Midway Atoll; c: The main island of Midway Atoll (Sand Island) was the studied areas (Red line); Red area in the figure: The paint and oil storage containers; Blue line in the figure: Fuel farm area).

$-25°C$. All samples were collected under the permit of the U.S. Fish and Wildlife Service.

Sample Preparation, Extraction and Cleanup

To determine PAHs in soils, an amount of 5 g soil was extracted with a supercritical fluid extractor SFX 220 (Isco, Inc., Lincoln, NE) according to the procedure previously described [14]. The extract was dried with anhydrous sodium sulfate (3 g) and rinsed with hexane (3 ml). The concentrated extract in hexane was cleaned up through an 8 mm i.d. aluminum/silica column. The column was packed with neutral silica (4.0 g, 3% deactivated), neutral alumina (2.0 g, 6% deactivated) and anhydrous sodium sulfate (1 cm) from the bottom to the top [15]. The column was eluted with 20 ml of solvent mixture (methylene chloride/hexane 1:1) to yield a fraction containing PAHs. The samples were concentrated to 20 μL under a gentle stream of high purity nitrogen.

Analysis of PAHs

The samples were analyzed on a Varian Saturn 2000 (Palo Alto, CA) gas chromatograph with mass spectrometric (ion trap) detection (GC/ITMS). The PAHs were separated by a capillary column DB-5MS (J and W Scientific Inc., 30 m, 0.25 mm i.d., 0.25 μm film thickness). The oven temperature was started at 50°C for 3 min, increased to 200°C at a rate of $10°C\ min^{-1}$, and increased to 280°C at a rate of $5°C\ min^{-1}$ and held for 8 min. The injector temperature was set at 280°C. Helium was used as the carrier gas at a constant flow rate of $1\ ml\ min^{-1}$. External calibration was done for each PAH using a certified mixture to determine 16 US-EPA priority PAHs.

Quality Assurance and Quality Control (QA/QC)

Average PAH recoveries and relative standard deviation (RSDs) were first obtained to evaluate the method performance by multiple analyses of 10 soil samples spiked with PAH standard (Accustandard, New Haven, CT), which contained 16 priority PAHs. The 16 PAHs were naphthalene (Nap), acenaphthylene (Acy), acenaphthene (Ace), fluorene (Flr), phenanthrene (Phn), anthracene (Ant), fluoranthene (Fla), pyrene (Pyr), benz[a]anthracene (BaA), chrysene (Chy), benzo[b]fluoranthene (BbF), benzo[k]fluoranthene (BkF), benzo[a]pyrene (BaP), dibenz[a,h]anthracene (DibA), benzo[ghi]perylene (BghiP) and indeno[1,2,3-cd]pyrene (InP). The spike level of each PAH was approximately 50–500 μg kg^{-1}. A solvent blank and matrix blank were analyzed through the entire procedure prior to and after every 10 samples. Standard solutions of PAHs were run at the beginning of sample analysis to determine the relative response factors and evaluate peak resolution. Each sample was analyzed in triplicate unless otherwise stated.

Limits of detection (LOD) were determined as signals 3 times the background signal. Peaks that were smaller than 3 times the signal-to-noise ratio were not considered. The LOD for PAHs ranged from 10 to 500 pg g^{-1}. The average recoveries of PAHs were 85–115% for 10 soil samples varying with the physicochemical properties of individual PAH.

PAH Diagnostic Ratios Analysis

PAH diagnostic ratios have recently come into common use as a tool for identifying and assessing pollution sources. These ratios distinguish PAH pollution originating from petroleum products, petroleum combustion and biomass or coal burning, such as Ant/

Table 1. Concentrations of PAHs in soils collected from Midway Atoll ($\mu g\ kg^{-1}$ dry weight).

PAHs	TEF	Minimum	Maximum	Mean	Median	Frequency	Type of PAHs	Soil guidelines
Nap	0.001	1.20	108	14.0	9.39	100%	LMW	5000
Acy	0.001	ND	10.0	0.61	0.05	60.4%	LMW	–
Ace	0.001	ND	5.30	0.40	0.09	60.4%	LMW	–
Flr	0.001	ND	6.83	0.84	0.66	88.3%	LMW	–
Phn	0.001	0.41	308	14.6	4.29	100%	LMW	5000
Ant	0.01	ND	41.6	2.61	0.41	66.7%	LMW	–
Fla	0.001	ND	649	35.4	3.57	99.1%	HMW	–
Pyr	0.001	ND	542	31.1	3.90	99.1%	HMW	10×10^3
BaA	0.1	ND	308	17.5	2.09	91.9%	HMW	1000
Chy	0.01	ND	363	23.3	3.10	94.6%	HMW	–
BbF	0.1	ND	339	20.5	2.26	89.2%	HMW	1000
BkF	0.1	ND	150	8.68	1.11	85.6%	HMW	1000
BaP	1.0	ND	197	13.6	3.20	92.8%	HMW	1000
InP	0.1	ND	170	6.58	0.16	55.0%	HMW	1000
DibA	1.0	ND	30.2	1.85	0.00	45.0%	HMW	1000
BghiP	0.01	ND	169	6.59	1.23	65.7%	HMW	–
LMW PAHs		2.86	374	33.1	19.2			
HMW PAHs		ND	2830	165	21.1			
$\sum$PAHs		3.55	3200	198	42.4			
LMW/HMW		0.05	16.9	1.52	0.62			
Total BaP$_{eq}$		ND	324	21.2	4.38			
$\sum$BaP$_{eq}$ of 10 PAHs		ND	262	17.2	3.79			

LMW PAHs denote low molecular weight 2–3 ring PAHs; HMW PAHs denote high molecular weight 4–6 ring PAHs; TEF denotes toxic equivalency factor [33]; BaPeq denotes Bap equivalent concentration. ND: not detected.
Soil guidelines: guidelines for residential and parkland soil, NOAA-National Oceanic and Atmospheric Administration.
$\sum$BaP$_{eq}$ of 10 PAHs: Nap, Phn, Ant, Fla, Chy, BaA, BaP, BkF, InP, BghiP.

(Phn+Ant), Fla/(Pyr+Fla), InP/(InP+BghiP) [16,17]. The compounds involved in each ratio have the same molar mass, so it is assumed they have similar physicochemical properties. Based on the PAH isomer ratios in source identification compiled by Yunker et al [18], the Fla/(Fla+Pyr) ratio <0.4 indicates petroleum input

as a source; 0.4–0.5 indicates petroleum (liquid fossil fuel, vehicle and crude oil); and >0.5 indicates combustion of biomass and coal. In addition, an Ant/(Phe+Ant) <0.1 implies a petroleum source, >0.1 implies combustion as a source [10,18].

Risk Assessment

Toxicity equivalent (TEQ) method was used to assess the ecotoxicological risk at a specific site. The total BaP equivalent concentration (BaP$_{eq}$) was calculated by the sum of BaP$_{eq}$ for each PAH using toxicity equivalent factors [19].

About 20 people live on Midway Island, but they form a complete and mutually interdependent community. Therefore, the incremental lifetime cancer risk (ILCR) was employed to evaluate the potential risk of PAHs in soils of Midway Atoll for human health in this study. The ILCRs for adults in terms of direct ingestion, dermal contact, and inhalation were calculated using the following equations [5]:

$$ILCRS_{Ingestion} = \frac{CS \times (CSF_{Ingestion} \times \sqrt[3]{(BW/70)}) \times IR_{soil} \times EF \times ED}{BW \times AT \times cf} \quad (1)$$

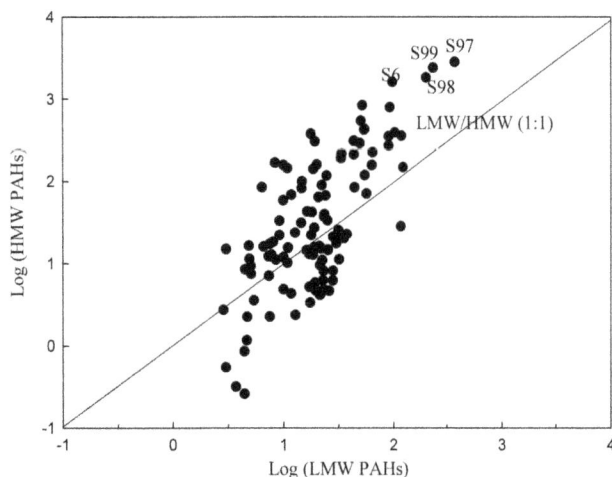

Figure 2. The logarithmic plot of LMW and HMW PAH concentrations of sampling sites on Midway Atoll.

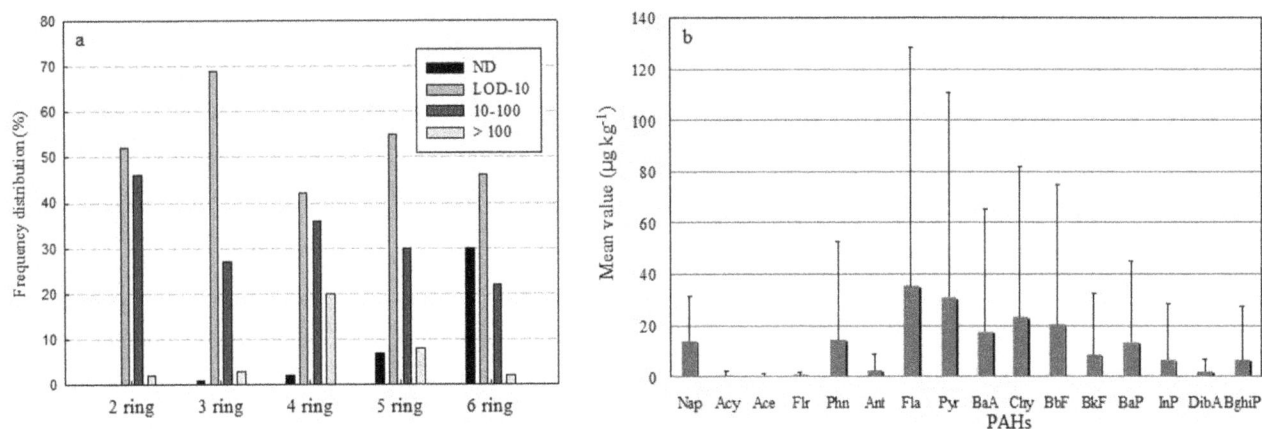

Figure 3. The frequency distribution of different ring PAHs (a) and mean concentrations of individual PAHs (b) on Midway Atoll.

$ILCRS_{Dermal}$

$$= \frac{CS \times (CSF_{Dermal} \times \sqrt[3]{(BW/70)}) \times SA \times AF \times ABS \times EF \times ED}{BW \times AT \times cf} \quad (2)$$

$ILCRS_{Inhalation}$

$$= \frac{CS \times (CSF_{Inhalation} \times \sqrt[3]{(BW/70)}) \times IR_{air} \times EF \times ED}{BW \times AT \times PEF \times cf} \quad (3)$$

where CS is the PAH concentration of soils ($\mu g\ kg^{-1}$), which was obtained by converting concentrations of PAHs according to toxic equivalents of BaP using the toxic equivalency factor (TEF in Table 1) [20]. The carcinogenic slope factor (mg kg^{-1} $day^{-1})^{-1}$ (CSF) was based on the cancer-causing ability of BaP: $CSF_{ingestion}$, CSF_{Dermal} and $CSF_{Inhalation}$ of BaP were 7.3, 25 and 3.85 (mg $kg^{-1}\ day^{-1})^{-1}$, respectively [13]. BW is body weight (kg): 70 kg; AT is average life span (year): 70 years; EF is exposure frequency (days $year^{-1}$): 350 days $year^{-1}$; ED is the exposure duration (year): 30 years; IR_{soil} is the soil intake rate (kg day^{-1}): 0.0001 kg day^{-1}; IR_{air} is the inhalation rate ($m^3\ day^{-1}$): 20 $m^3\ day^{-1}$; SA is the dermal surface exposure ($cm^2\ day^{-1}$): 5000 $cm^2\ day^{-1}$; cf is the conversion factor: 10^6; AF is the dermal adherence factor (kg cm^{-2}): 0.00001 kg cm^{-2}; ABS is the dermal adsorption fraction (unitless): 0.1; and PEF is the soil dust produce factor ($m^3\ kg^{-1}$): $1.32 \times 10^9\ m^3\ kg^{-1}$ [5,8]. The total risks were the sum of risks of ILCRs in terms of direct ingestion, dermal contact, and inhalation.

Results and Discussion

PAH Profiles in Soils of Midway Atoll

Table 1 shows the descriptive statistics for concentrations of PAHs in soils from Midway Atoll. The overall concentration of 16 US EPA priority PAHs in surface soils ranged from 3.55 to 3200 $\mu g\ kg^{-1}$ dry weight with a mean concentration of 198 μg kg^{-1}. The detection frequencies of Nap and Phn were the highest (100%), followed by Fla (99.1%) and Pyr (99.1%). The detection frequency of DibA was the lowest among the 16 PAHs at a detection rate of 45.0%. The concentrations of lower molecular weight PAHs (LMW, i.e., 2–3 ring PAHs) in soils ranged from 2.86 to 374 $\mu g\ kg^{-1}$ with a mean concentration of 33.1 $\mu g\ kg^{-1}$. The

concentrations of higher molecular weight PAHs (HMW, 4–6 ring PAHs) in soils ranged from ND (not detected) to 2830 $\mu g\ kg^{-1}$ with a mean concentration of 165 $\mu g\ kg^{-1}$. Most of the sampling sites (107 sites) had concentrations of LMW PAHs of <150 μg kg^{-1} and HMW PAHs of <1000 $\mu g\ kg^{-1}$ (Fig. 2). Only 4 sites (S6, S97, S98 and S99) had concentrations of LMW PAHs of >150 μg kg^{-1} and HMW PAHs of >1500 $\mu g\ kg^{-1}$. The high total PAH concentrations (S97, S98 and S99) were observed around the paint and oil storage on Midway Atoll. Site S6 having high total PAH concentrations located in fuel farm area (Fig. 1). Most of total PAH concentrations were distributed in the low concentration range with 50% of the samples showed concentrations less than 42.4 μg kg^{-1} (Median values in Table 1). The concentrations of individual PAH also showed a similar statistical characteristic for distribution, i.e., the median values were less than the average values. Compared with the established soil quality guidelines those from the National Oceanography and Atmospheric Administration (NOAA), concentrations of individual PAHs in soils of Midway Atoll were less than the guideline values (Table 1).

Fig. 3a shows frequency distribution for the concentration of different PAHs analogs in all soil samples from Midway Atoll, indicating that most of soil samples had levels of different PAH analogs ranging from LOD to 10 $\mu g\ kg^{-1}$. A two-ring PAH (Nap) was detected in all soil samples. The frequency distribution of three-rings PAHs in the range of LOD-10 $\mu g\ kg^{-1}$ even reached 69%. It was notable that six-ring PAHs was not detected in 30% of the samples. However, 20% of the samples had levels of four-ring PAHs more than 100 $\mu g\ kg^{-1}$. Fig. 3b shows that Fla, Pyr, Chy and BbF were found to be the main soil pollutants in Midway Atoll with mean value more than 20 $\mu g\ kg^{-1}$, which were four-ring PAHs except BbF.

Soils and sediments were considered as the primary steady sinks for PAHs in the environment. Table 2 summarizes PAHs concentrations ($\mu g\ kg^{-1}$ dry weight) in soils/sediments from islands and bays. Low contents of PAHs were found on the islands which were less disturbed by human activities, such as James Ross Island in Antarctica [21] and Admiralty Bay in King George Island [22]. LMW PAHs had a high prevalence in James Ross Island, indicating long-range atmospheric transport was the main source for PAHs contamination [21]. High content of PAHs were found in sediments and/or soils of the densely populated areas of islands, such as Coastal areas in the Shetland and Orkney Islands [23] and Island of Bermuda [24]. In this study, the concentrations of PAHs in Midway Atoll soils were found to be higher than those

Table 2. PAH concentrations (µg kg-1 dry weight) in soils/sediments from islands and bays.

Islands and bays	Soils/sediments	Number of PAHs	Range	Median/mean	Reference
Admiralty Bay, King George Island, Antarctica	Sediments	>16*	9.45–270	62.2	[22]
Coastal areas in the Shetland and Orkney Islands, Britain	Sediments	**	LOD-22600	**	[23]
James ROSS Island, Antarctica	Soils	16	34–171	**	[21]
Vasilievsky Island, Russia	Soils	11	0.197–8.20	1.97	[34]
Potter Cove, South Shetland Islands, Antarctica	Sediments	25	36.5–1910	484/90.4	[35]
Island of Bermuda, Britain	Sediments	13	33.0–10200	1910/1070	[24]
Midway Atoll, USA	Soils	16	3.55–3200	198/42.4	This study

*16 US EPA PAHs with alkyl-naphtalenes and methyl-phenanthrenes.
**No detail information.

of James Ross Island, but lower than those of densely populated Bermuda. Midway Atoll had been used as military bases. Human activities in Midway Atoll may play an important role in PAHs contamination. Through the Baseline Realignment and Closure process, the US Navy undertook a cleanup operation to remove many environmental contaminants that resulted from 90 years of military operations [25]. Contaminants included polychlorinated biphenyls (PCBs), PAHs, petroleum hydrocarbons, asbestos, pesticides such as dichlorodiphenyltrichloroethane (DDT) and dichlorodiphenyldichloroethylene (DDE), and numerous metals. The results of this study suggest that several areas require continued monitoring for possible further remediation, such as S6, S97, S98 and S99.

Potential Source of PAHs in Midway Atoll

The concentrations and patterns of PAHs in soils could reflect the source characteristics [26]. The ratios of LMW/HMW higher than 1 indicated the contaminations were mainly due to the petrogenic sources (hydrocarbon compounds associated with petroleum). On the other hand, the pyrogenic PAHs (hydrocarbon compounds associated with the combustion of petroleum, wood and coal) often showed to be at a LMW/HMW ratio less than 1.0 [27]. The LMW/HMW ratios in Midway Atoll soils ranged from 0.05 to 16.9 with a mean value 1.52. Among all 111 samples, 44 sites had a ratio greater than 1, indicating existence of petrogenic sources of PAHs (Fig. 2). Furthermore, recent pollution of petrogenic PAHs could occur at these 44 sites of predominance of low ring PAHs, because LMW PAHs were more biodegradable and less lipophilic than HMW PAHs. Similar results were also found for the PAHs in soils from Beijing, Tianjin and surrounding areas, North China [28]. Pyrogenic PAHs may be the main source at the other 67 sites with a ratio of LMW/HMW less than 1, such as S6, S97, S98 and S99 (Fig. 2).

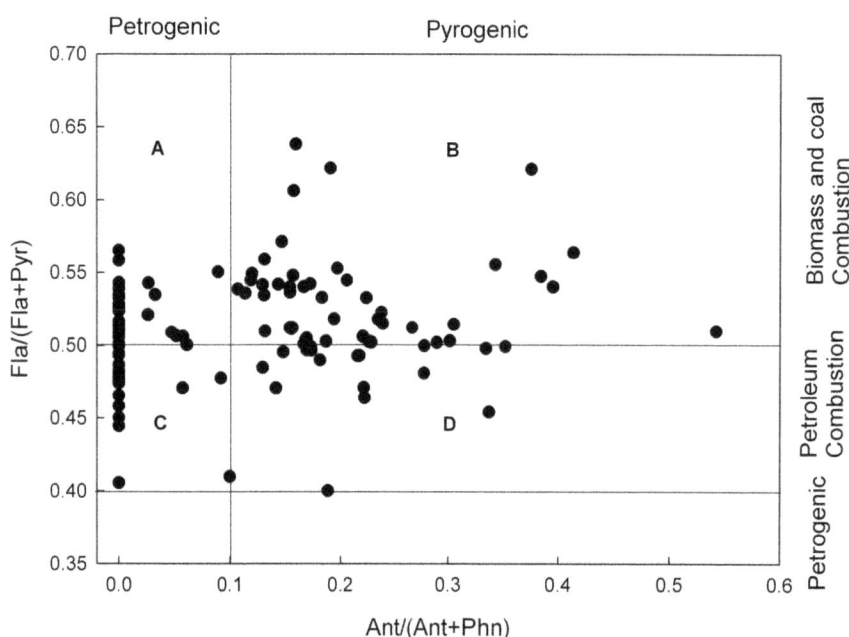

Figure 4. Cross plots for the ratios of Fla/(Pyr+Fla) and Ant/(Phn+Ant) (Fla: fluoranthene; Pyr: pyrene; Ant: anthracene; Phn:phenanthrene).

Table 3. Descriptive statistics of data on incremental lifetime cancer risks (ILCRs) in soils from Midway Atoll.

ILCRs	Minimum	Maximum	Mean	Median
Direct ingestion	5.56×10^{-12}	5.07×10^{-7}	3.31×10^{-8}	6.85×10^{-9}
Dermal contact	9.52×10^{-11}	8.69×10^{-6}	5.67×10^{-7}	1.17×10^{-7}
Inhalation	4.33×10^{-16}	4.05×10^{-11}	2.65×10^{-12}	5.47×10^{-13}
Total ILCRs	1.00×10^{-10}	9.20×10^{-6}	6.00×10^{-7}	1.24×10^{-7}

In the present study, the values of Fla/(Pyr+Fla) ranged from 0.40 to 0.64 and the values of Ant/(Phn+Ant) were between 0.0 to 0.54. Fig. 4 shows the cross plot of Fla/(Pyr+Fla) and Ant/(Phn+Ant), indicating that the sources of PAHs in soils could be classified into four distinct groups. About 15% and 45% of the sampling sites exhibited the typical characteristics of petroleum (liquid fossil fuel, vehicle and crude oil) combustion (Fig. 4D) and the signature of biomass and coal combustion (Fig. 4B), respectively. The remaining sites showed the signature of a mixture containinng petroleum and combustion (Fig. 4A and 4C). Hence, the primary source of PAHs in Midway Atoll could be considered as combustion. Midway Atoll is still an important military site and more than 90 years of military activities occurred on this island. The PAH sources of 55% of the sampling sites (Fig. 4A, 4C and 4D) were related with petroleum, indicating petroleum played an important role for energy and military activities in this island. This result was different from James Ross Island, which was in Antarctic Peninsula and far from human activities. The LMW PAHs dominated the PAH contamination in James Ross Island, indicating the long-range atmospheric transport was the primary source [21].

Risk Assessment of PAHs in Midway Atoll

Toxicity equivalent (TEQ) method was used to assess the ecotoxicological risk at a specific site. BaP_{eq} was calculated by the sum of BaP_{eq} for each PAH using toxicity equivalent factors [19]. In the present study, the total BaP_{eq} of 16 PAHs in soil samples were in the range of ND-324 µg kg^{-1} BaP_{eq} with a mean value of 21.2 µg kg^{-1} BaP_{eq} (Table 1). An half of all samples showed total BaP_{eq} concentrations less than 4.38 µg kg^{-1} BaP_{eq} and 86.5% of the samples had an exposure risk of less than 40 µg kg^{-1} BaP_{eq}. According to the Canadian soil quality guidelines, soils containing <0.1 mg kg^{-1} BaP are considered uncontaminated, soils containing 0.1–1.0 mg kg^{-1} BaP are considered slightly contaminated and soils containing 1–10 mg kg^{-1} BaP are considered to be significantly contaminated [29]. In the present study, 95.5% of sample sites contained less than 0.1 mg kg^{-1} BaP_{eq} and only 5 sites

had concentrations in the range 0.1–0.4 mg kg^{-1} BaP_{eq}, indicating most of soils in Midway Atoll could be considered uncontaminated. The Dutch standards are environmental pollutant reference values (i.e., concentrations in an environmental medium) used in environmental remediation, investigation and cleanup. The target values for the various substances are related to a national background concentration that was determined for the Netherlands [30]. The total BaP_{eq} concentrations of 10 PAHs of the Dutch standards(i.e., Dutch target value) were in the range of ND-262 µg kg^{-1} BaP_{eq}. Approximately one tenth of the sampling sites (11.3%) had values higher than the reference value 32.96 µg kg^{-1} BaP_{eq} [31,32], indicating that 13 sampling soils of Midway Atoll had potential risk to human health.

Incremental lifetime cancer risk is a carcinogenic risk used to evaluate the human health risk. Generally, an ILCR between 10^{-6} and 10^{-4} indicates a potential risk [8]. Table 3 shows the ILCRs levels calculated in the Midway Atoll soils, indicating a low human health risk from the exposure of direct ingestion and inhalation. The ILCRs values of dermal contact with soils ranged from 9.52×10^{-11} to 8.69×10^{-6} with a mean value of 5.67×10^{-7} and 13.5% of the sampling sites exhibited ILCRs values of dermal contact exceeding 10^{-6}, indicating a low potential carcinogenic risk via dermal contact at 15 sampling sites in Midway Atoll. The highest values of ILCRs was found in site S97, followed by S99, S98 and S6. This was accord with that highest concentrations of PAHs were found at these sites. Most of the total ILCRs were distributed in the low range and 86.5% of the samples showed the values less than 1.0×10^{-6}, indicating a negligible carcinogenic risk of PAHs in Midway Atoll.

Conclusion

PAHs are widely distributed in the soils collected from Midway Atoll, in which HMW PAH concentrations ranged from ND to 2830 µg kg^{-1} with a mean concentration of 165 µg kg^{-1}, accounting for 83.3% of the total PAH mass. The main PAH pollutants in Midway Atoll were found to be fluoranthene, pyrene, chrysene, benzo[b]fluoranthene and benzo[a]anthracene. Combustion of coal, petroleum and biomass was potentially the main source for PAH contamination in Midway Atoll. Majority of the sampling sites (95.5%) exhibited PAH concentrations less than 0.1 mg kg^{-1} BaP_{eq}, which could be considered uncontaminated. The ILCRs of PAHs showed that PAH concentrations in most of the sampling areas in Midway Atoll are likely harmless to human health. However, the soil sites that contain an exhibiting carcinogenic risk still need management strategies.

Author Contributions

Conceived and designed the experiments: LAW JW QXL. Performed the experiments: YY LAW JW. Analyzed the data: JW. Contributed reagents/materials/analysis tools: LAW JW QXL. Wrote the paper: YY JW QXL.

References

1. Masih A, Taneja A (2006) Polycyclic aromatic hydrocarbons (PAHs) concentrations and related carcinogenic potencies in soil at a semi-arid region of India. Chemosphere 65: 449–456.

2. Nam JJ, Song BH, Eom KC, Lee SH, Smith A (2003) Distribution of polycyclic aromatic hydrocarbons in agricultural soils in South Korea. Chemosphere 50: 1281–1289.

3. Seo JS, Keum YS, Harada RM, Li QX (2007) Isolation and characterization of bacteria capable of degrading polycyclic aromatic hydrocarbons (PAHs) and organophosphorus pesticides from PAH-contaminated soil in Hilo, Hawaii. Journal of Agricultural and Food Chemistry 55: 5383–5389.

4. Inomata Y, Kajino M, Sato K, Ohara T, Kurokawa JI, et al. (2012) Emission and Atmospheric Transport of Particulate PAHs in Northeast Asia. Environmental Science and Technology 46: 4941–4949.

5. Peng C, Chen WP, Liao XL, Wang ME, Ouyang ZY, et al. (2011) Polycyclic aromatic hydrocarbons in urban soils of Beijing: Status, sources, distribution and potential risk. Environmental Pollution 159: 802–808.

6. Zhao Z-Y, Chu Y-l, Gu J-D (2012) Distribution and sources of polycyclic aromatic hydrocarbons in sediments of the Mai Po Inner Deep Bay Ramsar Site in Hong Kong. Ecotoxicology 21: 1743–1752.

7. Olsson AC, Fevotte J, Fletcher T, Cassidy A, 't Mannetje A, et al. (2010) Occupational exposure to polycyclic aromatic hydrocarbons and lung cancer

risk: a multicenter study in Europe. Occupational and Environmental Medicine 67: 98–103.

8. USEPA (1993) Provisional Guidance for Quantitative Risk Assessment of PAHUS Environmental Protection Agency (1993) EPA/600/R-93/089.

9. Wild SR, Jones KC (1995) Polynuclear aromatic hydrocarbons in the United Kingdom environment: A preliminary source inventory and budget. Environmental Pollution 88: 91–108.

10. Jiang YF, Wang XT, Wu MH, Sheng GY, Fu JM (2011) Contamination, source identification, and risk assessment of polycyclic aromatic hydrocarbons in agricultural soil of Shanghai, China. Environmental Monitoring and Assessment 183: 139–150.

11. Kipopoulou AM, Manoli E, Samara C (1999) Bioconcentration of polycyclic aromatic hydrocarbons in vegetables grown in an industrial area. Environmental Pollution 106: 369–380.

12. Cachada A, Pato P, Rocha-Santos T, da Silva EF, Duarte AC (2012) Levels, sources and potential human health risks of organic pollutants in urban soils. Science of the Total Environment 430: 184–192.

13. Wang Z, Chen JW, Qiao XL, Yang P, Tian FL, et al. (2007) Distribution and sources of polycyclic aromatic hydrocarbons from urban to rural soils: A case study in Dalian, China. Chemosphere 68: 965–971.

14. Miao XS, Swenson C, Woodward LA, Li QX (2000) Distribution of polychlorinated biphenyls in marine species from French Frigate Shoals, North Pacific Ocean. Science of the Total Environment 257: 17–28.

15. Wang J, Caccamise SAL, Wu LJ, Woodward LA, Li QX (2011) Spatial distribution of organochlorine contaminants in soil, sediment, and fish in Bikini and Enewetak Atolls of the Marshall Islands, Pacific Ocean. Chemosphere 84: 1002–1008.

16. Tobiszewski M, Namieśnik J (2012) PAH diagnostic ratios for the identification of pollution emission sources. Environmental Pollution 162: 110–119.

17. Zhang XL, Tao S, Liu WX, Yang Y, Zuo Q, et al. (2005) Source Diagnostics of Polycyclic Aromatic Hydrocarbons Based on Species Ratios: A Multimedia Approach. Environmental Science and Technology 39: 9109–9114.

18. Yunker MB, Macdonald RW, Vingarzan R, Mitchell RH, Goyette D, et al. (2002) PAHs in the Fraser River basin: a critical appraisal of PAH ratios as indicators of PAH source and composition. Organic Geochemistry 33: 489–515.

19. Wickramasinghe AP, Karunaratne DGGP, Sivakanesan R (2012) PM10-bound polycyclic aromatic hydrocarbons: Biological indicators, lung cancer risk of realistic receptors and 'source-exposure-effect relationship' under different source scenarios. Chemosphere 87: 1381–1387.

20. Liao CM, Chiang KC (2006) Probabilistic risk assessment for personal exposure to carcinogenic polycyclic aromatic hydrocarbons in Taiwanese temples. Chemosphere 63: 1610–1619.

21. Klánová J, Matykiewiczová N, Máčka Z, Prošek P, Láska K, et al. (2008) Persistent organic pollutants in soils and sediments from James Ross Island, Antarctica. Environmental Pollution 152: 416–423.

22. Martins CC, Bicego MC, Taniguchi S, Montone RC (2004) Aliphatic and polycyclic aromatic hydrocarbons in surface sediments in Admiralty Bay, King George Island, Antarctica. Antarctic Science 16: 117–122.

23. Webster L, Fryer RJ, Dalgarno EJ, Megginson C, Moffat CF (2001) The polycyclic aromatic hydrocarbon and geochemical biomarker composition of sediments from voes and coastal areas in the Shetland and Orkney Islands. Journal of Environmental Monitoring 3: 591–601.

24. Jones RJ (2011) Spatial patterns of chemical contamination (metals, PAHs, PCBs, PCDDs/PCDFS) in sediments of a non-industrialized but densely populated coral atoll/small island state (Bermuda). Marine Pollution Bulletin 62: 1362–1376.

25. S.K. Taylor CSG, E Luciani, C Zeeman, J Gibson, A. Little Ecological risk assessment for lead in soil and Laysan albatross chicks on Sand Island. US Fish and Wildlife Service (2009).

26. Khairy MA, Kolb M, Mostafa AR, El-Fiky A, Bahadir M (2009) Risk assessment of polycyclic aromatic hydrocarbons in a Mediterranean semi-enclosed basin affected by human activities (Abu Qir Bay, Egypt). Journal of Hazardous Materials 170: 389–397.

27. Soclo HH, Garrigues P, Ewald M (2000) Origin of polycyclic aromatic hydrocarbons (PAHs) in coastal marine sediments: Case studies in Cotonou (Benin) and Aquitaine (France) areas. Marine Pollution Bulletin 40: 387–396.

28. Wang WT, Simonich SLM, Xue MA, Zhao JY, Zhang N, et al. (2010) Concentrations, sources and spatial distribution of polycyclic aromatic hydrocarbons in soils from Beijing, Tianjin and surrounding areas, North China. Environmental Pollution 158: 1245–1251.

29. CCME (2008) (Canadian Council of Ministers of the Environment) Guidelines for Carcinogenic and Other Polycyclic Aromatic Hydrocarbons (Environmental and Human Health Effects). Scientific Supporting Document. pp. 218.

30. Friday GP (1999) Ecological Screening Values for Surface Water, Sediment, and Soil. WSRC-TR-98–00110; United States WSRC-TR-98–00110; United States.

31. Crnkovic D, Ristic M, Jovanovic A, Antonovic D (2007) Levels of PAHs in the soils of Belgrade and its environs. Environmental Monitoring and Assessment 125: 75–83.

32. Van Brummelen TC, Verweij RA, Wedzinga SA, Van Gestel CAM (1996) Enrichment of polycyclic aromatic hydrocarbons in forest soils near a blast furnace plant. Chemosphere 32: 293–314.

33. Nisbet ICT, Lagoy PK (1992) Toxic Equivalency Factors (Tefs) for Polycyclic Aromatic-Hydrocarbons (Pahs). Regulatory Toxicology and Pharmacology 16: 290–300.

34. Lodygin ED, Chukov SN, Beznosikov VA, Gabov DN (2008) Polycyclic aromatic hydrocarbons in soils of Vasilievsky Island (St. Petersburg). Eurasian Soil Science 41: 1321–1326.

35. Curtosi A, Pelletier E, Vodopivez CL, Mac Cormack WP (2009) Distribution of PAHs in the water column, sediments and biota of Potter Cove, South Shetland Islands, Antarctica. Antarctic Science 21: 329–339.

18

Visualized Gene Network Reveals the Novel Target Transcripts Sox2 and Pax6 of Neuronal Development in Trans-Placental Exposure to Bisphenol A

Chung-Wei Yang[1], Wei-Chun Chou[1], Kuan-Hsueh Chen[1], An-Lin Cheng[2], I-Fang Mao[3], How-Ran Chao[4], Chun-Yu Chuang[1]*

1 Department of Biomedical Engineering and Environmental Sciences, National Tsing Hua University, Hsinchu, Taiwan, **2** Schools of Nursing and Health Studies, University of Missouri-Kansas City, Kansas City, Kansas, United States of America, **3** School of Occupational Safety and Health, Chung Shan Medical University, Taichung, Taiwan, **4** Emerging Compounds Research Center, Department of Environmental Science and Engineering, National Pingtung University of Science and Technology, Pingtung County, Taiwan

Abstract

Background: Bisphenol A (BPA) is a ubiquitous endocrine disrupting chemical in our daily life, and its health effect in response to prenatal exposure is still controversial. Early-life BPA exposure may impact brain development and contribute to childhood neurological disorders. The aim of the present study was to investigate molecular target genes of neuronal development in trans-placental exposure to BPA.

Methodology: A meta-analysis of three public microarray datasets was performed to screen for differentially expressed genes (DEGs) in exposure to BPA. The candidate genes of neuronal development were identified from gene ontology analysis in a reconstructed neuronal sub-network, and their gene expressions were determined using real-time PCR in 20 umbilical cord blood samples dichotomized into high and low BPA level groups upon the median 16.8 nM.

Principal Findings: Among 36 neuronal transcripts sorted from DAVID ontology clusters of 457 DEGs using the analysis of Bioconductor limma package, we found two neuronal genes, sex determining region Y-box 2 (Sox2) and paired box 6 (Pax6), had preferentially down-regulated expression (Bonferroni correction p-value $<10^{-4}$ and log2-transformed fold change ≤ -1.2) in response to BPA exposure. Fetal cord blood samples had the obviously attenuated gene expression of Sox2 and Pax6 in high BPA group referred to low BPA group. Visualized gene network of Cytoscape analysis showed that Sox2 and Pax6 which were contributed to neural precursor cell proliferation and neuronal differentiation might be down-regulated through sonic hedgehog (Shh), vascular endothelial growth factor A (VEGFA) and Notch signaling.

Conclusions: These results indicated that trans-placental BPA exposure down-regulated gene expression of Sox2 and Pax6 potentially underlying the adverse effect on childhood neuronal development.

Editor: Brian P. Chadwick, Florida State University, United States of America

Funding: This study was funded from Taiwan National Science Council grants NSC 100-2621-M-007-002 and NSC 102-2918-I-007 -011 (http://web1.nsc.gov.tw/). "The funders had no role in study design, data collection and analysis, decision to publish, or preparation of the manuscript."

Competing Interests: The authors have declared that no competing interests exist.

* Email: cychuang@mx.nthu.edu.tw

Introduction

Bisphenol A (BPA) has a frequent industrial use as a sealant or flux of plastic materials. Its application results in human exposure through the intake of foods and liquids in polycarbonate water bottles, food wraps, plastic bags, baby bottles, protective coatings on food containers, epoxy resin, and dental composites [1]. BPA is an endocrine-disrupting chemical that mimics hormones through estrogen or thyroid receptor mechanisms, and consequently causes adverse health effects on growth development in an intact organism and its progeny [2,3]. While some awareness exists about the possible adverse effects of BPA exposure, the broader knowledge of its effects on childhood neuronal development is still limited.

In mammalians, BPA can easily pass through placenta during pregnancy and has a latent effect on postnatal reproductive functions [4]. The reproductive toxicity of BPA is not as high as other environmental chemicals such as 2,3,7,8-Tetrachlorodibenzodioxin (TCDD) or nonylphenol (NP) [5]. However, maternal BPA exposure can cause metabolic and emotional disruption on offspring even in low dose exposure [6,7]. Urinary BPA has been detected in children in many developed countries e.g., Australia, United States, and Italy [8–10]. Our previous bio-monitoring study in Taiwan identified that prenatal BPA exposure concentration is negatively correlated with birth weight and affects gene expression of leptin and adiponectin in male neonates [11]. BPA also causes epigenetic disruptions. Low-dose prenatal BPA exposure alters mRNA expression of epigenetic regulators DNA methyltransferase (DNMT) 1 and DNMT3A in the brain [12]. Rat model indicated that perinatal BPA exposure is a potential causative agent of molar incisor hypomineralization (MIH) during a specific developmental time window [13].

Concerns have been raised about the effect of trans-placental BPA exposure on central nervous system and neuronal development. Early life BPA exposure has been associated with behavior problem such as anxiety, depression, and hyperactivity in children [14]. BPA has been reported in response to childhood behavioral and learning development at age 8–11 [15]. BPA (250 ng/kg/day) enhances fear memory and increases serotonin metabolites 5-hydroxyindoleacetic acid (5-HIAA) levels and 5-HIAA/serotonin (5-HT) in the hippocampus, striatum and midbrain in juvenile female mice [16], and delays perinatal chloride shift by significantly decreasing potassium chloride co-transporter 2 (Kcc2) mRNA expression in developing rat, mouse, and human cortical neurons [17]. BPA also causes adverse effects on neuronal morphology and functions as to interrupting neuronal dendritic and synaptic development in cultures of fetal rat hypothalamus cells at 10 and 100 nM [18]. BPA suppresses neurite extension by inhibiting phosphorylation of mitogen activated protein kinase (MAPK) in rat pheochromocytoma PC12 cells differentiated neuronal-like cells [19]. Furthermore, perinatal exposure to BPA causes GABAergic disinhibition and dopaminergic enhancement that is related to abnormal cortical basolateral amygdala synaptic transmission and plasticity; this effect may be responsible for hyperactivity and attention deficit in BPA-rats [20].

Microarray analysis is an effective way to explore possible mechanisms and has been used to study the molecular pathway of reproductive toxicity of BPA exposure in animal models [21–23]. However, few studies evaluated childhood neuronal development in exposure to BPA with human data. In human samples, umbilical cord blood is a postpartum placental remnant containing fetal blood which can be used as a surrogate for childhood study [24]. In this study, we used meta-analysis of publicly available microarray datasets to find the neuronal target genes in exposure to BPA, and explored whether trans-placental BPA exposure in mothers would alter gene expression on their progeny in human umbilical cord blood and the potential underlying mechanism from gene network analysis to childhood neuronal development.

Materials and Methods

Data Collection and Differentiated Expression Gene Analysis

Three microarray datasets of human cell models in exposure to low-dose BPA were selected from the ArrayExpress database (http://www.ebi.ac.uk/arrayexpress/) (Table 1). Data were analyzed using a Bioconductor package (http://www.bioconductor.org) implemented in R (http://cran.r-project.org), and were filtered by choosing probes having a standard deviation >0.15 over all the samples in the analysis. Two different platform chips (Agilent Whole Human Genome Microarray and Affymetrix HG-U133Plus2.0) were pre-processing using RMA algorithm (http://rmaexpress.bmbolstad.com/) prior to merging each other. We selected genes from all platforms based on the NIH Entrez Gene ID and used the median rank score method with the R package CONOR [25] for cross-platform normalization (Figure S1). The detailed procedure for performing the meta-analysis was provided in the Supporting Information (Information S1). In the analysis of Bioconductor limma package with t-statistic and false discovery rate (FDR) <0.1, differentially expressed genes (DEGs) with fold change greater or less than ±1.2 were identified individually among each BPA exposure groups (1 pM, 100 pM, 10 nM, 1 uM

and 10 uM) versus their corresponding controls between high (1 and 10 uM) and low (1 pM, 100 pM and 10 nM) BPA exposure groups. For gene ontology analysis, DEGs were imported into DAVID Bioinformatics Resources 6.7 to identify major function clusters. To investigate the major neuronal transcripts, Bonferroni correction was used as a screening method to account for multiple hypothesis testing. P-value of Bonferroni correction was $1.45*10^{-3}$ (0.05/36; adjusted p-value $=p/n$, where p is p-value, n is total number of neuronal transcripts) in this study. This study defined 10^{-4} as the cutoff p-value to make selected genes more specific. Therefore, the genes with expression values in the quadrant of greater or less than log-transformed fold change ±1.2 and p-value less than 10^{-4} (the absolute value of log10-transformed value >4) were selected for further network construction and pathway prediction.

Study Subjects and Sample Collection

This study randomly selected 20 umbilical cord blood samples of 157 healthy pregnant women in a previous birth cohort study [11] recruiting from January 2006 and August 2007 at an obstetrics and gynecology clinic in Hsinchu County, Taiwan. All pregnant women provided their written informed consent of genetic research to participate in this study, and the institutional review boards of National Tsing Hua University and Changhua Christian Hospital approved the bio-sampling process. When pregnant women gave birth, umbilical cord blood samples were collected in glass heparin tubes and delivered at 4°C to the lab within two days for RNA isolation. Plastics were excluded throughout the entire procedure to avoid BPA contamination. Twenty umbilical cord blood samples were dichotomized into high and low BPA level groups based on the median level of BPA exposure (16.8 nM) for determining gene expression of Sox2 and Pax6.

BPA Detection

The umbilical cord blood samples were centrifuged at 12,000 rpm for 10 min to separate the plasma and corpuscles, and stored at −80°C until analysis. Plasma fraction (500 μl) was mixed with 100 μl of 0.01 M ammonium acetate buffer (pH 4.5; Riedel-de Haen, Seelze, Germany), 4 ml mixture of n-hexane (HPLC grade; Echo Chemical, Miaoli, Taiwan) and diethyl ether (70:30 v/v, anhydrous; J.T. Baker, Phillipsburg, NJ). Then 8.71 μl of 9.187 M perchloric acid (purity 60–62%; Sigma-Aldrich, St. Louis, MO) was added into the plasma mixture and centrifuged with 3,000 rpm for 5 minutes. After centrifugation, the organic layer was evaporated to dryness and reconstituted with 100 μl of mobile phase (methanol:water 80:20 v/v) for BPA determination by a reverse-phase high performance liquid chromatography (HPLC) (D-7000) connected to a UV detector (L-7400) consisting of an autosampler (L-7200), a pump (L-2130) and a degasys (DG-2410) (Hitachi High Technologies America, Pleasanton, CA). The QA/QC materials were prepared from a plasma pool in analysis with standard, reagent blank and unknown samples. We performed external calibration using the chromatographic responses of seven standard concentrations in their corresponding solvent. The recovery rates of blanks extended from 96–103%. The relative standard deviation (RSD) among triplicate analyses were 1.99–7.53%, and the recovery percentage was 96.1% with an RSD of 7.53%. The limit of detection (LOD) was 1.75 nM.

Quantitative Real-Time PCR

The buffy coat of umbilical cord blood sample was pretreated using RBC lysis buffer to avoid the interference of red blood cells. RNA isolation was conducted using Trizol reagent with chloro-

Table 1. Microarray datasets used in the meta-analysis of gene expression in exposure to BPA.

Study	Accession	Control samples (n)	BPA treated samples (n)	Array Platform	BPA concentration
Qin et al., 2012	E-GEOD-35034	1	1	Agilent-028004 SurePrint G3 Human GE 8x60K Microarray	10 nM
Tiesman, 2011	E-GEOD-17624	4	4	Affymetrix GeneChip Human Genome U133 Plus 2.0	1 pM, 100 pM, 10 nM, 1uM
Huang, 2011	E-GEOD-32160	3	4	Affymetrix GeneChip Human Genome U133 Plus 2.0	1 uM, 10 uM

Source: http://www.ebi.ac.uk/arrayexpress/

form and isopropanol. The RNA pellet was washed using 75% alcohol and air dried. Total RNA was dissolved in DEPC contained water and stored under $-80°C$. Total RNA was reverse-transcribed to cDNA using the high capacity cDNA reverse transcription kits (ABI Inc., Foster, CA) for quantitative real-time analysis. PCR primers (Sox2: F-5'CAC ACT GCC CCT CTC ACA CA3', R-5'CCC ATT TCC CTC GTT TTT CTT3'; Pax6: F-5'TCG GGC ACC ACT TCA ACA3', R-5'CGG GAA CTT GAA CTG GAA CTG3') were designed using Primer Express V.3.0 software (ABI Inc, Foster, CA) according to the mRNA sequence from GenBank. Real-time PCR was performed with the FastStart SYBR Green Master (Roche Inc, Penzberg, Germany) and was analyzed using the 7300 Real-Time PCR system (ABI Inc, Foster, CA). The relative level of mRNA expression was analyzed using comparative method by SDS 1.4 software normalized to the endogenous housekeeping gene β-actin.

Network Construction and Pathway Prediction

The plug-in system "ClueGO+CluePedia" in the latest version of Cytoscape Software 3.0.2 was used to identify networks and functional pathways of DEGs in response to BPA exposure. ClueGO [26] performs an extensive database of functional interactions including GO, KEGG and Reactome. ClueGO creates the first binary gene-term matrix with the selected terms and their associated genes. Based on this matrix, a term-term similarity matrix is calculated using chance corrected kappa statistics to determine the association strength between the terms. For biological networks, CluePedia [27] calculates the correlation for DEGs based on four tests, Pearson correlation, Spearman's rank, distance correlation and maximal information coefficient (MIC), for investigating linear and non-linear dependencies between implemented variables. In this study, two-sided hypergeometric statistic was performed with kappa score threshold setting of 0.3. Enrichment/depletion was calculated based on Benjamini-Hochberg corrected p-value. GO groups with corrected p-value < 0.05 were denoted for different significant levels.

Results

Identification of Candidate Genes Relevant to Neuronal Development in Exposure to BPA

Figure 1 showed the framework in this study to explore candidate genes relevant to neuronal development in exposure to BPA exposure. This study combined three normalized public microarray datasets (Table 1) and analyzed using Bioconductor limma package to identify 457 DEGs (Table S1) among BPA exposure groups and their controls (1 pM n = 25, 100 pM n = 109, 10 nM n = 145, 1 uM n = 250, 10 uM n = 345). In the DAVID

ontology analysis of 457 DEGs, 36 transcripts were relevant to neuronal ontology and their up- and down-regulated expression were listed in Table 2. For exploring candidate genes of neuronal development in response to BPA exposure, 36 potential transcripts relevant to neuronal ontology were illustrated in a volcano plot (Figure 2). Two candidate genes, sex determining region Y-box 2 (Sox2) and paired box 6 (Pax6), had aberrant expression (Bonferroni correction p-value $<10^{-4}$ and ≤ -1.2 log2-transformed fold change) in comparison with control respectively responsible for developing neural tube cells and regulating neurogenesis in radial-glial-like neural stem cells [28]. Therefore, Sox2 and Pax6 were served as candidate neuronal genes for further investigation in human umbilical cord blood.

Comparison of Gene Expression between High and Low BPA Levels in Cord Blood Samples

Twenty human umbilical cord blood samples were randomly selected from our previous birth cohort [11] and their average BPA level were 23.6 nM. The gene expression of Sox2 and Pax6 were determined in low BPA group and high BPA group classified upon the median value (16.8 nM) of BPA in umbilical cord blood. Both gene expression of Sox2 and Pax6 significantly decreased in higher BPA group referred to lower BPA group (fold change 0.1 and 0.08, respectively) (Figure 3). Results suggested that trans-placental BPA exposure might affect childhood neuronal development underlying decreased Sox2 and Pax6 expression.

Neuronal Network Reconstruction with Sox2 and Pax6

This study imported 457 DEGs into Cytoscape plug-in ClueGO+CluePedia to investigate the gene network and functional pathway prediction in response to BPA exposure. The results of Cytoscape analysis presented that totally 959 genes constructed a gene network connecting with the 457 DEGs (Figure 4). The gene ontology enrichment of the 959 genes relevant to BPA exposure mapped for GO category presented in tetrapyrrole metabolic process, cellular amino acid biosynthetic process, endoplasmic reticulum unfolded protein response, and amino acid assembly. The further sub-network analysis in Figure 5 was visualized specifically to understand the neuronal functions that Sox2 and Pax6 were involved. Sox2 and Pax6 had similar neuronal functions such as regulation of neural precursor cell proliferation and forebrain neuron differentiation.

The results of ClueGO analysis presented a predicted pathway of Sox2 and Pax6 and their potential interaction genes resulted from BPA exposure in Figure 6. Sox2 and Pax6 acted in sonic hedgehog (Shh), Notch and vascular endothelial growth factor A (VEGFA) pathway for cell differentiation of spinal cord, forebrain neuron differentiation, and regulation of neural precursor cell proliferation. In Shh signaling, Shh is positively regulated by

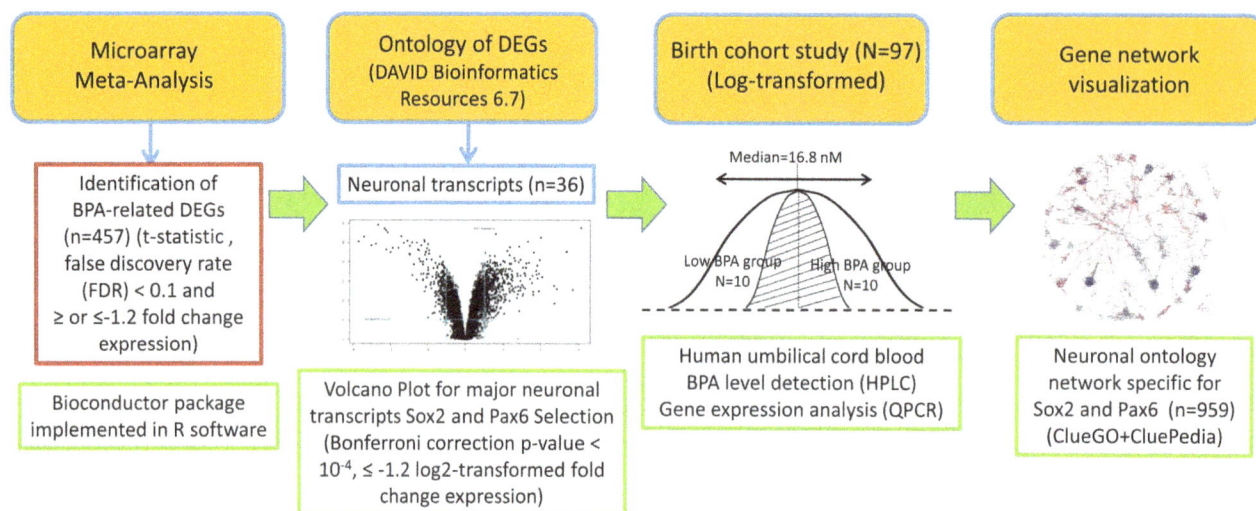

Figure 1. Schematic flow chart of study design. This study processed the meta-analysis of three microarray datasets to identify DEGs in exposure to various BPA levels compared to controls (n = 457). There were 36 neuronal transcripts sorted from 457 DEGs involved in ontology clusters, and top two down-regulated neuronal genes Sox2 and Pax6 were selected in response to BPA exposure. Gene expression of Sox2 and Pax6 were determined in 20 human umbilical cord blood samples randomly recruited from a previous birth cohort [11], and obviously attenuated in high BPA exposure group referred to low BPA group. The visualized gene network of Sox2 and Pax6 and their potential interaction genes specific for neuronal development was predicted in response to trans-placental BPA exposure. DEG: differentially expressed gene; Sox2: sex determining region Y-box 2; Pax6: paired box 6.

transcription factor encoded genes Gli1 and Gli2 and negatively modulated by Gli3. Regarding to BPA exposure, Shh down-regulates Sox2 through Pax6 antagonism. In VEGFA pathway, VEGFA generally activates prospero-related homeobox 1 (Prox1) to up-regulate Sox2 and Pax6, and suppresses Notch1. BPA exposure would cause insulin-like growth factor 1 (IGF1) to attenuate VEGFA expression and its down-stream genes. Notch signaling indirectly modulates Sox2 and Pax6 through Shh, VEGFA and Prox1 in response to BPA exposure.

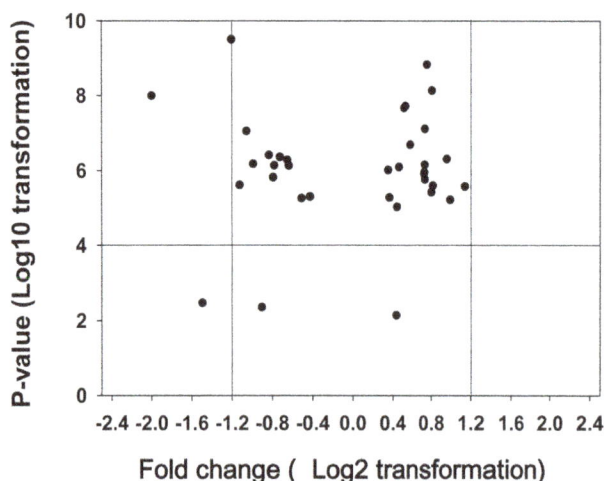

Figure 2. Volcano plot of gene expression relevant to neuronal development in exposure to BPA. X axis is log2-transformed fold change value of gene expression; Y axis is the absolute value of log10-transformed p-value. The genes Sox2 and Pax6 with log-transformed fold change less than -1.2 and p-value less than 10^{-4} were selected as the candidate target genes of neuronal development in trans-placental exposure to BPA.

Discussion

Human exposure to chemicals relevant to disease outcome is difficult to be determined and estimated, especially for the effect on child development. Although some studies have found that BPA affects growth and development of reproductive organ [29,30], potential adverse effects of BPA on childhood neuronal development are not fully understood yet. In this study, we identified Sox2 and Pax6 as neuronal development biomarkers whose gene expression was appeared in response to trans-placental BPA exposure. Such a biomarker holds promise in assessing BPA exposure and acts as a clinically relevant predictor for neurogenesis in children underlying maternal BPA exposure. In general, it's hard to explore the health effect of prenatal exposure in human subjects. A biomarker is found from a costly method because of the need for several gene chips with sufficient amount of samples; childhood neuronal development research takes time to prospectively follow up a cohort which generates additional challenges. The method described in this study offered an alternative strategy to examine the molecular effect of prenatal BPA exposure on child development. Candidate biomarkers were surveyed from the use of microarray meta-analysis and the gene expression of biomarkers were investigated in human samples from fetal umbilical cord blood for potential impact research.

In this study, the step-wise approach including the meta-analysis of microarray DEGs, GO and Bonferroni correction between high and low BPA exposure was used to investigate the neuronal candidate genes Sox2 and Pax6 in exposure to BPA. Additionally, the gene expression of Sox2 and Pax6 were determined in human fetal cord blood to evaluate whether it could be used as biomarkers for childhood neurogenesis deficiency resulting from trans-placental BPA exposure. This study found higher BPA exposure level in human fetal cord blood samples decreased gene expression of Sox2 and Pax6. The visualized network analysis showed that Sox2 and Pax6 were highly involved in the function of neurogenesis. This alternative method to evaluate the impact of

Table 2. Significantly 36 down-regulated and up-regulated genes of neuronal ontology in exposure to BPA.

Down-regulated genes				Up-regulated genes			
ID	Name	Fold change (log2 transformed)	p-value	ID	Name	Fold change (log2 transformed)	p-value
PAX6	paired box 6	−1.9937	1.05E−08	ESR1	estrogen receptor 1	1.1477	2.74E−06
UNC5B	unc-5 Homolog B	−1.4892	3.45E−06	ATP1A2	ATPase, Na+/K+ transporting, alpha 2 polypeptide	0.9995	6.13E−06
SOX2	SRY (Sex Determining Region Y)-Box 2	−1.20	3.22E−10	TGFB2	transforming growth factor, beta 2	0.9657	5.07E−07
PHGDH	phosphoglycerate Dehydrogenase	−1.1190	2.50E−06	TUBB2A	tubulin, beta 2A class IIa	0.8229	2.58E−06
MIB1	mindbomb E3 ubiquitin protein ligase 1	−1.0507	9.18E−08	BAIAP2	BAI1-associated protein 2	0.8133	7.60E−09
EIF2AK3	eukaryotic translation initiation factor 2-alpha kinase 3	−0.9850	6.85E−07	SEMA3A	sema domain, (semaphorin) 3A	0.8079	3.86E−06
KLHL24	Kelch-Like Family Member 24	−0.8977	0.004542	PLXNB2	plexin B2	0.7616	1.51E−09
TBCE	tubulin folding cofactor E	−0.8284	3.98E−07	FGFR1	fibroblast growth factor receptor 1	0.7410	1.77E−06
JMJD6	jumonji domain containing 6	−0.7862	1.56E−06	TUBB3	tubulin, beta 3 class III	0.7405	7.92E−08
NF1	Neurofibromin 1	−0.7749	7.51E−07	SLITRK5	SLIT and NTRK-like family, member 5	0.7387	7.19E−07
RUFY3	RUN and FYVE domain containing 3	−0.7190	4.47E−07	NCS1	neuronal calcium sensor 1	0.7352	1.13E−06
PPM1A	protein phosphatase, mg2+/mn2+ dependent, 1a	−0.6465	5.42E−07	ZNF488	zinc finger protein 488	0.7329	1.24E−06
PEX13	peroxisomal biogenesis factor 13	−0.6320	7.58E−07	BMP4	bone morphogenetic protein 4	0.5869	2.12E−07
DLC1	deleted in liver cancer 1	−0.5047	5.61E−06	ERBB2	v-erb-b2 avian erythroblastic leukemia viral oncogene homolog 2	0.5367	1.98E−08
GSK3B	glycogen synthase kinase 3 beta	−0.4210	5.05E−06	JAG2	jagged 2	0.5218	2.21E−08
PPARA	peroxisome proliferator-activated receptor alpha	−0.2660	5.24E−06	EPHA7	EPH receptor A7	0.4680	8.24E−07
				LRP8	low density lipoprotein receptor-related protein 8, apolipoprotein e receptor	0.4476	9.53E−06
				ADORA1	adenosine A1 receptor	0.4396	0.007408
				IQCB1	IQ motif containing B1	0.3704	5.36E−06
				EPHB2	EPH receptor B2	0.3587	9.93E−07

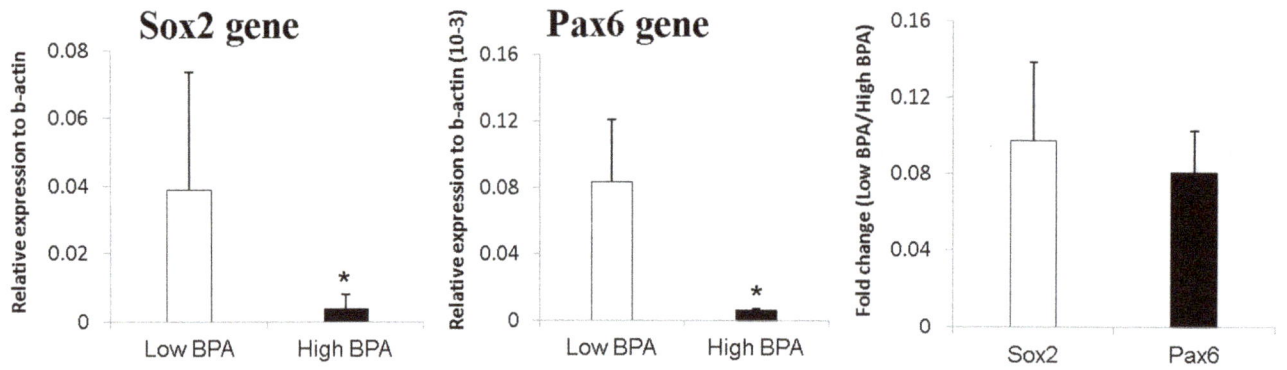

Figure 3. Gene expression and fold change of Sox2 and Pax6 between high and low BPA exposure groups in human umbilical cord blood samples. Both Sox2 and Pax6 were significantly down-regulated in high BPA exposure group. The expression of Sox2 and Pax6 were relative to house-keeping gene b-actin.

xenobiotics exposure on human is more effective than an animal model, and more convenient and time-saving than a cohort study for a long time follow-up.

Several but not many studies evaluated the impact of low-dose BPA exposure ($<$100 nM) on neuronal development. In human embryonic stem cells, BPA exposure (1, 10 or 100 nM) influences mammosphere area and down-regulate the expression of E-cadherin protein in the early differentiation stage of mammary epithelial cells [31]. In rattus experiment, maternal BPA exposure at 0.05, 0.5, 5 or 50 mg/kg per day from embryonic day 9 to day 20 in rats affects fetal growth (locomotor activity, exploratory habits and emotional behavior in open field test), synaptic structure (widened synaptic cleft, thinned postsynaptic density and unclear synaptic surface), and decreases mRNA and protein expressions of synaptophysin, PSD-95 (postsynaptic density protein 95), spinophilin, GluR1 (glutamate receptor 1) and NMDAR1 (N-methyl-D-aspartate receptor 1) in the hippocampus of male offspring on postnatal day 21 [32]. In neuronal or neuronal-like cells, neuronal differentiation is decreased in pheochromocytoma PC12 cells in pre-exposure to BPA (0.1, 1,

10 or 100 nM) for one week, longer or a week followed by a week's withdrawal [33]. PC12 cells treated with 43.8 nM BPA for 5 days suppress neurite extension through inhibition of phosphorylation of mitogen-activated protein kinase [19]. Rats exposed to 10 or 100 nM BPA for 7 days increases both MAP2 (microtubule associated protein-2) and synapsin I-positive areas for neuronal development as well as synaptic densities in hypothalamic neurons and glias [18]. In mouse purified astrocyte and neuron/glia co-cultures, exposure to low-dose BPA (100 fM, 1 pM, 10 pM, 100 pM, 1 nM, 10 nM or 100 nM) activates astrocytes, increases GFAP (glial fibrillary acidic protein) level and enhances Ca2+ responses to dopamine, which may contribute to potentiate the development of psychological dependence on supersensitivity to psychostimulant-induced pharmacological actions such as drugs of abuse [34]. Additionally, here are some BPA level studies in other countries for references. Serum BPA level of pregnant women (n = 61), fetuses (n = 61) and non-pregnant women (n = 26) in Eastern Townships of Canada ranged from non-detectable (ND) to 19.6 nM, ND to 20.2 nM, and 5.7 to 35.9 nM, respectively [35]. The cord blood BPA level of 106 boys in France was from 0.6 to 20.9 nM [36]. Also, BPA level in maternal blood and cord blood (n = 300) in Korea was from ND to 292.5 nM and ND to 38.9 nM, respectively [37]. These data supported that the measurement of BPA level in this study is similar with the range in other countries.

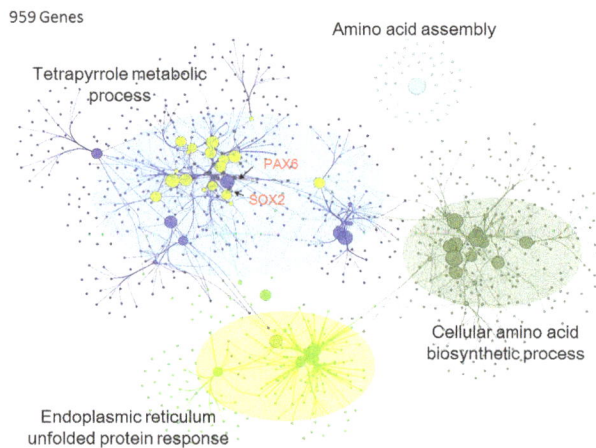

Figure 4. Gene regulatory network and gene ontology in exposure to BPA. Totally 959 genes connecting with the 457 DEGs relevant to BPA exposure constructed this gene network. The gene ontology enrichment of the 959 genes mapped for GO category presented in tetrapyrrole metabolic process (Sox2 and Pax6 were involved in), amino acid assembly, endoplasmic reticulum unfolded protein response, and cellular amino acid biosynthetic process.

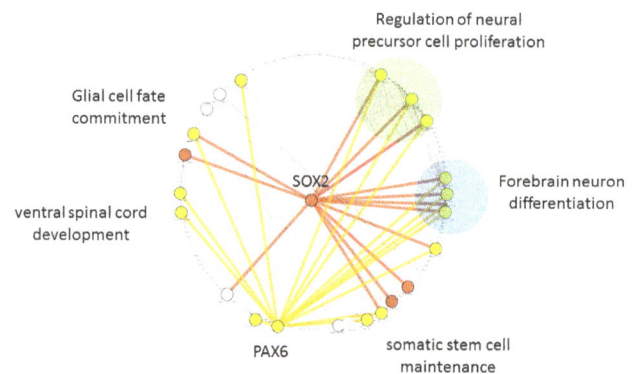

Figure 5. Sub-network illustration specific to Sox2 and Pax6 as hub genes. The visualized sub-network illustrated in the specific neuronal functions for Sox2 and Pax6 genes. Sox2 and Pax6 had similar neuronal functions such as regulation of neural precursor cell proliferation and forebrain neuron differentiation.

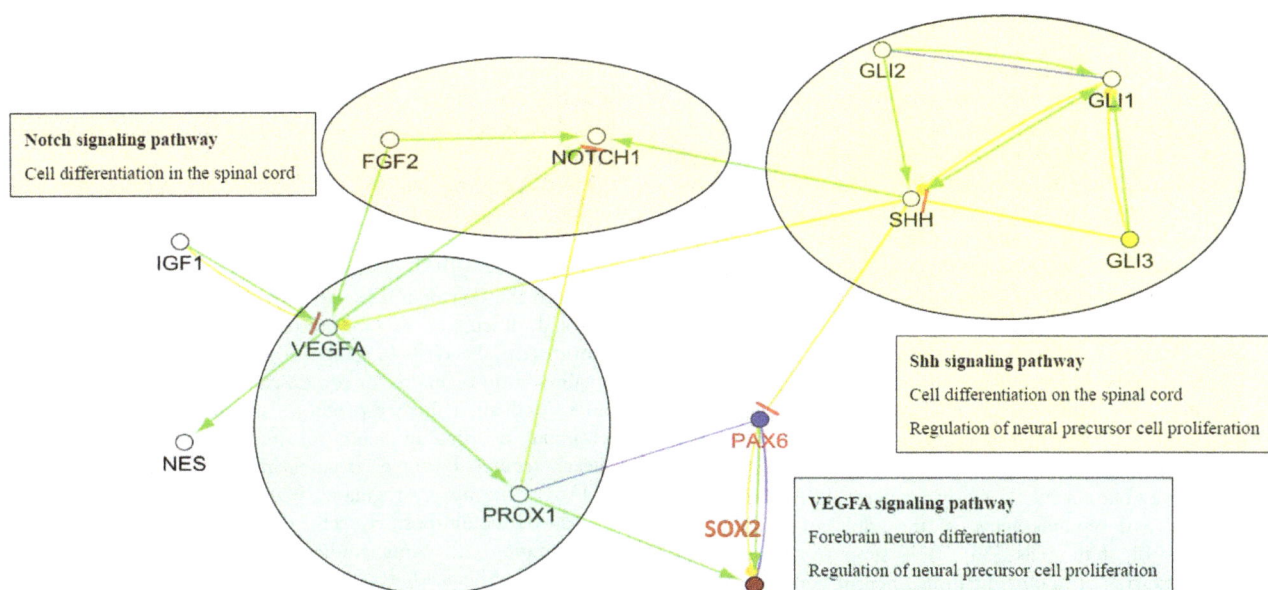

Figure 6. A predicted pathway of Sox2 and Pax6 and their potential interaction genes resulted from BPA exposure. Sox2 and Pax6 acted in Shh, Notch and VEGFA pathway for cell differentiation of spinal cord, forebrain neuron differentiation, and regulation of neural precursor cell proliferation. In Shh signaling, Shh is positively regulated by transcription factor encoded genes Gli1 and Gli2 and negatively modulated by Gli3. In VEGFA pathway, VEGFA generally activates Prox1 that up-regulated Sox2 and Pax6, and suppresses Notch1. Notch signaling indirectly modulates Sox2 and Pax6 through Shh, VEGFA and Prox1. In exposure to BPA, Shh down-regulated Sox2 through Pax6 antagonism, and IGF1 attenuated VEGFA expression and its down-stream genes. Shh: Sonic hedgehog; VEGFA: vascular endothelial growth factor A; IGF1: insulin-like growth factor 1; Green line: activation; blue line: binding; yellow line: expression; yellow line with a red bar: antagonizing.

Prenatal BPA exposure impairs murine fetal neocortical development by accelerating neuronal differentiation/migration during the early embryonic stage to affect neuronal plasticity and interferes corticosterone and corticosteroid receptors in the hippocampus [35–37]. According to mother report or behavior assessment system for children 2 (BASC-2) and behavior rating inventory of executive function-preschool (BRIEF-P), prenatal BPA exposure is significantly concerned in childhood behavioral problems such as anxiety, depression, inattention and hyperactivity in children less than age 7 [38,39]. However, these findings didn't further explore momentous biomarkers of neurogenesis relevant to trans-placental BPA exposure. The network analysis of this study presented that Sox2 and Pax6 were mainly involved in the regulation of neural precursor cell proliferation and forebrain neuron differentiation. A study using human embryonic stem cell-derived NPC for neurogenesis modeling found that Sox2 is required to maintain optimal levels of lin-28 homolog A (LIN28), a well-characterized suppressor of lethal-7 (let-7) microRNA biogenesis, and Sox2 loss causes proliferation deficit and differentiation inhibition in NPCs because of the abnormal expression of LIN28 and let-7 [40]. Sox2 is highly expressed in undifferentiated cells and declines with differentiation, and knockdown of Sox2 expression makes beta-tubulin-positive cells fail to progress to more mature neurons [41]. Aberrant expression of Sox2 is also related to childhood neuronal development. A study recruited 10 patients aged from 2 to 9 years screening for gene mutation found that Sox2 mutation is genetically correlated with inherited ocular phenotypes, anophthalmia and microphthalmia [42]. A review report presented that mutation of several transcription factors, including Sox2 gene, results in congenital hypopituitarism or septo-optic dysplasia in murine neonates [43]. Additionally, Pax6 knockdown reduces neurogenic capacity in embryonic stem cells [44]. Constitutive activation of Wnt signaling pathway in forebrain or brainstem precursor cells causes dramatic brain enlargement as well as in the formation of medulloblastoma, a malignant brain tumor in children [45]. Activated Wnt signaling leads to a virtually loss of Pax6 expression, and causes disruption in the proliferation and migration of neurons in mice [46]. Pax6 mutation affects downstream genes on mice cerebellar development for disturbed survival and migration and defects of neurite extension in producing granule cells [47]. Two case reports individually mentioned that mutated Pax6 results in reduced vision, photophobia and eyelid ptosis in an autistic child patient, and is responsible for impaired auditory sensory and higher order interhemispheric transfer in a 12 year old child [48,49]. Above studies indicated that Sox2 and Pax6 are relevant to childhood neurogenesis. In our study, we identified that BPA exposure affects Sox2 and Pax6 prenatally, and this might cause adverse effects on aberrant neuronal or behavioral development for children along with their increasing age.

This study found that BPA exposure decreased the gene expression of Sox2 and Pax6 in fetal cord blood, and Sox2 and Pax6 were involved in neuronal development according to network analysis. Findings in this study were consistent with a Xenopus laevis model that prenatal BPA exposure decreases the expression of Sox2 and Pax6 and disrupts Notch signaling to inhibit gamma-secretase activity for neurodegenerative abnormalities [50]. BPA also decreases the expression of Pax6, but not Sox2 to cause malformation of the head region in embryos through estrogen receptor 1 (ESR-1) and Notch signaling in Xenopus laevis [51]. Indirect evidence showed that decreased expressions of Sox2 and Pax6 cause adverse effects on neuronal development. Sox2 and Pax6 are lost and undergo glial differentiation after 5-bromo-2'-deoxyuridine (BrdU) exposure [52]. Conditionally, deleted Sox2 and Pax6 cause proliferative defects, alter morphology and reduce clonogenicity in forebrain-derived neural stem cells [28]. The results of fetal cord blood and network analysis in this study also supported that BPA exposure down-regulated Sox2 and Pax6 expression.

The visualized network analysis in response to BPA exposure was addressed in Figure 6 that illustrated that Shh, Notch and VEGFA pathways were in regulation to Sox2 and Pax6 for neuronal signaling including cell differentiation of spinal cord, forebrain neuron differentiation, and regulation of neural precursor cell proliferation. In exposure to BPA, Sox2 and Pax6 were down-regulated by Shh signaling and IGF1 attenuating VEGFA. Pax6 regulates the proliferation of neural progenitor cells in cortical subventricular zone through direct modulation of the Sox2 expression during the late developmental stage in mice [53]. Meis homeobox 2 (Meis2) activates Sox2 through the up-regulation of Pax6 for lens epithelial cell differentiation [54]. In human autopsy samples, the expressions of Sox2 and Pax6 are both higher in the radial glial cells and intermediate progenitors in the third trimester preterm birth, an important period for neurogenesis [55].

Shh pathway plays a role in many processes during embryonic development and remains active in the adult involving in the maintenance of stem cells [56]. Shh supports survival and stimulates growth of motor neurons, neurite outgrowth and neurosphere formation in murine primary embryonic spinal cord cell culture [57]. Shh pathway containing three cubitus interruptus (Ci) homologues, Gli1, Gli2 and Gli3, mediates Hedgehog-dependent cell fate specification in the developing spinal cord. The majority of slow-cycling neural stem cells (NSCs) express Gli2 and Gli3, whereas Gli1 is restricted ventrally, and all three genes are down-regulated when NSCs transition into proliferating progenitors [58]. Multiple ventral neuronal types can develop in the absence of Gli function in mice and chick embryos developing spinal cord models, but require balanced Gli protein activities for their correct patterning and differentiation [59]. Additionally, Shh/Gli1signaling is up-regulated in both mRNA and protein levels of the malignant glioma cells in U87-implanted nude mice, and down-regulated after curcumin treatment [60]. Shh signaling is activated accompanying with higher expression of miR-183~96~182 for medulloblastomas development in mice [61]. Taken above studies together, it suggested that BPA exposure might activate Shh to down-regulate Sox2 and Pax6 potentially for glioma and medulloblastoma brain cancers. However, prenatal exposure to BPA in mice produced significant decrease in the dopaminergic neuron development factors, Shh protein and glial cell line-derived neurotrophic factor [62], which is controversial to the prediction in this study after BPA exposure.

In general, VEGFA positively regulates Sox2 through Prox1 gene in embryonic eye development and lens differentiation [63]. VEGFA is required for normal forebrain development and downstream consequences for NSC fate decisions [64]. In this study, we proposed the induced expression of IGF1 to attenuate VEGFA in response to BPA exposure that led to down-regulated Sox2 and Pax6. IGF1 has been identified as a crucial factor in the central nervous system and is involved in cognitive functions, brain aging and development [65]. While circulating blood IGF1 and IGF2 levels exert trophic effects on neurogenesis and neuronal survival, CNS-derived or intrathecally derived IGF1 is also important in maintaining normal brain function [66]. IGF1 has been found to inhibit axon regeneration in aging C. elegans motor neurons [67]. The DNA methylation pattern of IGF1 and the ability of NSC differentiation are inhibited from alcohol consumption in human NSCs that means over-expression of IGF1 is harmful to neuronal development [68]. In an oxygen-induced retinopathy mice model, knockout of insulin or IGF1 receptor was associated with blunted elevation of VEGF, endothelial nitric oxide synthase and endothelin-1. According to these findings, we inferred that BPA exposure up-regulates IGF1 to antagonize VEGFA and down-regulates Sox2 through Prox1 gene that would be baneful to normal neuronal development and differentiation.

Notch pathway has been recognized as one of the main contributors in regulating neural development and has been proposed as a key mediator in neuroplasticity [69]. Notch intervenes with gp130 pathway at many stages to determine cell fate from the first neural lineage commitment and generation of neuronal precursors for the terminal specification of cells as neurons and glia [70]. In temporal lobe epilepsy patients, Notch signaling is up-regulated in response to epileptic seizure activity, and its activation further promotes neuronal excitation of CA1 pyramidal neurons in acute seizures [71]. In the developing chicken spinal cord, a disintegrin and metalloprotease 10 (ADAM10) negatively regulated Notch1 to increase the number of beta-III-tubulin-positive cells during neural progenitor cell differentiation [72]. Some studies started to focus on the pathway interactions. Shh mediated up-regulation of Notch1 is attenuated after cyclopamine treatment in both bovine retinal endothelial cells (BRECs) and pericytes (BRPs). As to anti-angiogenesis in cancer therapy, Notch signaling functions in the physiologic response to loss of VEGF signaling, and thus participates in tumor adaptation to VEGF inhibitors in human NGP neuroblastoma cells [73]. According to these findings, we addressed two explanations for Notch signaling on Sox2 and Pax6 expression in exposure to BPA. Notch1 down-regulates Sox2 and Pax6 in the process of Shh or VEGFA activation. Additionally, Notch1 was inhibited by VEGFA-activated Prox1 playing a feedback role for Sox2 and Pax6 expression. Based on above evidence, this study suggested BPA exposure would suppress neuronal transcripts Sox2 and Pax6 expression in fetal cord blood to cause neuronal development defect underlying the complicated regulations of Shh, VEGFA and Notch signaling pathways.

Conclusions

This study presented an alternative way to investigate the adverse effect of trans-placental xenobiotics exposure on child development in meta-analysis of public microarray datasets for exploring target genes and their potential gene network of neuronal functions. In the network-analysis of BPA exposure, we identified Sox2 and Pax6 as major neuronal transcript candidates and were involved in tetrapyrrole metabolic process. Sub-network analysis identified that Sox2 and Pax6 were highly involved in regulation of neuronal precursor cells and maintenance of stem cells, which supported the importance of Sox2 and Pax6 in neurogenesis. Furthermore, the decreased expression of Sox2 and Pax6 genes were determined in higher BPA level group of fetal umbilical cord blood. The findings suggested that trans-placental BPA exposure might down-regulate Sox2 and Pax6 expression and cause adverse effect on childhood neuronal development. It is necessary to pay attention in maternal BPA exposure during the period of pregnancy and its effect on child development.

Supporting Information

Figure S1   Process of matching probes among different microarray platforms. Microarrays from Affymetrix and Agilent platforms were originally normalized and summarized using RMA and 75th percentile normalization methods, respectively. Probes were combined to take the average measurements for genes with more than two probes. After the combination of genes from both platforms, correlation and median rank score (MRS) were used to select appropriate number of genes for further investigation.

Table S1   457 DEGs in response to BPA exposure and their corresponding ontology clusters.

Information S1   Description text of microarray data processing and differentially expressed analysis.

Author Contributions

Conceived and designed the experiments: CYC. Performed the experiments: KHC CWY. Analyzed the data: WCC ALC. Contributed reagents/materials/analysis tools: IFM HRC. Wrote the paper: CWY.

References

1. Lee BE, Park H, Hong YC, Ha M, Kim Y, et al. (2013) Prenatal bisphenol A and birth outcomes: MOCEH (Mothers and Children's Environmental Health) study. Int J Hyg Environ Health.
2. Matthews J, Celius T, Halgren R, Zacharewski T (2000) Differential estrogen receptor binding of estrogenic substances: a species comparison. J Steroid Biochem Mol Biol 74: 223–234.
3. Wang T, Lu J, Xu M, Xu Y, Li M, et al. (2013) Urinary bisphenol a concentration and thyroid function in Chinese adults. Epidemiology 24: 295–302.
4. Hong EJ, Choi KC, Jeung EB (2005) Maternal exposure to bisphenol a during late pregnancy resulted in an increase of Calbindin-D9k mRNA and protein in maternal and postnatal rat uteri. J Reprod Dev 51: 499–508.
5. Rankouhi TR, Sanderson JT, van Holsteijn I, van Leeuwen C, Vethaak AD, et al. (2004) Effects of natural and synthetic estrogens and various environmental contaminants on vitellogenesis in fish primary hepatocytes: comparison of bream (Abramis brama) and carp (Cyprinus carpio). Toxicol Sci 81: 90–102.
6. Angle BM, Do RP, Ponzi D, Stahlhut RW, Drury BE, et al. (2013) Metabolic disruption in male mice due to fetal exposure to low but not high doses of bisphenol a (BPA): Evidence for effects on body weight, food intake, adipocytes, leptin, adiponectin, insulin and glucose regulation. Reprod Toxicol.
7. Fujimoto T, Kubo K, Nishikawa Y, Aou S (2013) Postnatal exposure to low-dose bisphenol A influences various emotional conditions. J Toxicol Sci 38: 539–546.
8. Heffernan AL, Aylward LL, Toms LM, Eaglesham G, Hobson P, et al. (2013) Age-related trends in urinary excretion of bisphenol a in Australian children and adults: evidence from a pooled sample study using samples of convenience. J Toxicol Environ Health A 76: 1039–1055.
9. Khalil N, Ebert JR, Wang L, Belcher S, Lee M, et al. (2013) Bisphenol A and cardiometabolic risk factors in obese children. Sci Total Environ 470–471C: 726–732.
10. Nicolucci C, Rossi S, Menale C, Del Giudice EM, Perrone L, et al. (2013) A high selective and sensitive liquid chromatography-tandem mass spectrometry method for quantization of BPA urinary levels in children. Anal Bioanal Chem 405: 9139–9148.
11. Chou WC, Chen JL, Lin CF, Chen YC, Shih FC, et al. (2011) Biomonitoring of bisphenol A concentrations in maternal and umbilical cord blood in regard to birth outcomes and adipokine expression: a birth cohort study in Taiwan. Environ Health 10: 94.
12. Kundakovic M, Gudsnuk K, Franks B, Madrid J, Miller RL, et al. (2013) Sex-specific epigenetic disruption and behavioral changes following low-dose in utero bisphenol A exposure. Proc Natl Acad Sci U S A 110: 9956–9961.
13. Jedeon K, De la Dure-Molla M, Brookes SJ, Loiodice S, Marciano C, et al. (2013) Enamel defects reflect perinatal exposure to bisphenol A. Am J Pathol 183: 108–118.
14. Harley KG, Gunier RB, Kogut K, Johnson C, Bradman A, et al. (2013) Prenatal and early childhood bisphenol A concentrations and behavior in school-aged children. Environ Res.
15. Hong SB, Hong YC, Kim JW, Park EJ, Shin MS, et al. (2013) Bisphenol A in relation to behavior and learning of school-age children. J Child Psychol Psychiatry 54: 890–899.
16. Matsuda S, Matsuzawa D, Ishii D, Tomizawa H, Sajiki J, et al. (2013) Perinatal exposure to bisphenol A enhances contextual fear memory and affects the serotoninergic system in juvenile female mice. Horm Behav 63: 709–716.
17. Yeo M, Berglund K, Hanna M, Guo JU, Kittur J, et al. (2013) Bisphenol A delays the perinatal chloride shift in cortical neurons by epigenetic effects on the Kcc2 promoter. Proc Natl Acad Sci U S A 110: 4315–4320.
18. Yokosuka M, Ohtani-Kaneko R, Yamashita K, Muraoka D, Kuroda Y, et al. (2008) Estrogen and environmental estrogenic chemicals exert developmental effects on rat hypothalamic neurons and glias. Toxicol In Vitro 22: 1–9.
19. Seki S, Aoki M, Hosokawa T, Saito T, Masuma R, et al. (2011) Bisphenol-A suppresses neurite extension due to inhibition of phosphorylation of mitogen-activated protein kinase in PC12 cells. Chem Biol Interact 194: 23–30.
20. Zhou R, Bai Y, Yang R, Zhu Y, Chi X, et al. (2011) Abnormal synaptic plasticity in basolateral amygdala may account for hyperactivity and attention-deficit in male rat exposed perinatally to low-dose bisphenol-A. Neuropharmacology 60: 789–798.
21. Tainaka H, Takahashi H, Umezawa M, Tanaka H, Nishimune Y, et al. (2012) Evaluation of the testicular toxicity of prenatal exposure to bisphenol A based on

microarray analysis combined with MeSH annotation. J Toxicol Sci 37: 539–548.
22. Horstman KA, Naciff JM, Overmann GJ, Foertsch LM, Richardson BD, et al. (2012) Effects of transplacental 17-alpha-ethynyl estradiol or bisphenol A on the developmental profile of steroidogenic acute regulatory protein in the rat testis. Birth Defects Res B Dev Reprod Toxicol 95: 318–325.
23. Hwang KA, Park SH, Yi BR, Choi KC (2011) Gene alterations of ovarian cancer cells expressing estrogen receptors by estrogen and bisphenol a using microarray analysis. Lab Anim Res 27: 99–107.
24. Rajatileka S, Luyt K, El-Bokle M, Williams M, Kemp H, et al. (2013) Isolation of human genomic DNA for genetic analysis from premature neonates: a comparison between newborn dried blood spots, whole blood and umbilical cord tissue. BMC Genet 14: 105.
25. Rudy J, Valafar F (2011) Empirical comparison of cross-platform normalization methods for gene expression data. BMC Bioinformatics 12: 467.
26. Bindea G, Mlecnik B, Hackl H, Charoentong P, Tosolini M, et al. (2009) ClueGO: a Cytoscape plug-in to decipher functionally grouped gene ontology and pathway annotation networks. Bioinformatics 25: 1091–1093.
27. Bindea G, Galon J, Mlecnik B (2013) CluePedia Cytoscape plugin: pathway insights using integrated experimental and in silico data. Bioinformatics 29: 661–663.
28. Gomez-Lopez S, Wiskow O, Favaro R, Nicolis SK, Price DJ, et al. (2011) Sox2 and Pax6 maintain the proliferative and developmental potential of gliogenic neural stem cells In vitro. Glia 59: 1588–1599.
29. Snijder CA, Heederik D, Pierik FH, Hofman A, Jaddoe VW, et al. (2013) Fetal growth and prenatal exposure to bisphenol A: the generation R study. Environ Health Perspect 121: 393–398.
30. Suzuki A, Sugihara A, Uchida K, Sato T, Ohta Y, et al. (2002) Developmental effects of perinatal exposure to bisphenol-A and diethylstilbestrol on reproductive organs in female mice. Reprod Toxicol 16: 107–116.
31. Yang L, Luo L, Ji W, Gong C, Wu D, et al. (2013) Effect of low dose bisphenol A on the early differentiation of human embryonic stem cells into mammary epithelial cells. Toxicol Lett 218: 187–193.
32. Wang C, Niu R, Zhu Y, Han H, Luo G, et al. (2014) Changes in memory and synaptic plasticity induced in male rats after maternal exposure to Bisphenol A. Toxicology.
33. Nishimura Y, Nagao T, Fukushima N (2014) Long-term pre-exposure of pheochromocytoma PC12 cells to endocrine-disrupting chemicals influences neuronal differentiation. Neurosci Lett 570C: 1–4.
34. Miyatake M, Miyagawa K, Mizuo K, Narita M, Suzuki T (2006) Dynamic changes in dopaminergic neurotransmission induced by a low concentration of bisphenol-A in neurones and astrocytes. J Neuroendocrinol 18: 434–444.
35. Itoh K, Yaoi T, Fushiki S (2012) Bisphenol A, an endocrine-disrupting chemical, and brain development. Neuropathology 32: 447–457.
36. Masuo Y, Ishido M (2011) Neurotoxicity of endocrine disruptors: possible involvement in brain development and neurodegeneration. J Toxicol Environ Health B Crit Rev 14: 346–369.
37. Poimenova A, Markaki E, Rahiotis C, Kitraki E (2010) Corticosterone-regulated actions in the rat brain are affected by perinatal exposure to low dose of bisphenol A. Neuroscience 167: 741–749.
38. Braun JM, Yolton K, Dietrich KN, Hornung R, Ye X, et al. (2009) Prenatal bisphenol A exposure and early childhood behavior. Environ Health Perspect 117: 1945–1952.
39. Braun JM, Kalkbrenner AE, Calafat AM, Yolton K, Ye X, et al. (2011) Impact of early-life bisphenol A exposure on behavior and executive function in children. Pediatrics 128: 873–882.
40. Cimadamore F, Amador-Arjona A, Chen C, Huang CT, Terskikh AV (2013) SOX2-LIN28/let-7 pathway regulates proliferation and neurogenesis in neural precursors. Proc Natl Acad Sci U S A 110: E3017–3026.
41. Cavallaro M, Mariani J, Lancini C, Latorre E, Caccia R, et al. (2008) Impaired generation of mature neurons by neural stem cells from hypomorphic Sox2 mutants. Development 135: 541–557.
42. Schneider A, Bardakjian T, Reis LM, Tyler RC, Semina EV (2009) Novel SOX2 mutations and genotype-phenotype correlation in anophthalmia and microphthalmia. Am J Med Genet A 149A: 2706–2715.
43. Alatzoglou KS, Dattani MT (2009) Genetic forms of hypopituitarism and their manifestation in the neonatal period. Early Hum Dev 85: 705–712.

44. Quinn JC, Molinek M, Nowakowski TJ, Mason JO, Price DJ (2010) Novel lines of Pax6-/- embryonic stem cells exhibit reduced neurogenic capacity without loss of viability. BMC Neurosci 11: 26.

45. Lorenz A, Deutschmann M, Ahlfeld J, Prix C, Koch A, et al. (2011) Severe alterations of cerebellar cortical development after constitutive activation of Wnt signaling in granule neuron precursors. Mol Cell Biol 31: 3326–3338.

46. Poschl J, Grammel D, Dorostkar MM, Kretzschmar HA, Schuller U (2013) Constitutive activation of beta-catenin in neural progenitors results in disrupted proliferation and migration of neurons within the central nervous system. Dev Biol 374: 319–332.

47. Ha TJ, Swanson DJ, Kirova R, Yeung J, Choi K, et al. (2012) Genome-wide microarray comparison reveals downstream genes of Pax6 in the developing mouse cerebellum. Eur J Neurosci 36: 2888–2898.

48. Maekawa M, Iwayama Y, Nakamura K, Sato M, Toyota T, et al. (2009) A novel missense mutation (Leu46Val) of PAX6 found in an autistic patient. Neurosci Lett 462: 267–271.

49. Bamiou DE, Campbell NG, Musiek FE, Taylor R, Chong WK, et al. (2007) Auditory and verbal working memory deficits in a child with congenital aniridia due to a PAX6 mutation. Int J Audiol 46: 196–202.

50. Baba K, Okada K, Kinoshita T, Imaoka S (2009) Bisphenol A disrupts Notch signaling by inhibiting gamma-secretase activity and causes eye dysplasia of Xenopus laevis. Toxicol Sci 108: 344–355.

51. Imaoka S, Mori T, Kinoshita T (2007) Bisphenol A causes malformation of the head region in embryos of Xenopus laevis and decreases the expression of the ESR-1 gene mediated by Notch signaling. Biol Pharm Bull 30: 371–374.

52. Schneider L, d'Adda di Fagagna F (2012) Neural stem cells exposed to BrdU lose their global DNA methylation and undergo astrocytic differentiation. Nucleic Acids Res 40: 5332–5342.

53. Wen J, Hu Q, Li M, Wang S, Zhang L, et al. (2008) Pax6 directly modulate Sox2 expression in the neural progenitor cells. Neuroreport 19: 413–417.

54. Conte I, Carrella S, Avellino R, Karali M, Marco-Ferreres R, et al. (2010) miR-204 is required for lens and retinal development via Meis2 targeting. Proc Natl Acad Sci U S A 107: 15491–15496.

55. Malik S, Vinukonda G, Vose LR, Diamond D, Bhimavarapu BB, et al. (2013) Neurogenesis continues in the third trimester of pregnancy and is suppressed by premature birth. J Neurosci 33: 411–423.

56. Ingham PW, Nakano Y, Seger C (2011) Mechanisms and functions of Hedgehog signalling across the metazoa. Nat Rev Genet 12: 393–406.

57. Ma X, Turnbull P, Peterson R, Turnbull J (2013) Trophic and proliferative effects of Shh on motor neurons in embryonic spinal cord culture from wildtype and G93A SOD1 mice. BMC Neurosci 14: 119.

58. Petrova R, Garcia AD, Joyner AL (2013) Titration of GLI3 repressor activity by Sonic hedgehog signaling is critical for maintaining multiple adult neural stem cell and astrocyte functions. J Neurosci 33: 17490–17505.

59. Lei Q, Zelman AK, Kuang E, Li S, Matise MP (2004) Transduction of graded Hedgehog signaling by a combination of Gli2 and Gli3 activator functions in the developing spinal cord. Development 131: 3593–3604.

60. Du WZ, Feng Y, Wang XF, Piao XY, Cui YQ, et al. (2013) Curcumin Suppresses Malignant Glioma Cells Growth and Induces Apoptosis by Inhibition of SHH/GLI1 Signaling Pathway in Vitro and Vivo. CNS Neurosci Ther 19: 926–936.

61. Zhang Z, Li S, Cheng SY (2013) The miR-183 approximately 96 approximately 182 cluster promotes tumorigenesis in a mouse model of medulloblastoma. J Biomed Res 27: 486–494.

62. Miyagawa K, Narita M, Niikura K, Akama H, Tsurukawa Y, et al. (2007) Changes in central dopaminergic systems with the expression of Shh or GDNF in mice perinatally exposed to bisphenol-A. Nihon Shinkei Seishin Yakurigaku Zasshi 27: 69–75.

63. Saint-Geniez M, Kurihara T, D'Amore PA (2009) Role of cell and matrix-bound VEGF isoforms in lens development. Invest Ophthalmol Vis Sci 50: 311–321.

64. Cain JT, Berosik MA, Snyder SD, Crawford NF, Nour SI, et al. (2014) Shifts in the vascular endothelial growth factor isoforms result in transcriptome changes correlated with early neural stem cell proliferation and differentiation in mouse forebrain. Dev Neurobiol 74: 63–81.

65. Suh HS, Zhao ML, Derico L, Choi N, Lee SC (2013) Insulin-like growth factor 1 and 2 (IGF1, IGF2) expression in human microglia: differential regulation by inflammatory mediators. J Neuroinflammation 10: 37.

66. Russo VC, Gluckman PD, Feldman EL, Werther GA (2005) The insulin-like growth factor system and its pleiotropic functions in brain. Endocr Rev 26: 916–943.

67. Byrne AB, Walradt T, Gardner KE, Hubbert A, Reinke V, et al. (2014) Insulin/IGF1 Signaling Inhibits Age-Dependent Axon Regeneration. Neuron.

68. Zhou FC, Balaraman Y, Teng M, Liu Y, Singh RP, et al. (2011) Alcohol alters DNA methylation patterns and inhibits neural stem cell differentiation. Alcohol Clin Exp Res 35: 735–746.

69. Bonini SA, Ferrari-Toninelli G, Maccarinelli G, Bettinsoli P, Montinaro M, et al. (2013) Cytoskeletal Protection: Acting on Notch to Prevent Neuronal Dysfunction. Neurodegener Dis.

70. Kotasova H, Prochazkova J, Pachernik J (2013) Interaction of Notch and gp130 Signaling in the Maintenance of Neural Stem and Progenitor Cells. Cell Mol Neurobiol.

71. Sha L, Wu X, Yao Y, Wen B, Feng J, et al. (2013) Notch Signaling Activation Promotes Seizure Activity in Temporal Lobe Epilepsy. Mol Neurobiol.

72. Yan X, Lin J, Talabattula VA, Mussmann C, Yang F, et al. (2014) ADAM10 Negatively Regulates Neuronal Differentiation during Spinal Cord Development. PLoS One 9: e84617.

73. Hernandez SL, Banerjee D, Garcia A, Kangsamaksin T, Cheng WY, et al. (2013) Notch and VEGF pathways play distinct but complementary roles in tumor angiogenesis. Vasc Cell 5: 17.

19

Perinatal Exposure to Perfluorooctane Sulfonate Affects Glucose Metabolism in Adult Offspring

Hin T. Wan, Yin G. Zhao, Pik Y. Leung, Chris K. C. Wong*

Partner State Key Laboratory of Environmental and Biological Analysis, Croucher Institute for Environmental Sciences, Department of Biology, Hong Kong Baptist University, Hong Kong, People's Republic of China

Abstract

Perfluoroalkyl acids (PFAAs) are globally present in the environment and are widely distributed in human populations and wildlife. The chemicals are ubiquitous in human body fluids and have a long serum elimination half-life. The notorious member of PFAAs, perfluorooctane sulfonate (PFOS) is prioritized as a global concerning chemical at the Stockholm Convention in 2009, due to its harmful effects in mammals and aquatic organisms. PFOS is known to affect lipid metabolism in adults and was found to be able to cross human placenta. However the effects of *in utero* exposure to the susceptibility of metabolic disorders in offspring have not yet been elucidated. In this study, pregnant CD-1 mice (F_0) were fed with 0, 0.3 or 3 mg PFOS/kg body weight/day in corn oil by oral gavage daily throughout gestational and lactation periods. We investigated the immediate effects of perinatal exposure to PFOS on glucose metabolism in both maternal and offspring after weaning (PND 21). To determine if the perinatal exposure predisposes the risk for metabolic disorder to the offspring, weaned animals without further PFOS exposure, were fed with either standard or high-fat diet until PND 63. Fasting glucose and insulin levels were measured while HOMA-IR index and glucose AUCs were reported. Our data illustrated the first time the effects of the environmental equivalent dose of PFOS exposure on the disturbance of glucose metabolism in F_1 pups and F_1 adults at PND 21 and 63, respectively. Although the biological effects of PFOS on the elevated levels of fasting serum glucose and insulin levels were observed in both pups and adults of F_1, the phenotypes of insulin resistance and glucose intolerance were only evident in the F_1 adults. The effects were exacerbated under HFD, highlighting the synergistic action at postnatal growth on the development of metabolic disorders.

Editor: Julie A. Chowen, Hosptial Infantil Universitario Niño Jesús, CIBEROBN, Spain

Funding: This work is supported by the General Research Fund (HKBU 261812), University Grants Committee (CKC Wong). The funders had no role in study design, data collection and analysis, decision to publish, or preparation of the manuscript.

Competing Interests: The authors have declared that no competing interests exist.

* E-mail: ckcwong@hkbu.edu.hk

Introduction

The incidence of metabolic diseases (i.e. obesity, diabetes and fatty liver) has become highly prevalent globally [1,2]. Although genetic, nutrition and environmental factors have all been associated with the development of these diseases [3,4], epidemiological and laboratory animal studies have suggested the link between pollutant exposure (i.e. dioxin, bisphenol A (BPA), pesticides, heavy metals) and the impairment of glucose homeostasis and insulin resistance [5]. The two well-known anthropogenic pollutants, dioxins and BPA have been classified as the new diabetogenic factors [6] and are believed to affect glucose homeostasis via their actions on estrogen receptors (ER-α, -β) and/or aryl hydrocarbon receptor (AhR) [3]. However the presumed mechanisms still cannot explain the wide range of metabolic perturbations reported in both epidemiological and experimental studies. In addition to ERs and AhR, the lipid sensing and regulatory receptors, peroxisome proliferator-activated receptors PPARs are known to play pivotal roles in the regulation of insulin signaling, glucose/lipid metabolism [7] and the management of metabolic homeostasis. Of special interest is the emerging global pollutants perfluoroalkyl acids (PFAAs), which have been suggested to act on PPARs to modulate energy homeostasis [3] that warrants particular attention. In fact PFAAs

have been prioritized in the European research project OBELIX in 2009 as one of the risk factor in the alternation of development programming for metabolic diseases in life.

The three notorious family members of PFAAs, perfluorooctane sulfonate (PFOS), perfluorooctanoic acid (PFOA), and perfluorohexane sulfonate (PFHxS) are globally present in the environment and are widely distributed in human populations and wildlife [8–11]. The two most abundant PFAAs (i.e. PFOS and PFOA) are measurable in human plasma, umbilical blood, breast milk and liver [12–15] and have been related to developmental toxicity, immunotoxicity and hepatotoxicity in animals [10]. Due to the potential adverse health effects of PFOS, its use in industrial production was phased out in most countries in 2002, except in China where it's still manufactured and widely used today. In 2009, PFOS was listed under Annex B of the Stockholm Convention as one of the nine new persistent organic pollutants (POPs). Chronic PFOS exposure has been reported to cause effects on animal hepatic functions [16,17]. The major pathological manifestations include reduced body weights and loss of body fat, accompanied with increases of liver masses [10,18–20]. Histological examination of liver cells revealed peroxisome proliferation and lipid accumulation. Studies of hepatic gene expression profiles in PFOA/PFOS-treated rats highlighted that the up-regulation of genes are related to the metabolism and transport of lipids [18,20–

23]. Accordingly it has been postulated that its mode of action is via its pleiotropic interactions with multiple members of the nuclear hormone receptors, PPARs, constitutive androstane receptor (CAR) and pregnane X receptor (PXR) [24].

Maternal transfer of PFOS across the human placenta has been reported while the data suggest the potential harmful effects of in utero exposure to PFOS on fetus [25]. However there is no toxicological information regarding the perinatal PFOS exposure to susceptibility of metabolic disorders in the offspring. Our previous study revealed that PFOS exposure induced hepatic steatosis and altered lipid metabolisms in adult mice, suggesting the exposure may lead to non-alcoholic fatty liver disease (NAFLD) [26], which is strongly associated with type II diabetes, obesity and insulin resistance [27,28]. In this study we investigated the effects of perinatal exposure to PFOS on glucose metabolism in the offspring and demonstrated if the effects would be exacerbated under high fat diet.

Results

Oral Gavage Exposure to PFOS Interferes with Maternal Glucose Metabolism

Maternal mice (F_0) were sacrificed after weaning on PND 21. PFOS concentrations in serum and liver were remarkably greater in the PFOS-exposed groups (Tables 1 and 2). The relative liver weights were significantly increased in the maternal mice of the 3 mg PFOS/kg dosed-group (Fig. 1B). The maternal body and absolute liver weights were shown (Fig. S1A–B). There is an increasing trend in both the fasting serum glucose and insulin levels towards the high-dose treated group but the data are not statistically significant (Fig/1C–D). The HOMA-IR index was calculated which measures the one's tendency to develop insulin resistance [29]. The index was found to be significantly greater in both PFOS-treated groups as compared to the control (*p<0.02) (Table 3).

Gestation and Lactational Exposure to PFOS Affects Glucose Metabolism and Hepatic Gene Expression of the Pups at Postnatal Day 21

Perinatal exposures to PFOS led to a significant accumulation of the chemical in both serum and liver of the F_1 pups (Tables 4 and 5). The concentrations of PFOS in the liver and serum of pups are proportion to the dose of the exposure. A statistical difference between genders of pups was observed in serum PFOS contents while the level was significantly greater in the males. Although the gestation and lactational exposure did not cause noticeable effects on the body weight of the pups (Fig. 2A), the relative liver weights of perinatal PFOS-exposed pups of both sexes were significantly increased (p<0.05) (Fig. 2B). The absolute liver weights were shown in Fig. S1C. A modulation of hepatic gene expression was detected (Fig. 2C–F). The transcript levels of cytochrome P450 enzymes 4A14 (Cyp4a14), lipoprotein lipase (Lpl) and fatty acid translocase (Cd36) were induced in both male and female pups via perinatal exposure to 3 mg/kg PFOS (p<0.001 & 0.03 respectively) (Fig. 2C–D). The gene levels of hepatic membrane receptors were also altered. The transcript levels of insulin receptor (Ir) were up-regulated (p<0.001) while prolactin receptor (Prlr) were significantly down-regulated (p<0.03) in both sexes (Fig. 2E–F). The expression levels of the hepatic insulin-like growth factor (Igf-1) were significantly down-regulated while its receptor (Igf-1r) was up-regulated.

To determine if the perinatal PFOS exposure disturbed glucose metabolism of the pups, serum levels of insulin and glucose were measured (Fig. 3A–B). The serum insulin levels of the perinatal PFOS-exposed male pups were significantly higher than the control, however the effects weren't observed in the female pups. No noticeable difference in serum glucose levels between the control and perinatal PFOS-exposed groups were detected. No significant difference in the HOMA-IR index among the groups of F_1 pups was found (Table 6).

The Effects of Perinatal PFOS Exposure on Glucose Metabolism of STD- and HFD-fed F_1 Adults at PND 63

From PND 21 to PND 63, all weaned F_1 mice were kept without further PFOS exposure but were fed with either standard (STD) or high-fat diet (HFD). The consumption of standard diet and high fat diet did not show significant effects on their body weight (Fig. S2). They were sacrificed on PND 63 and the body levels of PFOS were measured (Tables 7 and 8). As compared to the F_1 pup's data at PND 21, notable reduction of PFOS levels from serum and liver in both STD- and HFD-fed F_1 adults were observed. Interestingly HFD-fed F_1 adults accumulated significantly greater levels of serum and liver PFOS than the STD-fed F_1 adults (P<0.05), indicating that the consumption of HFD led to a slower elimination rate of PFOS.

In the STD-fed group, the relative liver weights of male pups from high-dosed maternal group were significantly higher than the control pups (Fig. 4A). The absolute liver weights of the pups were also shown in Fig. S1D. The serum fasting glucose and insulin levels of pups of both sexes from PFOS-maternal groups were significantly higher than pups from the control-maternal group (Fig. 4B–C). OGTT was carried out to investigate the dynamic changes of glucose elimination in the STD-fed F_1 adults (Fig. 4D–E). No significant differences were observed as compared to the control F_1 adults.

In the HFD-fed group, a significant increase in the relative liver weights, fasting serum glucose and insulin levels of the male F_1 adults from high-dosed PFOS maternal groups was observed (Fig. 5A–C). The absolute liver weights were shown in figure S1E. Significant effects on fasting glucose and insulin levels were measured in the perinatal exposed female F_1 adults. OGTT experiments demonstrated that the blood glucose area under the curve (AUC) was significantly increased in the HFD-fed F_1 adults of both sexes from the high-dosed PFOS maternal group (*p<0.02) (Fig. 5D–E). Comparisons of the effects of the STD and HFD on perinatal PFOS-exposed F_1 adult offspring at PND63 were shown (Fig. S3). The relative liver weights and fasting blood glucose levels of the HFD-fed male adult offspring from the high-dose maternal group were significantly higher than that in the respective group in the STD. While the fasting blood insulin levels in the HFD-fed female adults (F_1) from the high-dosed exposed group were noticeably increased as compared to the respective group in the STD. Table 9 shows comparisons of HOMA-IR index among STD-/HFD-fed F_1 adults of both sexes. The index in HFD-fed F_1 adults was noticeably greater than the respective STD-fed pups. In the HFD-fed F_1 adults, the influence of perinatal PFOS exposure to HOMA-IR index was more prominent in the males than the females (P<0.001 for perinatal low-dose, P<0.05 for perinatal high-dose). In addition to the OGTT, insulin tolerance tests were performed. However no significant differences in glucose responses were observed among the control and the treatment groups (Fig. S4).

Discussion

The notorious member of PFAAs, perfluorooctane sulfonate (PFOS) is prioritized as a global concerning chemical. Although industrial production of PFOS in most countries has already been

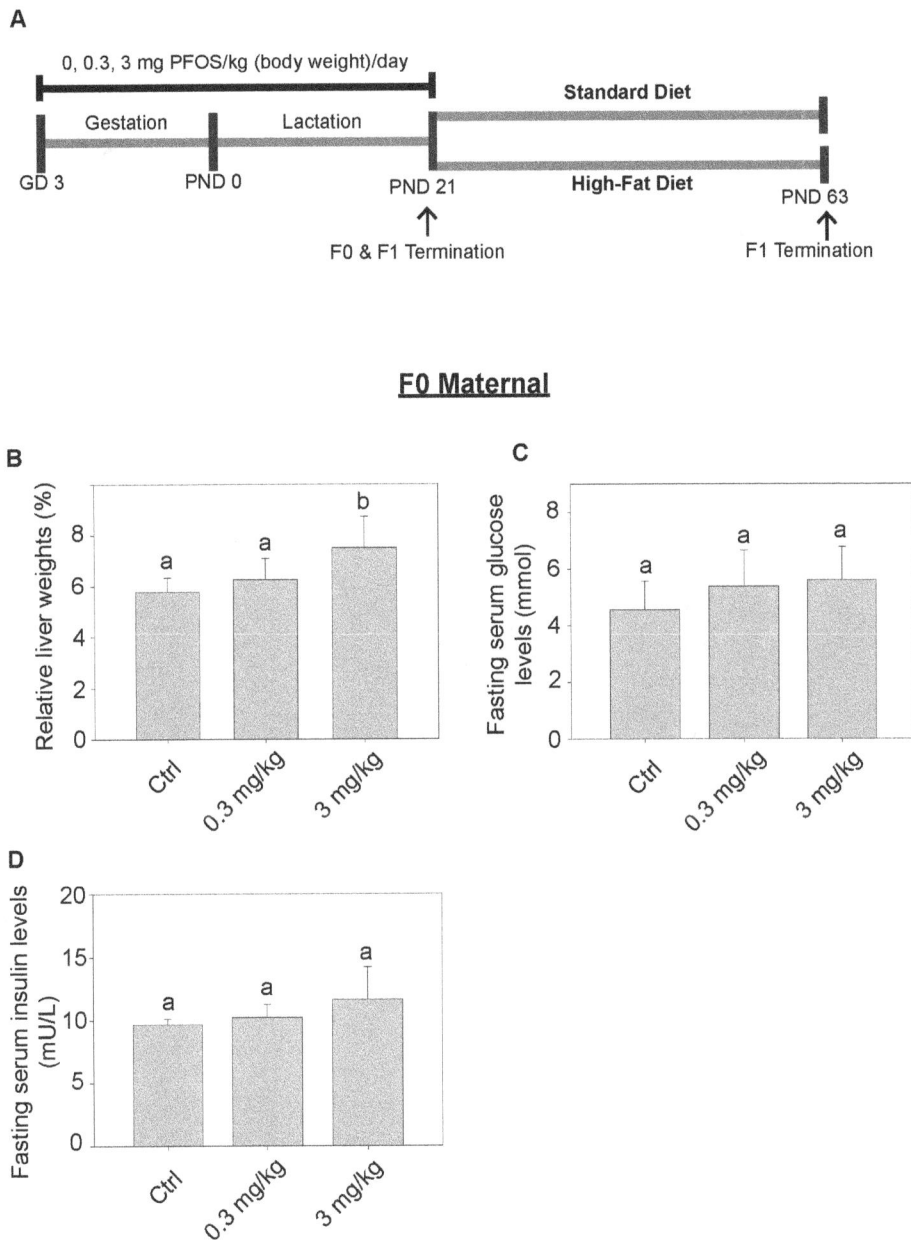

Figure 1. Experimental set-up and the effects of gestational and lactation PFOS exposure to maternal mice. (A) Pregnant CD-1 mice were administrated with corn oil as control, 0.3 or 3 mg PFOS/kg body weight daily by oral gavage from gestational day (GD) 3 to postnatal day (PND) 21. F_0 maternal and some F_1 pups were sacrificed on PND 21. The rest of the F_1 offspring were randomly separated into two groups, allowed freely access to either standard diet or high-fat diet until termination on PND 63. The relative liver weights (B), and fasting serum glucose (C) and insulin levels (D) of maternal mice were measured (n=6). The exposure to 3 mg PFOS/kg/day led to a significant increase in relative liver weight as compared to the control group. Bars with the same letter are not significantly different according to the results of one-way ANOVA followed by Tukey's test (p<0.05).

Table 1. The concentrations of PFOS (µg/ml) in the maternal (F_0) serum.

Maternal (F_0)	PFOS concentrations (µg/ml)
Ctrl	0.25±0.11
0.3 mg/kg	15.33±4.62
3 mg/kg	131.72±30.71

Table 2. The concentrations of PFOS (µg/g) in the maternal (F_0) livers.

Maternal (F_0)	PFOS concentrations (µg/g liver)
Ctrl	0.15±0.11
0.3 mg/kg	49.09±9.88
3 mg/kg	338.87±100.71

Table 3. Serum levels of fasting glucose-insulin and HOMA-IR index in the maternal mice.

Maternal (F$_0$)	Fasting glucose (mmol)	Fasting insulin (mU/L)	HOMA-IR index
Ctrl	4.55±1.03	9.63±0.46	1.90±0.50
0.3 mg/kg	6.08±1.28	10.23±1.01	3.05±0.97*
3 mg/kg	5.26±1.16	11.62±2.59	3.07±0.99*

*$p < 0.02$ as compared to the control.

F1 Postnatal day 21

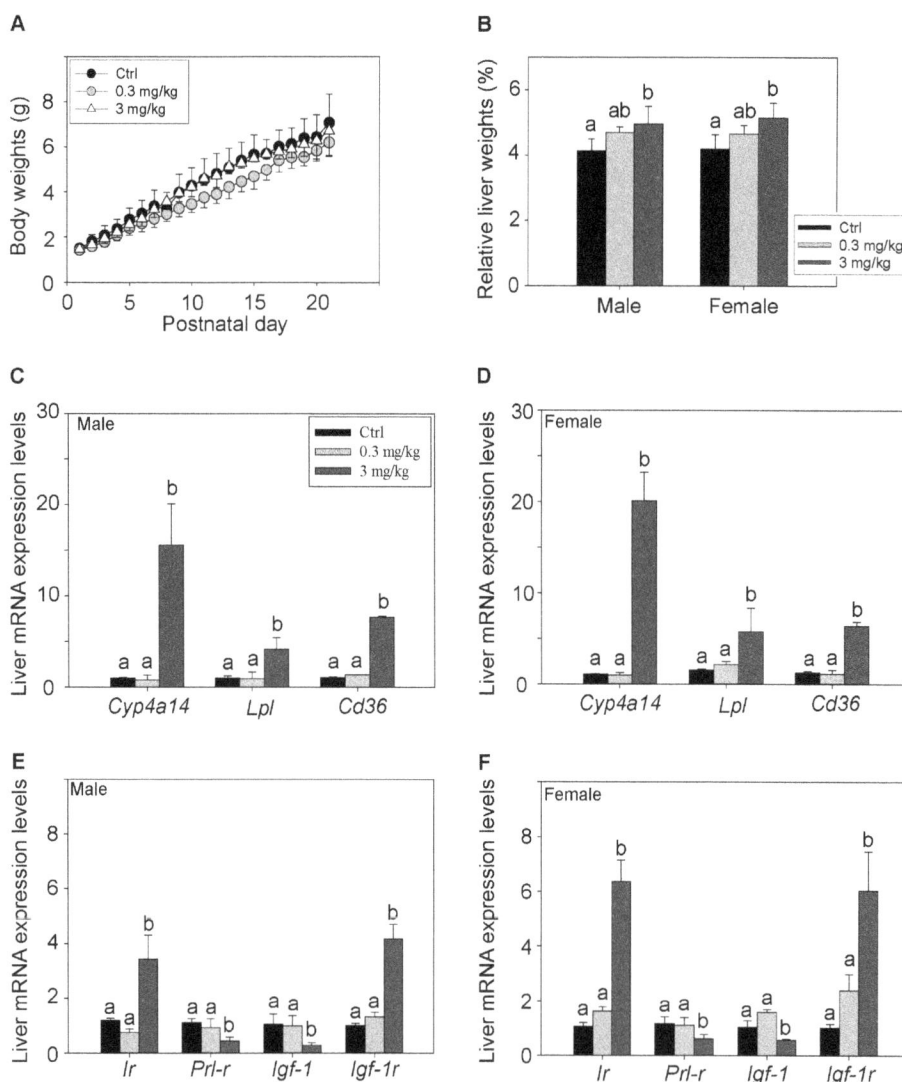

Figure 2. Effects of perinatal PFOS exposure on F1 pups at PND 21. (A) Body weights of the F$_1$ pups were measured daily from PND 1–21. No significant differences were observed between the control and the treatment groups (n = 6 per group). (B) Liver weights of the F$_1$ pups were determined at PND 21. Perinatal exposure to 3 mg PFOS/kg/day significantly increased the relative liver weights of both male and female pups as compared to the control group. Panels C–F, liver gene expression levels were determined on PND21. (C–D) Cyp4A14, lipoprotein lipase (*Lpl*) and fatty acid translocase (*Cd36*) gene expression levels were significantly up-regulated in dams of both sexes from the high-dose PFOS exposed maternal group as compared to the respective control groups. (E & F) The expression levels of the diabetic-related genes, hepatic insulin receptor (*Ir*) and insulin growth factor-1 receptor (*Igf-1r*) were significantly up-regulated in dams from the high-dose PFOS exposed groups while the expression levels of prolactin receptor (*Prl-r*) and insulin-like growth factor-1 (*Igf-1*) were decreased as compared to the respective control groups Bars with the same letter are not significantly different according to the results of one-way ANOVA followed by Tukey's test ($p < 0.05$).

Table 4. The concentrations of PFOS (μg/ml) in the F_1 pup serum.

F_1 pups (PND 21)	PFOS concentrations (μg/ml)	
	Male	Female
Ctrl	0	0
0.3 mg/kg	12.73$\pm$1.96	11.35$\pm$1.08
3 mg/kg	98.74$\pm$4.58* ($p<0.05$)	87.23$\pm$4.28

*$p<0.05$ as compared between gender of the same treatment group.

Table 5. The concentrations of PFOS (μg/g) in the F_1 pups livers.

F_1 pups (PND 21)	PFOS concentrations (μg/g liver)	
	Male	Female
Ctrl	0	0
0.3 mg/kg	20.14$\pm$4.06	17.96$\pm$6.38
3 mg/kg	242.98$\pm$55.62	178.44$\pm$79.03

ended since 2002, its contamination is still reported in our living environment and human blood samples, implying the exposure risk of our population to this compound. Since a number of reports have demonstrated PFOS-elicited metabolic syndromes in adult animals, we are interested in deciphering if perinatal PFOS exposure would increase susceptibility of metabolic diseases in the offspring. In this study, perinatal exposure to PFOS was performed to investigate the immediate and chronic effects of PFOS exposure on glucose metabolisms in offspring. In considering the uncertainty factors encountered in animal experiments, the oral gavage doses for maternal mice (F_0) were set at 0.3 to 3 mg/kg day, which is 10 to 100-fold higher than the human equivalent dose of exposure for general public but is comparable to the occupational exposure levels [30]. According to our data, the PFOS exposure dose to fetal and weaning pups (F_1), those indirectly receiving 6.2–10.7% of the maternal body load via placenta and lactation, would be equivalent to the exposure levels for general public. Our data illustrated the first time the effects of the environmental equivalent dose of PFOS exposure on the disturbance of glucose metabolisms in F_1 pups and adults at PND 21 and 63 respectively. Although the biological effects of PFOS on the elevated levels of fasting serum glucose and insulin levels were observed in both pups and adults of F_1, the phenotypes of insulin resistance and glucose intolerance were evident (i.e. HOMA-IR index and glucose AUC) in the F_1 adults. The effects were exacerbated under HFD, highlighting the synergistic action of chemical stressor and nutrition on the development of metabolic disorders.

Effects of PFOS Exposure on Glucose Metabolism in Maternal Mice

Similar to our previous findings, oral gavage exposure of maternal mice to PFOS caused significant increases in relative liver weights [26]. To relate the body compartment distribution of PFOS in liver and blood, an approximate 3:1 ratio was observed [10]. The remaining PFOS were probably partitioned in other body fluids (i.e. milk), tissues/organs, conceptus or excreted [31]. PFOS was known to bind with proteins [i.e. human serum albumin (HSA)] by interactions between the polar sulphonic groups to the hydrophilic residue of HSA to form a compact structure [32]. This structural interaction renders the relatively long human serum half-life (5.4 years) among other PFCs and chemical pollutants [8]. A previous toxicokinetic report demonstrated that the body elimination half-life of PFOS in rodents was about 1–2 months [33]. In contrast to the 5.4 years of the PFOS half-life in humans, it seems that the rodents displayed a relatively higher serum elimination rate.

Biochemical analysis of glucose metabolism illustrated that the chronic exposure had no noticeable effects on homeostatic regulation of blood glucose and insulin in the PFOS-exposed maternal mice as compared to the control. However the calculated

HOMA-IR index was found to be significantly greater in the high-dose maternal group. The HOMA-IR index is known to be a clinical parameter to evaluate hepatic insulin sensitivity and is useful to predict the risk of hypertension and type II diabetes in human diagnostics [34,35]. Therefore the observation indicated that PFOS exposure may be a potential chemical stressor to disrupt glucose metabolism during pregnancy.

Effects of Maternal Transfer and Lactational Exposure to PFOS on F_1 Pups at PND 21

It is known that PFOS can cross placenta and through lactation to impose developmental effects on fetus and neonates [25,31,36–40]. To investigate if the increase of liver weight might cause disturbance to liver functions as reported in the toxicological models in adult animals [22,23,26,41,42], the mRNA expression levels of hepatic genes targeting lipid metabolism were measured in the F_1 pups and were found to be modulated. Although the data showed perturbations of the gene expressions in related to lipid metabolism, further work are needed to elucidate if the changes at the transcript levels are linked with protein changes and so an alteration in liver function.

Recent studies on the perinatal exposure to BPA or phthalates (i.e. DEHP) were reported to illustrate negative impacts of the exposure on the regulation of blood glucose and insulin in rodents [43–45]. However data of metabolic effects of PFOS intoxication to fetal and neonates are largely not known. In this study, the measurement of fasting serum levels of glucose and insulin (indices of insulin resistance) indicated that the male pups from PFOS-exposed maternal groups showed significantly greater serum insulin levels than the control pups. A modulation of hepatic gene expression in glucose metabolism was observed in the perinatal PFOS-exposed pups. The transcript level of the hepatic insulin receptor (Ir) was found to be remarkably up-regulated while the prolactin receptor (Prlr) was down-regulated. The hepatic IGF-1/IGF-1r axis was also disrupted in the perinatal PFOS-exposed pups. Prolactin [46] and insulin are diabetic-related hormones. An alternation of their receptor gene expression levels suggested a modulation of their respective functions in glucose metabolism. The upregulation of Ir transcripts might reflect physiological changes to insulin responsiveness. Supposedly if the gene effects were recapitulated at the protein levels, the HOMA-IR would be lower. However our data showed that significant higher HOMA-IR indexes in the perinatal PFOS-exposed adult offspring were measured. It suggests that the Ir gene expression might not be linked to the protein level. Moreover PFOS-elicited effects might affect insulin receptor-signaling and so the increase in Ir expression couldn't rescue the effects of the perturbations. The pro-insulin factor, IGF-1 is known to confer insulin-like action to stimulate glucose uptake and its mal-regulation is recognized to associate with insulin resistance and diabetes [47]. The up-regulation of hepatic IGF-1 receptor in perinatal PFOS-exposed

F1 Postnatal day 21

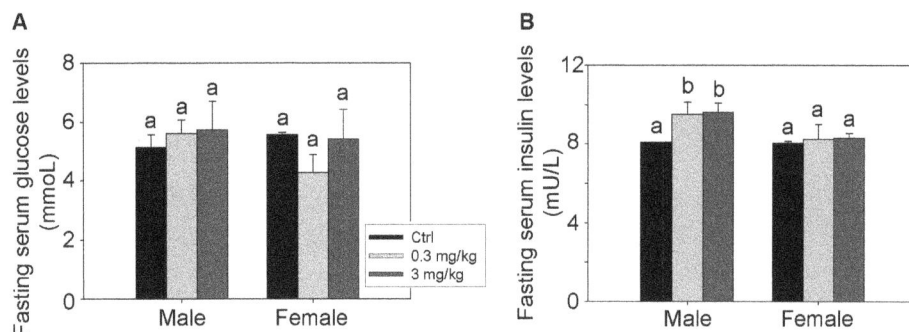

A

B

Figure 3. Effects of perinatal exposure to PFOS on fasting serum glucose and insulin levels of F_1 pups at PND 21. Blood glucose and insulin levels were measured after overnight fasting (n = 6 per group). (A) Glucose levels were comparable between the treatment and control groups. (B) The insulin levels of male pups were significantly greater than the control. Bars with the same letter are not significantly different according to the results of one-way ANOVA followed by Tukey's test (p<0.05).

pups may be due to the negative feedback from low level of hepatic IGF-1 biosynthesis. Patients with obesity, metabolic syndrome and diabetes showed significant lower plasma levels of IGF-1 [48]. The consistent metabolic effects demonstrated in both sexes of the perinatal PFOS-exposed pups rendered us to determine the long-term effects of the exposure on insulin resistance and glucose intolerance in the F_1 adult offspring.

Effects of Perinatal PFOS Exposure on Glucose Metabolism in F_1 Adult Offspring at PND 63

From PND 21 to 63, there were about 32–78% and 14–55% reductions in hepatic PFOS levels, respectively in the STD- and HFD-fed F_1 adults. The PFOS elimination rate was greater in the F_1 adults fed with the STD than the HFD. This observation indicates that dietary fat contents would modulate PFOS elimination from the animals. Although there is no study to indicate toxicokinetic difference in the elimination PFOS or related compounds, other study using a non-steroidal anti-inflammatory drug suggested that high dietary fat contents increased the half-life elimination of the drug in beagle's livers, owing to an alteration in the expression levels of efflux transporters [49]. The mechanism of HFD consumption in affecting PFOS toxicity is largely unknown, the possible effects on xenobiotic efflux transporters warrants further investigation. Male F_1 adults in general eliminated less and/or accumulated higher levels of PFOS in their livers. In the STD-fed F_1 adults, the ranges of hepatic PFOS accumulation in males (4.78–16.59 µg/liver) and females (3.48–14.91 µg/liver) while in the HFD-fed F_1 adults, the levels were ranging from (6.45–37.2 µg/liver) in the males and (4.15–28.75 µg/liver) females. The underlying mechanisms of the

gender-specific metabolism of PFOS is not known, however it is generally known that most drugs are cleared faster in females than males [50]. It may be probably due to sexual-dimorphic variations in the induction of hepatic cytochrome P450s, regulated by the pregnane X-receptor (PXR) and constitutive androstane receptor (CAR) in drug metabolism [51–53].

Both STD- and HFD-fed F_1 adults from PFOS-exposed maternal groups showed the elevated levels of fasting serum glucose and insulin as compared to the respective control groups. Consistently a significant greater in HOMA-IR index was observed in both STD- and HFD-fed PFOS-exposed F_1 adults. The data suggested the negative effects of the perinatal PFOS exposure to glucose metabolism in the F_1 adults. Moreover the significant greater values of HOMA-IR index in HFD-fed F_1 adults than the STD-fed groups underlined the synergistic effects of HFD on the development of insulin resistance in the animals. The synergistic effects were also exemplified in the oral glucose tolerance test (OGTT), while the glucose AUCs in HFD-fed F_1 adults were significantly greater than the STD-fed groups. Remarkably a human epidemiological study has demonstrated that serum levels of PFOS are associated with elevated serum insulin, HOMA-IR and altered β-cells functions [54]. Other than PFOS, in utero exposure to some toxicants such as arsenic, nicotine, organotins, phthalates, BPA, and pesticides were also shown to have positive association to the development of obesity, type II diabetes and insulin resistance [55]. Moreover nutrition is

Table 6. HOMA-IR index in the F_1 pups.

| F_1 pups (PND 21) | HOMA-IR index | |
	Male	Female
Ctrl	2.13±0.75	1.99±0.40
0.3 mg/kg	2.37±0.05	1.58±0.35
3 mg/kg	2.38±0.85	2.34±1.05

Table 7. The concentrations of PFOS (µg/ml) in the F_1 adult serum.

| F_1 adults (PND 63) | PFOS concentrations (µg/ml) | | | |
| | Male | | Female | |
	STD	HFD	STD	HFD
Ctrl	0	0	0	0
0.3 mg/kg	0.30±0.06	1.20±0.29*	0.51±0.11	1.50±0.27*
3 mg/kg	3.36±1.07	5.38±0.30*	3.40±1.08	5.76±1.24*

*p<0.05 as compared between STD and HFD diets under the same gender groups.

Table 8. The concentrations of PFOS (μg/g) in the F_1 adult livers.

| F_1 adults (PND 63) | PFOS concentrations (μg/g liver) | | | |
| | Male | | Female | |
	STD	HFD	STD	HFD
Ctrl	0	0	0	0
0.3 mg/kg	3.97±0.50	5.43±0.98*	3.34±0.50	4.27±1.75*
3 mg/kg	12.30±1.59	24.54±1.06*	13.77±4.05	21.34±3.36*

*$p<0.05$ as compared between STD and HFD diets under the same gender groups.

one of the non-chemical stressors which may contribute to disease progression in animals exposed to chemical toxicants [56]. The consumption of high-calorie, high-fat diet is one of the well-known risk factors to metabolic diseases [57]. Both nutritional and chemical stresses may exacerbate disease-associated biochemical factors in animals, such as induction of oxidative stress and inflammatory cytokines, to promote metabolic dysfunction in livers. Herein, we demonstrated that HFD can increase the susceptibility of perinatal PFOS-exposed adult offspring to metabolic disorders.

Increasing research studies have focused on the effects of perinatal exposure to determine if chemical pollutants predispose offspring to various kinds of metabolic disorders, underlining the risk of the exposure via maternal transfer. Our study indicated that perinatal exposure to PFOS caused disturbance to glucose metabolism in pups at PND 21. The development of insulin resistance and glucose intolerance was evident in adult offspring at PND 63. Further investigation of the effects on fetal liver and pancreas development would warrant a better understand of the underlying mechanistic actions of PFOS-induced metabolic disorders.

Materials and Methods

Experimental Animals and Chemicals

All experimental animals were housed and handled in accordance with the Guidelines and Regulations of Department of Health, the Government of Hong Kong Special Administrative Region. The protocol was approved by the Committee on the Use of Human and Animal Subjects of the Hong Kong Baptist University (Permit no. 261812). Female CD-1 mice (6–8 week old) were purchased from Laboratory Animal Service Centre of the Chinese University of Hong Kong (Hong Kong, China). The entire study was repeated for four times with mice that were received in separate batches. The animals were acclimatized for 1 week before experiments. Mice were allowed to mate for two consecutive nights and were randomly divided into three groups (about 6 individuals per group). Each group was housed in polypropylene cages with sterilized bedding and was maintained under controlled temperature (22°C) and 12L:12D cycles (0600–1800 h). The mice were weighted by an electronic balance (Shiamdzu, Tokyo, Japan) and orally administered 0.3 and 3 mg PFOS/kg body weight by gavage in corn oil in the afternoon from the last day of mating, then daily throughout gestation until the end of weaning period (PND 21) (Fig. 1A). Perfluorooctane sulfonate (98% purity, Sigma-Aldrich, US) was dissolved in dimethyl sulfoxide (DMSO, Sigma-Aldrich, US) before mixing

with corn oil (the final concentration of DMSO is less than 0.05% in all group). The control group was given corn oil with 0.05% DMSO. The F_0 maternal mice were fed with standard food (Rodent Diet 5001; LabDiet) and water *ad libitum*. After weaning, 2 pups per dam and all F_0 mice were sacrificed at postnatal day 21 for follow-up experiments. The rest of the pups were randomly divided into two groups, and were fed with either standard diet (STD) (Rodent Diet 5001; LabDiet) or high fat diet (HFD) (MP Biomedicals, US) and were sacrificed at PND63.

All the mice were fasted overnight (16 h) and killed by cervical dislocation in the morning on the designed dates. Blood sample was collected by cardiocenthesis, and serum was prepared by centrifugation at 3000×g for 15 min. The sera were stored at −20°C immediately until further analysis.

Liver and Blood Serum PFOS Analysis

A mass-labeled standard solution for PFOS (used as the internal standard) was purchased from Wellington Laboratories (Ontario, Canada). Purities of the analytical standard were greater than 98%. The method for the extraction and analysis of PFOS was performed as previously described [58]. Briefly, liver sample was homogenized in MilliQ water, while serum sample was diluted in 1 ml of MilliQ water (n = 4 for each group per experiment). One ml of liver homogenate or the diluted serum sample was then mixed with 1 ml tetra-n-butyl ammonium hydrogen sulfate (TBA), 2 ml TBA buffer and 5 ml methyl-tert-butyl ether (MTBE), followed by shaking for 30 min at 300 mot/min at room temperature. After centrifugation at 3,500 rpm for 15 min, the supernatant (organic phase) was transferred to a clean 50 ml polypropylene tube. The remaining aqueous phase was subjected to extraction twice with 5 ml MTBE. All three organic phases were pooled and were concentrated to dryness under a gentle stream of nitrogen and reconstituted with 1 mL of 10 mM Ammonium/Acetate:Acetonitrile (6:4) prior to LC/MS/MS analysis. Standards of PFOS and labeled-PFOS used for calibration were both prepared in methanol. The detection of PFOS was performed using an Agilent 1200 high-performance liquid chromatograph coupled with tandem mass spectrometry (HPLC-MS/MS, Aglient 1200 series, Aglient Technologies, California, US). A 30 μL aliquot of the extract was injected into a guard column (Zobrax Eclipse Plus-C8, 2.1 mm i.d. ×12.5 mm length, 5-μm; Agilent Technologies), which was connected to a Zorbax Eclipse Plus C8 column (2.1 mm i.d. x100 mm length, 3.5-μm; Agilent Technologies). Instrumental parameters for analysis were described in Zhao and coworkers [59]. LOD was defined as 3-fold higher than the signal-to-noise ratio and 0.4 ng/ml for PFOS. LOQ was defined as 10-fold higher than the signal-to-noise ratio. The vales of matrix recovery were 99.95%.

RNA Isolation and Real-Time PCR

RNA isolation was carried out by TriReagent according to manufacturer's instructions. Total RNA with A260:A280 ratio above 1.85 was used for real-time PCR analyses. Complementary DNA was synthesized from 150 ng of total cellular RNA using High Capacity RNA-to-cDNA Master Mix (Applied Biosystem, Foster City, CA). Gene-specific primers were designed from published sequences (Table S1). Real-time PCR was conducted with a program of 3 min at 95°C followed by 40 cycles of 95°C for 15 sec, 56°C for 20 sec, and 72°C for 30 s. Standards and cDNAs from samples were quantified using StepOne Real-Time PCR system using SYBR Green Master mix (Applied Biosystems). By applying the comparative CT method [60], the data were presented as relative to the mouse actin and normalized to the control. The error bars in the control group displayed the

F1 Postnatal day 63

Standard Diet

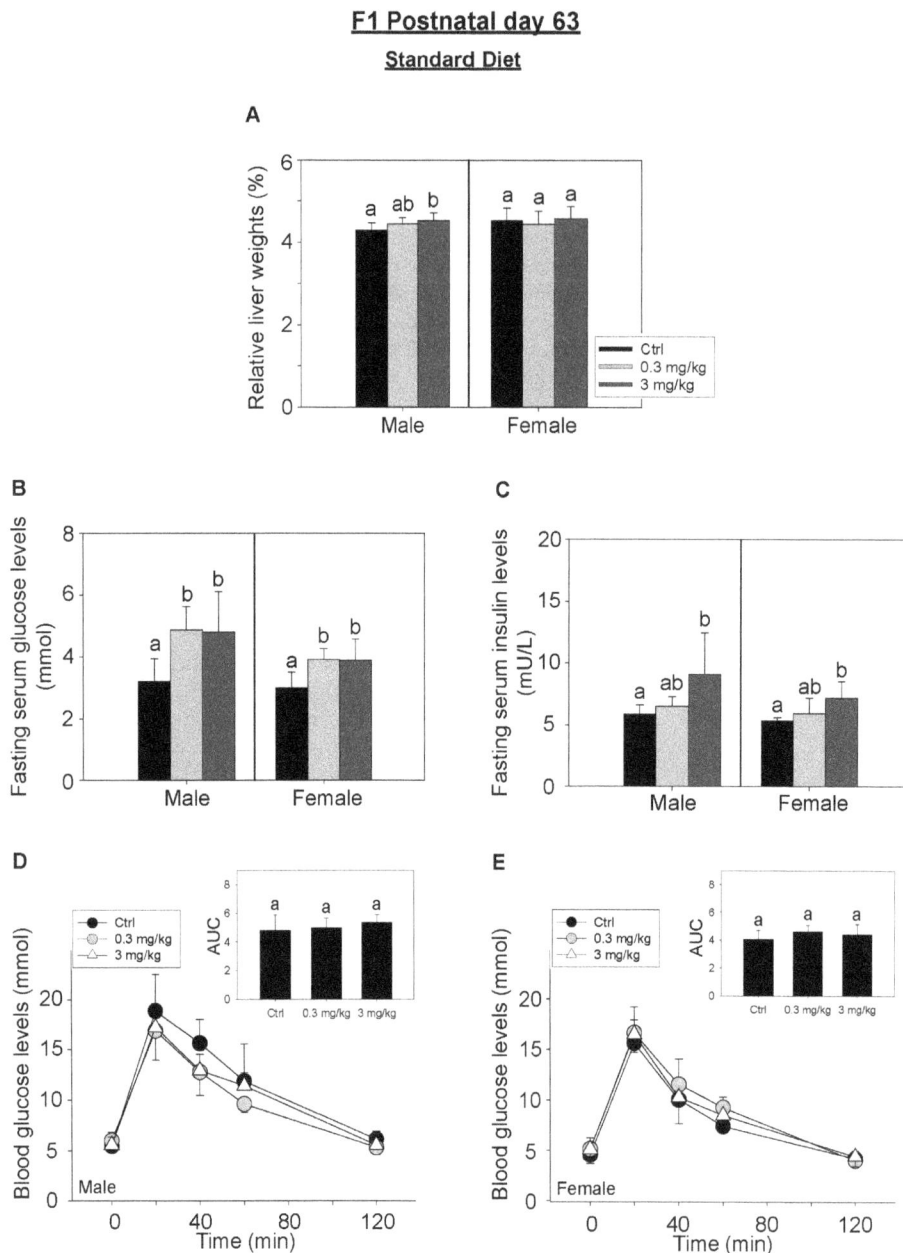

Figure 4. The effects perinatal PFOS exposure to STD-fed F$_1$ adult offspring at PND63. F$_1$ offspring were fed with the standard diet (STD) and grown without further PFOS exposure. The F$_1$ adults were sacrificed on PND 63. The relative liver weights, fasting blood glucose and insulin levels were measured and OGTT was performed. (A) The relative liver weights of the male adults (F$_1$) from the high-dose maternal group were significantly increased (n = 8) as compared to the control group. (B) Fasting blood glucose levels were increased in the F$_1$ adults of both sexes from PFOS-exposed maternal groups (n = 6) as compared to the respective control groups. (C) Fasting blood insulin levels in the F$_1$ adults of both sexes from the high-dosed exposed group were noticeably increased (n = 6) as compared to the respective control groups. In panels D–E, F$_1$ adults were given 2 g glucose/kg body weight by oral gavage at time 0 (min) and blood glucose levels were measured by the glucometer at specific time intervals. The blood glucose levels reached maximum at time 20 (min) and gradually drop to baseline levels at time 120 (min). OGTT results of the F$_1$ male (D) and female adults (E) were shown (n = 4). The data at the same time point from the control and the perinatal PFOS treated groups were compared. No statistical differences were detected. The area under curve (AUC) analysis showed that there were no significant differences between the respective control and the treatment groups. Bars with the same letter are not significantly different according to the results of one-way ANOVA followed by Tukey's test (p<0.05).

variation of different mice within the group. Statistical analysis was performed using normalized data by Sigma Stat (version 3.5) while simple t-test analysis was conducted. Occurrences of primer-dimers and secondary products were evaluated using melting curve analysis. Control amplifications were done either without RT or without RNA. All glassware and plastic ware were treated with diethyl pyrocarbonate and autoclaved.

F1 Postnatal day 63
High-Fat Diet

A

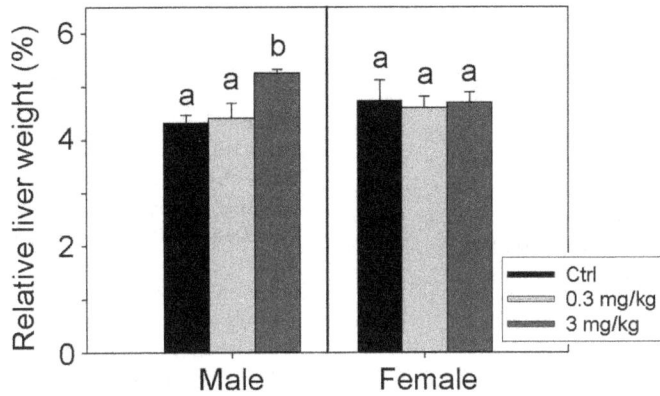

B

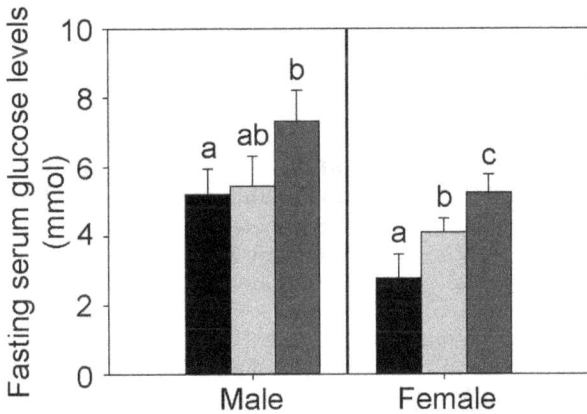

C

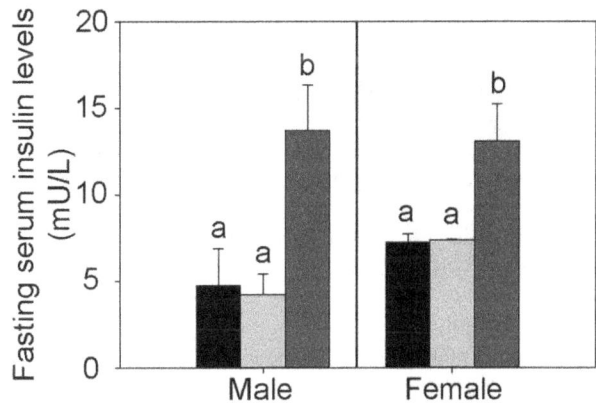

D

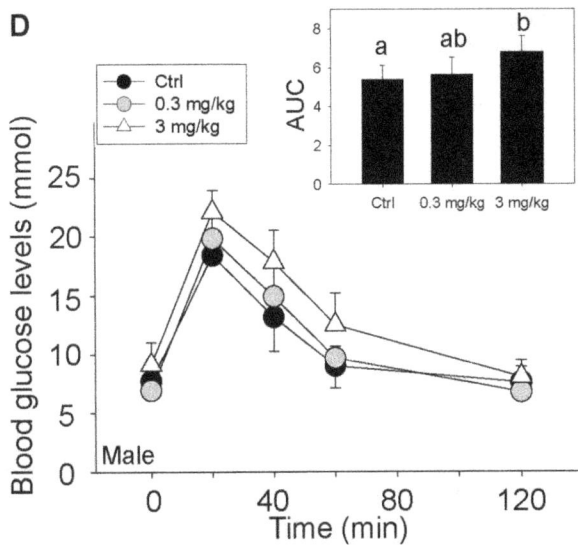

E

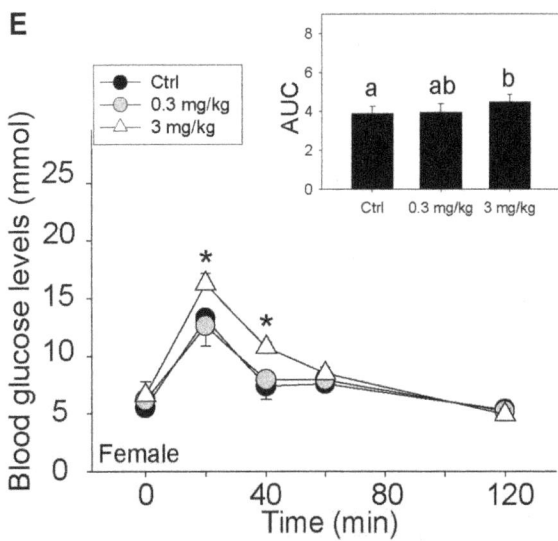

Figure 5. The effects perinatal PFOS exposure to HFD-fed F$_1$ adult offspring at PND63. F$_1$ offspring were fed with the high-fat diet (HFD) and grown without further PFOS exposure. The F$_1$ adults were sacrificed on PND 63. The relative liver weights, fasting blood glucose and insulin levels were measured and OGTT was performed. (A) The relative liver weights of the male adults (F$_1$) from the high-dose maternal group were significantly increased (n = 8) as compared to the control group. (B) Fasting blood glucose levels were increased in the F$_1$ adults of both sexes from PFOS-exposed maternal groups (n = 6) as compared to the respective control groups. (C) Fasting blood insulin levels in the F$_1$ adults of both sexes from the high-dosed exposed group were noticeably increased (n = 6) as compared to the respective control groups. The data of the OGTT from the F$_1$ male (D) and female adults (E) were shown. The data at the same time point from the control and the perinatal PFOS treated groups were compared. The area under curve (AUC) analysis showed that there were significant increases in the F$_1$ adults of both sexes as compared to the respective control groups. Bars with the same letter are not significantly different according to the results of one-way ANOVA followed by Tukey's test (p<0.05).

Fasting Serum Levels of Glucose and Insulin

Serum fasting glucose (n = 6 per group) was measured by StanBio Glucose Liquicolor (StanBio Laboratory, Boerne, US) according the manufacturer's manual. Serum insulin (n = 6 per group) was determined by a mouse insulin ELISA kit (Mercodia, Sweden) according to manufacturer's protocol. A Homeostatic Model Assessment for Insulin Resistance (HOMA-IR index) was calculated by blood glucose (mmol/L) × insulin (mU/L)/22.5 [29].

Oral Glucose Tolerance Test (OGTT)

Mice were fasted for 16 h before the experiment (n = 4 per group). The fasting glucose level was measured by Accu-chek Glucometer (Roche, US). After the measurement, 2 g/kg body weight glucose solution was given to the mice orally by gavage and blood glucose was measured at time 15, 30, 60 and 120 min. Glucose response during the OGTT was calculated by the area under the curve (AUC) using the trapezoidal method [61].

Intraperitoneal Insulin Tolerance Test (ITT)

Mice were fasted for 16 h before the experiment (n = 4 per group). The fasting glucose level was measured by Accu-chek Glucometer (Roche, US). After the initial measurement, intraperitoneal injections of insulin (0.5 U/kg body weight, Sigma) were given to the mice and blood glucose levels were measured at time 15, 30, 45 and 60 min.

Statistical Analysis

Statistical evaluations were conducted by SigmaStat 3.5. All data were tested to be normally distributed and independent with significance of 0.05. Differences between treatment groups and corresponding control groups were tested for statistical significance by one-way ANOVA followed by Tukey's test (significance at p<0.05) or Student's t-tests as appropriate. Results are presented as the mean ± SD.

Supporting Information

Figure S1   Pregnant CD-1 mice were administrated with corn oil as control, 0.3 or 3 mg PFOS/kg body weight daily by oral gavage from gestational day (GD) 3 to postnatal day (PND) 21. F$_0$ maternal (n = 6 per group) were sacrificed on PND 21. The body weight (A) and absolute liver weights (B) were measured. No significant differences were observed between the control and the treatment groups. The absolute liver weights of F$_1$ pups on PND 21 (C) and PND 63 (D–E) were shown. Bars with the same letter are not significantly different according to the results of one-way ANOVA followed by Tukey's test (p<0.05).

Figure S2   F$_1$ adult offspring were fed with either the standard diet (STD) or high fat diet (HFD) after weaning (PND 21). The body weights were measured weekly (n = 11 per each group). No significant differences were observed between the control and the perinatal PFOS exposed groups from either the STD or HFD groups.

Figure S3   Comparisons of the effects of STD and HFD on perinatal PFOS-exposed F$_1$ adult offspring at PND63. F$_1$ offspring were fed with either standard diet (STD) or high fat diet (HFD) and grown without further PFOS exposure. The relative liver weights, fasting blood glucose and insulin levels of the STD- and HFD-fed F$_1$ were shown. The relative liver weights (n = 8) (A) and fasting blood glucose levels (n = 6) (C) of the HFD-fed male adults (F$_1$) from the high-dose maternal group were significantly higher than that in the respective group in the STD ([#]p<0.05, student's t test). (F) Fasting blood insulin levels in the HFD-fed female adults (F$_1$) from the high-dosed exposed group were noticeably increased as compared to the respective group in the STD (n = 6, [#]p<0.05, Student's t test). For the comparison of the control and treatment groups under either the STD or HFD, bars with the same letter are not significantly different according to the results of one-way ANOVA followed by Tukey's test (p<0.05).

Figure S4   Effects of perinatal PFOS exposure on glucose responses in the insulin tolerance test (ITT). F$_1$ adult offspring were given intraperitoneal injection of insulin (0.5 U insulin/kg body weight) at time 0 (min). Blood glucose levels were measured by the glucometer at the designated time intervals (15, 30, 45 and 60 min). The ITT data of the adult offspring from the STD (A & B) and HFD (C & D) were shown (n = 4). The data at the same time point from the control and the perinatal PFOS treated groups were compared using one-way ANOVA (p<0.05). Statistical analysis showed that there were no significant differences among the control and the perinatal PFOS-exposed groups.

Table 9. The HOMA-IR index in STD- and HFD-fed F$_1$ adults.

F$_1$ adults (PND 63)	HOMA-IR index			
	Male		Female	
	STD	HFD	STD	HFD
Ctrl	0.99±0.52	1.11±0.41	0.71±0.13	0.93±0.16
0.3 mg/kg	1.40±0.23	3.01±0.33#	1.02±0.17	1.46±0.13#
3 mg/kg	1.90±0.76*	5.01±2.22*#	1.17±0.43*	2.99±1.40*#

*p<0.01 as compared with the respective control groups.
#p<0.01 as compared between STD and HFD diets under the same gender groups.

Table S1 Nucleotide sequences of primers used in the present study.

Author Contributions

Conceived and designed the experiments: CKCW. Performed the experiments: HTW YGZ PYL. Analyzed the data: HTW YGZ. Wrote the paper: HTW.

References

1. Ismail MH (2011) Nonalcoholic fatty liver disease and type 2 diabetes mellitus: the hidden epidemic. Am J Med Sci 341: 485–492.

2. Chiang DJ, Pritchard MT, Nagy LE (2011) Obesity, diabetes mellitus, and liver fibrosis. Am J Physiol Gastrointest Liver Physiol 300: G697–G702.

3. Casals-Casas C, Desvergne B (2011) Endocrine disruptors: from endocrine to metabolic disruption. Annu Rev Physiol 73: 135–162.

4. Rinaudo P, Wang E (2011) Fetal Programming and Metabolic Syndrome. Annu Rev Physiol.

5. Neel BA, Sargis RM (2011) The paradox of progress: environmental disruption of metabolism and the diabetes epidemic. Diabetes 60: 1838–1848.

6. Alonso-Magdalena P, Quesada I, Nadal A (2011) Endocrine disruptors in the etiology of type 2 diabetes mellitus. Nat Rev Endocrinol 7: 346–353.

7. Sugden MC, Holness MJ (2008) Role of nuclear receptors in the modulation of insulin secretion in lipid-induced insulin resistance. Biochem Soc Trans 36: 891–900.

8. Olsen GW, Burris JM, Ehresman DJ, Froehlich JW, Seacat AM, et al. (2007) Half-life of serum elimination of perfluorooctanesulfonate,perfluorohexanesulfonate, and perfluorooctanoate in retired fluorochemical production workers. Environ Health Perspect 115: 1298–1305.

9. Seacat AM, Thomford PJ, Hansen KJ, Olsen GW, Case MT, et al. (2002) Subchronic toxicity studies on perfluorooctanesulfonate potassium salt in cynomolgus monkeys. Toxicol Sci 68: 249–264.

10. Lau C, Anitole K, Hodes C, Lai D, Pfahles-Hutchens A, et al. (2007) Perfluoroalkyl acids: a review of monitoring and toxicological findings. Toxicol Sci 99: 366–394.

11. Calafat AM, Wong LY, Kuklenyik Z, Reidy JA, Needham LL (2007) Polyfluoroalkyl chemicals in the U.S. population: data from the National Health and Nutrition Examination Survey (NHANES) 2003–2004 and comparisons with NHANES 1999–2000. Environ Health Perspect 115: 1596–1602.

12. Karrman A, Mueller JF, van Bavel B, Harden F, Toms LM, et al. (2006) Levels of 12 perfluorinated chemicals in pooled australian serum, collected 2002–2003, in relation to age, gender, and region. Environ Sci Technol 40: 3742–3748.

13. Apelberg BJ, Goldman LR, Calafat AM, Herbstman JB, Kuklenyik Z, et al. (2007) Determinants of fetal exposure to polyfluoroalkyl compounds in Baltimore, Maryland. Environ Sci Technol 41: 3891–3897.

14. Olsen GW, Hansen KJ, Stevenson LA, Burris JM, Mandel JH (2003) Human donor liver and serum concentrations of perfluorooctanesulfonate and other perfluorochemicals. Environ Sci Technol 37: 888–891.

15. So MK, Yamashita N, Taniyasu S, Jiang Q, Giesy JP, et al. (2006) Health risks in infants associated with exposure to perfluorinate compounds in human breast milk from Zhoushan, China. Environ Sci Technol 40: 2924–2929.

16. Beach SA, Newsted JL, Coady K, Giesy JP (2006) Ecotoxicological evaluation of perfluorooctanesulfonate (PFOS). Rev Environ Contam Toxicol 186: 133–174.

17. Jensen AA, Leffers H (2008) Emerging endocrine disrupters: perfluoroalkylated substances. Int J Androl 31: 161–169.

18. Martin MT, Brennan RJ, Hu W, Ayanoglu E, Lau C, et al. (2007) Toxicogenomic study of triazole fungicides and perfluoroalkyl acids in rat livers predicts toxicity and categorizes chemicals based on mechanisms of toxicity. Toxicol Sci 97: 595–613.

19. Cui L, Zhou QF, Liao CY, Fu JJ, Jiang GB (2009) Studies on the toxicological effects of PFOA and PFOS on rats using histological observation and chemical analysis. Arch Environ Contam Toxicol 56: 338–349.

20. Zhang H, Shi Z, Liu Y, Wei Y, Dai J (2008) Lipid homeostasis and oxidative stress in the liver of male rats exposed to perfluorododecanoic acid. Toxicol Appl Pharmacol 227: 16–25.

21. Guruge KS, Yeung LW, Yamanaka N, Miyazaki S, Lam PK, et al. (2006) Gene expression profiles in rat liver treated with perfluorooctanoic acid (PFOA). Toxicol Sci 89: 93–107.

22. Rosen MB, Schmid JE, Das KP, Wood CR, Zehr RD, et al. (2009) Gene expression profiling in the liver and lung of perfluorooctane sulfonate-exposed mouse fetuses: comparison to changes induced by exposure to perfluorooctanoic acid. Reprod Toxicol 27: 278–288.

23. Bjork JA, Lau C, Chang SC, Butenhoff JL, Wallace KB (2008) Perfluorooctane sulfonate-induced changes in fetal rat liver gene expression. Toxicology 251: 8–20.

24. Vanden Heuvel JP, Thompson JT, Frame SR, Gillies PJ (2006) Differential activation of nuclear receptors by perfluorinated fatty acid analogs and natural fatty acids: a comparison of human, mouse, and rat peroxisome proliferator-activated receptor-alpha, -beta, and -gamma, liver X receptor-beta, and retinoid X receptor-alpha. Toxicol Sci 92: 476–489.

25. Kim S, Choi K, Ji K, Seo J, Kho Y, et al. (2011) Trans-placental transfer of thirteen perfluorinated compounds and relations with fetal thyroid hormones. Environ Sci Technol 45: 7465–7472.

26. Wan HT, Zhao YG, Wei X, Hui KY, Giesy JP, et al. (2012) PFOS-induced hepatic steatosis, the mechanistic actions on beta-oxidation and lipid transport. Biochim Biophys Acta 1820: 1092–1101.

27. Takamura T, Misu H, Ota T, Kaneko S (2012) Fatty liver as a consequence and cause of insulin resistance: lessons from type 2 diabetic liver. Endocr J 59: 745–763.

28. Targher G, Byrne CD (2013) Clinical Review: Nonalcoholic fatty liver disease: a novel cardiometabolic risk factor for type 2 diabetes and its complications. J Clin Endocrinol Metab 98: 48 3–495.

29. Matthews DR, Hosker JP, Rudenski AS, Naylor BA, Treacher DF, et al. (1985) Homeostasis model assessment: insulin resistance and beta-cell function from fasting plasma glucose and insulin concentrations in man. Diabetologia 28: 412–419.

30. Clarke DB, Bailey VA, Routledge A, Lloyd AS, Hird S, et al. (2010) Dietary intake estimate for perfluorooctanesulphonic acid (PFOS) and other perfluorocompounds (PFCs) in UK retail foods following determination using standard addition LC-MS/MS. Food Addit Contam Part A Chem Anal Control Expo Risk Assess 27: 530–545.

31. Loccisano AE, Campbell JL Jr, Butenhoff JL, Andersen ME, Clewell HJ III (2012) Evaluation of placental and lactational pharmacokinetics of PFOA and PFOS in the pregnant, lactating, fetal and neonatal rat using a physiologically based pharmacokinetic model. Reprod Toxicol 33: 468–490.

32. Salvalaglio M, Muscionico I, Cavallotti C (2010) Determination of energies and sites of binding of PFOA and PFOS to human serum albumin. J Phys Chem B 114: 14860–14874.

33. Chang SC, Noker PE, Gorman GS, Gibson SJ, Hart JA, et al. (2012) Comparative pharmacokinetics of perfluorooctanesulfonate (PFOS) in rats, mice, and monkeys. Reprod Toxicol 33: 428–440.

34. Borai A, Livingstone C, Kaddam I, Ferns G (2011) Selection of the appropriate method for the assessment of insulin resistance. BMC Med Res Methodol 11: 158.

35. Sarafidis PA, Lasaridis AN, Nilsson PM, Pikilidou MI, Stafilas PC, et al. (2007) Validity and reproducibility of HOMA-IR, 1/HOMA-IR, QUICKI and McAuley's indices in patients with hypertension and type II diabetes. J Hum Hypertens 21: 709–716.

36. Lien GW, Huang CC, Wu KY, Chen MH, Lin CY, et al. (2013) Neonatal-maternal factors and perfluoroalkyl substances in cord blood. Chemosphere 92: 843–850.

37. Beesoon S, Webster GM, Shoeib M, Harner T, Benskin JP, et al. (2011) Isomer profiles of perfluorochemicals in matched maternal, cord, and house dust samples: manufacturing sources and transplacental transfer. Environ Health Perspect 119: 1659–1664.

38. Liu J, Li J, Liu Y, Chan HM, Zhao Y, et al. (2011) Comparison on gestation and lactation exposure of perfluorinated compounds for newborns. Environ Int 37: 1206–1212.

39. Kato K, Basden BJ, Needham LL, Calafat AM (2011) Improved selectivity for the analysis of maternal serum and cord serum for polyfluoroalkyl chemicals. J Chromatogr A 1218: 2133–2137.

40. Yu WG, Liu W, Jin YH, Liu XH, Wang FQ, et al. (2009) Prenatal and postnatal impact of perfluorooctane sulfonate (PFOS) on rat development: a cross-foster study on chemical burden and thyroid hormone system. Environ Sci Technol 43: 8416–8422.

41. Bjork JA, Wallace KB (2009) Structure-activity relationships and human relevance for perfluoroalkyl acid-induced transcriptional activation of peroxisome proliferation in liver cell cultures. Toxicol Sci 111: 89–99.

42. Takacs ML, Abbott BD (2007) Activation of mouse and human peroxisome proliferator-activated receptors (alpha, beta/delta, gamma) by perfluorooctanoic acid and perfluorooctane sulfonate. Toxicol Sci 95: 108–117.

43. Ryan KK, Haller AM, Sorrell JE, Woods SC, Jandacek RJ, et al. (2010) Perinatal exposure to bisphenol-a and the development of metabolic syndrome in CD-1 mice. Endocrinology 151: 2603–2612.

44. Lin Y, Wei J, Li Y, Chen J, Zhou Z, et al. (2011) Developmental exposure to di(2-ethylhexyl) phthalate impairs endocrine pancreas and leads to long-term adverse effects on glucose homeostasis in the rat. Am J Physiol Endocrinol Metab 301: E527–E538.

45. Wei J, Lin Y, Li Y, Ying C, Chen J, et al. (2011) Perinatal exposure to bisphenol A at reference dose predisposes offspring to metabolic syndrome in adult rats on a high-fat diet. Endocrinology 152: 3049–3061.

46. Ben Jonathan N, Hugo ER, Brandebourg TD, LaPensee CR (2006) Focus on prolactin as a metabolic hormone. Trends Endocrinol Metab 17: 110–116.

47. Clemmons DR (2006) Involvement of insulin-like growth factor-I in the control of glucose homeostasis. Curr Opin Pharmacol 6: 620–625.

48. Clemmons DR (2012) Metabolic actions of insulin-like growth factor-I in normal physiology and diabetes. Endocrinol Metab Clin North Am 41: 425-viii.

49. Homer LM, Clarke CR, Weingarten AJ (2005) Effect of dietary fat on oral bioavailability of tepoxalin in dogs. J Vet Pharmacol Ther 28: 287–291.

50. Meibohm B, Beierle I, Derendorf H (2002) How important are gender differences in pharmacokinetics? Clin Pharmacokinet 41: 329–342.

51. Hernandez JP, Chapman LM, Kretschmer XC, Baldwin WS (2006) Gender-specific induction of cytochrome P450s in nonylphenol-treated FVB/NJ mice. Toxicol Appl Pharmacol 216: 186–196.

52. Wolbold R, Klein K, Burk O, Nussler AK, Neuhaus P, et al. (2003) Sex is a major determinant of CYP3A4 expression in human liver. Hepatology 38: 978–988.

53. Sierra-Santoyo A, Hernandez M, Albores A, Cebrian ME (2000) Sex-dependent regulation of hepatic cytochrome P-450 by DDT. Toxicol Sci 54: 81–87.

54. Lin CY, Chen PC, Lin YC, Lin LY (2009) Association among serum perfluoroalkyl chemicals, glucose homeostasis, and metabolic syndrome in adolescents and adults. Diabetes Care 32: 702–707.

55. Thayer KA, Heindel JJ, Bucher JR, Gallo MA (2012) Role of environmental chemicals in diabetes and obesity: a National Toxicology Program workshop review. Environ Health Perspect 120: 779–789.

56. Hennig B, Ormsbee L, McClain CJ, Watkins BA, Blumberg B, et al. (2012) Nutrition can modulate the toxicity of environmental pollutants: implications in risk assessment and human health. Environ Health Perspect 120: 771–774.

57. Misra A, Singhal N, Khurana L (2010) Obesity, the metabolic syndrome, and type 2 diabetes in developing countries: role of dietary fats and oils. J Am Coll Nutr 29: 289S–301S.

58. Wan HT, Leung PY, Zhao YG, Wei X, Wong MH, et al. (2013) Blood plasma concentrations of endocrine disrupting chemicals in Hong Kong populations. J Hazard Mater 261: 763–769.

59. Zhao YG, Wan HT, Law AY, Wei X, Huang YQ, et al. (2011) Risk assessment for human consumption of perfluorinated compound-contaminated freshwater and marine fish from Hong Kong and Xiamen. Chemosphere 85: 277–283.

60. Schmittgen TD, Livak KJ (2008) Analyzing real-time PCR data by the comparative C(T) method. Nat Protoc 3: 1101–1108.

61. Purves RD (1992) Optimum numerical integration methods for estimation of area-under-the-curve (AUC) and area-under-the-moment-curve (AUMC). J Pharmacokinet Biopharm 20: 211–226.

20

Cultivation-Independent Screening Revealed Hot Spots of IncP-1, IncP-7 and IncP-9 Plasmid Occurrence in Different Environmental Habitats

Simone Dealtry[1], Guo-Chun Ding[1], Viola Weichelt[1], Vincent Dunon[2], Andreas Schlüter[3], María Carla Martini[4], María Florencia Del Papa[4], Antonio Lagares[4], Gregory Charles Auton Amos[5], Elizabeth Margaret Helen Wellington[5], William Hugo Gaze[5], Detmer Sipkema[6], Sara Sjöling[7], Dirk Springael[2], Holger Heuer[1], Jan Dirk van Elsas[8], Christopher Thomas[9], Kornelia Smalla[1]*

1 Julius Kühn-Institut – Federal Research Centre for Cultivated Plants (JKI), Institute for Epidemiology and Pathogen Diagnostics, Braunschweig, Germany, 2 Division of Soil and Water Management, KU Leuven, Heverlee, Belgium, 3 Center for Biotechnology (CeBiTec), Institute for Genome Research and Systems Biology, Bielefeld University, Bielefeld, Germany, 4 IBBM (Instituto de Biotecnología y Biología Molecular), CCT-CONICET-La Plata, Departamento de Ciencias Biológicas, Facultad de Ciencias Exactas, Universidad Nacional de La Plata, La Plata, Argentina, 5 School of Life Sciences, University of Warwick, Warwick, United Kingdom, 6 Laboratory of Microbiology, Wageningen University, Wageningen, The Netherlands, 7 Södertörns högskola (Sodertorn University), Inst. för Naturvetenskap, Miljö och medieteknik (School of Natural Sciences, Environmental Studies and media tech), Huddinge, Sweden, 8 University of Groningen, Groningen, The Netherlands, 9 School of Biosciences, University of Birmingham, Edgbaston, Birmingham, Warwick, United Kingdom

Abstract

IncP-1, IncP-7 and IncP-9 plasmids often carry genes encoding enzymes involved in the degradation of man-made and natural contaminants, thus contributing to bacterial survival in polluted environments. However, the lack of suitable molecular tools often limits the detection of these plasmids in the environment. In this study, PCR followed by Southern blot hybridization detected the presence of plasmid-specific sequences in total community (TC-) DNA or fosmid DNA from samples originating from different environments and geographic regions. A novel primer system targeting IncP-9 plasmids was developed and applied along with established primers for IncP-1 and IncP-7. Screening TC-DNA from biopurification systems (BPS) which are used on farms for the purification of pesticide-contaminated water revealed high abundances of IncP-1 plasmids belonging to different subgroups as well as IncP-7 and IncP-9. The novel IncP-9 primer-system targeting the *rep* gene of nine IncP-9 subgroups allowed the detection of a high diversity of IncP-9 plasmid specific sequences in environments with different sources of pollution. Thus polluted sites are "hot spots" of plasmids potentially carrying catabolic genes.

Editor: Axel Cloeckaert, Institut National de la Recherche Agronomique, France

Funding: This study was funded by the EU 7th Framework Programme (MetaExplore 222625) and the Inter-University Attraction Pole (IUAP) "μ-manager" of the Belgian Science Policy (BELSPO, P7/25). The funders had no role in study design, data collection and analysis, decision to publish, or preparation of the manuscript.

Competing Interests: The authors have declared that no competing interests exist.

* E-mail: kornelia.smalla@jki.bund.de

Introduction

The search for novel enzymes able to degrade recalcitrant natural contaminants such as chitins and lignins and man-made pollutants such as halogenated aliphatic and aromatic compounds motivated metagenomic explorations of various environments. It has been observed that the microbial metagenomes of open ecosystems, including soils and aquatic habitats, clearly represent rich reservoirs of genes that determine the desired enzymatic reactions in which chitinases, ligninases and dehalogenases are involved [1,2]. By anthropogenic activities, recalcitrant compounds have also been released as environmental pollutants. Typical metagenomic approaches employ genetic or activity screens of cloned large DNA fragments from various environments [3]. However, the idea of capturing complete mobile genetic elements (MGE) into suitable recipients might be an alternative and complementary approach to access the genes coding for novel enzymes or even complete degradative pathways. Mobile genetic elements such as plasmids are often found to play an important role in the adaptation of bacterial communities to changing and, due to pollutants, often challenging environmental conditions. For example, partial or complete degradative pathways were previously reported to be localized on plasmids belonging to the IncP-1, IncP-7 or IncP-9 group [1]. The present study aimed to monitor various environments for the abundance of these plasmids by using a cultivation-independent total community (TC-) DNA based approach to select the most promising habitats for mining plasmids potentially carrying genes coding for novel enzymes. We hypothesized that the frequency of occurrence of genes encoding the desired enzymatic activities is increased in the MGE gene pool. In particular, plasmids belonging to the incompatibility groups (Inc) P-1, P-7 and P-9 often carry genes responsible for the degradation of xenobiotic (man-made) and natural organic pollutants, being essential players in the adaptation of bacterial communities to new toxic compounds released in the environment [1]. Therefore, selected natural or treated environments were analyzed for the prevalence of plasmids belonging to the IncP-1, IncP-7 and IncP-9 groups by a cultivation-independent approach. Some of these environments were enriched for the desired degradation function by adding the relevant substrates, i.e. chitin, lignin and/or organohalogens. The habitats sampled included a variety of soils (one soil sample amended with chitin, peat bogs), biopurification systems (BPS) for pesticide removal from contaminated water, biogas production plants, wastewater, as well as aquatic (river bank sediments, sponges) environments from a wide range of geographic regions. Total community DNA was analyzed for the presence of IncP-1, IncP-7 and IncP-9 plasmids by means of PCR and subsequent Southern blot hybridization. A novel primer system for the specific amplification of IncP-9 plasmids was developed and tested in the present study. Southern blot hybridization using probes derived from reference plasmids belonging to different subgroups of IncP-1 plasmids provided new insights into their environmental dissemination. Our results showed a particularly widespread dissemination of IncP-1 plasmid-specific sequences. Different hot spots of plasmid occurrence were identified.

Materials and Methods

Ethics Statement of Provided Samples

None of the samples used in the present work involved any endangered or protected species. The marine sponges were obtained under legal permits from competent authorities: *Halichondria panicea* was obtained under a permit from authorities given to Wageningen University, while *Corticium candelabrum* and *Petrosia ficiformis* marine sponges were sampled under a Spanish permit to CEAB-CSIC. The sediments and soil originated from UK were taken from a river bed accessed from a public right of way and therefore no permissions were needed. Landsort Deep was sampled from a national environmental monitoring site (BY31) in conjunction with the Baltic Sea monitoring programme. The sampling permission was provided by the Stockholm University marine research Center. The Askö samples were sampled at the Stockholm Marine Research Center (now the Stockholm University Marine Research Center. The sediment and soils from Argentina were obtained from public locations as part of fundamental studies performed through a collaborative project with the agreement of the Facultad de Ciencias Exactas, Universidad Nacional de La Plata, and did not require any specific permission. The biopurification systems (BPS) samples were obtained from private land with permission from the local farmers in Kortrijk, Leefdaal, Lierde and Koksijde, located in Belgium.

Extraction of Total Community DNA (TC-DNA) and Metagenomic DNA (Pooled Fosmid Library) from Different Environmental Samples

The TC-DNA and/or metagenomic DNA (metagenomic DNA represented by the metagenomic pooled fosmid library from Baltic Sea) from different environmental samples originating from various geographic regions were extracted using different methods. The protocols used for TC-DNA extraction of each sample type are given in Table 1.

16S rRNA Gene PCR Amplification and Quantification

16S rRNA gene PCR amplification reaction was done as previously described by Heuer *et al.* (2009) [5] (product size of 1506 bp). The quality of the PCR product was determined by electrophoresis in 1% agarose gel and visualized with ethidium bromide staining and under UV light by comparison with the 1-kb gene-rulerTM DNA ladder (Fermentas, St Leon-Rot, Germany). Quantitative PCR (qPCR) targeting the 16S rRNA gene was performed with the TaqMan system as described by Suzuki *et al.* (2000) [6]. The 16S rRNA gene qPCR standard was made from cloned 16S rRNA gene amplicons (1467 bp) of *E. coli.* and 10^9, 10^8, 10^7 16S rRNA gene copy numbers were used.

Southern Blot-PCR Based Detection of IncP-1 Plasmids

IncP-1 plasmids belonging to the α, β, γ, δ and ε subgroups were detected based on the amplification of the *trfA* region (product size of 281 bp) from TC-DNA and metagenomic DNA using the primers described by Bahl *et al.* (2009) [2]. Digoxygenin-labeled probes targeting different IncP-1 plasmids subgroups were generated from reference plasmids belonging to the IncP-1α, β, γ, δ and ε plasmids (Table 2). The IncP-1 mixed probe was prepared by mixing probes generated for the different subgroups. The random primed digoxigenin labeling of PCR amplicons excised from preparative agarose gels was done according to the Roche manufacturer's protocol (Roche Diagnostics Deutschland GmbH, Mannheim, Germany).

Southern Blot-PCR Based Detection of IncP-7

PCR amplification of the *rep* region of IncP-7 plasmids (product size of 524 bp) from TC-DNA was performed as previously described by Izmalkova *et al.* (2005) [4]. Southern blotted PCR amplicons were hybridized at medium stringency with the dig-labeled IncP-7 probe generated from the reference plasmid pCAR1 isolated from *Pseudomonas resinovorans* according to the manufacturer's instructions (QIAGEN® Plasmid Mini Kit) (Table 2). The randomly primed digoxigenin labeling of PCR amplicons was done as described above.

Analyzing the Diversity and Abundance of IncP-9 Plasmids by a Novel PCR System Targeting the *oriV-rep* Region

To study the abundance and diversity of IncP-9 plasmids, a novel PCR system targeting the *oriV-rep* regions was developed and applied to detect IncP-9 plasmids in TC- and metagenomic DNA from all samples analyzed (Table 1).

Multiple alignments of 28 sequences of *oriV* (EU499619-EU499641, AF078924, AB237655, AJ344068, AB257759 and AF491307) and *rep* (EU499644-EU499666, AF078924, AB237655, AJ344068, AB257760 and AF491307) were performed with Molecular Evolutionary Genetics Analysis (MEGA 4). Conserved regions of sequences belonging to nine IncP-9 subgroups [7] were used for the primer design. The selected primer system consists of 21-mer degenerate forward primer (5-GAG GGT TTG GAG ATC ATW AGA-3) and reverse primer (5-GGT CTG TAT CCA GTT RTG CTT-3). *In silico* analysis showed no mismatch for at least 12 bp at the 3′ end of each primer and 1–4 mismatches for each sequence type at the 5′ end (Fig. S1). The expected amplicon size is 610–637 bp. The primers were further tested with plasmid DNA from the reference plasmids summarized in Fig. S1. None of the plasmids belonging to other incompatibility groups was amplified while the reference plasmids were amplified. The reaction mixture (25 µl) contained 1 µl template DNA (1–5 ng), 1× Stoffel buffer (Applied Biosystems, Foster, CA), 0,2 mM dNTPs, 2,5 mM $MgCl_2$, 2 µg/µl bovine serum albumin, 0.2 µM of each primer, and 2.5 U TrueStartTaq DNA polymerase (Stoffel fragment, Applied Biosystems). Denaturation was carried out at 94°C for 5 min, followed by 35 cycles

Table 1. Description of environmental samples analyzed and TC-DNA extraction applied.

Samples	Description of samples	TC-DNA extraction method
A	Biogas production plant fermentation sample from Bielefeld, Germany	[24]
B.1	*Biopurification system (BPS) from Leefdaal, Belgium	[17]
B.2	BPS from Leefdaal, Belgium	[17]
B.3	BPS from Leefdaal, Belgium	[17]
C.1	*BPS from Belgium (Pcfruit)	[17]
C.2	BPS from Belgium (Pcfruit)	[17]
C.3	BPS from Belgium (Pcfruit)	[17]
C.4	BPS from Belgium (Pcfruit)	[17]
C.5	BPS from Belgium (Pcfruit)	[17]
C.6	BPS from Belgium (Pcfruit)	[17]
D.1	*BPS from Lierde, Belgium	[17]
D.2	BPS from Lierde, Belgium	[17]
D.3	BPS from Lierde, Belgium	[17]
E.1	*BPS from Kortrijk, Belgium	[17]
E.2	BPS from Kortrijk, Belgium	[17]
E.3	BPS from Kortrijk, Belgium	[17]
F.1	*BPS from Koksijde, Belgium	[17]
F.2	BPS from Koksijde, Belgium	[17]
F.3	BPS from Koksijde, Belgium	[17]
G.1	Soil from La Plata, Argentina polluted with industrial residues and petrol	[18]
G.2	Soil from La Plata, Argentina polluted with industrial residues and petrol	[18]
G.3	Soil from La Plata, Argentina polluted with industrial residues and petrol	[18]
H.1	Sediments from La Plata, Argentina polluted with pesticides and petrol	[18]
H.2	Bordering soil from a water channel in La Plata, Argentina polluted with pesticides, residues from paper industry	[18]
H.3	Bordering soil from a water channel in La Plata, Argentina polluted with pesticides, residues from paper industry	[18]
J	Marginal river forest soil from La Plata, Argentina polluted with industrial residues	[18]
L.1	Bordering soil from a water channel in Buenos Aires, Argentina polluted with industrial residues	[18]
L.2	Bordering soil from a water channel in Buenos Aires, Argentina polluted with industrial residues	[18]
L.3	Bordering soil from a water channel in Buenos Aires, Argentina polluted with industrial residues	[18]
M	*Halichondria panicea* (marine sponge) from Oosterschelde, Netherlands	[25]
N	*Corticium candelabrum* (marine sponge) from Punta Santa Anna (Blanes), Spain	[25]
O	*Petrosia ficiformis* (marine sponge) from Punta Santa Anna (Blanes), Spain	[25]
P.1	Askö sediment from Baltic Sea Sweden (bottom fraction - anoxic)	[26]
P.2	Askö sediment from Baltic Sea Sweden (middle fraction - mixed anoxic/oxic)	[26]
P.3	Askö sediment from Baltic Sea Sweden (top fraction - oxic)	[26]
Q	Pooled fosmid library, Askö sediment, Baltic Sea	[3]
R	Landsort in Sweden	[26]
S.1	Sediment from a river in Warwickshire, UK	[27]
S.2	Sediment from a river in Warwickshire, UK	[28]
T	Soil from UK amended with chitin (Test site 1)	[28]

*BPS samples received water contaminated with different types of pesticides from spillage and residue water collected when cleaning the spraying equipment such as ethofumesate, fenpropimorf, fluroxypyr, glyphosate, linuron, metamitron and S-metalochlor (information provided by the farmers).

of 1 min at 94°C, 1 min at 53°C (primer annealing) and 2 min at 72°C and a final extension of 10 min at 72°C.

PCR amplicons of *oriV-rep* regions of nine IncP-9 subgroups IncP-9 plasmids (Table 2) were gel-purified and digoxigenin-labeled as described above. Southern blot hybridization of *oriV-rep* amplicons from different environmental samples listed above was performed with a mixture of these probes under medium

stringency following the manufacturer's instructions (Roche Diagnostics Deutschland GmbH, Mannheim, Germany). Clone libraries were generated for these three BPS to confirm primer specificity. *oriV-rep* amplicons were gel-purified, ligated into pGEM vectors, and transformed into *E. coli* JM109 competent cells according to the instructions of the manufacturer. Clones containing the correct inserts were selected for sequencing.

Table 2. Generation of probes for Southern blot hybridization.

Probe	Reference plasmid	Plasmids host strain	Primers
IncP-1α	RP4	E. coli	[2]
IncP-1β	R751	E. coli CM544	[2]
IncP-1γ	pQKH54	E. coli DH10B	[2]
IncP-1δ	pEST4011	Alcaligenes xylosoxidans EST4002	[2]
IncP-1ε	p3-408	E. coli cv601-GFP	[2]
IncP-7	pCAR1,	Pseudomonas resinovorans CA10	[4]
IncP-9 α	pM3	Pseudomonas putida	This study
IncP-9 β	pBS2	Pseudomonas putida BS268	This study
IncP-9 γ	pSN11	Pseudomonas putida BS349	This study
IncP-9 δ	pSN11	Pseudomonas putida SN11	This study
IncP-9 ε	pMG18	Pseudomonas putida AC34	This study
IncP-9 ζ	pNL60	Pseudomonas spp. 18d/1	This study
IncP-9 η	pNL15	E. coli C600	This study
IncP-9 θ	pSVS15	Pseudomonas fluorescens SVS15	This study
IncP-9 ι	pNL22	Pseudomonas spp. 41a/2	This study

BLAST-N analysis was used to identify *oriV-rep* sequences of IncP-9. All sequences analyzed share high similarity with IncP-9 *oriV* or *rep* sequences in NCBI. The sequences and those of known *oriV-rep* sequences in the data base were aligned and phylogenetic tree was calculated according to the neighbor-joining method and bootstrapping analysis using MEGA 4.

Nucleotide Sequence Accession Numbers of Cloned IncP-9 *oriV-rep* Gene Amplicons

Amplicon sequences have been submitted to NCBI SRA with IncP-9 *oriV-rep* gene amplicons under accession numbers KF706553 - KF706633.

Results

Determination of Bacterial 16S rRNA Gene Copies by qPCR

To estimate the bacterial density of the different environmental samples analyzed, 16S rRNA gene copies were determined by quantitative real-time PCR from the TC-DNA. Most of the samples (Table 3) showed a high abundance of bacterial populations ranging from 10^8 to 10^9 16S rRNA gene copy numbers per gram of material. For a few samples significantly lower 16S rRNA gene copy numbers per gram of material (Tukey's test $p > 0.05$) were detected (Table 3).

Distribution of IncP-1 Plasmids in Different Environments

To investigate the presence of IncP-1 plasmids in different habitats a detection system based on Southern blot-PCR was applied. Using the IncP-1 mixed probe from PCR products hybridization signals of the expected size (251 bp) were detected in a very wide range of different habitats (Table 3), indicating that IncP-1 plasmids of different subgroups are widely distributed. By using probes specific for the five different IncP-1 different subgroups (α, β, γ, δ and ε), differences in the composition of IncP-1 plasmids according to the geographic area and sample type were observed. Strong hybridization signals of IncP-1α plasmids

were only observed in one TC-DNA from Askö sediment (Sweden), in TC-DNA from a biogas production plant (Germany) and fosmid DNA from Baltic Sea sediments. Strong hybridization signals were observed using the IncP-1β specific probe in the TC-DNAs of all biopurification system (BPS) samples from Belgium (Table 3, Fig. 1) and most of the sediment samples from Argentina, indicating that in these environments bacterial populations carrying IncP-1β plasmids were highly abundant. The highest IncP-1γ hybridization signal was observed in the TC-DNA of the BPS located in Kortrijk. Less intense IncP-1γ hybridization signals were detected in the TC-DNAs of other BPS from Belgium and in TC-DNA of sediments from Argentina. In all TC-DNAs of BPS from Belgium, strong IncP-1δ hybridization signals were observed and a weaker hybridization signal, compared to BPS TC-DNA, was detected in TC-DNA from sediments in Argentina. Very strong IncP-1ε hybridization signals were again detected in all BPS TC-DNAs from Belgium (Table 3, Fig. 2) and most of the sediments from Argentina. Using IncP-1 mixed-probe, strong hybridization signals were detected in soils from Argentina and soil treated with chitin from the UK, indicating a high abundance of IncP-1 plasmids.

Distribution of IncP-7 Plasmids in Different Environments

To investigate the occurrence of IncP-7 plasmids in different environments, a PCR-based detection approach was applied in combination with Southern blot hybridization. Strong hybridization signals were observed in all TC-DNAs from BPS analyzed (Fig. 3), indicating a high abundance of bacterial populations carrying in BPS IncP-7 plasmid. Less intense hybridization signals were observed in the TC-DNAs of seven sediment river samples from Argentina and in the TC-DNA from soil amended with chitin from the UK. Hybridization signals using the amplicon probe specific for IncP-7 plasmids were not detected in any of the other environmental samples analyzed (Table 3).

IncP-9 Plasmid Occurrence and Diversity in Different Environmental Samples

In order to verify the occurrence and diversity of IncP-9 plasmids in different habitats the new IncP-9 primer system developed in the present work was applied. Very strong hybridization signals were detected in all TC-DNAs of BPS samples, indicating that BPS are reservoirs of bacteria carrying IncP-9 plasmids. Less intense hybridization signals were observed in the TC-DNA of sediment samples from Argentina. A weaker hybridization signal was detected in the soil amended with chitin from the UK (Table 3).

To verify primer specificity and to gain insights into the IncP-9 plasmid diversity from BPS samples (indicated as a "hot spot" of IncP-9 plasmids), a clone library was generated with amplicons from PCR using primers targeting the IncP-9 *oriV-rep* region in TC-DNA of three different BPS. Sequencing revealed the presence of different IncP-9 subgroups while phylogenetic analysis (Fig. 4) showed IncP-9 plasmid types similar to *oriV-rep* sequences of pWWO and pM3 as well as several sequences that could not be affiliated to previously known IncP-9 plasmid groups indicating an undiscovered diversity of this plasmid group.

Discussion

In the present study a PCR-based screening combined with Southern blot hybridization allowed the detection of IncP-1, IncP-7 and IncP-9 plasmids in a wide range of different geographic areas and sample types. The results indicated a high abundance of these plasmids in environments with different sources of pollution.

Table 3. Bacterial densities and PCR-Southern blot hybridization detection of plasmid replicon-specific sequences belonging to the five IncP-1 subgroups, IncP-7 and IncP-9.

Sample	Description of samples	P-1	α	β	ε	γ	δ	P-7	P-9	16S log10/g
A	Biogas production plant from Bielefeld, Germany	+++	++	++	++	–	++	–	–	9,34
B.1	Biopurification system (BPS) from Leefdaal, Belgium	+++	–	++	+++	++	+++	+++	+++	9,32
B.2	BPS from Leefdaal, Belgium	+++	–	++	+++	++	+++	+++	+++	9,25
B.3	BPS from Leefdaal, Belgium	+++	–	++	+++	++	++	+++	+++	8,43
C.1	BPS from Belgium (Pcfruit)	+++	+	+++	+++	++	++	+++	+++	9,32
C.2	BPS from Belgium (Pcfruit)	+++	+	+++	++	++	+++	+++	+++	8,28
C.3	BPS from Belgium (Pcfruit)	++	–	++	++	–	+++	++	+	8,36
C.4	BPS from Belgium (Pcfruit)	+	–	++	+	+	+++	+++	+++	8,54
C.5	BPS from Belgium (Pcfruit)	+++	(+)	+++	+++	++	+++	+++	+++	8,66
C.6	BPS from Belgium (Pcfruit)	+	–	++	++	+	++	++	–	8,15
D.1	BPS from Lierde, Belgium	+++	–	++	–	–	+++	++	++	8,61
D.2	BPS from Lierde, Belgium	+++	–	++	++	–	+++	++	++	8,59
D.3	BPS from Lierde, Belgium	+++	–	++	+++	++	+++	++	++	8,31
E.1	BPS from Kortrijk, Belgium	+++	–	+++	+++	+++	+	+++	++	9,2
E.2	BPS from Kortrijk, Belgium	+++	–	+++	+++	+++	++	+++	+++	9,03
E.3	BPS from Kortrijk, Belgium	+++	–	++	++	+++	–	+++	+++	9,11
F.1	BPS from Koksijde, Belgium	++	(+)	++	+	–	+++	(+)	+++	9,01
F.2	BPS from Koksijde, Belgium	++	+++	++	–	–	++	–	+++	8,9
F.3	BPS from Koksijde, Belgium	++	(+)	++	+	–	+++	–	+++	8,95
G.1	Soil from La Plata, Argentina	+++	(+)	+	+	–	+++	+++	+++	8,55
G.2	Soil from La Plata, Argentina	+++	–	(+)	+	++	+	+++	+++	8,53
G.3	Soil from La Plata, Argentina	+++	–	++	–	++	+	–	(+)	8,22
H.1	Sediments from La Plata, Argentina	+++	++	+++	++	++	+++	+++	+++	8,96
H.2	Bordering soil from a water channel in La Plata, Argentina	+++	++	++	++	++	+++	+	+++	8,49
Sample	Description of samples	P-1	α	β	ε	γ	δ	P-7	P-9	16S log10/g
H.3	Bordering soil from a water channel in La Plata, Argentina	+++	+	++	++	++	+++	–	+++	8,7
I	Sweet-water soil from a river in La Plata, Argentina	+++	+	++	++	++	++	+++	+++	7,91
J	Marginal river forest soil from La Plata, Argentina	++	–	–	–	–	–	–	–	8,32
L.1	Bordering soil from a water channel in Buenos Aires, Argentina	+++	–	++	+	–	++	–	–	8,29
L.2	Bordering soil from a water channel in Buenos Aires, Argentina	+++	+	+++	++	++	+++	–	(+)	8,6
L.3	Bordering soil from a water channel in Buenos Aires, Argentina	+++	–	++	++	++	+++	++	+++	7,66
M	*Halichondria panicea* (marine sponge) from Oosterschelde, Netherlands	++	–	++	–	–	+++	–	–	7,32
N	*Corticium candelabrum* (marine sponge) from Punta Santa Anna (Blanes), Spain	++	–	+	–	–	+	–	–	8,18
O	*Petrosia ficiformis* (marine sponge) from Punta Santa Anna (Blanes), Spain	++	–	++	–	–	+	–	–	8,4
P.1	Askö sediment from Baltic Sea Sweden (bottom fraction - anoxic)	++	–	–	–	–	+	–	–	8,34
P.2	Askö sediment from Baltic Sea Sweden (middle fraction mixed anoxic/oxic)	++	++	+++	++	–	+	–	–	8,43
P.3	Askö sediment from Baltic Sea Sweden (top fraction - oxic)	+++	–	++	+++	–	+	–	–	8,09
Q	Pooled fosmid library, Askö sediment, Baltic Sea	+++	++	+	+	–	+	–	–	5,01
R	Landsort in Sweden	+++	++	–	–	–	–	–	–	8,16
S.1	Sediment from a river in Warwick, UK	+++	/	/	/	/	/	–	(+)	5,78
S.2	Sediment from a river in Warwick, UK	+++	/	/	/	/	/	++	–	6,26
T	Soil from Cuba amended with chitin (Test site 1)	+++	/	/	/	/	/	+++	++	6,95
negative control		–						–	–	

Table 3. Cont.

Sample	Description of samples	P-1	α	β	ε	γ	δ	P-7	P-9	16S log10/g
RP4 (IncP-1α)		+++	+++							
R751 (IncP-1β)		+++		+++						
pKJK5 (IncP-1ε)		++			++					
pQKH54 (IncP-1γ)		+++				+++				
pEST4011 (IncP-1δ)		+++					+++			
pCAR1 (IncP-7)								+++		
pNF 142 (IncP-9)									+++	

Hybridization signal: (+++) very strong, with exposure time up to five minutes; (++) strong, with exposure time up to one hour; (+) weak, with exposure time up to three hours; (−) none, with exposure time of more than three hours; (/) not analyzed.

It is tempting to speculate that degradative genes localized on the plasmid of these groups might contribute to the bacterial degradation of a variety of pollutants such as pesticides, due to the "metabolic complementation" resulting from the combination of different genes brought together by different plasmids. While IncP-1 plasmids typically host genes associated with the degradation of man-made pollutants (xenobiotics) [8], IncP-7 and IncP-9 plasmids often carry genes responsible for degradation of natural contaminants, such as polyaromatic hydrocarbons [9]. Screening TC-DNA revealed that IncP-1, IncP-7 and IncP-9 specific sequences vary according to sample type and degree of pollution. IncP-1 plasmid specific sequences were detected in a wide range of environments: marine sponges, soils and sediments, Baltic Sea sediment fosmid library, biogas production plant, river sediments, chitin-treated soils and BPS contaminated with pesticides. Very strong hybridization signals for all different IncP-1 subgroups tested except for IncP-1α plasmids were especially observed in the BPS samples heavily contaminated with pesticides, indicating an

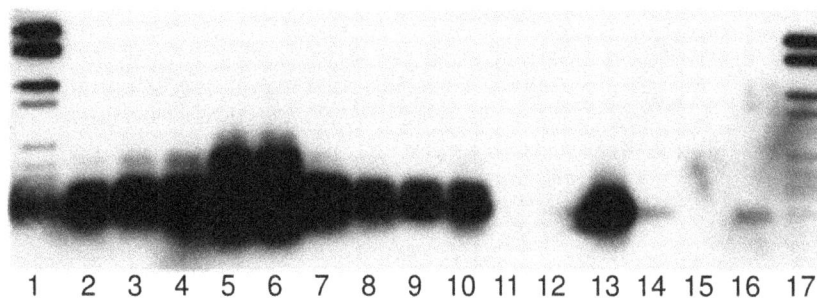

Figure 1. Biopurification systems (BPS). Hybridization of Southern-blotted PCR products obtained with *trfA* primer system from TC-DNA of BPS (IncP-1β specific group). Lanes: 1 and 17, dig ladder; lanes 2 to 4, BPS from Lierde, Belgium; lanes 5 to 7, BPS from Kortrijk, Belgium; lanes 8–10, BPS from Koksijde, Belgium; lane 11, negative control; lanes 12–16, IncP-1 positive controls RP4 (α), R751 (β), pKJK5 (ε), pQKH54 (γ) and pEST4011 (δ). Exposure time of 5 min.

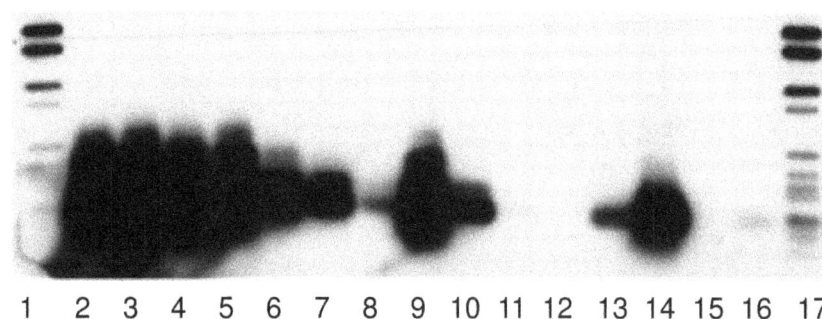

Figure 2. Biopurification systems (BPS). Hybridization of Southern-blotted PCR products obtained with *trfA* primer system from TC-DNA of BPS with the IncP-1ε specific probe. Lanes: 1 and 17, dig ladder; lanes 2 to 4 BPS from Lierde, Belgium; lanes 5 to 7, BPS from Kortrijk, Belgium; lanes 8 to 10, BPS from Koksijde, Belgium; lane 11, negative control; lanes 12 to 15, IncP-1 positive controls RP4 (α), R751 (β), pKJK5 (ε), pQKH54 (γ) and pEST4011 (δ). Exposure time of 5 min.

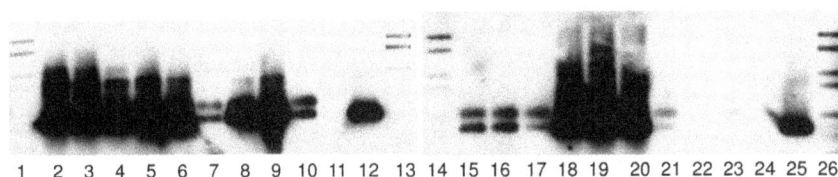

Figure 3. Biopurification systems (BPS). Hybridization of Southern-blotted PCR products obtained with *rep* primer system from TC-DNA of BPS with the IncP-7 probe generated from pCAR1. Lanes: 1, 13 and 26, dig ladder; lanes 2 to 4, BPS from Leefdaal, Belgium; lanes 5 to 10, BPS from Belgium (Pcfruit); lanes 15 to 17, BPS from Lierde, Belgium; lanes 18 to 20, BPS from Kortrijk, Belgium; lanes 21 to 23, BPS from Koksijde, Belgium; lanes 11 and 24, negative control; lanes 12 and 25 IncP-7 positive control pCAR-1. Exposure time of 5 min.

unusual high abundance of bacterial populations carrying IncP-1 plasmids. Indeed, the use of BPS, defined as a pollution control technique employing microorganisms to degrade pesticides through biodegradation processes [10], in on-farm treatment of water contaminated with pesticides has substantially increased and enhanced the degradation rates [11]. Strong hybridization signals of IncP-1β and IncP-1ε plasmids observed in all BPS samples and in some sediments from Argentina contaminated with oil, suggested that IncP-1β and IncP-1ε plasmids might be important in the local adaptation of bacteria to changing environmental conditions [12,13]. Strong IncP-1 plasmid hybridization signals observed in sediments from different regions: Warwick (UK), La Plata (Argentina) and sediments from Sweden indicated that IncP-1 plasmids might also have an important ecological role in the adaptation and biodegradation processes in sediments as previously reported already for mercury-contaminated sediments in Kazakhstan [14]. The apparently high abundance of IncP-1 plasmids in soils from different regions contaminated with different pollutants, such as soils from Argentina polluted with oil and soils from the UK enriched with chitin, also suggested that IncP-1 plasmids might substantially contribute to the adaptation and survival of the soil bacterial communities in response to wide range of environmental pollutants [8,15–17]. The results from several studies suggested a correlation between IncP-1 plasmid abundance and pollution as hypothesized by Smalla *et al.* (2006) and confirm

previously published quantitative data on the abundance of IncP-1 plasmids in BPS samples from one BPS site by means of a qPCR targeting the *korB* gene. Obviously, the relative abundance of IncP-1 plasmids can only be precisely quantified by quantitative real-time PCR. However, the recently developed *korB* quantitative PCR system [18] cannot indicate the relative abundance of the different IncP-1 subgroups which was achieved with specific probes for different IncP-1 groups used in the present study in a semi-quantitative manner.

The study by Sevastsyanovich *et al.* (2008) already showed that IncP-9 plasmid diversity is much broader than previously imagined. In view of this huge plasmid diversity, a novel IncP-9 primer system was developed and established in the present work. Typically, IncP-9 plasmids are related to the degradation of natural pollutants as polyaromatic hydrocarbons [19]. However, the detection of very strong IncP-9 hybridization signals mainly in BPS indicated that populations carrying IncP-9 plasmids are also important players in the degradation of man-made pollutants or wood-derived aromatic compounds. IncP-9 plasmids often possess different aromatic-ring degradation genes. BPS typically contain wood chips but also various aromatic ring-containing pesticides such as bentazon, epoxiconazol and diflufencian [20], which could explain the high abundance of IncP-9 plasmids observed in BPS. Cloning and sequencing of amplicons obtained with the novel IncP-9 primers from BPS TC-DNA confirmed not only the specificity of the primers but also showed the presence of plasmids with high similarity to pWWO, that were previously reported to carry degradative genes (Fig. 4) [21,22]. The presence of several sequences with high similarity to the *oriV-rep* sequence of pM3, an antibiotic resistance plasmid belonging to the IncP-9α subgroup, in BPS 2 might be caused by manure addition in the beginning of every year (on March) by the farmers as a C-source in BPS material. Therefore, the addition of manure in BPS as nutrient source for the microorganisms might be reconsidered and replaced for an alternative one.

The indication of high abundance of IncP-9 plasmids in soils from Argentina contaminated with oil is not too surprising. IncP-9 plasmids are important vehicles for the dissemination of genes coding for enzymes involved in the degradation of polycyclic aromatic hydrocarbons (PAH) and are very often found in environments polluted with oil [23] (Flocco *et al.*, unpublished).

PCR-Southern blot hybridization results showed that bacteria hosting IncP-7 plasmids were also highly abundant in BPS, indicating a role of these plasmids in the degradation of man-made pollutants such as pesticides. It can be concluded that PCR-Southern blot hybridization detection of plasmid-specific sequences from TC-DNA is a suitable and specific but semi-quantitative approach to investigate the occurrence of plasmid-specific sequences in different environments and in a large number of samples. The detection of plasmids was possible independently of

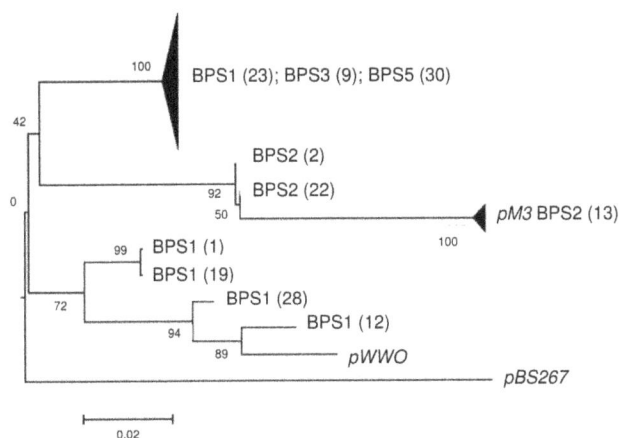

Figure 4. Neighbor-Joining phylogenetic tree based on the multiple alignment of cloned amplicon sequences of the *oriV-rep* IncP-9 gene. Sequences from known IncP-9 plasmids have been included as references. Value at each node is percent bootstrap support of 1,000 replicates. BPS1; BPS2 and BPS5 correspond to three different biopurification systems (BPS), located in Belgium. Numbers in brackets correspond to number of clones and numbers without brackets correspond to the clone designation.

the cultivation of their original hosts [1] and indicated "hot spots" of IncP-1, IncP-7 and IncP-9 plasmids, such as BPS.

Author Contributions

Performed the experiments: SD VW. Analyzed the data: SD GCD HH KS. Contributed reagents/materials/analysis tools: HH CT. Wrote the paper: SD. Contributed sending samples: VD AS MCM MFP AL GCA EMHW WHG D Sipkema SS D Springael JDE. Corrected the writing: KS.

References

1. Heuer H, Smalla K (2012) Plasmids foster diversification and adaptation of bacterial populations in soil. FEMS Microbiology Reviews 36: 1083–1104.
2. Bahl MI, Burmølle M, Meisner A, Hansen LH, Sørensen SJ (2009) All IncP-1 plasmid subgroups, including the novel ε subgroup, are prevalent in the influent of a Danish wastewater treatment plant. Plasmid 62: 134–139.
3. Hardeman F, Sjoling S (2007) Metagenomic approach for the isolation of a novel low-temperature-active lipase from uncultured bacteria of marine sediment. FEMS Microbiology Ecology 59: 524–534.
4. Izmalkova TY, Sazonova OI, Sokolov SL, Kosheleva IA, Boronin AM (2005) The P-7 incompatibility group plasmids responsible for biodegradation of naphthalene and salicylate in fluorescent pseudomonads. Microbiology 74: 290–295.
5. Schauss K, Focks A, Heuer H, Kotzerke A, Schmitt H, et al. (2009) Analysis, fate and effects of the antibiotic sulfadiazine in soil ecosystems. Trac-Trends in Analytical Chemistry 28: 612–618.
6. Suzuki MT, Taylor LT, DeLong EF (2000) Quantitative analysis of small-subunit rRNA genes in mixed microbial populations via 5'-nuclease assays. Applied and Environmental Microbiology 66: 4605–4614.
7. Sevastsyanovich YR, Krasowiak R, Bingle LEH, Haines AS, Sokolov SL, et al. (2008) Diversity of IncP-9 plasmids of Pseudomonas. Microbiology-Sgm 154: 2929–2941.
8. Krol JE, Penrod JT, McCaslin H, Rogers LM, Yano H, et al. (2012) Role of IncP-1 beta plasmids pWDL7::rfp and pNB8c in chloroaniline catabolism as determined by genomic and functional analyses. Applied and Environmental Microbiology 78: 828–838.
9. Jutkina JHE, Vedler E, Juhanson J, Heinaru A (2011) Occurrence of plasmids in the aromatic degrading Bacterioplankton of the Baltic Sea. Genes 2: 853–868.
10. Castillo MdP, Torstensson L, Stenström J (2008) Biobeds for environmental protection from pesticide use - a review. Journal of Agricultural and Food Chemistry 56: 6206–6219.
11. Omirou M, Dalias P, Costa C, Papastefanou C, Dados A, et al. (2012) Exploring the potential of biobeds for the depuration of pesticide-contaminated wastewaters from the citrus production chain: Laboratory, column and field studies. Environmental Pollution 166: 31–39.
12. Trefault N, De la Iglesia R, Molina AM, Manzano M, Ledger T, et al. (2004) Genetic organization of the catabolic plasmid pJP4 from Ralstonia eutropha JMP134 (pJP4) reveals mechanisms of adaptation to chloroaromatic pollutants and evolution of specialized chloroaromatic degradation pathways. Environmental Microbiology 6: 655–668.
13. Oliveira CS, Lazaro B, Azevedo JSN, Henriques I, Almeida A, et al. (2012) New molecular variants of epsilon and beta IncP-1 plasmids are present in estuarine waters. Plasmid 67: 252–258.
14. Smalla K, Haines AS, Jones K, Krögerrecklenfort E, Heuer H, et al. (2006) Increased abundance of IncP-1 beta plasmids and mercury resistance genes in mercury-polluted river sediments: First discovery of IncP-1 beta plasmids with a complex mer transposon as the sole accessory element. Applied and Environmental Microbiology 72: 7253–7259.
15. Top EM, Springael D (2003) The role of mobile genetic elements in bacterial adaptation to xenobiotic organic compounds. Current Opinion in Biotechnology 14: 262–269.
16. Sen DY, Brown CJ, Top EM, Sullivan J (2013) Inferring the Evolutionary History of IncP-1 Plasmids Despite Incongruence among Backbone Gene Trees. Molecular Biology and Evolution 30: 154–166.
17. Dunon VSK, Bers K, Lavigne R, Smalla K, Springael D (2013) High prevalence of IncP-1 plasmids and IS1071 insertion sequences in on-farm biopurification systems and other pesticide polluted environments. FEMS Microbiology Ecology.
18. Jechalke S, Dealtry S, Smalla K, Heuer H (2013) Quantification of IncP-1 plasmid prevalence in environmental samples. Applied and Environmental Microbiology 79: 1410–1413.
19. Gomes NCM, Flocco CG, Costa R, Junca H, Vilchez R, et al. (2010) Mangrove microniches determine the structural and functional diversity of enriched petroleum hydrocarbon-degrading consortia. FEMS Microbiology Ecology 74: 276–290.
20. Fetzner S, Lingens F (1994) Bacterial Dehalogenases: Biochemistry, Genetics, and Biotechnological Applications. Microbiological Reviews 58: 641–685.
21. Greated A, Lambertsen L, Williams PA, Thomas CM (2002) Complete sequence of the IncP-9 TOL plasmid pWW0 from Pseudomonas putida. Environmental Microbiology 4: 856–871.
22. Sota M, Yano H, Ono A, Miyazaki R, Ishii H, et al. (2006) Genomic and functional analysis of the IncP-9 naphthalene-catabolic plasmid NAH7 and its transposon Tn4655 suggests catabolic gene spread by a tyrosine recombinase. Journal of Bacteriology 188: 4057–4067.
23. Izmalkova TY, Mavrodi DV, Sokolov SL, Kosheleva IA, Smalla K, et al. (2006) Molecular classification of IncP-9 naphthalene degradation plasmids. Plasmid 56: 1–10.
24. Zhou Y (1996) Two-phase anaerobic digestion of water hyacinth pretreated with dilute sulphuric acid. Huanjing Kexue 17: 13–92.
25. Sipkema D, Blanch HW (2010) Spatial distribution of bacteria associated with the marine sponge Tethya californiana. Marine Biology 157: 627–638.
26. Edlund A, Hardeman F, Jansson JK, Sjoling S (2008) Active bacterial community structure along vertical redox gradients in Baltic Sea sediment. Environmental Microbiology 10: 2051–2063.
27. Gaze WH, Zhang LH, Abdouslam NA, Hawkey PM, Calvo-Bado L, et al. (2011) Impacts of anthropogenic activity on the ecology of class 1 integrons and integron-associated genes in the environment. ISMEJournal 5: 1253–1261.
28. Byrne-Bailey KG, Gaze WH, Kay P, Boxall ABA, Hawkey PM, et al. (2009) Prevalence of sulfonamide resistance genes in bacterial isolates from manured agricultural soils and pig slurry in the United Kingdom. Antimicrobial Agents and Chemotherapy 53: 696–702.

21

Spatial Interpolation of Fine Particulate Matter Concentrations Using the Shortest Wind-Field Path Distance

Longxiang Li[1], Jianhua Gong[2], Jieping Zhou[1]*

1 Institute of Remote Sensing and Digital Earth, Chinese Academy of Sciences, Olympic Science & Technology Park of CAS, Beijing, China, **2** Institute of Remote Sensing and Digital Earth, Chinese Academy of Sciences, Olympic Science & Technology Park of CAS, Beijing, China and Zhejiang-CAS Application Center for Geoinformatics, Jiashan, Zhejiang, China

Abstract

Effective assessments of air-pollution exposure depend on the ability to accurately predict pollutant concentrations at unmonitored locations, which can be achieved through spatial interpolation. However, most interpolation approaches currently in use are based on the Euclidean distance, which cannot account for the complex nonlinear features displayed by air-pollution distributions in the wind-field. In this study, an interpolation method based on the shortest path distance is developed to characterize the impact of complex urban wind-field on the distribution of the particulate matter concentration. In this method, the wind-field is incorporated by first interpolating the observed wind-field from a meteorological-station network, then using this continuous wind-field to construct a cost surface based on Gaussian dispersion model and calculating the shortest wind-field path distances between locations, and finally replacing the Euclidean distances typically used in Inverse Distance Weighting (IDW) with the shortest wind-field path distances. This proposed methodology is used to generate daily and hourly estimation surfaces for the particulate matter concentration in the urban area of Beijing in May 2013. This study demonstrates that wind-fields can be incorporated into an interpolation framework using the shortest wind-field path distance, which leads to a remarkable improvement in both the prediction accuracy and the visual reproduction of the *wind-flow effect*, both of which are of great importance for the assessment of the effects of pollutants on human health.

Editor: Qinghua Sun, The Ohio State University, United States of America

Funding: This work was partially supported by the Plans of National Sci-Tech Major Special Item 2014ZX10003002, the National Science Foundation of China 41301437, the Foundation of the State Key Laboratory of Remote Sensing Sciences in China and the National Science Foundation of China 41371387. The funders had no role in study design, data collection and analysis, decision to publish, or preparation of the manuscript.

Competing Interests: The authors have declared that no competing interests exist.

* E-mail: jpzhou@irsa.ac.cn

Introduction

Public health studies of air-pollution exposure require accurate predictions of concentrations at unmonitored locations to minimize the misclassification of exposure levels [1]. Recent studies have reported that intra-urban-scale variations in air-pollution concentration may exceed the differences between cities [2,3], suggesting the potential importance of predicting air-pollution at fine spatial scales [4,5]. Correspondingly, estimates of exposure to pollutants on small temporal scales are necessary to study the short-term or acute impacts of air-pollution [6]. Spatial interpolation models, landuse regression (LUR) models, remote-sensing-based models and diffusion models are robust tools for intra-urban air-pollution prediction [7,8]. However, spatial interpolation techniques, which generate concentration surfaces from *in situ* observations, are preferred for the estimation of real-time concentrations when data availability and software and hardware costs are taken into account [3,9,10].

Wind is a key meteorological factor that has major impacts on the movement and distribution of air pollutants in a region. When the wind-speed is relatively high, local wind-field exert substantial influence on the horizontal transport of air-pollution; this phenomenon is known as the *wind-flow effect* [11,12]. For example, areas downwind of highways are more heavily exposed to traffic-related pollutants than are upwind areas. This effect illustrates the necessity of incorporating wind-field into spatial interpolation. In a number of recent studies, the consideration of a negative correlation between air-pollution concentration and wind-speed has led to the application of the wind-speed as an auxiliary variable in multi-variable interpolation methods [9,13,14]. Although there have been several attempts to incorporate long-term, large-scale wind-fields into corresponding air-pollution estimations, short-term, small-scale wind-fields have not been extensively used for this purpose, because no direct numerical relations exist between the angle of the wind-direction and the concentration level in such cases. As a result, these approaches fail to capture the expected short-term effects of the wind flow.

By including the wind-fields indirectly, some regression-based methods are able to capture the complex features of pollutant distributions [2,3,14]. A recent study assessed the use of the wind-direction in LUR to improve predictions of nitrogen dioxide levels in Toronto-Hamilton area [11,15]. This method shows great potential, as it quantifies the influence of the wind-direction with the downwind distances from highways. However, real-time air-pollution assessment using this model is economically infeasible because of the cost of collecting sufficiently diverse data sets. Therefore, one objective of the present study is to incorporate wind-fields directly into interpolation frameworks.

Most interpolation techniques depend on Euclidean or straight-line distances to compute spatial dependency. However, the complex features of certain spatial phenomena impede the ability to obtain accurate dependency descriptions using Euclidean distances [16,17]. An appropriate non-Euclidean distance may outperform the Euclidean distance in determining such types of spatial dependency and in capturing complex features [18]. The shortest path distance (SPD) is an important subclass of non-Euclidean distance and has exhibited great potential in diverse interpolation studies. The hydrological distance, a derivative of the SPD, has been used to characterize the spatial configurations, connectivity and directionality of the water temperatures and chemical pollutants in stream networks [17,19–22]. Accounting for geological anisotropy has led to the interpolation of deposits in conjunction with shortest anisotropy path distances [23]. Along-road continuity has been described in carbon dioxide estimations after replacing the Euclidean distance with the SPD [24]. The inclusion of the effect of topographical factors in simulations of the genetic dispersion path has led to the development of the concept of effective distance, which is the length of a virtual movement route. Although the quantitation of these factors remains unclear, this metric exhibits a greater correlation with genetic variance than does the straight-line distance and has been used to characterize the nonlinear features of genetic dispersal [25]. Road-network connectivity has also been incorporated into the interpolations of urban travel speeds using the approximate road-network distance, another derivative of the SPD [26]. The works listed above are important references for the methodology presented in this paper. However, the characteristics of the wind-flow effect differ from those investigated in these previous works. Current SPD techniques are insufficient to successfully capture such features. To address this shortcoming, another derivation of SPD, the shortest wind-field path distance (SWPD), is proposed to determine the spatial dependency effected by a wind-field and is exploited to integrate wind-fields into interpolation frameworks.

In this study, a new interpolation method based on the SPD is developed to describe the influence of the wind-field. This technique is then applied to generate concentration surfaces for fine particulate matter of less than $2.5 \mu m$ in diameter ($PM_{2.5}$) on the experimental dates in the study area. Comparisons are performed between this technique and the conventional methodology to illustrate the improvements achieved by including the wind-field.

Materials and Methods

Data collection and processing

Study area and context. Beijing, the capital of the People's Republic of China, is an international metropolis and has experienced a rapid increase in urban population, energy consumption and vehicle numbers over the past several decades [27]. An urban area inside the surrounding ring road (Fig. 1) was selected as the study area (approximately 30×30 km) because of the relatively dense air-pollution monitoring network that is present in this area. Six non-consecutive days in May 2013 with major air-pollution in terms of PM2.5 and daily wind-speeds above 1.5 m/s were selected as the experimental dates because no accurate wind-direction measurements were available for low-speed wind conditions.

Observed PM2.5 concentration. To improve air-pollution monitoring, a network of 35 automated stations has been established by the Beijing Environmental Protection Bureau (BJEPB). Each station measures hourly $PM_{2.5}$ concentrations and releases real-time data to the public through the Beijing Municipal Environmental Monitoring Center (www.bjmemc.com.cn). 13 urban sampling sites compose a dense urban monitoring subsystem across the study area (Fig. 1). This monitoring network enables the detection of real-time, small-area variations in the $PM_{2.5}$ concentration. We collected an experimental data set from all 13 sites for six selected dates. The concentration data are given in units of $\mu g/m^3$. Daily average concentrations were calculated from the hourly data.

Observed wind-fields. Hourly wind-field observations were obtained through the Chinese Meteorological Data Service Platform (cdc.bjmb.gov.cn). The measurements were collected over a network of 16 weather stations throughout the study area operated by the Beijing Meteorological Bureau (Fig. 1). The daily average wind-speed and wind-direction were calculated from the hourly real-time data. Influenced by the complex urban morphology, the urban wind-fields exhibit dramatic small-scale variations that cannot be captured by model-simulated fields with overlarge grid sizes [11].

Methodology

At the heart of the proposed method is the shortest-path analysis. An appropriate simulation of air-pollution movement using this method entails the construction of a cost surface, which lays the foundation for the shortest-path analysis. After the SWPD between every pair of unmonitored locations and sampling locations is obtained, this distance metric is used to determine the spatial dependency. Inverse Distance Weighting (IDW) in conjunction with SWPD is then implemented to calculate the concentration surface. Therefore, this method consists of generating a continuous wind-field, then implementing the shortest-path analysis and finally creating the estimation surface using SWPD-based IDW.

Generation of continuous wind-field. The creation of the cost surface for shortest-path analysis first requires the generation of continuous wind-field. This process also involves spatial interpolation. Unlike other scalar weather variables, a wind-field is a vector quantity whose interpolation is unique in meteorology. Typically, one wind-vector is decomposed into two Cartesian wind components (an east-west component and a north-south component). Each component is then interpolated separately into a corresponding surface using multiquadric (MQ) radial basis functions (RBF) [28]. The wind-field is then constructed backward from the two Cartesian-component surfaces using trigonometry (Fig. 2). This methodology has been widely used to interpolate diverse vector-type data since its proposal and is considered to be a robust approach for various meteorological studies [28]. In this study, a continuous wind-field is established using a grid size of 0.5 km, which is an appropriate resolution for urban air-pollution research [29].

Shortest-path analysis. The shortest-path analysis includes two stages. First, a continuous wind-field is modeled onto a cost surface that depicts the movement cost between adjacent cells. Second, a shortest-path algorithm is implemented to acquire the SWPDs between locations.

Creating a cost surface using wind-field data. The movement of PM2.5 from one location to another may be facilitated or impeded by the local wind-field. A cost surface must be well defined based on the properties of the air-pollution movement to ensure that each shortest path acquired represents

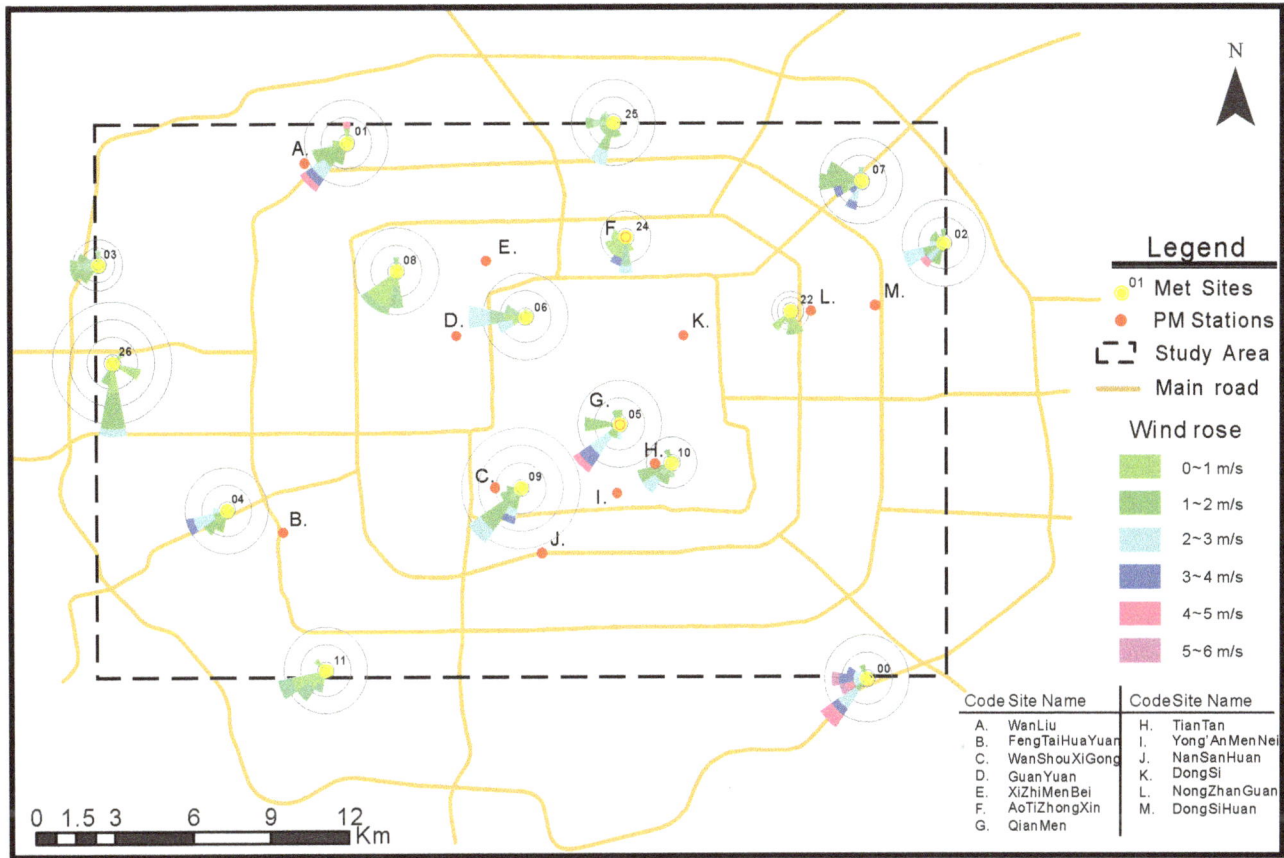

Figure 1. Study area and locations of PM$_{2.5}$ monitoring stations and meteorological sites with daily wind roses for May 18th.

the true movement trajectories and thus reveals the path along which the two locations are related.

The grid-based representation of field data is sufficient to depict the cost of traversing each cell but is incapable of representing the movement cost associated with not only the distance between cells but also the relative positions of adjacent cells. An alternate methodology is to reform the grid raster (Fig. 3A) into a graph (Fig. 3C) with pixels as vertices and virtual connecting lines as edges, where each edge has an associated cost value that indicates the cost of traveling along this edge.

The calculation of the edge cost that depicts the movement difficulty between adjacent cells is performed based on Gaussian dispersion model, which is the standard model for the study of the transport of airborne contaminants under the influence of wind-field. This model simulates a cross section of the air-pollution dispersion and assumes that both the horizontal and vertical concentration distributions are normal [30,31]. The basic formula for this model can be written as follows:

$$C_0(x,y,z,u) = \frac{Q}{\pi u \sigma_y \sigma_z} \exp\left(-\frac{y^2}{2\sigma_y^2} - \frac{z^2}{2\sigma_z^2}\right) \quad (1)$$

where $C_0(x,y,z,u)$ is the concentration at ground level, x is the downwind distance, y is the horizontal distance between the point of interest and the centerline, z is the height of the emission source, u is the horizontal wind-speed, σ_y is the standard deviation of horizontal dispersion and σ_z is the standard deviation of vertical dispersion.

The Cartesian coordinates (x and y) in the model can be transformed into polar coordinates (γ and θ) as shown below [32]:

$$C_0(\gamma,\theta,z,u) = \frac{Q}{\pi u \sigma_\theta \sigma_z} \exp\left(-\frac{z^2}{2\sigma_z^2} - \frac{\gamma^2 \theta^2}{2\sigma_\theta^2} - \Omega \frac{\theta^2}{2}\right) \quad (2)$$

where Ω is a correction term given by

$$\Omega = \alpha(\gamma) + \beta(\gamma)\frac{z^2}{\sigma_z^2} \quad (3)$$

with

$$\alpha(\gamma) = -2 + \frac{p\gamma}{a+\gamma} + \frac{q\gamma}{a+\gamma} \quad (4)$$

$$\beta(\gamma) = 1 - \frac{q\gamma}{a+\gamma} \quad (5)$$

where a, p and q are dispersion parameters that depend on the atmospheric stability. Further information can be found in the cited references. When the focus is placed only on horizontal diffusion and downwind advection, the formula can be written as

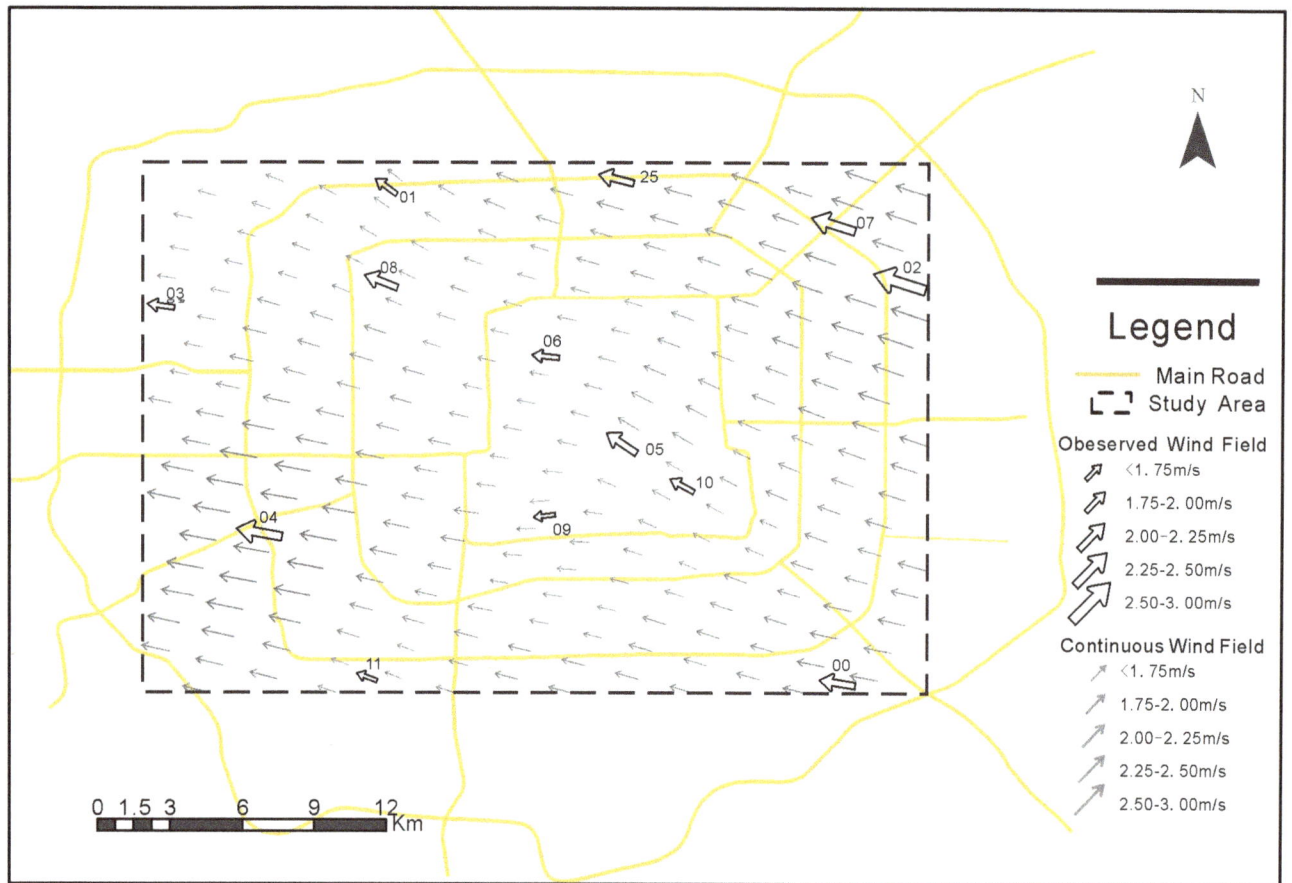

Figure 2. Observed wind-field and continuous wind-field generated from the monitoring data of May 28th.

$$C_0(\gamma,\theta,u) = \frac{Q}{\pi u \sigma_\theta}\exp(-\frac{\gamma^2\theta^2}{2\sigma_\theta^2}-\Omega\frac{\theta^2}{2}) \qquad (6)$$

which indicates that the concentration will decrease with increasing downwind distance (γ) and azimuth (θ), both of which are used to determine the movement cost in the following steps. However, this formula is too complicated to be directly used in calculation, and the following simplified version is applied instead:

$$Cost(E_{AB}) = [\mathbb{F}(D_A,D_M)+\mathbb{F}(D_B,D_M)] \times L_{AB} \qquad (7)$$

where A and B are two points in the wind-field, E_{AB} is the edge connecting these two points, D_A and D_B are the wind-directions at these two points and D_M is the direction of E_{AB}, namely, the potential movement direction. L_{AB} is the length of E_{AB}, which is functionally equivalent to γ. The function $\mathbb{F}$ is used to calculate the azimuth which is functionally identical to θ. It is important to note that the Gaussian dispersion model is designed to simulate the diffusion of contaminants from definite sources and cannot be directly applied in this study because there are no stable emission sources. However, this model illustrates the origin of the movement cost, which serves as the foundation of this section.

Implementing shortest-path analysis. Based on the establishment of the cost surface, shortest-path algorithms can be used to calculate the paths with the minimum accumulated movement cost between point pairs, indicating the most likely path along

which the two points are related (Fig. 3d). In this study, the classic Dijkstra algorithm [33] is employed. The shortest paths and their associated costs are computed as the output of this step. However, SWPD is not measured as a summation of movement costs but as the total length of the shortest path segments between location pairs.

Interpolation based on SWPD. After the SWPD between each pair of prediction locations and measurements is acquired, SWPD-based interpolation can be used to generate estimation surfaces. IDW in conjunction with the SWPD was selected as the technique used to incorporate wind-field in this study. Although the total sample size in this study is small, the density of the observation network permits the estimation of small-area variations in the air-pollution concentration using the proposed method [34]. The feasibility of the method is confirmed by the low mean squared error and mean absolute error in the following sections.

The IDW approach aims to predict the pollutant concentration at a given location based on a weighted average of the measurements obtained at surrounding stations. As a direct application of Tobler's First Law (TFL), the relations between the point of interest and the nearby stations are determined by the distances between them. The method takes the following form:

$$z^*(u) = \sum_{\alpha=1}^{n} \lambda_\alpha z(u_\alpha) \qquad (8)$$

where $z^*(u)$ is the estimate at location u, $z(u_\alpha)$ is the measurement at

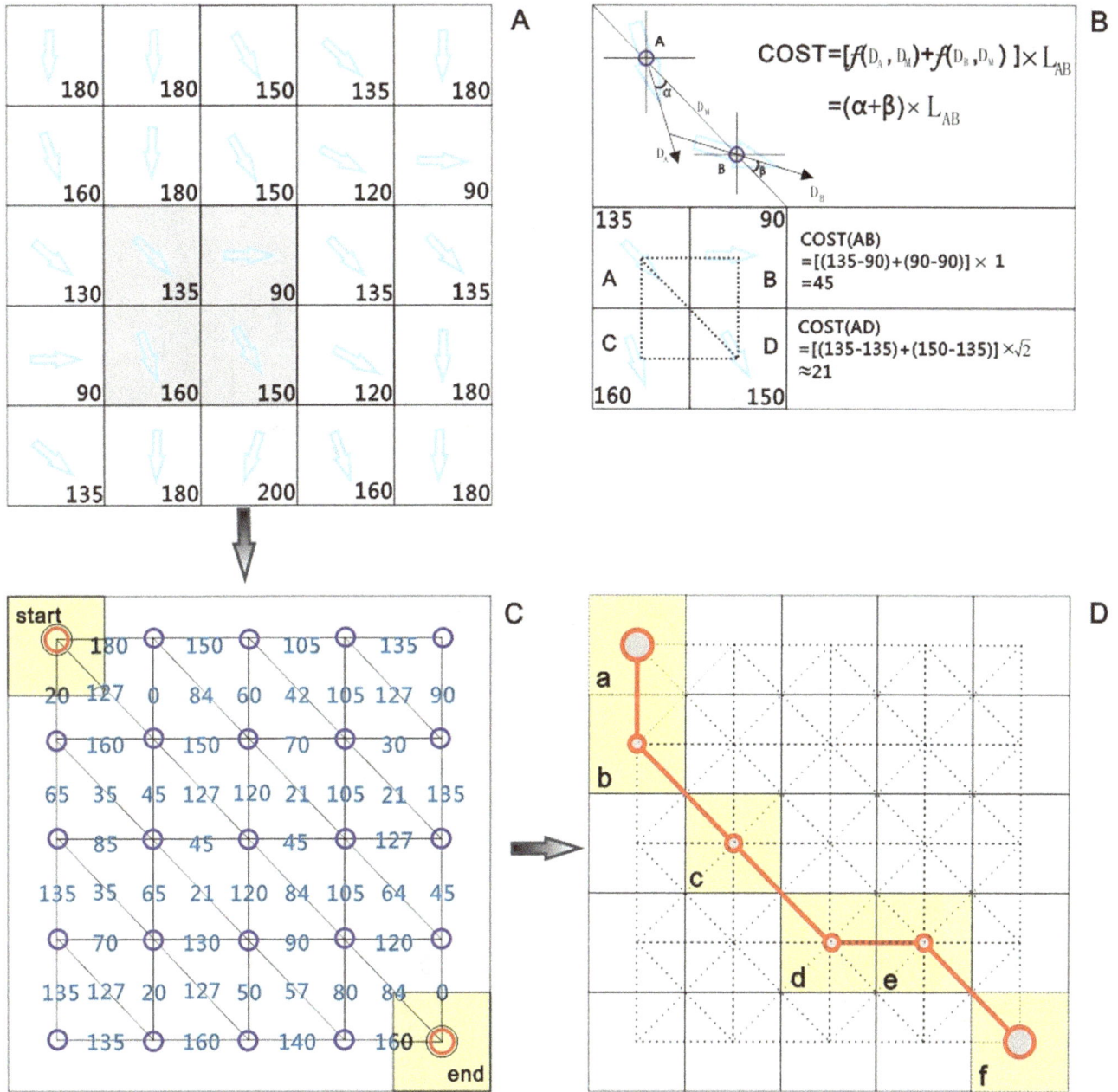

Figure 3. Steps for calculating SWPDs using a wind-field. (A) Grid-based representation of a wind-field. (B) Computing the cost associated with the edge between adjacent cells. (C) Reforming a grid-based wind-field into a graph. (D) SWPD calculated between starting point A and ending point F.

locationu_α, n is the number of stations used for the estimation and$\lambda_\alpha$is the interpolation weight of the measurement atu_α. The calculation of$\lambda_\alpha$takes the following form:

$$\lambda_\alpha = \frac{1/D_\alpha^\omega}{\sum_{\alpha=1}^n 1/D_\alpha^\omega} \qquad (9)$$

where $D_\alpha$is the distance between the monitoring station numbered$\alpha$and the point of interest; ω is the exponent of the distance and is set to 2 by default. The incorporation of the SWPD can be achieved by replacing the distances with the SWPD values, as

demonstrated below:

$$\lambda_\alpha = \frac{1/SWPD_\alpha^\omega}{\sum_{\alpha=1}^n 1/SWPD_\alpha^\omega} \qquad (10)$$

where$SWPD_\alpha$is the shortest path distance in the wind-field between the monitoring station numbered$\alpha$and the point of interest.

Results

Here, comparisons are made on various temporal scales between IDW based on the SWPD (IDWS), as proposed in this paper, and IDW based on the Euclidean distance (IDWE). First, a cross-validation by "leaving one out" is performed to assess the estimation accuracy, as described below. Second, the abilities of the two methods to visually reproduce the wind-flow effect in the interpolation results are also compared. The method used in cross-validation involves temporarily removing one $PM_{2.5}$ measurement from the data set and then predicting the concentration at this location based on the remaining measurements using the same methodology.

The three comparison criteria below are used to assess the performance of the interpolations:

$$MSE = \frac{1}{n}\sum_{\alpha=1}^{n}[z(u_\alpha) - z^*(u_\alpha)]^2 \qquad (11)$$

$$MAE = \frac{1}{n}\sum_{\alpha=1}^{n}|z(u_\alpha) - z^*(u_\alpha)| \qquad (12)$$

$$MRE = \frac{1}{n}\sum_{\alpha=1}^{n}\frac{z(u_\alpha) - z^*(u_\alpha)}{z(u_\alpha)} \qquad (13)$$

The mean squared error (MSE) measures the average squared difference between the removed true $PM_{2.5}$ measurement $z(u_\alpha)$ and its estimate $z^*(u_\alpha)$. The mean absolute error (MAE) measures the average absolute difference between $z(u_\alpha)$ and $z^*(u_\alpha)$. The mean relative error (MRE) measures the average relative deviation between $z(u_\alpha)$ and $z^*(u_\alpha)$. In the case of reasonably accurate estimation, the values of all three statistics should be close to zero [26,34,35].

Interpolation of the daily PM$_{2.5}$ concentration

Daily estimation surfaces for the $PM_{2.5}$ concentration on the experimental dates were calculated using IDWS and IDWE. The local prediction is improved when wind-fields are incorporated, as evidenced by the average decrease of 15.66% in the MSE, the average decrease of 6.46% in the MAE and the evident decline in the MRE obtained for the IDWS estimation compared to the IDWE estimation (Table 1). The spatial distributions of the relative errors are presented in Table 2.

In addition to the benefit of lowering these three statistics, improvements are evident when the estimation surfaces obtained using two methods are compared visually (Fig. 4). The two methods produce different distributions when the $PM_{2.5}$ value measured at a single location is much higher than those measured at surrounding stations. In the results obtained using IDWS, greater continuity is apparent on the downwind side of the downtown area and there is a shorter dispersion distance on the upwind side, as would be expected from the wind-flow effect (Fig. 4A). IDWE methods always produce an eye-shaped pattern in such cases, which is commonly considered to be a major shortcoming of this interpolation method (Fig. 4B). Under the complex local wind-field northwest of the urban area, the results of the interpolation method proposed in this paper also exhibit an

Table 1. Comparison of the interpolation accuracies achieved using MSE, MAE and MRE based on a cross-validation analysis.

Date	PM$_{2.5}$($\mu g/m^3$)	wind speed (m/s)	wind direction(°)	MSE of IDWS	MSE of IDWE	MAE of IDWS	MAE of IDWE	MRE of IDWS	MRE of IDWE	Improvement in MSE (%)	Improvement in MAE (%)
4th May	92.2	2.19	206.2	49.33	62.51	5.81	6.24	−0.36%	−1.21%	21.07%	7.40%
18th May	118.8	1.96	194.2	105.86	129.71	8.69	8.91	0.93%	2.53%	18.39%	2.53%
21st May	110.8	1.65	93.4	695.81	781.08	18.54	20.99	1.48%	5.03%	10.92%	13.21%
26th May	106.9	1.71	101.9	85.78	90.23	8.17	8.09	−0.27%	0.41%	4.93%	−0.98%
28th May	17.7	1.99	285.3	96.71	113.19	8.22	9.09	0.22%	0.35%	14.56%	10.58%
29th May	78.8	1.53	251.7	13.83	18.22	2.82	2.99	−1.63%	−2.59%	24.11%	6.03%

Table 2. Distributions of the relative errors of the two methods for the daily $PM_{2.5}$ estimation on 21st May.

PM station Number	Estimation of IDWS	Estimation of IDWE	Measured value	Relative error of IDWS	Relative error of IDWE
A	100.9	100.2	129	−21.78%	−22.33%
B	116.8	115.9	100	16.80%	15.90%
C	115.7	121.6	104	11.25%	16.92%
D	113.7	112.6	95	19.68%	18.53%
E	100.7	106.1	104	−3.17%	2.02%
F	106.2	107.6	98	8.37%	9.80%
G	101.3	103.2	179	−43.41%	−42.35%
H	123.8	136.3	98	26.33%	39.08%
I	111.2	117.5	112	−0.71%	4.91%
J	111.9	113.6	118	−5.17%	−3.73%
K	111	118.4	100	11.00%	18.40%
L	105.6	112.5	91	16.04%	23.63%
M	94.9	95.6	113	−16.02%	−15.40%

accordingly complex anisotropy (Fig. 4A). However, the estimation surface obtained using IDWE fails to capture this feature (Fig. 4B).

Interpolation of the hourly $PM_{2.5}$ concentration

Here, we consider the process of interpolating the $PM_{2.5}$ concentration on a smaller temporal scale. Although IDWS outperforms IDWE on most experimental dates, the improvement in the MSE on the 26th of May is only 4.93%, and the MAE of IDWS is larger than that of IDWE (Table 1). On May 26th, the $PM_{2.5}$ concentration increased gradually during the diurnal hours, reached a peak at approximately 4 pm and then decreased dramatically because of the washout caused by a moderate rainfall event that offset the impact of increasing traffic volume during the evening rush hour (Fig. 5). The hourly measured $PM_{2.5}$ concentrations were interpolated from 6 am to 8 pm. Twelve of the 15 experimental hours exhibit smaller MSE values in the IDWS estimation than in the IDWE estimation, whereas the remaining three hours exhibit larger MSE values, suggesting that

the incorporation of the wind-fields had a negative influence on the interpolation accuracy during these three hours. The hours immediately preceding and immediately following these three hours also exhibit a limited improvement of less than 10%. Moreover, the improvement-ratio curve of the MAE follows a similar trend. Thus, two valleys appear in the curves: at noon and at sunset.

The prevailing wind-direction was approximately 65° NE in the morning (Fig. 6A) and changed to 120° SE in the afternoon (Fig. 6C). From 11 am to 1 pm, corresponding to the first valley in both improvement-ratio curves, the prevailing direction experienced dramatic variations from northeast to southeast (Fig. 6B), which limited the accuracy of the wind-field measurements. Because modest direction errors on the order of 10 degrees can lead to large errors in the estimation of air-pollution trajectories, the effectiveness of the proposed methodology no longer holds in cases of strongly varying wind-direction [36]. As is also indicated by the meteorological data set, the study area experienced rainfall from 4 pm to 6 pm, corresponding to the second valley. Because

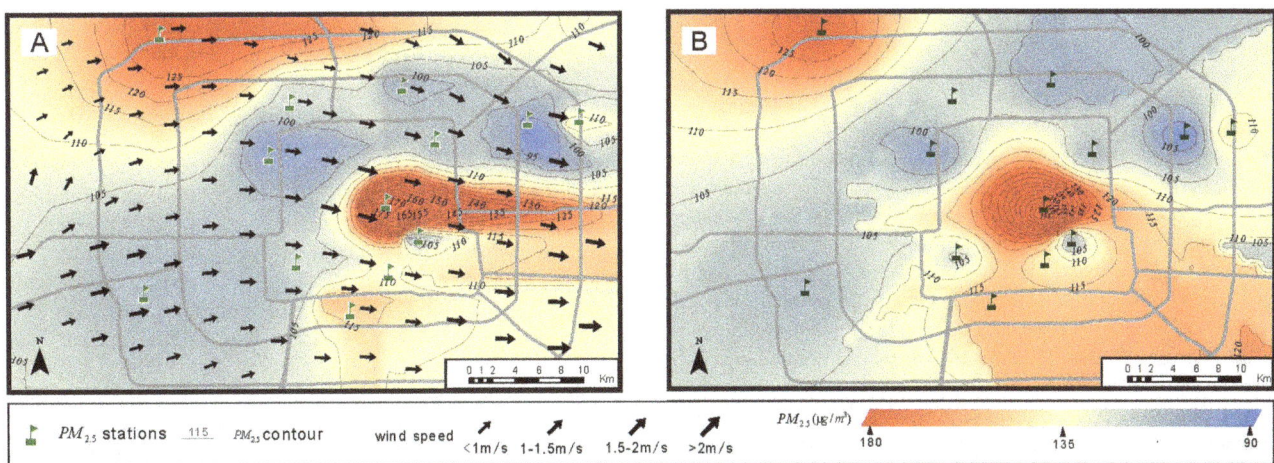

Figure 4. Comparison between the daily $PM_{2.5}$ (May 21st) estimations obtained using IDW based on the SWPD and the Euclidean distance. (A)Interpolation results obtained using IDWS. (B) Interpolation results obtained using IDWE.

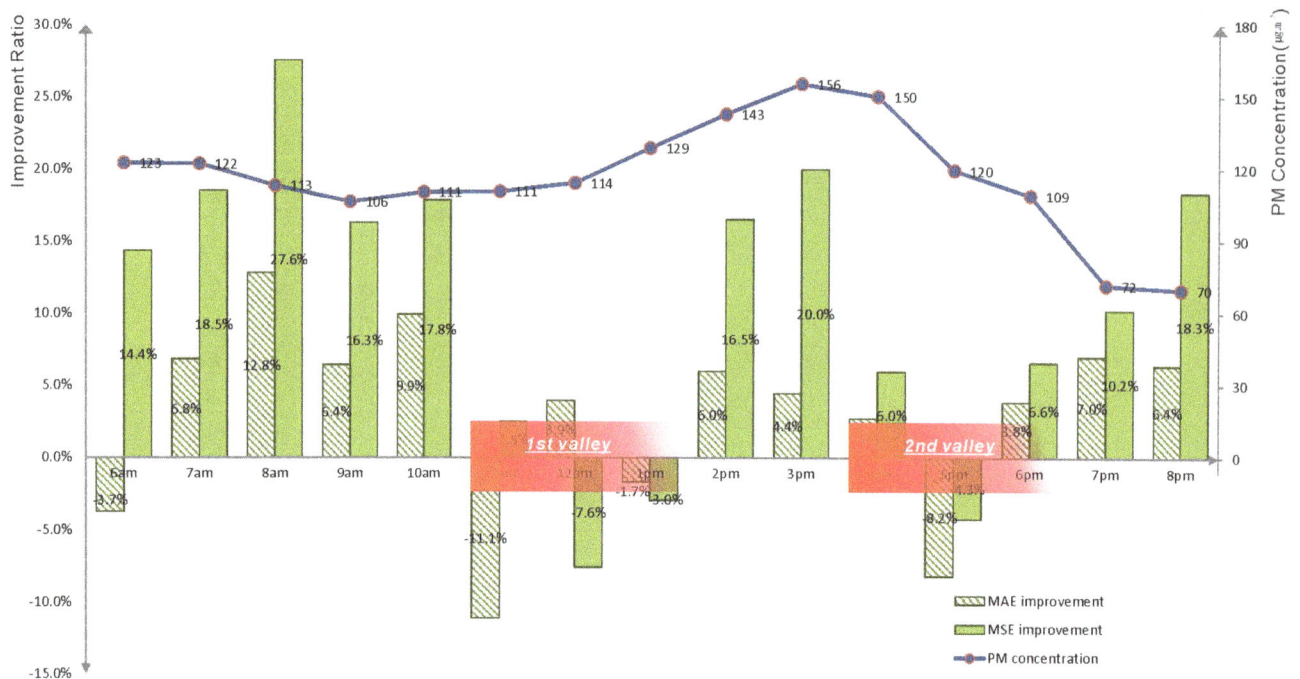

Figure 5. Temporal variation in the hourly PM$_{2.5}$ concentration and the improvement ratio of IDWS with respect to IDWE on May 26th.

precipitation accelerates the deposition of particulate matter, the transport effects of the wind-field were reduced (Fig. 6D). However, under weather conditions with fewer or weaker variations, the methodology proposed in this paper yielded better results on an hourly temporal scale than did the conventional interpolation method.

Discussion

This study demonstrates the potential of incorporating wind-field into interpolation using the IDWS approach. In addition to minimizing estimation errors, a major advantage of this approach stems from its ability to reproduce complex nonlinear features caused by the wind-flow effect. This capability deserves further investigation for its potential use in studies of air-pollution and the negative health effects thereof. As shown in Fig. 4A, the asymmetric distribution of PM$_{2.5}$ on the two sides of the downtown area suggests that residents living east of downtown were exposed to higher concentrations, whereas those living to the west were protected by the wind. By contrast, the symmetric distribution predicted by IDWE (Fig. 4B) may overestimate the PM$_{2.5}$ exposure of upwind residents. Furthermore, the estimation surface obtained using IDWS exhibits greater downwind continuity. Ignoring the wind-flow effect will lead to the underestimation of downwind dispersion distance and the overestimation of dispersion distances in other directions. The IDWS technique also enables the modeling of smaller-scale variations, which can reduce prediction uncertainties in exposure assessments.

As demonstrated in the previous sections, the proposed method produced relatively inaccurate estimates on certain experimental dates. A combination of the dramatically changing wind-direction at noon and the precipitation that occurred at sunset led to poor performance on May 26th. The daily PM$_{2.5}$ on May 20th was also interpolated, and the improvement in the MSE with respect to the

results obtained using IDWE was only 3.31%. Although the daily average wind-speed was greater than 1.5 m/s on May 20th, the major pollutant was dust particles caused by blowing sand, which was assumed to be the major source of the prediction uncertainty. These results suggest the need for careful evaluation of the specific weather conditions prior to including the wind-field using IDWS.

The basic version of IDW was applied to test the feasibility of the novel distance metric proposed in this study. Neither the problem of the influence radius nor the problem of zero distance was considered. Some variants of the classic method may be used to verify the validity of the SWPD or even to achieve more accurate estimations. Although it offers a number of advantages, IDW always achieves poorer performance than kriging or other more sophisticated interpolation methods. Thus, now that the effectiveness of the SWPD has been demonstrated and the metric has been shown to offer improved accuracy and realistic visual representation, a need exists to combine this distance metric with a more robust technique to obtain prediction surfaces with higher accuracy [23].

The incorporation of secondary information, instead of relying solely on station measurements, enables the estimation surfaces obtained to reflect localized variations and thus improves the predictive capacity of the analysis [34]. Most currently used auxiliary data are scalar, and little weight is given to vector data. One objective of this paper is to propose a methodology for incorporating vector-type secondary information into interpolation. Now that the feasibility of this methodology has been confirmed, a more general method that is capable of including both scalar-type data and vector-type data is desirable.

The primary intent of this study was to verify the effectiveness of the proposed method, so little attention was directed toward improving the computing efficiency. All algorithms were implemented using C++ with no optimization, and the visualization was performed using ArcGIS (ESRI). At the current stage of

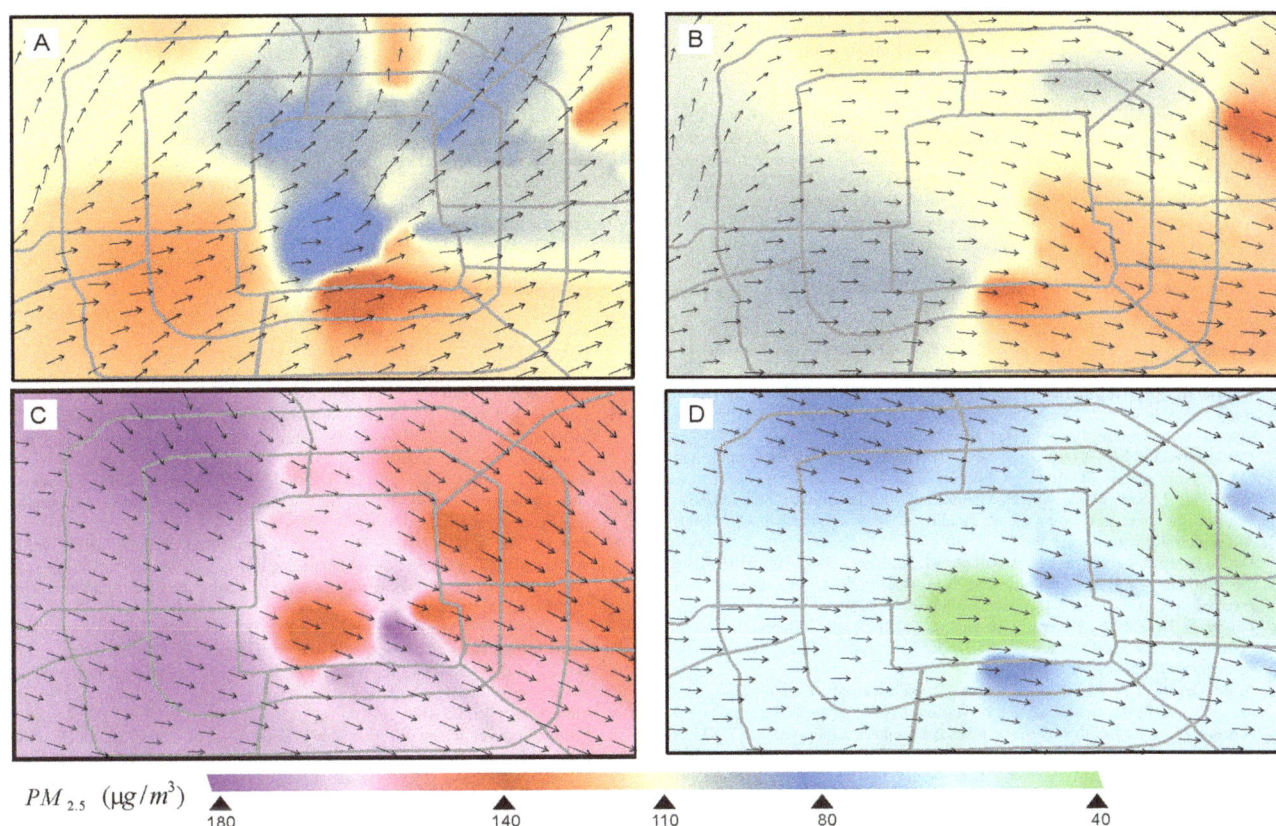

Figure 6. Interpolation results of the hourly PM$_{2.5}$ concentrations on May 26th at various times: (A) 8 am, (B) 12 pm, (C) 4 pm and (D) 8 pm.

development, the time required to generate one estimation surface for the study area is approximately 10 minutes. Further research is necessary to accelerate the calculation and to allow this technique to be used for real-time estimation.

The results of cross-validation and visual assessment have demonstrated that including wind-field in the interpolation of the PM$_{2.5}$ concentration improves the predictive performance. However, the experiments were conducted only in the core area of Beijing during six selected days in May. The method should be assessed over much longer monitoring periods spanning all four seasons to confirm its year-round effectiveness. Additional measurements are also required to confirm the usefulness of the SWPD or even to discover a better distance metric. The interpolation of other types of air-pollution, such as nitrogen dioxide and coarse particulate matter, should also be performed to verify that the model has general applicability. Furthermore, wind-fields with higher resolution may have the potential to improve the predictive capability of the technique and deserve further research.

Conclusions

Three major conclusions can be drawn from this study:

(1) Wind-fields are of great importance to studies of the negative effects of airborne pollutants. Incorporating wind-fields into the spatial interpolation of air-pollution distributions serves to enhance the predictive capability of such interpolation.

(2) The shortest wind-field path distance (SWPD) shows great potential for determining the spatial dependence and enables SWPD-based interpolations to capture complex features of air-pollution distributions with higher accuracy than methods based on the Euclidean distance.

(3) The workflow proposed in this paper, which consists of wind-field generation, shortest-path analysis and IDW in conjunction with the SWPD, has been demonstrated to be a robust technique for predicting urban-scale PM$_{2.5}$ concentrations.

Acknowledgments

We thank Professor Dongping Ming, Professor Jinfeng Wang and Professor Tao Pei for their advice. We also thank Zhuoqin Song for her assistance.

Author Contributions

Conceived and designed the experiments: LL JG. Performed the experiments: LL JZ. Analyzed the data: LL JG. Contributed reagents/materials/analysis tools: LL JZ. Wrote the paper: LL JZ.

References

1. Ryan PH, LeMasters GK (2007) A review of land-use regression models for characterizing intraurban air pollution exposure. Inhal Toxicol 19 Suppl 1: 127–133.

2. Lee SJ, Serre ML, van Donkelaar A, Martin RV, Burnett RT, et al. (2012) Comparison of geostatistical interpolation and remote sensing techniques for estimating long-term exposure to ambient PM2.5 concentrations across the continental United States. Environ Health Perspect 120: 1727–1732.

3. Jerrett M, Arain A, Kanaroglou P, Beckerman B, Potoglou D, et al. (2004) A review and evaluation of intraurban air pollution exposure models. Journal of Exposure Science and Environmental Epidemiology 15: 185–204.

4. Liao D, Peuquet DJ, Duan Y, Whitsel EA, Dou J, et al. (2006) GIS approaches for the estimation of residential-level ambient PM concentrations. Environmental health perspectives 114: 1374.

5. Ensor KB, Raun LH, Persse D (2013) A Case-Crossover Analysis of Out-of-Hospital Cardiac Arrest and Air PollutionClinical Perspective. Circulation 127: 1192–1199.

6. Sullivan J, Sheppard L, Schreuder A, Ishikawa N, Siscovick D, et al. (2005) Relation between short-term fine-particulate matter exposure and onset of myocardial infarction. Epidemiology 16: 41–48.

7. Bowman KW (2013) Toward the next generation of air quality monitoring: Ozone. Atmospheric Environment 80: 571–583.

8. Engel-Cox J, Kim Oanh NT, van Donkelaar A, Martin RV, Zell E (2013) Toward the next generation of air quality monitoring: Particulate Matter. Atmospheric Environment 80: 584–590.

9. Pearce JL, Rathbun SL, Aguilar-Villalobos M, Naeher LP (2009) Characterizing the spatiotemporal variability of PM2.5 in Cusco, Peru using kriging with external drift. Atmospheric Environment 43: 2060–2069.

10. Janssen S, Dumont G, Fierens F, Mensink C (2008) Spatial interpolation of air pollution measurements using CORINE land cover data. Atmospheric Environment 42: 4884–4903.

11. Arain MA, Blair R, Finkelstein N, Brook JR, Sahsuvaroglu T, et al. (2007) The use of wind fields in a land use regression model to predict air pollution concentrations for health exposure studies. Atmospheric Environment 41: 3453–3464.

12. Seaman NL (2000) Meteorological modeling for air-quality assessments. Atmospheric Environment 34: 2231–2259.

13. Liu Y, Paciorek CJ, Koutrakis P (2009) Estimating regional spatial and temporal variability of PM2. 5 concentrations using satellite data, meteorology, and land use information. Environmental Health Perspectives 117: 886.

14. Tian J, Chen D (2010) A semi-empirical model for predicting hourly ground-level fine particulate matter (PM2.5) concentration in southern Ontario from satellite remote sensing and ground-based meteorological measurements. Remote Sensing of Environment 114: 221–229.

15. Hoek G, Beelen R, de Hoogh K, Vienneau D, Gulliver J, et al. (2008) A review of land-use regression models to assess spatial variation of outdoor air pollution. Atmospheric Environment 42: 7561–7578.

16. Curriero F (2006) On the Use of Non-Euclidean Distance Measures in Geostatistics. Mathematical Geology 38: 907–926.

17. Greenberg JA, Rueda C, Hestir EL, Santos MJ, Ustin SL (2011) Least cost distance analysis for spatial interpolation. Computers & Geosciences 37: 272–276.

18. Boisvert JB, Manchuk JG, Deutsch CV (2009) Kriging in the Presence of Locally Varying Anisotropy Using Non-Euclidean Distances. Mathematical Geosciences 41: 585–601.

19. Peterson E, Merton A, Theobald D, Urquhart NS (2006) Patterns of Spatial Autocorrelation in Stream Water Chemistry. Environmental Monitoring and Assessment 121: 569–594.

20. Peterson EE, Theobald DM, Ver Hoef JM (2007) Geostatistical modelling on stream networks: developing valid covariance matrices based on hydrologic distance and stream flow. Freshwater Biology 52: 267–279.

21. Hoef JV, Peterson E, Theobald D (2006) Spatial statistical models that use flow and stream distance. Environmental and Ecological Statistics 13: 449–464.

22. Gardner B, Sullivan PJ, Lembo JAJ (2003) Predicting stream temperatures: geostatistical model comparison using alternative distance metrics. Canadian Journal of Fisheries and Aquatic Sciences 60: 344–351.

23. Boisvert JB, Deutsch CV (2011) Programs for kriging and sequential Gaussian simulation with locally varying anisotropy using non-Euclidean distances. Computers & Geosciences 37: 495–510.

24. Boisvert J, Deutsch C (2011) Modeling locally varying anisotropy of CO2 emissions in the United States. Stochastic Environmental Research and Risk Assessment 25: 1077–1084.

25. Broquet T, Ray N, Petit E, Fryxell J, Burel F (2006) Genetic isolation by distance and landscape connectivity in the American marten (Martes americana). Landscape Ecology 21: 877–889.

26. Zou H, Yue Y, Li Q, Yeh AG (2012) An improved distance metric for the interpolation of link-based traffic data using kriging: a case study of a large-scale urban road network. International Journal of Geographical Information Science 26: 667–689.

27. Zhang A, Qi Q, Jiang L, Zhou F, Wang J (2013) Population Exposure to PM2. 5 in the Urban Area of Beijing. PloS one 8: e63486.

28. Goodin WR, McRae GJ, Seinfeld JH (1980) An objective analysis technique for constructing three-dimensional urban-scale wind fields. Journal of Applied Meteorology 19: 98–108.

29. Kanaroglou PS, Jerrett M, Morrison J, Beckerman B, Arain MA, et al. (2005) Establishing an air pollution monitoring network for intra-urban population exposure assessment: A location-allocation approach. Atmospheric Environment 39: 2399–2409.

30. Turner DB (1973) Workbook of atmospheric dispersion estimates: US Government Printing Office.

31. de Mesnard L (2013) Pollution models and inverse distance weighting: Some critical remarks. Computers & Geosciences 52: 459–469.

32. Green A, Singhal R, Venkateswar R (1980) Analytic extensions of the Gaussian plume model. Journal of the Air Pollution Control Association 30: 773–776.

33. Dijkstra EW (1959) A note on two problems in connexion with graphs. Numerische mathematik 1: 269–271.

34. Goovaerts P (2000) Geostatistical approaches for incorporating elevation into the spatial interpolation of rainfall. Journal of Hydrology 228: 113–129.

35. Picard RR, Cook RD (1984) Cross-validation of regression models. Journal of the American Statistical Association 79: 575–583.

36. Warner TT, Fizz RR, Seaman NL (1983) A Comparison of Two Types of Atmospheric Transport Models—Use of Observed Winds Versus Dynamically Predicted Winds. Journal of Applied Meteorology 22: 394–406.

22

Association between Air Pollution and General Outpatient Clinic Consultations for Upper Respiratory Tract Infections in Hong Kong

Wilson W. S. Tam[1,2], Tze Wai Wong[1]*, Lorna Ng[3], Samuel Y. S. Wong[1], Kenny K. L. Kung[1], Andromeda H. S. Wong[1]

1 Jockey Club School of Public Health and Primary Care, The Chinese University of Hong Kong, Sha Tin, Hong Kong, **2** Shenzhen Municipal Key Laboratory for Health Risk Analysis, Shenzhen Research Institute of The Chinese University of Hong Kong, Shenzhen, Guangdong Province, China, **3** General Outpatient Clinics, Kwong Wah Hospital, Yaumatei, Hong Kong

Abstract

Background and Objectives: Many studies have shown the adverse effects of air pollution on respiratory health, but few have examined the effects of air pollution on service utilisation in the primary care setting. The aim of this study was to examine the association between air pollution and the daily number of consultations due to upper respiratory tract infections (URTIs) in general outpatient clinics (GOPCs) in Hong Kong.

Methods: Daily data on the numbers of consultations due to URTIs in GOPCs, the concentrations of major air pollutants, and the mean values of metrological variables were retrospectively collected over a 3-year period (2008–2010, inclusive). Generalised additive models were constructed to examine the association between air pollution and the daily number of consultations, and to derive the relative risks and 95% confidence intervals (95% CI) of GOPC consultations for a unit increase in the concentrations of air pollutants.

Results: The mean daily consultations due to URTIs in GOPCs ranged from 68.4 to 253.0 over the study period. The summary relative risks (and 95% CI) of daily consultations in all GOPCs for the air pollutants PM_{10}, NO_2, O_3, and SO_2 were 1.005 (1.002, 1.009), 1.010 (1.006, 1.013), 1.009 (1.006, 1.012), and 1.004 (1.000, 1.008) respectively, per 10 $\mu g/m^3$ increase in the concentration of each pollutant.

Conclusion: Significant associations were found between the daily number of consultations due to URTIs in GOPCs and the concentrations of air pollutants, implying that air pollution incurs a substantial morbidity and increases the burden of primary health care services.

Editor: Qinghua Sun, The Ohio State University, United States of America

Funding: The authors have no support or funding to report.

Competing Interests: The authors have declared that no competing interests exist.

* E-mail: twwong@cuhk.edu.hk

Introduction

Most epidemiological time series studies on air pollution focus on hospital admissions and mortality as health outcomes. [1–6] These outcomes represent, respectively, serious morbidity and the ultimate health consequence of air pollution. Illnesses seen in primary health care settings, by contrast, form the much wider base of the 'pyramid' of air pollution-related diseases, but are much less studied.

Respiratory diseases are very common in all age groups and generate a major demand on health care services worldwide. Acute respiratory infections represent one of the most common reasons for seeking medical attention in the primary health care setting. [7] A recent cross-sectional morbidity study in Hong Kong revealed that 26.4% of outpatient consultations were due to upper respiratory tract infections [8].

Gaseous air pollutants such as sulphur dioxide (SO_2), ozone (O_3), and nitrogen dioxide (NO_2) have been shown to cause irritation and constriction of the large airways. [9] However, only a few studies have examined the association between air pollution and the daily number of consultations in the primary health care sector. [10–13] One reason was the lack, in most countries, of routinely collected data on primary care service utilisation in both private and public sectors; hence, data might need to be collected prospectively. For example, in a previous study we conducted in Hong Kong, we invited 13 private general practitioners (GPs) to manually record the daily number of patients consulted for respiratory diseases. We used these data to examine the association between the number of GP consultations and the concentration of air pollutants [13].

In Hong Kong, primary health care is provided by both the private general practitioner clinics and the public general outpatient clinics (GOPCs). The GOPCs network covers all districts in Hong Kong, and provides primary care services mainly for the socially disadvantaged, while middle- to high-income groups often prefer private GP clinics for their primary care needs. Proportionately, more elderly patients and those with chronic diseases attend the GOPCs, whereas more patients with acute health problems consult GPs in private clinics. The GOPCs are heavily subsidised by the government; the charge per consultation in these clinics (HK$45 or ~US$6) is around a quarter of the median charges in private clinics, which range from US$19 to US$26. [14] The Hospital Authority (HA), a Hong Kong government-funded 'statutory board', manages the public GOPC network, and since 1999, medical records in these clinics have progressively been computerised [15].

The aim of this study is to examine the effects of air pollution on the frequency of consultations for upper respiratory tract infections (URTIs) in the public primary care clinics. We retrospectively collected data on daily consultations due to URTIs, at the GOPCs located in two of seven geographically-defined hospital and clinic 'clusters' under the Hospital Authority, from 2008 to 2010. We used a time-series approach to investigate the association between air pollutant concentrations and GOPCs consultations.

Materials and Methods

Data

This is a retrospective time series study. We identified 'upper respiratory tract infection' by the diagnostic code R74, defined as 'URI (head cold)/rhinitis, not otherwise classified' in the Revised Edition of the International Classification of Primary Care 2 (ICPC-2), prepared by the World Organization of National Colleges, Academies, and Academic Associations of General Practitioners/Family Physicians (WONCA). [16] Under this classification system, R74 excludes specific diagnostic labels such as acute and chronic sinusitis (R75), acute tonsillitis (R76), acute laryngitis/tracheitis/croup (R77), acute bronchitis/bronchiolitis (R78), and influenza (R80). Data on the number of daily visits for URTIs (R74) were extracted from the Clinical Data Analysis & Reporting System of the Hospital Authority, for five of the HA's GOPCs. Three of the GOPCs were in the Kowloon Central Cluster, which covers the Kowloon peninsula – a flat, densely populated urban area. The other two, part of the Kowloon West Cluster, were located in the north-western 'new towns' of Kwai Tsing and Tsuen Wan which, besides being residential, also have a container terminal and an industrial estate, respectively. Data were retrospectively collected from 1 January 2008 to 31 December 2010. [15] Four monitoring stations of the Hong Kong Government's Environmental Protection Department were matched to the five GOPCs according to their geographic locations; one of the monitoring stations served both GOPC 2 and GOPC 3 (Figure 1).

Four pollutants were regularly monitored in all of the monitoring stations. These were: nitrogen dioxide (NO_2), sulphur dioxide (SO_2), ozone (O_3), and particulates with an aerodynamic diameter less than 10 μm (PM_{10}). Data regarding hourly concentration of these pollutants were obtained for the corresponding 3-year period as the clinical data (2008–2010) from the Environmental Protection Department (www.epd.gov.hk) and the collection method has been described in our previous study. [13] The daily mean concentration of each pollutant was computed as the average of the hourly data, if more than two-thirds of the hourly data were available for that day; otherwise, the daily mean was deemed as missing. Missing daily concentrations in one station were predicted by a regression of data from that station on the corresponding data from the nearest neighbouring station. [17]

The daily mean temperatures and relative humidity at different districts were acquired from the Hong Kong Observatory (www.hko.gov.hk).

Statistical Modelling

A generalised additive model, using the Poisson distribution with a log-link function, was used to construct a core model for each of the five GOPCs. [18] In brief, each model regressed the daily numbers of the URTIs consultations on several variables: time (day), day of the week, daily mean temperature and humidity, and a holiday indicator. To adjust for the potential confounding effect of influenza on the number of URTIs consultations, we used a dichotomous variable to indicate the weeks during which the number of influenza consultations exceeded the 75th percentile for the study period. This method has been used in time series studies of air pollution and hospital admissions for respiratory illnesses. [19] The estimated weekly numbers of influenza were obtained from the Centre for Health Protection, Hong Kong SAR Government (www.chp.gov.hk). We used penalised smoothing splines [20,21] to adjust for seasonal patterns and long-term trends in daily consultations, temperature, and relative humidity, with degrees of freedom (df) selected a priori, based on previous studies. [22,23] Specifically, we used 7 df per year for time trends, 6 df for temperature and 3 df for humidity [19].

The quasi-likelihood method was used to correct for over-dispersion. [18] To minimise autocorrelation, which would bias the standard errors, we specified that the absolute values of the partial autocorrelation function for the model residuals had to be <0.1 for the first 2 lag days. [24] When these criteria were not met, we added autoregressive terms for the outcome variable to the core model, up to a maximum of three autoregressive terms.

Linear effects of all the air pollutants for the same day (lag 0) up to 3 lag days (lag 3) were tested in each model and the best lag for each pollutant at the GOPC was chosen as the one that yielded the largest t-value.

We conducted all analyses using the MGCV package in R [25] and expressed the results as the relative risks (RRs) of visits for URTIs in each GOPC for every 10 μg/m^3 increase in the concentrations of all four pollutants. The best lag RRs for each pollutant, obtained from individual GOPCs, were then combined using random effects models. [26] Ethics approval was obtained for conducting this study from the Research Ethics Committee, Kowloon West Cluster, Hospital Authority (Ref: KW/EX-11–137). Our study did not involve any patient's personal information.

Results

Daily Consultations at GOPC

Table 1 shows summary statistics of the consultations due to URTIs, at the five GOPCs, over the study period. From 2008 to 2010, the total number of consultations was 817,240, with mean daily consultations ranging from 68.4 to 253.0. The district-specific demographic information for each GOPC is also shown.

Descriptive Statistics for Air Pollutants

Table 2 shows the mean, median, and quartile values of the daily air pollutant concentrations measured at each monitoring station throughout the study. During this period, the mean daily temperature and humidity in Hong Kong were 23.3°C (SD = 5.2) and 78.1% (SD = 11.1) respectively. Table 3 shows the Pearson's correlations between the air pollutants, listed by each monitoring station.

Figure 1. A location map of GOPCs and the air monitoring stations.

Relative Risk of Daily Consultation

The individual relative risks (RRs) of consultation due to URTIs by GOPC, and the summary RRs for all GOPCs (per 10 $\mu g/m^3$ increase in the concentrations of each air pollutant) are shown in Table 4. The RRs for NO_2 in individual clinics were consistently significant. For PM_{10}, the RRs were significant in three out of five clinics while the RRs for O_3 were significant in four out of five clinics. For SO_2, none of the RRs in individual clinics were significant, although all were above unity. Statistically significant summary RRs were found for NO_2, PM_{10}, and O_3, while the combined RR for SO_2 was marginally significant (p = 0.060). The summary relative risks (with 95% confidence intervals) from all GOPC for NO_2, PM_{10}, O_3, and SO_2 were 1.005 (1.002, 1.009), 1.010 (1.006, 1.013), 1.009 (1.006, 1.012) and 1.004 (1.000, 1.008) respectively, per 10 $\mu g/m^3$ increase in the concentration of each air pollutant.

Discussion

While air pollution has been extensively researched with regard to its association with mortality and hospital admissions, studies

Table 1. Daily number of URTIs consultations at the five GOPCs, and district specific statistics, from 1 Jan 2008 to 31 Dec 2010.

	GOPC 1	GOPC 2	GOPC 3	GOPC 4	GOPC 5
Mean (SD)	179.5 (97.3)	115.8 (70.4)	68.4 (48.6)	253.0 (149.2)	130.1 (68.5)
Median	197	132	78	288	136
Quartile	99, 253	68, 165	24, 107	163, 360	71.5, 175
District specific statistics*					
Area (km²)	9.36	9.48	6.85	21.82	60.70
Population	420,183	380,855	307,878	511,167	304,637
Mean age (years)	44.5	43.2	41.1	41.9	41.3
Male gender	46.7%	46.7%	46.4%	47.3%	46.6%
Never married**	33.7%	31.2%	29.5%	32.6%	28.9%
Post- secondary educated or above***	17.9%	22.8%	31.8%	18.7%	29.8%
Mean number of persons in household	2.9	2.7	2.7	3.0	2.9
Mean household income	$17,000	$16,280	$22,070	$17,000	$24,100

*Information extracted from 2011 Population Census by the Census of Statistics Department, Hong Kong Special Administrative Region Government. The two districts served by GOPC 4 and 5 are much larger than the others, as they include uninhabited country park areas. The populations in these districts are concentrated in urban areas, much like the other districts.
**aged 15 or above.
***non-student population aged 20 or above.

Table 2. Summary statistics of the four air pollutants in the four monitoring stations from 1 January 2008 to 31 December 2010.

Station	Pollutant	Mean (SD) ($\mu g/m^3$)	Median ($\mu g/m^3$)	Quartile ($\mu g/m^3$)
Station 1 for GOPC 5	NO_2	62.7 (20.1)	58.5	48.2, 73.2
(Missing: 39 days)	PM_{10}	48.7 (28.9)	41.9	28.8, 62.9
	O_3	30.8 (19.6)	26.4	14.1, 43.8
	SO_2	19.8 (13.5)	17.2	10.7, 25.1
Station 2 for GOPC 4	NO_2	64.9 (21.8)	59.7	49.6, 75.7
(Missing: 44 days)	PM_{10}	47.8 (27.4)	42.0	29.3, 62.1
	O_3	30.5 (20.3)	27.2	13.2, 43.6
	SO_2	23.6 (19.9)	15.3	8.1, 36.9
Station 3 for GOPC 2 & 3	NO_2	67.5 (22.0)	66.6	50.8, 79.7
(Missing: 23 days)	PM_{10}	49.1 (29.9)	42.8	27.9, 64.0
	O_3	28.3 (19.1)	23.6	12.8, 40.0
	SO_2	16.6 (14.4)	11.9	7.3, 20.8
Station 4 for GOPC 1	NO_2	60.8 (20.3)	56.2	47.1, 70.9
(Missing: 43 days)	PM_{10}	49.0 (31.8)	42.6	28.9, 63.9
	O_3	33.4 (21.9)	30.5	14.0, 48.5
	SO_2	16.4 (16.3)	10.3	6.9, 17.8

Remark: 'Missing' indicates that at least one pollutant measurement was missing that day.

reporting its effects on primary care consultations are scarce.[11,13,27–30] This time series study is one of the few that investigate air pollution's effects on primary health care service utilisation in the public sector. We found significant positive associations between three air pollutants – namely, NO_2, O_3, and PM_{10} (from zero to three lag days) – and the daily number of consultations for URTIs at outpatient clinics in Hong Kong.

Table 3. Correlation between pollutants by monitoring station.

Station	Pollutant	NO_2	PM_{10}	O_3	SO_2
Station 1 for GOPC 5	NO_2	1	0.58**	0.24**	0.37**
(Missing: 39 days)	PM_{10}		1	0.45**	0.21**
	O_3			1	−0.22**
	SO_2				1
Station 2 for GOPC 4	NO_2	1	0.60**	0.14**	0.27**
(Missing: 44 days)	PM_{10}		1	0.39*	0.06*
	O_3			1	−0.46*
	SO_2				1
Station 3 for GOPC 2 & 3	NO_2	1	0.61**	0.34**	0.38**
(Missing: 23 days)	PM_{10}		1	0.46**	0.26**
	O_3			1	−0.07*
	SO_2				1
Station 4 for GOPC 1	NO_2	1	0.50**	0.14**	0.36**
(Missing: 43 days)	PM_{10}		1	0.42**	0.15**
	O_3			1	−0.22**
	SO_2				1

*$p<0.05$;
**$p<0.01$.

The combined RRs for NO_2, O_3, and PM_{10} (at 1.010, 1.009 and 1.005 respectively) for visits to public clinics herein investigated were substantially lower than their corresponding RRs reported for private GP visits in Hong Kong (at 1.030, 1.024 and 1.020 respectively) in our previous study in 2006. [13] One possible explanation is that the GOPCs services required patients to register either by walk-in or by phone, with daily quotas placed on the number of consultations. Because of the much lower prices charged by the GOPCs compared to that charged by the private sector GPs, the waiting time for patients attending GOPCs could be much longer than that for the latter. [31] Therefore, it is plausible that to reduce waiting time, patients with more severe URTIs may have consulted private GPs or even the accident and emergency departments of hospitals. This spillage of patients into alternative services could then result in the smaller RRs found at the GOPCs in this study. This would explain the findings in this study that the 'response' or 'health outcome' to a unit increase in the concentration of air pollutants, as GOPCs visits, was lower than consultations to private GPs. The lack of a significant association with SO_2 could be explained by the relatively low concentrations of SO_2 in Hong Kong. Levels of SO_2 in three of the four stations were within the air quality guidelines of 20 $\mu g/m^3$ recommended by the World Health Organization. [9] By contrast, the mean concentrations of NO_2 and PM_{10} in all stations were much higher than their annual air quality guidelines,

The lag times in both Hong Kong studies were similar, ranging from zero to three lag days. In a GP study in London, significant RRs for SO_2 and PM_{10} were reported for specific diseases (allergic rhinitis and asthma) and in different age groups for URTIs, as the 10th to the 90th percentile change in air pollutant concentrations, from zero to 3 lag days. [27,28] This makes comparison of RRs difficult. Ostro reported similar associations for PM_{10} and O_3 with URTIs or lower respiratory tract infections (LRTIs) among children of different age groups in Santiago, Chile. [29] In an ambulatory care study in Atlanta, Georgia, in the USA, Sinclair and Tolsma found weak but significant associations between PM and consultations for asthma and LRTIs. They reported a longer

Table 4. Relative risks and 95% CI of the visits to five GOPCs for upper respiratory tract infections (URTIs), per 10 μg/m^3 increase in the concentration of air pollutants.

Clinic	NO$_2$	PM$_{10}$	O$_3$	SO$_2$
GOPC 1	1.007 [lag 0 day]	1.007 [lag 1 day]	1.010 [lag 1 day]	1.001 [lag 0 day]
	(1.000, 1.013)*	(1.004, 1.010)*	(1.004, 1.016)*	(0.992, 1.009)
GOPC 2	1.014 [lag 0 day]	1.005 [lag 0 day]	1.003 [lag 1 day]	1.014 [lag 0 day]
	(1.000, 1.029)*	(0.998, 1.013)	(0.988, 1.018)	(0.998, 1.031)
GOPC 3	1.013 [lag 0 day]	1.011 [lag 3 days]	1.015 [lag 1 day]	1.010 [lag 3 days]
	(1.004, 1.023)*	(1.006, 1.017)*	(1.004, 1.025)*	(0.999, 1.022)
GOPC 4	1.010 [lag 3 days]	1.005 [lag 3 days]	1.010 [lag 1 days]	1.002 [lag 0 day]
	(1.004, 1.015)*	(1.001, 1.009)*	(1.004, 1.016)*	(0.996, 1.009)
GOPC 5	1.010 [lag 0 day]	0.999 [lag 3 days]	1.010 [lag 3 day]	1.003 [lag 0 day]
	(1.002, 1.017)*	(0.995, 1.003)	(1.003, 1.018)*	(0.993, 1.014)
Combined	1.010	1.005	1.009	1.004
	(1.006, 1.013)*	(1.002, 1.009)*	(1.006, 1.012)*	(1.000, 1.008)#

*p<0.05;
#p = 0.06.

lag time of 3 to 5 days. [30] Differences in lag days may have been due to differences in the primary health care settings and in consultation behaviour across different cities.

Besides time series studies, spatial studies have also reported the association between air pollution and primary care utilisation. Hwang et al. reported that carbon monoxide (CO), NO$_2$, SO$_2$, and PM$_{10}$ all had significant effects on daily visits to clinics due to LRTIs, based on small area design and hierarchical modelling of data from 50 communities in Taiwan. [12] Oiamo et al. used a community health survey on GP access and utilisation in Sarnia, Ontario, in Canada, and data on spatial differences in air pollutant concentrations, to demonstrate their relationships. [32] A time series study that took place over the 2008 Olympic Games in Beijing showed a significant reduction in outpatient visits for asthma during the air pollution control period, during which traffic and industrial emissions were restricted. [33] This provided strong evidence of a cause-effect relationship between air pollution and respiratory illnesses in an outpatient setting.

Ciencewicki and Jaspers have suggested a potential mechanism linking exposure to environmental air pollutants to their adverse health effects – including those effects related to respiratory infections. Pollutants could induce oxidative stress, resulting in the production of free radicals. These, in turn, could have damaging effects on the respiratory system, thereby lowering the resistance of the tissues to viral and bacterial infections. [34] Also, the oxidative stress induced by exposure to air pollutants may enhance the morbidity of an infection through an increased inflammatory response. Potential mechanisms of individual pollutants have been reported elsewhere [35–38].

A major strength of our study is the large number of consultations (which increases the study's power) and the reliability of the data sources, as all GOPCs data of the Hospital Authority are stored in one computerized database. In addition, we have a comprehensive network of air quality monitoring stations in different districts of the city. There are several limitations to our study results. First, as our study area was limited to one region of the city, and our GOPCs visitors belonged to the lower social class, selection bias is a possibility. Secondly, misclassification of the respiratory infections is a potential problem. For instance, in an outpatient setting, some LRTIs such as acute bronchitis or

influenza might be wrongly coded as URTIs. This would have resulted in wider confidence intervals of the RRs.

Upper respiratory tract infections should include R74 to R77 in the ICPC-2 codes. Owing to logistical constraints in the acquisition of daily data, we were only able to download one diagnostic code from the system. A check on the annual data (which could be accessed more easily) showed that the total number of cases coded as R75–77 was about 1% of the cases coded as R74. The mean daily number of the former group ranged from 0.9 to 2,6 cases in each of the five clinics, whereas the corresponding mean daily number of cases coded as R74 ranged from 68–253. Hence, we believe that our findings would not deviate much if we had included all the URTI codes.

Furthermore, all patients attending one clinic were assigned the same level of exposure to air pollutants according to data from the nearest monitoring station. In reality, individual exposure could be quite different from the ambient pollution concentration. We cannot rule out the possibility that some patients living in one district would use GOPCs in another district, because the distances between some clinics are fairly short. Hence, as in other ecological studies, the misclassification of patients' true exposure to air pollution is a potential source of error that is difficult to assess. Nonetheless, the data on air pollutants show little variation between districts (Table 2), as our study areas were adjacent to each other, and consisted largely of flat terrain. In addition, the possible biases discussed above may be assumed as about constant over the study period, and therefore they should not have remarkably confounded the positive relationship found between changes in the concentration of air pollutants and the GOPCs visits for URTIs.

In our choice of health outcomes, we only focused on URTIs, the only illnesses with a sufficiently large number of mean daily consultations for time series analysis. By contrast, the average daily number of visits for asthma in our clinics ranged from 6 to 11. Many existing studies focused on PM$_{2.5}$, which have more damaging effects on the lungs. However, it was not routinely measured in all monitoring stations. Moreover, PM$_{10}$ are more relevant because of their effects on the upper respiratory tract.

Conclusion

Our results showed a significant association between the concentrations of several air pollutants and the daily number of GOPCs consultations due to URTIs in Hong Kong. These findings provide further evidence that air pollution is a major public health problem. The number of consultations in primary care represents much higher than hospital admissions and deaths. Air pollution thus incurs a substantial burden to health care services in an urban community.

Author Contributions

Conceived and designed the experiments: WWST TWW. Performed the experiments: WWST TWW LN SYSW KKLK. Analyzed the data: WWST TWW LN SYSW KKLK AHSW. Contributed reagents/materials/analysis tools: WWST TWW LN. Wrote the paper: WWST TWW LN SYSW KKLK AHSW.

References

1. Atkinson RW, Anderson HR, Sunyer J, Ayres J, Baccini M, et al. (2001) Acute effects of particulate air pollution on respiratory admissions: results from APHEA 2 project. Air Pollution and Health: a European Approach. Am J Respir Crit Care Med 164: 1860–1866.
2. Samet JM, Dominici F, Curriero FC, Coursac I, Zeger SL (2000) Fine particulate air pollution and mortality in 20 U.S. cities, 1987–1994. N Engl J Med 343: 1742–1749.
3. Katsouyanni K, Touloumi G, Spix C, Schwartz J, Balducci F, et al. (1997) Short-term effects of ambient sulphur dioxide and particulate matter on mortality in 12 European cities: results from time series data from the APHEA project. Air Pollution and Health: a European Approach. BMJ 314: 1658–1663.
4. Schwartz J (1999) Air pollution and hospital admissions for heart disease in eight U.S. counties. Epidemiology 10: 17–22.
5. Bremner SA, Anderson HR, Atkinson RW, McMichael AJ, Strachan DP, et al. (1999) Short-term associations between outdoor air pollution and mortality in London 1992–4. Occup Environ Med 56: 237–244.
6. Kan H, Chen B (2003) Air pollution and daily mortality in Shanghai: a time-series study. Arch Environ Health 58: 360–367.
7. WHO WHO (2004) Respiratory Care in Primary Care Services – A Survey in 9 counties. World Health Organization: 121.
8. Lo YYC LC, Mercer SW, Fong DYT, Lee A, Lam TP (2011) Patient morbidity and management patterns of community-based primary health care services in Hong Kong. Hong Kong Medical Journal 17: 7.
9. WHO WHO (2006) WHO Air quality guidelines for particulate matter, ozone, nitrogen dioxide and sulfur dioxide. Global update 2005. Summary of risk assessment. Geneva, Switzerland: World Health Organization.
10. Chang CJ, Yang HH, Chang CA, Tsai HY (2012) Relationship between air pollution and outpatient visits for nonspecific conjunctivitis. Invest Ophthalmol Vis Sci 53: 429–433.
11. Hajat S, Anderson HR, Atkinson RW, Haines A (2002) Effects of air pollution on general practitioner consultations for upper respiratory diseases in London. Occup Environ Med 59: 294–299.
12. Hwang JS, Chan CC (2002) Effects of air pollution on daily clinic visits for lower respiratory tract illness. Am J Epidemiol 155: 1–10.
13. Wong TW, Tam W, Tak Sun Yu I, Wun YT, Wong AH, et al. (2006) Association between air pollution and general practitioner visits for respiratory diseases in Hong Kong. Thorax 61: 585–591.
14. Yam CH, Liu S, Huang OH, Yeoh EK, Griffiths SM (2011) Can vouchers make a difference to the use of private primary care services by older people? Experience from the healthcare reform programme in Hong Kong. BMC Health Serv Res 11: 255.
15. Cheung NT, Fung KW, Wong KC, Cheung A, Cheung J, et al. (2001) Medical informatics–the state of the art in the Hospital Authority. Int J Med Inform 62: 113–119.
16. World Organization of National Colleges Academies and Academic Associations of General Practitioners/Family Physicians (1998) ICPC-2 : international classification of primary care. Oxford; New York: Oxford University Press. x, 190 p.
17. Duffy ME (2006) Handling missing data: a commonly encountered problem in quantitative research. Clin Nurse Spec 20: 273–276.
18. Hastie T, Tibshirani R (1995) Generalized additive models for medical research. Stat Methods Med Res 4: 187–196.
19. Qiu H, Yu IT, Tian L, Wang X, Tse LA, et al. (2012) Effects of coarse particulate matter on emergency hospital admissions for respiratory diseases: a time-series analysis in Hong Kong. Environ Health Perspect 120: 572–576.
20. Host S, Larrieu S, Pascal L, Blanchard M, Declercq C, et al. (2008) Short-term associations between fine and coarse particles and hospital admissions for cardiorespiratory diseases in six French cities. Occup Environ Med 65: 544–551.
21. Kan H, London SJ, Chen G, Zhang Y, Song G, et al. (2007) Differentiating the effects of fine and coarse particles on daily mortality in Shanghai, China. Environ Int 33: 376–384.
22. Bell ML, Ebisu K, Peng RD, Walker J, Samet JM, et al. (2008) Seasonal and regional short-term effects of fine particles on hospital admissions in 202 US counties, 1999–2005. Am J Epidemiol 168: 1301–1310.
23. Peng RD, Chang HH, Bell ML, McDermott A, Zeger SL, et al. (2008) Coarse particulate matter air pollution and hospital admissions for cardiovascular and respiratory diseases among Medicare patients. JAMA 299: 2172–2179.
24. Wong CM, Vichit-Vadakan N, Kan H, Qian Z (2008) Public Health and Air Pollution in Asia (PAPA): a multicity study of short-term effects of air pollution on mortality. Environ Health Perspect 116: 1195–1202.
25. Wood SN (2006) Generalized additive models : an introduction with R. Boca RatonFL: Chapman & Hall/CRC. xvii, 391 p.
26. Fleiss JL (1993) The statistical basis of meta-analysis. Stat Methods Med Res 2: 121–145.
27. Hajat S, Haines A, Atkinson RW, Bremner SA, Anderson HR, et al. (2001) Association between air pollution and daily consultations with general practitioners for allergic rhinitis in London, United Kingdom. Am J Epidemiol 153: 704–714.
28. Hajat S, Haines A, Goubet SA, Atkinson RW, Anderson HR (1999) Association of air pollution with daily GP consultations for asthma and other lower respiratory conditions in London. Thorax 54: 597–605.
29. Ostro BD, Eskeland GS, Sanchez JM, Feyzioglu T (1999) Air pollution and health effects: A study of medical visits among children in Santiago, Chile. Environ Health Perspect 107: 69–73.
30. Sinclair AH, Tolsma D (2004) Associations and lags between air pollution and acute respiratory visits in an ambulatory care setting: 25-month results from the aerosol research and inhalation epidemiological study. J Air Waste Manag Assoc 54: 1212–1218.
31. Ng LV, Kam CW, Ng PTK, Chong SYC, Shek JKP, et al. (2009) What are the major factors influencing patients' decision in queuing up for registration in a General Out-Patient Clinic attached to a regional hospital? The Hong Kong Practitioner 31: 10.
32. Oiamo TH, Luginaah IN, Atari DO, Gorey KM (2011) Air pollution and general practitioner access and utilization: a population based study in Sarnia, 'Chemical Valley,' Ontario. Environ Health 10: 71.
33. Li Y, Wang W, Kan H, Xu X, Chen B (2010) Air quality and outpatient visits for asthma in adults during the 2008 Summer Olympic Games in Beijing. Sci Total Environ 408: 1226–1227.
34. Ciencewicki J, Jaspers I (2007) Air pollution and respiratory viral infection. Inhal Toxicol 19: 1135–1146.
35. Rahman I, MacNee W (2000) Oxidative stress and regulation of glutathione in lung inflammation. Eur Respir J 16: 534–554.
36. Reist M, Jenner P, Halliwell B (1998) Sulphite enhances peroxynitrite-dependent alpha1-antiproteinase inactivation. A mechanism of lung injury by sulphur dioxide? FEBS Lett 423: 231–234.
37. Scapellato ML, Lotti M (2007) Short-term effects of particulate matter: an inflammatory mechanism? Crit Rev Toxicol 37: 461–487.
38. Uysal N, Schapira RM (2003) Effects of ozone on lung function and lung diseases. Curr Opin Pulm Med 9: 144–150.

PERMISSIONS

The contributors of this book come from diverse backgrounds, making this book a truly international effort. This book will bring forth new frontiers with its revolutionizing research information and detailed analysis of the nascent developments around the world.

We would like to thank all the contributing authors for lending their expertise to make the book truly unique. They have played a crucial role in the development of this book. Without their invaluable contributions this book wouldn't have been possible. They have made vital efforts to compile up to date information on the varied aspects of this subject to make this book a valuable addition to the collection of many professionals and students.

This book was conceptualized with the vision of imparting up-to-date information and advanced data in this field. To ensure the same, a matchless editorial board was set up. Every individual on the board went through rigorous rounds of assessment to prove their worth. After which they invested a large part of their time researching and compiling the most relevant data for our readers.

The editorial board has been involved in producing this book since its inception. They have spent rigorous hours researching and exploring the diverse topics which have resulted in the successful publishing of this book. They have passed on their knowledge of decades through this book. To expedite this challenging task, the publisher supported the team at every step. A small team of assistant editors was also appointed to further simplify the editing procedure and attain best results for the readers.

Apart from the editorial board, the designing team has also invested a significant amount of their time in understanding the subject and creating the most relevant covers. They scrutinized every image to scout for the most suitable representation of the subject and create an appropriate cover for the book.

The publishing team has been an ardent support to the editorial, designing and production team. Their endless efforts to recruit the best for this project, has resulted in the accomplishment of this book. They are a veteran in the field of academics and their pool of knowledge is as vast as their experience in printing. Their expertise and guidance has proved useful at every step. Their uncompromising quality standards have made this book an exceptional effort. Their encouragement from time to time has been an inspiration for everyone.

The publisher and the editorial board hope that this book will prove to be a valuable piece of knowledge for researchers, students, practitioners and scholars across the globe.

LIST OF CONTRIBUTORS

Siobán D. Harlow
Department of Epidemiology, University of Michigan School of Public Health, Ann Arbor, Michigan, United States of America

Sung Kyun Park
Department of Epidemiology, University of Michigan School of Public Health, Ann Arbor, Michigan, United States of America
Department of Environmental Health Sciences, University of Michigan School of Public Health, Ann Arbor, Michigan, United States of America

John D. Meeker
Department of Environmental Health Sciences, University of Michigan School of Public Health, Ann Arbor, Michigan, United States of America

Yebin Tao and Bhramar Mukherjee
Department of Biostatistics, University of Michigan School of Public Health, Ann Arbor, Michigan, United States of America

Hangjun Zhang, Xiaojun Jiang, Wenfeng Xiao and Liping Lu
College of Life and Environmental Sciences, Hangzhou Normal University, Hangzhou, Zhejiang, China

Jun Ying
Department of Environmental Health, University of Cincinnati College of Medicine, Cincinnati, Ohio, United States of America
Center for Environmental Genetics, University of Cincinnati College of Medicine, Cincinnati, Ohio, United States of America

Pheruza Tarapore and Bin Ouyang
Department of Environmental Health, University of Cincinnati College of Medicine, Cincinnati, Ohio, United States of America
Center for Environmental Genetics, University of Cincinnati College of Medicine, Cincinnati, Ohio, United States of America
Cincinnati Cancer Center, University of Cincinnati College of Medicine, Cincinnati, Ohio, United States of America

Barbara Burke and Bruce Bracken
Department of Surgery, University of Cincinnati College of Medicine, Cincinnati, Ohio, United States of America

Shuk-Mei Ho
Department of Environmental Health, University of Cincinnati College of Medicine, Cincinnati, Ohio, United States of America
Center for Environmental Genetics, University of Cincinnati College of Medicine, Cincinnati, Ohio, United States of America
Cincinnati Cancer Center, University of Cincinnati College of Medicine, Cincinnati, Ohio, United States of America
Department of Surgery, University of Cincinnati College of Medicine, Cincinnati, Ohio, United States of America
Cincinnati Veteran Affairs Hospital Medical Center, Cincinnati, Ohio, United States of America

Taotao Mu, Siying Chen, Yinchao Zhang, Pan Guo, He Chen and Fandong Meng
School of Optoelectronics, Beijing Institute of Technology, Beijing, China

Sher Bahadar Khan, Mohammed M. Rahman, Abdullah M. Asiri and Malik Abdul Rub
Center of Excellence for Advanced Materials Research (CEAMR), King Abdulaziz University, Jeddah, Saudi Arabia
Chemistry Department, Faculty of Science, King Abdulaziz University, Jeddah, Saudi Arabia

Kalsoom Akhtar
Division of Nano Sciences and Department of Chemistry, Ewha Womans University, Seoul, Korea

Alessandra Ghiani, Pietro Fumagalli, Tho Nguyen Van, Rodolfo Gentili and Sandra Citterio
Department of Earth and Environmental Sciences, University of Milano-Bicocca, Milan, Italy

Chao Chai, Hongzhen Cheng, Dong Ma and Yanxi Shi
College of Resources and Environment, Qingdao Agricultural University, Qingdao, China

Wei Ge
College of Life Sciences, Qingdao Agricultural University, Qingdao, China

Ming Wu, Ren-Qiang Han, Jin-Yi Zhou and Jin-Kou Zhao
Department of Non-communicable Chronic Disease Control, Jiangsu Provincial Center for Disease Control and Prevention, Nanjing, Jiangsu, China

Zi-Yi Jin
Jiangyin Center for Disease Control and Prevention, Jiangyin, Jiangsu, China

Xiao-Feng Zhang and Xu-Shan Wang
Ganyu Center for Disease Control and Prevention, Ganyu, Jiangsu, China

Ai-Ming Liu
Dafeng Center for Disease Control and Prevention, Dafeng, Jiangsu, China

Qing-Yi Lu
Center for Human Nutrition, David Geffen School of Medicine, University of California Los Angeles (UCLA), Los Angeles, California, United States of America

Claire H. Kim and Zuo-Feng Zhang
Department of Epidemiology, Fielding School of Public Health, University of California Los Angeles (UCLA), Los Angeles, California, United States of America

Lina Mu
Department of Social and Preventive Medicine, State University of New York at Buffalo, Buffalo, New York, United States of America

Amanda Ode, Karin Källén, Lars Rylander, Christian H. Lindh and Anna Rignell-Hydbom
Division of Occupational and Environmental Medicine, Lund University, Lund, Sweden

Peik Gustafsson
Child and Adolescent Psychiatry Unit, Department of Clinical Sciences, Lund University, Lund, Sweden

Per Olofsson
Obstetrics and Gynecology Unit, Department of Clinical Sciences, Skåne University Hospital, Lund University, Malmö, Sweden

Bo A. G. Jönsson and Sten A. Ivarsson
Department of Clinical Sciences, Unit of Pediatric Endocrinology, Lund University/Clinical Research Centre (CRC), Malmö, Sweden

Emily G. I. Payne, Ana Deletic and Belinda E. Hatt
Monash Water for Liveability, Department of Civil Engineering, Monash University, Victoria, Australia

Tim D. Fletcher
Department of Resource Management and Geography, Melbourne School of Land and Environment, The University of Melbourne, Victoria, Australia

Douglas G. Russell, Michael R. Grace, Victor Evrard and Perran L. M. Cook
Water Studies Centre, School of Chemistry, Monash University, Victoria, Australia

Timothy R. Cavagnaro
School of Biological Sciences, Monash University, Victoria, Australia

Rong Yin and Tongcheng Cao
Department of Chemistry, Tongji University, Shanghai, China

Liang Gu, Min Li, Cizhong Jiang and Xiaobai Zhang
Shanghai Key Laboratory of Signaling and Disease Research, the School of Life Sciences and Technology, Tongji University, Shanghai, China

Duk-Hee Lee
Department of Preventive Medicine, School of Medicine, Kyungpook National University, Daegu, Korea
BK21 Plus KNU Biomedical Convergence Program, Department of Biomedical Science, Kyungpook National University, Daegu, Korea

Lars Lind
Department of Medical Sciences, Cardiovascular Epidemiology, Uppsala University Hospital, Uppsala, Sweden

David R. Jacobs Jr.
Division of Epidemiology and Community Health, School of Public Health, University of Minnesota, Minneapolis, Minnesota, United States of America
Department of Nutrition, University of Oslo, Oslo, Norway

Samira Salihovic and Bert van Bavel
MTM Research Center, School of Science and Technology, Örebro University, Örebro, Sweden

P. Monica Lind
Department of Medical Sciences, Occupational and Environmental Medicine, Uppsala University, Uppsala, Sweden

Gilbert Cadelis and Rachel Tourres
Department of Pulmonary Medicine, Universitary Hospital of Pointe-a-Pitre, Pointe-a-Pitre, Guadeloupe, French West Indies

Jack Molinie
Laboratory of Research in Geoscience and Energy, University of Antilles and Guyane, Pointe-a-Pitre, Guadeloupe, French West Indies

Denis Bard, Wahida Kihal, Christophe Fermanian and Sophie Glorion
Department of Epidemiology and Biostatistics, École des Hautes Études en Santé Publique, Rennes and Sorbonne Paris Cité, Paris, France

Charles Schillinger
Association pour la Surveillance de la Qualité de l'Air en Alsace-ASPA, Schiltigheim, France

Claire Ségala
SEPIA-Santé, Baud, France

Dominique Arveiler
Department of Epidemiology and Public Health (EA3430), University of Strasbourg, Strasbourg, France

Christiane Weber
Laboratoire Image, Ville, Environnement (LIVE UMR7362 CNRS), Faculté de géographie et d'aménagement, University of Strasbourg, Strasbourg, France

Mengling Tang, Fangxing Yang and Weiping Liu
MOE Key Laboratory of Environmental Remediation and Ecosystem Health, College of Environmental and Resource Sciences, Zhejiang University, Hangzhou, China

Kun Chen
Department of Epidemiology & Health Statistics, School of Public Health, Zhejiang University, Hangzhou, China

Hung-Chieh Lee and Huai-Jen Tsai
Institute of Molecular and Cellular Biology, National Taiwan University, Taipei, Taiwan

Po-Nien Lu and Hui-Lan Huang
Institute of Molecular and Cellular Biology, National Taiwan University, Taipei, Taiwan
Liver Disease Prevention & Treatment Research Foundation, Taipei, Taiwan

Hong-Ping Li
Taiwan Agricultural Chemicals and Toxic Substances Research Institute Council of Agriculture, Executive Yuan, Taichung, Taiwan

Chien Chu
Taiwan Agricultural Chemicals and Toxic Substances Research Institute Council of Agriculture, Executive Yuan, Taichung, Taiwan
Department of Soil and Environmental Sciences, National Chung Hsing University, Taichung, Taiwan

Yuyi Yang
Key Laboratory of Aquatic Botany and Watershed Ecology, Wuhan Botanical Garden, Chinese Academy of Sciences, Wuhan, Hubei Province, China

Qing X. Li and Jun Wang
Key Laboratory of Aquatic Botany and Watershed Ecology, Wuhan Botanical Garden, Chinese Academy of Sciences, Wuhan, Hubei Province, China
Department of Molecular Biosciences and Bioengineering, University of Hawaii at Manoa, Honolulu, Hawaii, United States of America

Lee Ann Woodward
U.S. Fish and Wildlife Service, Pacific Reefs NWRC, Honolulu, Hawaii, United States of America

Chung-Wei Yang, Wei-Chun Chou, Kuan-Hsueh Chen and Chun-Yu Chuang
Department of Biomedical Engineering and Environmental Sciences, National Tsing Hua University, Hsinchu, Taiwan

An-Lin Cheng
Schools of Nursing and Health Studies, University of Missouri-Kansas City, Kansas City, Kansas, United States of America

I-Fang Mao
School of Occupational Safety and Health, Chung Shan Medical University, Taichung, Taiwan

How-Ran Chao
Emerging Compounds Research Center, Department of Environmental Science and Engineering, National Pingtung University of Science and Technology, Pingtung County, Taiwan

Hin T. Wan, Yin G. Zhao, Pik Y. Leung and Chris K. C. Wong
Partner State Key Laboratory of Environmental and Biological Analysis, Croucher Institute for Environmental Sciences, Department of Biology, Hong Kong Baptist University, Hong Kong, People's Republic of China

Simone Dealtry, Guo-Chun Ding, Viola Weichelt, Holger Heuer and Kornelia Smalla
Julius Kühn-Institut – Federal Research Centre for Cultivated Plants (JKI), Institute for Epidemiology and Pathogen Diagnostics, Braunschweig, Germany

Dirk Springael and Vincent Dunon
Division of Soil and Water Management, KU Leuven, Heverlee, Belgium

Andreas Schlüter
Center for Biotechnology (CeBiTec), Institute for Genome Research and Systems Biology, Bielefeld University, Bielefeld, Germany

María Carla Martini, María Florencia Del Papa and Antonio Lagares
IBBM (Instituto de Biotecnología y Biología Molecular), CCT-CONICET-La Plata, Departamento de Ciencias Biológicas, Facultad de Ciencias Exactas, Universidad Nacional de La Plata, La Plata, Argentina

Gregory Charles Auton Amos, Elizabeth Margaret Helen Wellington and William Hugo Gaze
School of Life Sciences, University of Warwick, Warwick, United Kingdom

Detmer Sipkema
Laboratory of Microbiology, Wageningen University, Wageningen, The Netherlands

Sara Sjöling
Södertörns högskola (Sodertorn University), Inst. för Naturvetenskap, Miljöoch medieteknik (School of Natural Sciences, Environmental Studies and media tech), Huddinge, Sweden

Jan Dirk van Elsas
University of Groningen, Groningen, The Netherlands

Christopher Thomas
School of Biosciences, University of Birmingham, Edgbaston, Birmingham, Warwick, United Kingdom

Longxiang Li and Jieping Zhou
Institute of Remote Sensing and Digital Earth, Chinese Academy of Sciences, Olympic Science & Technology Park of CAS, Beijing, China

Jianhua Gong
Institute of Remote Sensing and Digital Earth, Chinese Academy of Sciences, Olympic Science & Technology Park of CAS, Beijing, China and Zhejiang-CAS Application Center for Geoinformatics, Jiashan, Zhejiang, China

Tze Wai Wong, Samuel Y. S. Wong, Kenny K. L. Kung and Andromeda H. S. Wong
Jockey Club School of Public Health and Primary Care, The Chinese University of Hong Kong, Sha Tin, Hong Kong

Wilson W. S. Tam
Jockey Club School of Public Health and Primary Care, The Chinese University of Hong Kong, Sha Tin, Hong Kong

Shenzhen Municipal Key Laboratory for Health Risk Analysis, Shenzhen Research Institute of The Chinese University of Hong Kong, Shenzhen, Guangdong Province, China

Lorna Ng
General Outpatient Clinics, Kwong Wah Hospital, Yaumatei, Hong Kong

Index

www.ingramcontent.com/pod-product-compliance
Lightning Source LLC
Chambersburg PA
CBHW080637200326
41458CB00013B/4661